Selected Titles in This Series

21 **Michel Fortin, Editor,** Plates and shells, 1999

20 **Katie Coughlin, Editor,** Semi-analytic methods for the Navier-Stokes equations, 1999

19 **Rajiv Gupta and Kenneth S. Williams, Editors,** Number theory, 1999

18 **Serge Dubuc and Gilles Deslauriers, Editors,** Spline functions and the theory of wavelets, 1999

17 **Olga Kharlampovich, Editor,** Summer school in group theory in Banff, 1996, 1998

16 **Alain Vincent, Editor,** Numerical methods in fluid mechanics, 1998

15 **François Lalonde, Editor,** Geometry, topology, and dynamics, 1998

14 **John Harnad and Alex Kasman, Editors,** The bispectral problem, 1998

13 **Michel C. Delfour, Editor,** Boundaries, interfaces, and transitions, 1998

12 **Peter C. Greiner, Victor Ivrii, Luis A. Seco, and Catherine Sulem, Editors,** Partial differential equations and their applications, 1997

11 **Luc Vinet, Editor,** Advances in mathematical sciences: CRM's 25 years, 1997

10 **Donald E. Knuth,** Stable marriage and its relation to other combinatorial problems: An introduction to the mathematical analysis of algorithms, 1997

9 **D. Levi, L. Vinet, and P. Winternitz, Editors,** Symmetries and integrability of difference equations, 1996

8 **J. Feldman, R. Froese, and L. M. Rosen, Editors,** Mathematical quantum theory II: Schrödinger operators, 1995

7 **J. Feldman, R. Froese, and L. M. Rosen, Editors,** Mathematical quantum theory I: Field theory and many-body theory, 1994

6 **Guido Mislin, Editor,** The Hilton Symposium 1993: Topics in topology and group theory, 1994

5 **D. A. Dawson, Editor,** Measure-valued processes, stochastic partial differential equations, and interacting systems, 1994

4 **Hershy Kisilevsky and M. Ram Murty, Editors,** Elliptic curves and related topics, 1994

3 **Rémi Vaillancourt and Andrei L. Smirnov, Editors,** Asymptotic methods in mechanics, 1993

2 **Philip D. Loewen,** Optimal control via nonsmooth analysis, 1993

1 **M. Ram Murty, Editor,** Theta functions: From the classical to the modern, 1993

Volume 21

CRM PROCEEDINGS & LECTURE NOTES

Centre de Recherches Mathématiques
Université de Montréal

Plates and Shells

Michel Fortin
Editor

The Centre de Recherches Mathématiques (CRM) of the Université de Montréal was created in 1968 to promote research in pure and applied mathematics and related disciplines. Among its activities are special theme years, summer schools, workshops, postdoctoral programs, and publishing. The CRM is supported by the Université de Montréal, the Province of Québec (FCAR), and the Natural Sciences and Engineering Research Council of Canada. It is affiliated with the Institut des Sciences Mathématiques (ISM) of Montréal, whose constituent members are Concordia University, McGill University, the Université de Montréal, the Université du Québec à Montréal, and the Ecole Polytechnique. The CRM may be reached on the Web at www.crm.umontreal.ca.

American Mathematical Society
Providence, Rhode Island USA

The production of this volume was supported in part by the Fonds pour la Formation de Chercheurs et l'Aide à la Recherche (Fonds FCAR) and the Natural Sciences and Engineering Research Council of Canada (NSERC).

1991 *Mathematics Subject Classification.* Primary 73Kxx; Secondary 65–XX.

Library of Congress Cataloging-in-Publication Data

Plates and shells / Michel Fortin, editor.
p. cm. — (CRM proceedings & lecture notes ; v. 21)
Includes bibliographical references.
ISBN 0-8218-0950-4 (alk. paper)
1. Shells (Engineering) 2. Plates (Engineering) 3. Elastic plates and shells. I. Fortin, Michel. II. Series.
TA660.S5P56 1999
624.1′776—dc21 99-22844
CIP

∞ The paper used in this book is acid-free and falls within the guidelines
established to ensure permanence and durability.
This volume was typeset using $\mathcal{A}_{\mathcal{M}}\mathcal{S}$-TeX,
the American Mathematical Society's TeX macro system,
and submitted to the American Mathematical Society in camera ready
form by the Centre de Recherches Mathématiques.
Visit the AMS home page at URL: http://www.ams.org/

10 9 8 7 6 5 4 3 2 1 04 03 02 01 00 99

Contents

Preface vii

Derivation and Justification of Plate Models by Variational Methods
Stephen M. Alessandrini, Douglas N. Arnold, Richard S. Falk, and Alexandre L. Madureira 1

Simple Triangular Shell Elements for Large Strain Estimations of Sheet Metal Forming Parts
Jean-Louis Batoz, Ying Qiao Guo, and Frédéric Mercier 21

Some Approximation Methods for Linear Thin Shell Problems
Michel Bernadou 37

Numerical Analysis of Piezoelectric Shells
Michel Bernadou and Christophe Haenel 55

Étude intrinsèque d'un problème de coque en grandes déformations
M. Carrive and P. Le Tallec 65

Locking-Free Mixed Stabilized Finite Element Methods for Bending-Dominated Shells
Dominique Chapelle and Rolf Stenberg 81

Un élément de coque simple à trois nœuds en grandes rotations et élastoplasticité
F. Dammak, S. Chamlal, A. Gakwaya, and G. Dhatt 95

Intrinsic Nonlinear Models of Shells
Michel C. Delfour and Jiabin Zhao 109

Convergence of the Linear $P(1,1)$ and $P(2,1)$ Thin Shells to Asymptotic Shells
Michel C. Delfour and Jean-Paul Zolésio 125

Two-Level Overlapping Schwarz Methods for Plate Elements on Unstructured Meshes Using Non-Matching Coarse Grids
Qingping Deng and Xiaobing Feng 159

Large Elasto-Plastic Deformations of Thin Shells with Application to the Stamping of Anisotropic Sheet Metal Parts
J.-C. Gelin and L. Boubakar 171

Justification of Two-Dimensional Linear Shell Models by the Use of Γ-Convergence Theory
Karine Genevey 185

Numerical Simulation of High-Rate Stamping of Tubes and Sheets
S. F. Golovashchenko and N. M. Bessonov 199

An Intrinsic Form of the Stress Resultant Geometrically Exact Shell Theory
Adnan Ibrahimbegović 209

Numerical Inf-Sup Analysis of MITC Plate Bending Elements
Alexander Iosilevich, Klaus-Jürgen Bathe, and Franco Brezzi 225

Optimisation des plaques et coques
E. Pagnacco and J. E. Souza de Cursi 243

Vibroacoustic Behavior of a Rectangular Box
M. A. Tournour and N. Atalla 255

Error Analysis of Mixed Finite Elements for Cylindrical Shells
Geng Yang, Michel C. Delfour, and Michel Fortin 267

Preface

The 1996 Summer Seminar of the Canadian Mathematical Society took place in Québec city from July 22 to July 27, in the building of l'École d'Architecture at the Université Laval. The scientific committee consisted of

- Michel Fortin (chair), Département de mathématiques et de statistique, Laval,
- Mario Fafard, Département de Génie civil, Laval,
- Michel Delfour, Centre de recherches mathématiques, Montréal.

The objective of this Seminar was to bring together both mathematicians and engineers interested in the theory or application of plates and shells, or more generally in the modelisation of thin structures. From this it was hoped that there could emerge a better understanding of the problem: mathematicians could become more aware of how actual engineering applications were performed and engineers would know better what information of theoretical results could provide.

Thin structures are met everywhere in applications. Their approximation by a lower-dimensional model is natural and almost compulsory . It is also full of pitfalls and some engineering catastrophes can be attributed to bad numerical schemes for shells.

From the mathematical point of view, many studies try to make clearer the convergence of full three-dimensional models to models of plates or shells. In particular, the underlying assumptions are made more explicit.

Numerical methods are the key to actual computations. Thin structures are known to need special methods, often inspired by mixed finite element methods. The goal is to avoid numerical locking while providing a good accuracy.

New applications, such as the simulation of crash in the automobile industry, bring new challenges as models and method are stretched to their extreme limit.

All these aspects are reported in these Proceedings which give a snapshot of the state of the art of a fast evolving subject.

Michel Fortin
Montréal, February 1999

Centre de Recherches Mathématiques
CRM Proceedings and Lecture Notes
Volume **21**, 1999

Derivation and Justification of Plate Models by Variational Methods

Stephen M. Alessandrini, Douglas N. Arnold,
Richard S. Falk, and Alexandre L. Madureira

ABSTRACT. We consider the derivation of two-dimensional models for the bending and stretching of a thin three-dimensional linearly elastic plate using variational methods. Specifically we consider restriction of the trial space in two different forms of the Hellinger-Reissner variational principle for 3-D elasticity to functions with a specified polynomial dependence in the transverse direction. Using this approach many different plate models are possible and we classify and investigate the most important. We study in detail a method which leads naturally not only to familiar plate models, but also to error bounds between the plate solution and the full 3-D solution.

1. Introduction

Let Ω be a smoothly bounded domain in $\mathbb{R}^2$ and $t \in (0,1]$. We consider an isotropic, homogeneous, linearly elastic plate occupying the region $P_t = \Omega \times (-t/2, t/2)$. Denote the union of the top and bottom surfaces of the plate by $\partial P_t^{\pm} = \Omega \times \{-t/2, t/2\}$ and the lateral boundary by $\partial P_t^{\mathrm{L}} = \partial\Omega \times (-t/2, t/2)$. See Fig. 1. We suppose that the plate is loaded by a surface force density $\underline{g}\colon \partial P_t^{\pm} \to \mathbb{R}^3$ and a volume force density $\underline{f}\colon P_t \to \mathbb{R}^3$, and is clamped along its lateral boundary. The resulting stress $\underline{\underline{\sigma}}^*\colon P_t \to \mathbb{R}^{3\times 3}_{\mathrm{sym}}$ and displacement $\underline{u}^*\colon P_t \to \mathbb{R}^3$ then satisfy the boundary-value problem

$$\begin{aligned} \underline{\underline{\underline{\underline{A}}}}\,\underline{\underline{\sigma}}^* &= \underline{\underline{\varepsilon}}(\underline{u}^*), \quad -\underline{\operatorname{div}}\,\underline{\underline{\sigma}}^* = \underline{f} \text{ in } P_t, \\ \underline{\underline{\sigma}}^*\underline{n} &= \underline{g} \text{ on } \partial P_t^{\pm}, \quad \underline{u}^* = 0 \text{ on } \partial P_t^{\mathrm{L}}. \end{aligned} \tag{1}$$

Here $\underline{\underline{\varepsilon}}(\underline{u}^*)$ denotes the infinitesimal strain tensor associated to the displacement vector $\underline{u}^*$, namely the symmetric part of its gradient, and $\underline{\operatorname{div}}\,\underline{\underline{\sigma}}$ denotes the vector divergence of the symmetric matrix $\underline{\underline{\sigma}}$ taken by rows. The compliance tensor $\underline{\underline{\underline{\underline{A}}}}$ is

1991 *Mathematics Subject Classification.* Primary: 73K10; Secondary: 73C02.

The second author was supported NSF grant DMS-9500672 and by the Institute for Mathematics and its Applications. The third author was supported by NSF grant DMS-9403552. The fourth author was supported by a fellowship from CNPq-Brazil.

This is the final form of the paper.

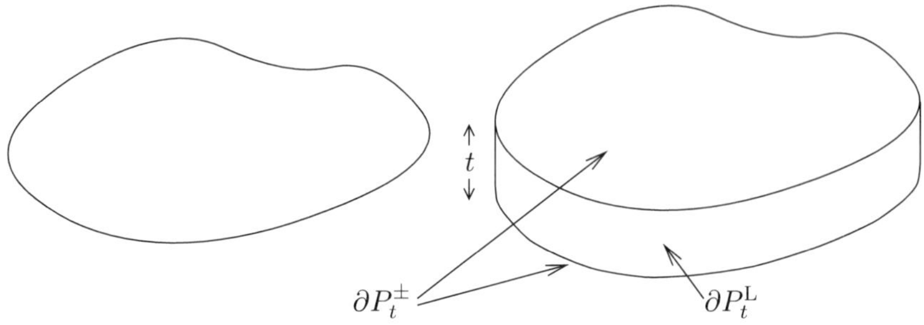

FIGURE 1. The two-dimensional domain Ω and plate domain P_t.

given by $\underline{\underline{\underline{\underline{A}}}}\,\underline{\underline{\tau}} = (1+\nu)\underline{\underline{\tau}}/E - \nu\,\mathrm{tr}(\underline{\underline{\tau}})\underline{\underline{\delta}}/E$, with $E > 0$ Young's modulus, $\nu \in [0, 1/2)$ Poisson's ratio, and $\underline{\underline{\delta}}$ the 3×3 identity matrix.

A plate model seeks to approximate the solution of the elasticity problem (1) in terms of the solution of a system of partial differential equations on the two-dimensional domain Ω without requiring the solution of a three-dimensional problem. The passage from the 3-D problem to a plate model is known as *dimensional reduction*.

In this paper we discuss systematic procedures of dimensional reduction to plate models based on variational formulations of the three-dimensional problem (1). In addition to the derivation of plate models, we are very concerned with their rigorous justification in the sense of proving convergence and rates of convergence, in an appropriate sense, of the plate model solution to the 3-D solution as the plate thickness tends to zero.

Before proceeding, we introduce some notation. As has already been seen, we indicate tensors in three variables with underbars. A first-order tensor (or 3-vector) is written with one underbar, a second-order tensor (or 3×3 matrix) with two underbars, etc. For tensors in two variables we use undertildes in the same way. By way of illustration, any 3-vector may be expressed in terms of a 2-vector giving its in-plane components and a scalar giving its transverse component, and any 3×3 symmetric matrix may be expressed in terms of a 2×2 symmetric matrix, a 2-vector, and a scalar thus:

$$\underline{v} = \begin{pmatrix} \underset{\sim}{v} \\ v_3 \end{pmatrix}, \quad \underline{\underline{\tau}} = \begin{pmatrix} \underset{\approx}{\tau} & \underset{\sim}{\tau} \\ \underset{\sim}{\tau}^T & \tau_{33} \end{pmatrix}.$$

Underbars and undertildes will be used for tensor-valued functions, operators yielding such functions, and spaces of such operators, as well. Even without explicit mention, all second-order tensors arising in this paper will be assumed symmetric. Thus, for example, the notation $\underset{\approx}{H}^s(\Omega)$ denotes the Sobolev space of order s which consists of all functions on Ω with values in $\mathbb{R}^{2\times 2}_{\mathrm{sym}}$ whose partial derivatives of order at most s are square integrable. We denote the norm on a (scalar or tensor-valued) Sobolev space of order s on a set Q by $\|\cdot\|_{s,Q}$. In the case where $Q = \Omega$ we just write $\|\cdot\|_s$.

By taking odd and even parts with respect to the variable x_3, the three-dimensional plate problem splits into two decoupled problems which correspond to *stretching* and *bending* of the plate. More precisely, for a function $k(\underline{x})$ defined on P_t or $\partial P_t^{\pm}$, define its odd and even parts with respect to x_3:

$$k^{\text{even}}(\underline{x}) = \frac{k(\underset{\sim}{x}, x_3) + k(\underset{\sim}{x}, -x_3)}{2}, \quad k^{\text{odd}}(\underline{x}) = \frac{k(\underset{\sim}{x}, x_3) - k(\underset{\sim}{x}, -x_3)}{2}.$$

We then decompose the loads

$$\underline{g} = \underline{g}^{\mathrm{s}} + \underline{g}^{\mathrm{b}}, \quad \underline{f} = \underline{f}^{\mathrm{s}} + \underline{f}^{\mathrm{b}},$$

with

$$\underline{g}^{\mathrm{s}} = \begin{pmatrix} \underset{\sim}{g}^{\text{even}} \\ g_3^{\text{odd}} \end{pmatrix}, \quad \underline{g}^{\mathrm{b}} = \begin{pmatrix} \underset{\sim}{g}^{\text{odd}} \\ g_3^{\text{even}} \end{pmatrix}, \quad \underline{f}^{\mathrm{s}} = \begin{pmatrix} \underset{\sim}{f}^{\text{even}} \\ f_3^{\text{odd}} \end{pmatrix}, \quad \underline{f}^{\mathrm{b}} = \begin{pmatrix} \underset{\sim}{f}^{\text{odd}} \\ f_3^{\text{even}} \end{pmatrix}.$$

Define the stretching portion $(\underline{\underline{\sigma}}^{*\mathrm{s}}, \underline{u}^{*\mathrm{s}})$ of the solution by the system (1) with $\underline{g}$ replaced by $\underline{g}^{\mathrm{s}}$ and $\underline{f}$ replaced by $\underline{f}^{\mathrm{s}}$, and define the bending portion of the solution analogously. It is then easy to see that

$$\underline{u}^{*\mathrm{s}} = \begin{pmatrix} \underset{\sim}{u}^{*\text{even}} \\ u_3^{*\text{odd}} \end{pmatrix}, \quad \underline{\underline{\sigma}}^{*\mathrm{s}} = \begin{pmatrix} \underset{\approx}{\sigma}^{*\text{even}} & \underset{\sim}{\sigma}^{*\text{odd}} \\ (\underset{\sim}{\sigma}^{*\text{odd}})^T & \sigma_{33}^{*\text{even}} \end{pmatrix},$$

$$\underline{u}^{*\mathrm{b}} = \begin{pmatrix} \underset{\sim}{u}^{*\text{odd}} \\ u_3^{*\text{even}} \end{pmatrix}, \quad \underline{\underline{\sigma}}^{*\mathrm{b}} = \begin{pmatrix} \underset{\approx}{\sigma}^{*\text{odd}} & \underset{\sim}{\sigma}^{*\text{even}} \\ (\underset{\sim}{\sigma}^{*\text{even}})^T & \sigma_{33}^{*\text{odd}} \end{pmatrix},$$

and that

$$\underline{u} = \underline{u}^{\mathrm{s}} + \underline{u}^{\mathrm{b}}, \quad \underline{\underline{\sigma}} = \underline{\underline{\sigma}}^{\mathrm{s}} + \underline{\underline{\sigma}}^{\mathrm{b}}.$$

Thus, the three-dimensional elasticity problem (1) splits into two decoupled problems, one for the stretching portion of the solution and one for the bending portion.

All the common two-dimensional plate models admit a similar splitting. The most common plate stretching models are variants of the equations of generalized plane stress. The most common plate bending models are variants of the Kirchhoff-Love biharmonic plate model or of the Reissner-Mindlin plate model. We speak of variants here, because the specification of the forcing functions for the 2-D differential equations in terms of the 3-D loads $\underline{g}$ and $\underline{f}$ differs for different models to be found in the literature, as does the specification of the approximate 3-D stresses and displacements in terms of the solutions of the 2-D boundary-value problems. Moreover, there is a coefficient in the Reissner-Mindlin model, the so-called shear correction factor, which is given different values in the literature. So there is no universally accepted basic two-dimensional model of plate stretching or bending.

2. A Variational Approach to Dimensional Reduction

The Hellinger-Reissner principle, or HR for short, gives a variational characterization of the solution to the three-dimensional problem (1). To state this principle we define

$$\underline{\underline{\Sigma}}^{\bullet} = \underline{\underline{L}}^2(P_t), \quad \underline{V}^{\bullet} = \{\underline{v} \in \underline{H}^1(P_t) : \underline{v} = 0 \text{ on } \partial P_t^{\mathrm{L}}\}.$$

Then HR characterizes $(\underline{\underline{\sigma}}^*, \underline{u}^*)$ as the unique critical point (namely a saddle point) of the HR functional

$$J(\underline{\underline{\tau}}, \underline{v}) = \frac{1}{2} \int_{P_t} \underline{\underline{\underline{\underline{A}}}}\underline{\underline{\tau}} : \underline{\underline{\tau}} \, d\underline{x} - \int_{P_t} \underline{\underline{\tau}} : \underline{\underline{\varepsilon}}(\underline{v}) \, d\underline{x} + \int_{P_t} \underline{f} \cdot \underline{v} \, d\underline{x} + \int_{\partial P_t^{\pm}} \underline{g} \cdot \underline{v} \, d\underset{\sim}{x},$$

TABLE 1. Summary of the principle plate models based on the Hellinger-Reissner principle.

model	$\deg_3 \underset{\approx}{\sigma}$	$\deg_3 \underset{\sim}{\sigma}$	$\deg_3 \sigma_{33}$	$\deg_3 \underset{\sim}{u}$	$\deg_3 u_3$
$\mathrm{HR}_1(p)$	p	$p-1$	$p-2$	p	$p-1$
$\mathrm{HR}_2(p)$	p	$p-1$	p	p	$p-1$
$\mathrm{HR}_3(p)$	p	$p+1$	p	p	$p+1$

on $\underline{\underline{\Sigma}}^\bullet \times \underline{V}^\bullet$. Equivalently, $(\underline{\underline{\sigma}}^*, \underline{u}^*)$ is the unique element of $\underline{\underline{\Sigma}}^\bullet \times \underline{V}^\bullet$ satisfying the weak equations

$$\int_{P_t} \underline{\underline{\underline{\underline{A}}}}\,\underline{\underline{\sigma}}^* : \underline{\underline{\tau}}\, d\underline{x} - \int_{P_t} \underline{\underline{\varepsilon}}(\underline{u}) : \underline{\underline{\tau}}\, d\underline{x} = 0 \quad \text{for all } \underline{\underline{\tau}} \in \underline{\underline{\Sigma}}^\bullet,$$

$$\int_{P_t} \underline{\underline{\sigma}} : \underline{\underline{\varepsilon}}(\underline{v})\, d\underline{x} = \int_{P_t} \underline{f} \cdot \underline{v}\, d\underline{x} + \int_{\partial P_t^\pm} \underline{g} \cdot \underline{v}\, d\underset{\sim}{x} \quad \text{for all } \underline{v} \in \underline{V}^\bullet.$$

Plate models may be derived by replacing $\underline{\underline{\Sigma}}^\bullet$ and $\underline{V}^\bullet$ in HR with subspaces $\underline{\underline{\Sigma}}$ and $\underline{V}$ which admit only a specified polynomial dependence on x_3 and then defining $(\underline{\underline{\sigma}}, \underline{u})$ as the unique critical point of J over $\underline{\underline{\Sigma}} \times \underline{V}$. This is equivalent to restricting the trial and test spaces in the weak formulation to $\underline{\underline{\Sigma}} \times \underline{V}$. Now if the subspaces $\underline{\underline{\Sigma}}$ and $\underline{V}$ are chosen carelessly, there may not exist any such critical point or it may not be unique. We can, and shall, insure a unique solution by insisting that $\underline{\underline{\varepsilon}}(\underline{V}) \subset \underline{\underline{\Sigma}}$. To describe the spaces we consider, we write $\deg_3 v = p$, with $v \in L^2(P_t)$ and $p \in \mathbb{Z}$, to mean that v is a polynomial in x_3 of degree p with coefficients in $L^2(\Omega)$. If $p < 0$, we take this to mean that $v \equiv 0$.

To describe one family of plate models derived in this way, fix a positive integer p, and let

$$\underline{\underline{\Sigma}} = \{\underline{\underline{\tau}} \in \underline{\underline{\Sigma}}^\bullet : \deg_3 \underset{\approx}{\tau} \le p,\ \deg_3 \underset{\sim}{\tau} \le p-1,\ \deg_3 \tau_{33} \le p-2\},$$

$$\underline{V} = \{\underline{v} \in \underline{V}^\bullet : \deg_3 \underset{\sim}{v} \le p,\ \deg_3 v_3 \le p-1\}.$$

We refer to this plate model as the $\mathrm{HR}_1(p)$ model.

This family of models and two others, called $\mathrm{HR}_2(p)$ and $\mathrm{HR}_3(p)$, are specified in Table 1. The most interesting plate models based on HR seem to be those in Table 1 and those which consist of one method from Table 1 applied to the stretching problem and another applied to the bending problem. Thus, for example, with p odd, we may apply $\mathrm{HR}_3(p-1)$ to the stretching problem and $\mathrm{HR}_2(p)$ to the bending problem. This corresponds to seeking a critical point of the Hellinger-Reissner functional over fields with all components of degree at most p.

Each of these choices of space fulfils the condition $\underline{\underline{\varepsilon}}(\underline{V}) \subset \underline{\underline{\Sigma}}$, and so defines a unique $\underline{\underline{\sigma}}$ and $\underline{u}$. It is not hard to see that the coefficients of $\underline{\underline{\sigma}}$ and $\underline{u}$ (as polynomials in x_3) are determined by boundary-value problems posed on Ω (we shall illustrate a specific example below).

The $\mathrm{HR}_2(p)$ and $\mathrm{HR}_3(p)$ models satisfy the condition $\underline{\underline{\underline{\underline{A}}}}^{-1}\underline{\underline{\varepsilon}}(\underline{V}) \subset \underline{\underline{\Sigma}}$ as well. For these two families of models it then follows that $\underline{\underline{\sigma}} = \underline{\underline{\underline{\underline{A}}}}^{-1}\underline{\underline{\varepsilon}}(\underline{u})$—i.e., the three-dimensional constitutive equation is satisfied exactly—and that $\underline{u}$ is determined as the minimizer in $\underline{V}$ of the potential energy

$$E(\underline{v}) = \frac{1}{2}\int_{P_t} \underline{\underline{\underline{A}}}^{-1}\underline{\underline{\varepsilon}}(\underline{v}):\underline{\underline{\varepsilon}}(\underline{v})\,d\underline{x} - \int_{P_t} \underline{f}\cdot\underline{v}\,d\underline{x} - \int_{\partial P_t^{\pm}} \underline{g}\cdot\underline{v}\,d\underset{\sim}{x}.$$

Thus the $\mathrm{HR}_2(p)$ and $\mathrm{HR}_3(p)$ models are minimum energy models (or projection models), which have been discussed by many authors, for example [**1, 4, 13**]. The $\mathrm{HR}_1(p)$ models are *not* minimum energy models.

We now consider some specific low-order models. For this we introduce notations for the odd and even parts of $\underset{\sim}{g}$ and g_3 and for the first two moments of $\underset{\sim}{f}$ and f_3 with respect to x_3, all viewed as functions on Ω. We set

$$g_3^0(\underset{\sim}{x}) = \frac{1}{2}\big[g_3(\underset{\sim}{x},t/2) + g_3(\underset{\sim}{x},-t/2)\big], \quad g_3^1(\underset{\sim}{x}) = \frac{1}{2}\big[g_3(\underset{\sim}{x},t/2) - g_3(\underset{\sim}{x},-t/2)\big],$$

$$f_3^0(\underset{\sim}{x}) = \int_{-t/2}^{t/2} f_3(\underset{\sim}{x},x_3)\,dx_3, \quad f_3^1(\underset{\sim}{x}) = \int_{-t/2}^{t/2} f_3(\underset{\sim}{x},x_3)\frac{x_3}{t}\,dx_3,$$

and define $\underset{\sim}{g}^0$, $\underset{\sim}{g}^1$, $\underset{\sim}{f}^0$, and $\underset{\sim}{f}^1$ analogously.

The $\mathbf{HR_1(1)}$ method. Define the two-dimensional analogue of the compliance tensor by $\underset{\approx}{\underset{\approx}{A}}\underset{\approx}{\tau} = (1+\nu)\underset{\approx}{\tau}/E - \nu\,\mathrm{tr}(\underset{\approx}{\tau})\underset{\approx}{\delta}/E$. It can be shown that the $\mathrm{HR}_1(1)$ solution is given by

$$\underline{u}(\underline{x}) = \begin{pmatrix} \underset{\sim}{\eta}(\underset{\sim}{x}) \\ 0 \end{pmatrix} + \begin{pmatrix} -\underset{\sim}{\phi}(\underset{\sim}{x})x_3 \\ \omega(\underset{\sim}{x}) \end{pmatrix}, \tag{2}$$

$$\underline{\underline{\sigma}}(\underline{x}) = \begin{pmatrix} \underset{\approx}{\underset{\approx}{A}}^{-1}\underset{\approx}{\varepsilon}(\underset{\sim}{\eta}) & 0 \\ 0 & 0 \end{pmatrix} + \begin{pmatrix} -\underset{\approx}{\underset{\approx}{A}}^{-1}\underset{\approx}{\varepsilon}(\underset{\sim}{\phi})x_3 & \frac{E}{2(1+\nu)}(\underset{\sim}{\nabla}\omega - \underset{\sim}{\phi}) \\ \frac{E}{2(1+\nu)}(\underset{\sim}{\nabla}\omega - \underset{\sim}{\phi})^T & 0 \end{pmatrix}, \tag{3}$$

where $\underset{\sim}{\eta}$ is determined by a classical generalized plane stress problem and $\underset{\sim}{\phi}$ and ω by a Reissner-Mindlin problem. Specifically,

$$-t\,\underset{\sim}{\mathrm{div}}\,\underset{\approx}{\underset{\approx}{A}}^{-1}\underset{\approx}{\varepsilon}(\underset{\sim}{\eta}) = 2\underset{\sim}{g}^0 + \underset{\sim}{f}^0 \text{ in } \Omega, \tag{4}$$

$$\underset{\sim}{\eta} = 0 \text{ on } \partial\Omega, \tag{5}$$

and

$$-\frac{t^3}{12}\,\underset{\sim}{\mathrm{div}}\,\underset{\approx}{\underset{\approx}{A}}^{-1}\underset{\approx}{\varepsilon}(\underset{\sim}{\phi}) + t\frac{E}{2(1+\nu)}(\underset{\sim}{\phi} - \underset{\sim}{\nabla}\omega) = -t(\underset{\sim}{g}^1 + \underset{\sim}{f}^1) \text{ in } \Omega, \tag{6}$$

$$t\frac{E}{2(1+\nu)}\,\mathrm{div}(\underset{\sim}{\phi} - \underset{\sim}{\nabla}\omega) = 2g_3^0 + f_3^0 \text{ in } \Omega, \tag{7}$$

$$\underset{\sim}{\phi} = 0, \quad \omega = 0 \text{ on } \partial\Omega. \tag{8}$$

The verification of these equations is straightforward, but tedious. A similar, but more involved, computation (that for the $\mathrm{HR}_4'(1)$ model defined in §3) will be given in the appendix.

Note that in the case of a purely transverse bending load, the system (6)–(8) is the classical Reissner-Mindlin system with shear correction factor 1. When the bending is also affected by nonzero $\underset{\sim}{g}^1$ or $\underset{\sim}{f}^1$, then these appear as an applied couple in the Reissner-Mindlin system.

Thus we see that the $\mathrm{HR}_1(1)$ method is a simple approach to deriving the classical generalized plane stress and Reissner-Mindlin models.

The $\mathbf{HR_2(1)}$ model. The conclusion is altogether different for the $\mathrm{HR}_2(1)$ model, the simplest minimum energy model. In this case it can be shown that $\underline{u}$ is again given by (2) (and $\underline{\underline{\sigma}}$ is given by an expression a bit more complicated than(3) which we don't report here), except that the equations (4) and (6) are replaced by

$$-t\big[\underset{\sim}{\mathrm{div}}\, \underset{\approx}{A}^{-1} \underset{\approx}{\varepsilon}(\underset{\sim}{\eta}) + c\underset{\sim}{\nabla}\, \mathrm{div}\, \underset{\sim}{\eta}\big] = 2\underset{\sim}{g}^0 + \underset{\sim}{f}^0 \text{ in } \Omega,$$

and

$$-\frac{t^3}{12}\big[\underset{\sim}{\mathrm{div}}\, \underset{\approx}{A}^{-1} \underset{\approx}{\varepsilon}(\underset{\sim}{\phi}) + c\underset{\sim}{\nabla}\, \mathrm{div}\, \underset{\sim}{\phi}\big] + t\frac{E}{2(1+\nu)}(\underset{\sim}{\phi} - \underset{\sim}{\nabla}\omega) = -t(\underset{\sim}{g}^1 + \underset{\sim}{f}^1) \text{ in } \Omega,$$

respectively, with $c = -E\nu^2/[(1-\nu^2)(2\nu-1)]$. These additional terms are spurious, and cause the $\mathrm{HR}_2(1)$ model to be divergent as t tends to 0 (in a sense which will be made precise in §4).

Thus the $\mathrm{HR}_2(1)$ model is incorrect. For $p \geq 3$ it can be shown that the $\mathrm{HR}_2(p)$ model is convergent. For $p = 3$, it can be shown to be identical to a method of Lo, Christensen, and Wu [**8**]. However, we feel it possible that even for larger p, the $\mathrm{HR}_2(p)$ method is both more complicated than and less accurate than the $\mathrm{HR}_1(p)$ method.

The $\mathbf{HR_3(1)}$ model. For lack of space we will not discuss this model here, except to note that it is a convergent model, but that the stretching system it yields is more complicated than the generalized plane stress system and the bending system it yields is more complicated than the Reissner-Mindlin system. In particular, the stretching boundary-value problem is a coupled system in three scalar unknowns in contrast to the two of generalized plane stress, and the bending boundary-value problem is a coupled system in four scalar unknowns in contrast to the three unknowns of the Reissner-Mindlin system. In the case of the bending problem this is referred to as the $(1, 1, 2)$ model in the terminology of Babuška and Li [**4, 13**].

3. An Alternative Variational Approach

A variant of the Hellinger-Reissner principle, which we shall call HR$'$, leads to somewhat different plate models. We shall discuss one of these models which is both more accurate and more amenable to rigorous justification than the methods based on HR discussed above. For HR$'$ we define

$$\underline{\underline{\Sigma}}_g^* = \{\underline{\underline{\sigma}} \in \underline{\underline{H}}(\underline{\mathrm{div}}, P_t) \mid \underline{\underline{\sigma}}\underline{n} = \underline{g} \text{ on } \partial P_t^{\pm}\}, \quad \underline{V}^* = \underline{L}^2(P).$$

The space $\underline{\underline{H}}(\underline{\mathrm{div}}, P_t)$ is the space of square integrable symmetric matrix-valued functions on P_t with square integrable divergence. Then HR$'$ characterizes $(\underline{\underline{\sigma}}^*, \underline{u}^*)$ as the unique critical point (again a saddle point) of the HR$'$ functional

$$J'(\underline{\underline{\tau}}, \underline{v}) = \frac{1}{2}\int_{P_t} \underline{\underline{\underline{\underline{A}}}}\underline{\underline{\tau}} : \underline{\underline{\tau}}\, d\underline{x} + \int_{P_t} \underline{\mathrm{div}}\, \underline{\underline{\tau}} \cdot \underline{v}\, d\underline{x} + \int_{P_t} \underline{f} \cdot \underline{v}\, d\underline{x},$$

TABLE 2. Summary of the principle plate models based on the second form of the Hellinger-Reissner principle.

model	$\deg_3 \underset{\approx}{\sigma}$	$\deg_3 \underset{\sim}{\sigma}$	$\deg_3 \sigma_{33}$	$\deg_3 \underset{\sim}{u}$	$\deg_3 u_3$
$\mathrm{HR}_1'(p)$	p	$p-1$	p	p	$p-1$
$\mathrm{HR}_2'(p)$	p	$p+1$	p	p	$p-1$
$\mathrm{HR}_3'(p)$	p	$p+1$	p	p	$p+1$
$\mathrm{HR}_4'(p)$	p	$p+1$	$p+2$	p	$p+1$

on $\underline{\underline{\Sigma}}_g^* \times \underline{V}^*$. Equivalently, $(\underline{\underline{\sigma}}^*, \underline{u}^*)$ is the unique element of $\underline{\underline{\Sigma}}_g^* \times \underline{V}^*$ satisfying the weak equations

$$\int_{P_t} \underline{\underline{\underline{\underline{A}}}}\underline{\underline{\sigma}}^* : \underline{\underline{\tau}}\, d\underline{x} + \int_{P_t} \underline{u} \cdot \underline{\operatorname{div}}\, \underline{\underline{\tau}}\, d\underline{x} = 0 \quad \text{for all } \underline{\underline{\tau}} \in \underline{\underline{\Sigma}}_0^*, \tag{9}$$

$$\int_{P_t} \underline{\operatorname{div}}\, \underline{\underline{\sigma}} \cdot \underline{v}\, d\underline{x} = -\int_{P_t} \underline{f} \cdot \underline{v}\, d\underline{x} \quad \text{for all } \underline{v} \in \underline{V}^*. \tag{10}$$

Here $\underline{\underline{\Sigma}}_0^* = \{\underline{\underline{\sigma}} \in \underline{\underline{H}}(\underline{\operatorname{div}}, P_t) \mid \underline{\underline{\sigma}}\underline{n} = 0 \text{ on } \partial P_t^{\pm}\}$. Note that the displacement boundary conditions, which were essential to the first form of the Hellinger-Reissner principle, are natural in this setting, while the reverse situation holds for the traction boundary conditions.

By restricting J' to subspaces of $\underline{\underline{\Sigma}}_g^*$ and $\underline{V}^*$ with a specified polynomial dependence on x_3, we obtain a variety of plate models. Table 2 lists four families of models of this type. All but the second were studied (in the case of bending only) by Alessandrini in his thesis [**1**].

Here we shall consider only the model $\mathrm{HR}_4'(1)$. For this model (and a number of others from Table 2), we have $\underline{\operatorname{div}}\, \underline{\underline{\Sigma}}_0 = \underline{V}$. This not only implies that there is a unique solution $(\underline{\underline{\sigma}}, \underline{u})$ but also that

$$\underline{\operatorname{div}}\, \underline{\underline{\sigma}} = -P_{\underline{V}}\underline{f}, \tag{11}$$

the orthogonal projection of $\underline{f}$ in $\underline{V}^*$ onto $\underline{V}$. It follows that $\underline{\underline{\sigma}}$ minimizes the complementary energy

$$E_c(\underline{\underline{\tau}}) = \frac{1}{2}\int_{P_t} \underline{\underline{\underline{\underline{A}}}}\underline{\underline{\tau}} : \underline{\underline{\tau}}\, d\underline{x},$$

over all $\underline{\underline{\tau}} \in \underline{\underline{\Sigma}}_g$ satisfying the equilibrium condition $\underline{\operatorname{div}}\, \underline{\underline{\tau}} = -P_{\underline{V}}\underline{f}$. That is, the $\mathrm{HR}_4'(1)$ model is a *minimum complementary energy model.*

We remark that one of Reissner's own derivation of the Reissner-Mindlin plate bending model [**11**] used a complementary energy approach close to $\mathrm{HR}_4'(1)$. However, the trial subspace of $\underline{\underline{\Sigma}}_g^*$ which Reissner used is not determined simply by bounding the polynomial degrees of the various components. The resulting plate bending model is very close to what we shall obtain, but the approach is not as amenable to error analysis or to generalization to higher formal order.

The $\mathrm{HR}_4'(1)$ model gives

$$\underline{u}(\underline{x}) = \begin{pmatrix} \underset{\sim}{\eta}(\underset{\sim}{x}) \\ \rho(\underset{\sim}{x})x_3 \end{pmatrix} + \begin{pmatrix} -\underset{\sim}{\phi}(\underset{\sim}{x})x_3 \\ \omega(\underset{\sim}{x}) + \omega_2(\underset{\sim}{x})r(x_3) \end{pmatrix}, \tag{12}$$

and

$$(13)\quad \underline{\underline{\sigma}}(\underset{\sim}{x}) = \begin{pmatrix} \underset{\approx}{\sigma}^0(\underset{\sim}{x}) & \frac{2x_3}{t}\underset{\sim}{g}^0(\underset{\sim}{x}) \\ \frac{2x_3}{t}\underset{\sim}{g}^0(\underset{\sim}{x})^T & g_3^1(\underset{\sim}{x}) + \sigma_{33}^0(\underset{\sim}{x})q(x_3) \end{pmatrix} + \begin{pmatrix} \underset{\approx}{\sigma}^1(\underset{\sim}{x})\frac{x_3}{t} & \underset{\sim}{g}^1(\underset{\sim}{x}) + \underset{\sim}{\sigma}^0(\underset{\sim}{x})q(x_3) \\ \underset{\sim}{g}^1(\underset{\sim}{x})^T + \underset{\sim}{\sigma}^0(\underset{\sim}{x})^T q(x_3) & g_3^0(\underset{\sim}{x})\frac{2x_3}{t} + \sigma_{33}^1(x_3)s(x_3) \end{pmatrix},$$

where the coefficient functions $\underset{\sim}{\eta}$, ρ, $\underset{\sim}{\phi}$, ω, ω_2, $\underset{\approx}{\sigma}^0$, σ_{33}^0, $\underset{\approx}{\sigma}^1$, $\underset{\sim}{\sigma}^0$, and σ_{33}^1 are functions of $\underset{\sim}{x}$ which we shall describe, and the polynomials q, r, and s are given by $q(z) = 3/2 - 6z^2/t^2$, $r(z) = 6z^2/t^2 - 3/10$, and $s(z) = (5/2)z/t - 10z^3/t^3$. In the appendix we verify that the coefficient functions are determined by the following 2-D boundary-value problems.

The stretching solution. The stretching portion of the solution is determined by the solution to the boundary-value problem

$$(14)\qquad -t \,\underset{\sim}{\operatorname{div}}\, \underset{\approx\!\approx}{A}^{-1} \underset{\approx}{\varepsilon}(\underset{\sim}{\eta}) = \underset{\sim}{l}_1 + t\frac{\nu}{1-\nu}\underset{\sim}{\nabla} l_2 \text{ in } \Omega,$$

$$(15)\qquad \underset{\sim}{\eta} = 0 \text{ on } \partial\Omega,$$

where

$$(16)\qquad \underset{\sim}{l}_1 = 2\underset{\sim}{g}^0 + \underset{\sim}{f}^0, \quad l_2 = g_3^1 + \frac{t}{6}\operatorname{div}\underset{\sim}{g}^0 + f_3^1.$$

Note that only the loading term $\underset{\sim}{l}_1$ appeared in the $\mathrm{HR}_1(1)$ model described earlier. This model takes into account the effect of an odd transverse load on stretching, which is ignored in $\mathrm{HR}_1(1)$.

With $\underset{\sim}{\eta}$ uniquely determined by (14)–(15), the remaining solution quantities are given by

$$(17a)\qquad \underset{\approx}{\sigma}^0 = \underset{\approx\!\approx}{A}^{-1}\underset{\approx}{\varepsilon}(\underset{\sim}{\eta}) + \frac{\nu}{1-\nu} l_2 \underset{\approx}{\delta},$$

$$(17b)\qquad \sigma_{33}^0 = \frac{t}{6}\operatorname{div}\underset{\sim}{g}^0 + f_3^1,$$

$$(17c)\qquad \rho = \frac{1}{E}\big[-\nu \operatorname{tr}(\underset{\approx}{\sigma}^0) + \frac{6}{5}\sigma_{33}^0 + g_3^1\big].$$

The bending solution. The bending portion of the solution is determined by the solution to the boundary-value problem

$$(18)\qquad -\frac{t^3}{12}\underset{\sim}{\operatorname{div}}\, \underset{\approx\!\approx}{A}^{-1}\underset{\approx}{\varepsilon}(\underset{\sim}{\phi}) + t\frac{5}{6}\frac{E}{2(1+\nu)}(\underset{\sim}{\phi} - \underset{\sim}{\nabla}\omega) = t\underset{\sim}{k}_1 - \frac{t^2}{12}\underset{\sim}{\nabla} k_2 \text{ in } \Omega,$$

$$(19)\qquad t\frac{5}{6}\frac{E}{2(1+\nu)}\operatorname{div}(\underset{\sim}{\phi} - \underset{\sim}{\nabla}\omega) = k_3 \text{ in } \Omega,$$

$$(20)\qquad \underset{\sim}{\phi} = 0, \quad \omega = 0 \text{ on } \partial\Omega,$$

where

$$(21)\qquad \underset{\sim}{k}_1 = -\frac{5}{6}\underset{\sim}{g}^1 - \underset{\sim}{f}^1, \quad k_2 = \frac{\nu}{1-\nu}\Big[\frac{t}{5}\operatorname{div}\underset{\sim}{g}^1 + \frac{12}{5}g_3^0 + f_3^2\Big], \quad k_3 = \frac{t}{6}\operatorname{div}\underset{\sim}{g}^1 + 2g_3^0 + f_3^0.$$

Here we have introduced the quantity $f_3^2(\underset{\sim}{x}) = \int_{-t/2}^{t/2} f_3(\underline{x}) r(x_3)\, dx_3$.

The equations (18)–(19) are a somewhat different version of the Reissner-Mindlin equations than (6)–(7), which arose from the $\mathrm{HR}_1(1)$ model. Not only are the formulas for the applied load and couple more involved, but a shear correction factor of 5/6 has been introduced.

With $\underset{\sim}{\phi}$ and ω determined by (18)–(20) we find

$$\underset{\approx}{\sigma}^1 = -t\underset{\approx}{A}^{-1}\underset{\approx}{\varepsilon}(\underset{\sim}{\phi}) + k_2\underset{\approx}{\delta}, \tag{22a}$$

$$\underset{\sim}{\sigma}^0 = \frac{5}{6}\left[\frac{E}{2(1+\nu)}(-\underset{\sim}{\phi} + \underset{\sim}{\nabla}\omega) - \underset{\sim}{g}^1\right], \tag{22b}$$

$$\sigma_{33}^1 = \frac{t}{5}\operatorname{div}\underset{\sim}{g}^1 + \frac{2}{5}g_3^0 + f_3^2, \tag{22c}$$

$$\omega_2 = \frac{t}{E}\left[\frac{1}{6}g_3^0 + \frac{5}{42}\sigma_{33}^1 - \frac{\nu}{12}\operatorname{tr}(\underset{\approx}{\sigma}^1)\right]. \tag{22d}$$

4. Bounding the Modelling Error

An important advantage of the $\mathrm{HR}_4'(1)$ method is that it is amenable to rigorous error analysis. To state our main result, we make the following assumptions on the data $\underline{f}$ and $\underline{g}$. We assume that $\underline{f} \in \underline{V}$, i.e., that $\underset{\sim}{f}$ varies linearly with x_3 and f_3 quadratically, and that $f_3^i \in H^1(\Omega)$, $i = 1, 2, 3$. We assume that $\underline{g} \in \underline{L}^2(\partial P_t^{\pm})$ and also that $\operatorname{div}\underset{\sim}{g}, g_3 \in H^1(\partial P_t^{\pm})$. For such loading functions we define the norms

$$N(\underset{\sim}{f}^0) = \|\underset{\sim}{f}^0\|_0, \quad N(\underset{\sim}{f}^1) = \|\underset{\sim}{f}^1\|_0,$$
$$N(f_3^0, f_3^2) = \|f_3^0\|_{-1} + t^{3/2}\|f_3^2\|_0 + t^2\|f_3^2\|_1, \quad N(f_3^1) = \|f_3^1\|_0 + t^{1/2}\|f_3^1\|_1,$$
$$N(\underset{\sim}{g}^0) = \|\underset{\sim}{g}^0\|_0 + t^{3/2}\|\operatorname{div}\underset{\sim}{g}^0\|_0 + t^2\|\operatorname{div}\underset{\sim}{g}^0\|_1,$$
$$N(\underset{\sim}{g}^1) = \|\underset{\sim}{g}^1\|_0 + t^{3/2}\|\operatorname{div}\underset{\sim}{g}^1\|_0 + t^2\|\operatorname{div}\underset{\sim}{g}^1\|_1,$$
$$N(g_3^0) = \|g_3^0\|_{-1} + t^{3/2}\|g_3^0\|_0 + t^2\|g_3^0\|_1, \quad N(g_3^1) = \|g_3^1\|_1.$$

The precise form of these t-dependent norms is unessential. The key point is that if the functions $\underset{\sim}{f}^i$, f_3^i, $\underset{\sim}{g}^i$, g_3^i on Ω are nonzero smooth functions independent of t, then the norms given tend to a finite nonzero limit as t tends to zero.

THEOREM 1. *Let $t \in (0, 1]$ and suppose that $\underline{f}: P_t \to \mathbb{R}^3$ and $\underline{g}: \partial P_t^{\pm} \to \mathbb{R}^3$ satisfy the conditions just stated. Define $(\underline{\underline{\sigma}}^*, \underline{u}^*)$ as the solution to the three-dimensional elasticity problem* (1) *and define $(\underline{\underline{\sigma}}, \underline{u})$ as the solution to the* $\mathrm{HR}_4'(1)$ *model* (12)–(22). *Then we have*

$$\|\underline{\underline{\varepsilon}}(\underline{u}^* - \underline{u})\|_{0,P_t} + \|\underline{\underline{\sigma}}^* - \underline{\underline{\sigma}}\|_{0,P_t} \le C\big[N(\underset{\sim}{g}^0) + N(\underset{\sim}{f}^0) + tN(g_3^1) + t^{1/2}N(f_3^1)$$
$$+ t^{-1}N(g_3^0) + t^{-1}N(f_3^0, f_3^2) + N(\underset{\sim}{g}^1) + N(\underset{\sim}{f}^1)\big],$$

where C is a constant independent of t, $\underline{g}$, and $\underline{f}$.

Before proceeding to the proof of the theorem, we discuss its significance. First we consider a simple case of pure stretching, namely that where the surface load is purely in-plane and even in x_3, and the volume load vanishes. That is, suppose that for each t,

$$\underline{g}(\underset{\sim}{x}, t/2) = \underline{g}(\underset{\sim}{x}, -t/2) = \begin{pmatrix} \underset{\sim}{g}^0(\underset{\sim}{x}) \\ 0 \end{pmatrix},$$

with $\underset{\sim}{g}^0$ a smooth function on Ω independent of t, and that $\underline{f}$ is identically zero. Then Theorem 1 gives the estimate

$$\|\underline{\underline{\varepsilon}}(\underline{u}^* - \underline{u})\|_{0,P_t} + \|\underline{\underline{\sigma}}^* - \underline{\underline{\sigma}}\|_{0,P_t} \le \text{const.},$$

with the constant depending on $\underset{\sim}{g}$ but not on t. However, from (14)–(17), we easily see that $\|\underset{\approx}{\varepsilon}(\underset{\sim}{\eta})\|_0$ and $\|\underset{\approx}{\sigma}^0\|_0$ behave as $O(t^{-1})$ as t tends to zero, and consequently $\|\underline{\underline{\varepsilon}}(\underline{u})\|_{0,P_t}$ and $\|\underline{\underline{\sigma}}\|_{0,P_t}$ behave as $O(t^{-1/2})$. Thus we find that the relative error

$$\frac{\|\underline{\underline{\varepsilon}}(\underline{u}^* - \underline{u})\|_{0,P_t}}{\|\underline{\underline{\varepsilon}}(\underline{u})\|_{0,P_t}} + \frac{\|\underline{\underline{\sigma}}^* - \underline{\underline{\sigma}}\|_{0,P_t}}{\|\underline{\underline{\sigma}}\|_{0,P_t}} \le Ct^{1/2}.$$

In other words, the plate model converges with order $t^{1/2}$. Note that we could have as well assumed that the surface load was proportional to a function of t, rather than independent of t, and still have obtained $O(t^{1/2})$ convergence. Indeed, for the relative error, any such factor cancels out.

In the case of a purely in-plane and even volume load and vanishing surface load, we obtain the same result in essentially the same way.

Next consider the simplest case of pure bending, namely a transverse, even surface load and no volume load:

$$\underline{g}(\underset{\sim}{x}, t/2) = \underline{g}(\underset{\sim}{x}, -t/2) = \begin{pmatrix} 0 \\ g_3^0(\underset{\sim}{x}) \end{pmatrix}.$$

In this case the theorem gives

$$\|\underline{\underline{\varepsilon}}(\underline{u}^* - \underline{u})\|_{0,P_t} + \|\underline{\underline{\sigma}}^* - \underline{\underline{\sigma}}\|_{0,P_t} \le Ct^{-1}.$$

But from (18)–(22) we see that $\|\underset{\sim}{\phi}\|_1$ behaves as $O(t^{-3})$ and $\|\underset{\approx}{\sigma}^1\|_0$ as $O(t^{-2})$. Thus $\|\underline{\underline{\varepsilon}}(\underline{u})\|_{0,P_t}$ and $\|\underline{\underline{\sigma}}\|_{0,P_t}$ behave as $O(t^{-3/2})$, and again we have a convergence rate of $O(t^{1/2})$ in relative energy norm. The same results holds for bending induced by a transverse, even volume load.

A less common situation is when bending is induced by an in-plane surface load which acts in opposite directions on the top and bottom surfaces:

$$\underline{g}(\underset{\sim}{x}, t/2) = -\underline{g}(\underset{\sim}{x}, -t/2) = \begin{pmatrix} \underset{\sim}{g}^1(\underset{\sim}{x}) \\ 0 \end{pmatrix}.$$

Once again we obtain a relative error convergence rate of $O(t^{1/2})$. The same result holds for bending induced by an odd in-plane volume loading, and for stretching induced by an odd transverse surface loading.

We are aware of one case where $O(t^{1/2})$ convergence does not hold. Suppose that the surface load vanishes and the volume load is transverse and odd in x_3:

$$\underline{f}(\underline{x}) = \begin{pmatrix} 0 \\ f(\underset{\sim}{x})x_3 \end{pmatrix},$$

where f is nonconstant. Then $f_3^1 = t^2 f/12$, while the other load quantities vanish. Therefore Theorem 1 gives

$$\big\|\underline{\underline{\varepsilon}}(\underline{u}^* - \underline{u})\big\|_{0,P_t} + \|\underline{\underline{\sigma}}^* - \underline{\underline{\sigma}}\|_{0,P_t} \leq Ct^{5/2},$$

but $\big\|\underline{\underline{\varepsilon}}(\underline{u})\big\|_{0,P_t}$ and $\|\underline{\underline{\sigma}}\|_{0,P_t}$ are $O(t^{5/2})$ in this case as well. Thus, in this loading case, we don't see any convergence. (We remark that this exceptional situation doesn't pertain to a transverse volume load due to gravity, since then its odd part vanishes.)

5. Proof of the Modelling Error Bounds

The proof of Theorem 1 will be based on the two energies principle or Prager-Synge theorem [**10**]. To state this principle, we define the energy norm of $\underline{v} \in \underline{H}^1(P_t)$ by

$$\|\underline{v}\|_{\mathrm{e}}^2 = \int_{P_t} \underline{\underline{\underline{A}}}^{-1}\underline{\underline{\varepsilon}}(\underline{v}) : \underline{\underline{\varepsilon}}(\underline{v})\, d\underline{x},$$

and the complementary energy norm of $\underline{\underline{\tau}} \in \underline{\underline{L}}^2(P_t)$ by

$$\|\underline{\underline{\tau}}\|_{\mathrm{c}}^2 = \int_{P_t} \underline{\underline{\underline{A}}}\underline{\underline{\tau}} : \underline{\underline{\tau}}\, d\underline{x}.$$

Clearly, $\|\underline{v}\|_{\mathrm{e}}$ is equivalent to $\big\|\underline{\underline{\varepsilon}}(\underline{v})\big\|_{0,P_t}$ and $\|\underline{\underline{\tau}}\|_{\mathrm{c}}$ is equivalent to $\|\underline{\underline{\tau}}\|_{0,P_t}$ with equivalence constants independent of t. The form of the two energies principle we use states that if a stress field and displacement field satisfy the equilibrium equation and boundary conditions of the three-dimensional elasticity problem, then the energy norm of their difference from the true solution of the 3D problem can be written in terms of a norm of the residual of the constitutive equation.

THEOREM 2. *Suppose that* $\bar{\underline{\underline{\sigma}}} \in \underline{\underline{H}}(\underline{\mathrm{div}}, P_t)$ *satisfies*

$$-\underline{\mathrm{div}}\,\bar{\underline{\underline{\sigma}}} = \underline{f} \text{ in } P_t, \quad \bar{\underline{\underline{\sigma}}}\underline{n} = \underline{g} \text{ on } \partial P_t^{\pm}, \tag{23}$$

and that $\bar{\underline{u}} \in \underline{H}^1(P_t)$ *satisfies*

$$\bar{\underline{u}} = 0 \text{ on } \partial P_t^{\mathrm{L}}. \tag{24}$$

Then

$$\|\underline{u}^* - \bar{\underline{u}}\|_{\mathrm{e}}^2 + \|\underline{\underline{\sigma}}^* - \bar{\underline{\underline{\sigma}}}\|_{\mathrm{c}}^2 = \|\bar{\underline{\underline{\sigma}}} - \underline{\underline{\underline{A}}}^{-1}\underline{\underline{\varepsilon}}(\bar{\underline{u}})\|_{\mathrm{c}}^2.$$

Since $\underline{\underline{\sigma}}$ belongs to $\underline{\underline{\Sigma}}_g$ and satisfies (11), and because we assumed that $\underline{f} \in \underline{V}$, we may choose $\bar{\underline{\underline{\sigma}}} = \underline{\underline{\sigma}}$ and (23) is satisfied. We may not, however, take $\bar{\underline{u}} = \underline{u}$, because, in general, the third component of $\underline{u}$ does not vanish on the lateral boundary and so (24) would not be satisfied.

Therefore we define

$$\bar{\underline{u}}(\underline{x}) = \bar{\underline{u}}^{\mathrm{s}}(\underline{x}) + \bar{\underline{u}}^{\mathrm{b}}(\underline{x}) := \begin{pmatrix} \underset{\sim}{\eta}(\underset{\sim}{x}) \\ \bar{\rho}(\underset{\sim}{x})x_3 \end{pmatrix} + \begin{pmatrix} -\underset{\sim}{\phi}(\underset{\sim}{x})x_3 \\ \omega(\underset{\sim}{x}) + \bar{\omega}_2(\underset{\sim}{x})r(x_3) \end{pmatrix},$$

where $\bar{\rho}$ and $\bar{\omega}_2$ are functions in $\mathring{H}^1(\Omega)$ which are yet to be defined. We will use the Prager-Synge theorem to bound $\big\|\underline{\underline{\varepsilon}}(\underline{u}^* - \bar{\underline{u}})\big\|_{0,P_t}$ and $\|\underline{\underline{\sigma}}^* - \underline{\underline{\sigma}}\|_{0,P_t}$, and then bound $\big\|\underline{\underline{\varepsilon}}(\underline{u}^* - \underline{u})\big\|_{0,P_t}$ via the triangle inequality.

A priori estimates. The first step is to derive some a priori estimates for the plate model.

Stretching. The following lemma summarizes the regularity of quantities associated to the stretching problem.

LEMMA 3. *Let $s \geq 0$ be an integer. There exist a constant C depending only on s, Ω, E, and ν such that*

$$\|\sigma_{33}^0\|_s \leq C\big(t\|\operatorname{div} \underset{\sim}{g}^0\|_s + \|f_3^1\|_s\big),$$

and

$$\|\underset{\sim}{\eta}\|_{s+1} + \|\underset{\approx}{\sigma}^0\|_s + \|\rho\|_s \leq C\big(t^{-1}\|\underset{\sim}{g}^0\|_{s-1} + t^{-1}\|\underset{\sim}{f}^0\|_{s-1} + \|g_3^1\|_s + \|f_3^1\|_s + t\|\operatorname{div} \underset{\sim}{g}^0\|_s\big).$$

PROOF. The estimate for σ_{33}^0 is immediate from its definition (17) (with $C = 1$). The estimate for $\underset{\sim}{\eta}$ follows directly from standard regularity estimates for the plane elasticity problem (14)–(15), and then the estimates for $\underset{\approx}{\sigma}$ and ρ follow from their definitions. □

Bending. For the bending problem, the situation is more complicated, because the boundary-value problem (18)–(20) involves a boundary layer. The following result is proved in [**2**].

THEOREM 4. *Let Ω be a convex polygon or a smoothly bounded domain in the plane. For any $t \in (0, 1]$, $\underset{\sim}{F} \in \underset{\sim}{H}^{-1}(\Omega)$, and $G \in H^{-1}(\Omega)$, there exists a unique $(\underset{\sim}{\phi}, \omega) \in \underset{\sim}{H}^1(\Omega) \times H^1(\Omega)$ such that*

$$-\frac{1}{12}\operatorname{div} \underset{\approx}{A}^{-1}\underset{\approx}{\varepsilon}(\underset{\sim}{\phi}) + t^{-2}\frac{5}{6}\frac{E}{2(1+\nu)}(\underset{\sim}{\phi} - \underset{\sim}{\nabla}\omega) = \underset{\sim}{F} \text{ in } \Omega,$$
$$t^{-2}\frac{5}{6}\frac{E}{2(1+\nu)}\operatorname{div}(\underset{\sim}{\phi} - \underset{\sim}{\nabla}\omega) = G \text{ in } \Omega,$$
$$\underset{\sim}{\phi} = 0, \quad \omega = 0 \text{ on } \partial\Omega.$$

Moreover, if $\underset{\sim}{F} \in \underset{\sim}{L}^2(\Omega)$, then $\underset{\sim}{\phi} \in \underset{\sim}{H}^2(\Omega)$ and there exists a constant C independent of t, $\underset{\sim}{F}$, and G, such that

$$\|\underset{\sim}{\phi}\|_2 + \|\omega\|_1 + t^{-2}\|\underset{\sim}{\nabla}\omega - \underset{\sim}{\phi}\|_0 \leq C\big(\|\underset{\sim}{F}\|_0 + \|G\|_{-1}\big).$$

We note that the estimate for $\underset{\sim}{\nabla}\omega - \underset{\sim}{\phi}$ follows directly from that for $\underset{\sim}{\phi}$ and the first differential equation. We also remark that this result does not hold for some other common boundary conditions for the Reissner-Mindlin plate, due to the presence of a stronger boundary layer. For such problems a more refined analysis, based on the asymptotic analysis of [**3**], is needed.

LEMMA 5.

(1) *For any s*

$$\|\sigma_{33}^1\|_s \leq t\|\operatorname{div} \underset{\sim}{g}^1\|_s + \|g_3^0\|_s + \|f_3^2\|_s.$$

(2) *There exist a constant C depending only on Ω, E, and ν such that*

$$\|\underset{\sim}{\phi}\|_2 + \|\omega\|_1 + t^{-2}\|\underset{\sim}{\nabla}\omega - \underset{\sim}{\phi}\|_0 + t^{-1}\|\underset{\approx}{\sigma}^1\|_1 + t^{-2}\|\omega_2\|_1$$
$$\le Ct^{-3}\big(\|g_3^0\|_{-1} + t^2\|g_3^0\|_1 + \|f_3^0\|_{-1} + t^2\|f_3^2\|_1 + t\|\underset{\sim}{g}^1\|_0 + t^3\|\operatorname{div}\underset{\sim}{g}^1\|_1 + t\|\underset{\sim}{f}^1\|_0\big).$$

PROOF. The estimate for σ_{33}^1 is immediate from its definition (22c). The estimates for $\underset{\sim}{\phi}$, ω, and $\underset{\sim}{\nabla}\omega - \underset{\sim}{\phi}$ follow directly from Theorem 4 by taking $\underset{\sim}{F} = t^{-2}\underset{\sim}{k}_1 - t^{-1}\underset{\sim}{\nabla}k_2/12$ and $G = t^{-3}k_3$, and the estimates for $\underset{\approx}{\sigma}^1$ and ω_2 then follow directly from (22). □

Estimation of the residual in the constitutive equation. We now proceed to the estimation of $\|\underline{\underline{\underline{A}}}\underline{\underline{\sigma}} - \underline{\underline{\varepsilon}}(\underline{\bar{u}})\|_{0,P_t}$ (which is equivalent to $\|\underline{\underline{\sigma}} - \underline{\underline{\underline{A}}}^{-1}\underline{\underline{\varepsilon}}(\underline{\bar{u}})\|_c$).

Stretching. The form of $\underline{\underline{\sigma}}^s$ and $\underline{\bar{u}}^s$ give

$$\underline{\underline{\underline{A}}}\underline{\underline{\sigma}}^s = \begin{pmatrix} \underset{\approx}{\underset{\approx}{A}}\underset{\approx}{\sigma}^0 - \frac{\nu}{E}(g_3^1 + \sigma_{33}^0 q)\underset{\approx}{\delta} & \frac{1+\nu}{E}\frac{2x_3}{t}\underset{\sim}{g}^0 \\ \frac{1+\nu}{E}\frac{2x_3}{t}\underset{\sim}{g}^{0T} & \frac{1}{E}(g_3^1 + \sigma_{33}^0 q) - \frac{\nu}{E}\operatorname{tr}(\underset{\approx}{\sigma}^0) \end{pmatrix},$$

and

$$\underline{\underline{\varepsilon}}(\underline{\bar{u}}^s) = \begin{pmatrix} \underset{\approx}{\varepsilon}(\underset{\sim}{\eta}) & (\underset{\sim}{\nabla}\bar{\rho})\frac{x_3}{2} \\ (\underset{\sim}{\nabla}\bar{\rho})^T\frac{x_3}{2} & \bar{\rho} \end{pmatrix}, \tag{25}$$

which, combined with the formulas for $\underset{\approx}{\sigma}^0$, σ_{33}^0, and ρ above, give

$$\underline{\underline{\underline{A}}}\underline{\underline{\sigma}}^s - \underline{\underline{\varepsilon}}(\underline{\bar{u}}^s) = \begin{pmatrix} \frac{\nu}{E}\sigma_{33}^0(1-q)\underset{\approx}{\delta} & \frac{1+\nu}{E}\frac{2x_3}{t}\underset{\sim}{g}^0 - (\underset{\sim}{\nabla}\bar{\rho})\frac{x_3}{2} \\ \frac{1+\nu}{E}\frac{2x_3}{t}\underset{\sim}{g}^{0T} - (\underset{\sim}{\nabla}\bar{\rho})^T\frac{x_3}{2} & \rho - \bar{\rho} + \frac{1}{E}\sigma_{33}^0(q - 6/5) \end{pmatrix}.$$

It follows that

$$\|\underline{\underline{\underline{A}}}\underline{\underline{\sigma}}^s - \underline{\underline{\varepsilon}}(\underline{\bar{u}}^s)\|_{0,P_t}^2 \le Ct\big(\|\underset{\sim}{g}^0\|_0^2 + \|\sigma_{33}^0\|_0^2 + t^2\|\underset{\sim}{\nabla}\bar{\rho}\|_0^2 + \|\bar{\rho} - \rho\|_0^2\big). \tag{26}$$

In view of this estimate we choose $\bar{\rho}$ as the $\mathring{H}^1(\Omega)$ function which minimizes

$$t^2\|\underset{\sim}{\nabla}\bar{\rho}\|_0^2 + \|\bar{\rho} - \rho\|_0^2,$$

or, equivalently, as the solution of the boundary-value problem

$$-t^2\Delta\bar{\rho} + \bar{\rho} = \rho \text{ in } \Omega, \quad \bar{\rho} = 0 \text{ on } \partial\Omega. \tag{27}$$

We then have the following estimate.

LEMMA 6. *Let $\rho \in H^1(\Omega)$ and let $\bar{\rho}$ be the solution of (27). Then*

$$t^2\|\underset{\sim}{\nabla}\bar{\rho}\|_0^2 + \|\bar{\rho} - \rho\|_0^2 \le C\big(t\|\rho\|_{0,\partial\Omega}^2 + t^2\|\rho\|_1^2\big).$$

PROOF. Multiplying the differential equation by $-\Delta\bar{\rho}$ and integrating by parts gives

$$t^2\|\Delta\bar{\rho}\|_0^2 + \|\underset{\sim}{\nabla}\bar{\rho}\|_0^2 = \int_\Omega \underset{\sim}{\nabla}\rho\cdot\underset{\sim}{\nabla}\bar{\rho}\,d\underset{\sim}{x} - \int_{\partial\Omega}\rho\frac{\partial\bar{\rho}}{\partial n}\,ds.$$

Now

$$\|\partial\bar{\rho}/\partial n\|_{0,\partial\Omega}^2 \le C\|\bar{\rho}\|_1\|\bar{\rho}\|_2 \le C\big(t^{-1}\|\bar{\rho}\|_1^2 + t\|\bar{\rho}\|_2^2\big),$$

so, for any $\epsilon > 0$ we have

$$\left| \int_{\partial\Omega} \rho \frac{\partial \bar\rho}{\partial n}\, ds \right| \le C_\epsilon t^{-1} \|\rho\|_{0,\partial\Omega}^2 + \epsilon\big(\|\bar\rho\|_1^2 + t^2\|\bar\rho\|_2^2\big).$$

Clearly also

$$\big|(\underset{\sim}{\nabla}\rho, \underset{\sim}{\nabla}\bar\rho)\big| \le C_\epsilon \|\rho\|_1^2 + \epsilon\|\bar\rho\|_1^2.$$

It follows easily from these estimates that

$$t^2\|\Delta\bar\rho\|_0^2 + \|\underset{\sim}{\nabla}\bar\rho\|_0^2 \le C\big(t^{-1}\|\rho\|_{0,\partial\Omega}^2 + \|\rho\|_1^2\big).$$

Multiplying by t^2 and making the substitution $t^2\Delta\bar\rho = \bar\rho - \rho$ gives the lemma. □

Applying the lemma to (26) and bounding $\|\rho\|_{0,\partial\Omega}$ by $\|\rho\|_1$, we get

$$\begin{aligned}(28)\quad \big\|\underline{\underline{A}}\underline{\underline{\sigma}}^{\mathrm{s}} - \underline{\underline{\varepsilon}}(\underline{\bar u}^{\mathrm{s}})\big\|_{0,P_t} &\le Ct^{1/2}\big(\|\underset{\sim}{g}^0\|_0 + \|\sigma_{33}^0\|_0 + t^{1/2}\|\rho\|_1\big)\\ &\le C\big(\|\underset{\sim}{g}^0\|_0 + t^{3/2}\|\operatorname{div}\underset{\sim}{g}^0\|_0 + t^2\|\operatorname{div}\underset{\sim}{g}^0\|_1\\ &\qquad + \|\underset{\sim}{f}^0\|_0 + t\|g_3^1\|_1 + t^{1/2}\|f_3^1\|_0 + t\|f_3^1\|_1\big),\end{aligned}$$

where we have used Lemma 3 in the last step.

REMARK. If we use the formula (17) for ρ we can compute

$$\begin{aligned}\big\|\underline{\underline{A}}\underline{\underline{\sigma}}^{\mathrm{s}} - \underline{\underline{\varepsilon}}(\underline{u}^{\mathrm{s}})\big\|_{0,P_t} \le Ct^{1/2}\big(&\|\underset{\sim}{g}^0\|_0 + t\|\operatorname{div}\underset{\sim}{g}^0\|_0 + t^2\|\operatorname{div}\underset{\sim}{g}^0\|_1 + \|\underset{\sim}{f}^0\|_0\\ &+ t\|g_3^1\|_1 + \|f_3^1\|_0 + t\|f_3^1\|_1\big).\end{aligned}$$

Thus this quantity is (for most loading conditions) one half order of t smaller than $\big\|\underline{\underline{A}}\underline{\underline{\sigma}}^{\mathrm{s}} - \underline{\underline{\varepsilon}}(\underline{\bar u}^{\mathrm{s}})\big\|_{0,P_t}$. This suggests, correctly, that if we consider a different boundary-value problem, such as that for a periodic plate, in which the plate model does not need to be corrected to satisfy the conditions on the lateral boundary, then the model will approximate the three dimensional solution to higher order.

For later use we note one other consequence of Lemma 6. From (25) and the analogous formula for $\underline{\underline{\varepsilon}}(\underline{u}^{\mathrm{s}})$ we have

$$\begin{aligned}\big\|\underline{\underline{\varepsilon}}(\underline{\bar u}^{\mathrm{s}}) - \underline{\underline{\varepsilon}}(\underline{u}^{\mathrm{s}})\big\|_{0,P_t} &\le Ct^{1/2}\big[t\big\|\underset{\sim}{\nabla}(\bar\rho - \rho)\big\|_0 + \|\bar\rho - \rho\|_0\big]\\ &\le Ct^{1/2}\big(t\|\underset{\sim}{\nabla}\bar\rho\|_0 + t\|\underset{\sim}{\nabla}\rho\|_0 + \|\bar\rho - \rho\|_0\big)\\ &\le Ct\|\rho\|_1.\end{aligned}$$

In view of Lemma 3 we conclude

$$(29)\quad \big\|\underline{\underline{\varepsilon}}(\underline{\bar u}^{\mathrm{s}}) - \underline{\underline{\varepsilon}}(\underline{u}^{\mathrm{s}})\big\|_{0,P_t} \le C\big(\|\underset{\sim}{g}^0\|_0 + t^2\|\operatorname{div}\underset{\sim}{g}^0\|_1 + \|\underset{\sim}{f}^0\|_0 + t\|g_3^1\|_1 + t\|f_3^1\|_1\big).$$

Bending. Since

$$\underline{\underline{A}}\underline{\underline{\sigma}}^{\mathrm{b}} = \begin{pmatrix} \underset{\approx}{A}\underset{\approx}{\sigma}^1\frac{x_3}{t} - \frac{\nu}{E}\big(g_3^0\frac{2x_3}{t} + \sigma_{33}^1 s\big)\underset{\approx}{\delta} & \frac{1+\nu}{E}\big(\underset{\sim}{g}^1 + \underset{\sim}{\sigma}^0 q\big)\\ \frac{1+\nu}{E}\big(\underset{\sim}{g}^1 + \underset{\sim}{\sigma}^0 q\big)^T & \frac{1}{E}\big(g_3^0\frac{2x_3}{t} + \sigma_{33}^1 s\big) - \frac{\nu}{E}\operatorname{tr}(\underset{\approx}{\sigma}^1)\frac{x_3}{t}\end{pmatrix},$$

and

$$\underline{\underline{\varepsilon}}(\underline{\bar u}^{\mathrm{b}}) = \begin{pmatrix} -\underset{\approx}{\varepsilon}(\underset{\sim}{\phi})x_3 & \frac12\big[\underset{\sim}{\nabla}\omega + (\underset{\sim}{\nabla}\bar\omega_2)r - \underset{\sim}{\phi}\big]\\ \frac12\big[\underset{\sim}{\nabla}\omega + (\underset{\sim}{\nabla}\bar\omega_2)r - \underset{\sim}{\phi}\big]^T & \bar\omega_2\frac{12x_3}{t^2}\end{pmatrix},$$

we obtain, in view of the formulas for $\underset{\approx}{\sigma}^1$, $\underset{\sim}{\sigma}^0$, and ω_2,

$$\underline{\underline{\underline{A}}}\underline{\underline{\sigma}}^{\mathrm{b}} - \underline{\underline{\varepsilon}}(\underline{\bar{u}}^{\mathrm{b}}) = \begin{pmatrix} \frac{\nu}{E}\sigma^1_{33}(\frac{x_3}{t} - s)\underset{\approx}{\delta} & \underset{\sim}{\zeta} \\ \underset{\sim}{\zeta}^T & \zeta \end{pmatrix}.$$

with

$$\underset{\sim}{\zeta} = \left(\frac{1}{2} - \frac{5}{12}q\right)\left[\frac{2(1+\nu)}{E}\underset{\sim}{g}^1 - \underset{\sim}{\nabla}\omega + \underset{\sim}{\phi}\right] - \frac{1}{2}\underset{\sim}{\nabla}\bar{\omega}_2 r,$$

$$\begin{aligned}\zeta &= \frac{1}{E}\left(g_3^0\frac{2x_3}{t} + \sigma^1_{33}s\right) - \frac{\nu}{E}\operatorname{tr}(\underset{\approx}{\sigma}^1)\frac{x_3}{t} - \bar{\omega}_2\frac{12x_3}{t^2}, \\ &= \frac{12x_3}{t^2}(\omega_2 - \bar{\omega}_2) - \frac{1}{E}\left[\frac{10}{7}\frac{x_3}{t} - s(x_3)\right]\sigma^1_{33}.\end{aligned}$$

Thus

$$\begin{aligned}(30)\quad &\left\|\underline{\underline{\underline{A}}}\underline{\underline{\sigma}}^{\mathrm{b}} - \underline{\underline{\varepsilon}}(\underline{\bar{u}}^{\mathrm{b}})\right\|^2_{0,P_t} \\ &\qquad \le Ct\big(\|\underset{\sim}{g}^1\|_0^2 + \|\sigma^1_{33}\|_0^2 + \|\underset{\sim}{\nabla}\omega - \underset{\sim}{\phi}\|_0^2 + \|\underset{\sim}{\nabla}\bar{\omega}_2\|_0^2 + t^{-2}\|\bar{\omega}_2 - \omega_2\|_0^2\big).\end{aligned}$$

If we define $\bar{\omega}_2$ by

$$-t^2\Delta\bar{\omega}_2 + \bar{\omega}_2 = \omega_2 \text{ in } \Omega, \quad \bar{\omega}_2 = 0 \text{ on } \partial\Omega,$$

we obtain from Lemma 6 that

$$\|\underset{\sim}{\nabla}\bar{\omega}_2\|_0^2 + t^{-2}\|\bar{\omega}_2 - \omega_2\|_0^2 \le Ct^{-1}\|\omega_2\|_1^2.$$

Substituting this estimate in (30) and invoking Lemma 5 we get

$$\begin{aligned}(31)\quad \left\|\underline{\underline{\underline{A}}}\underline{\underline{\sigma}}^{\mathrm{b}} - \underline{\underline{\varepsilon}}(\underline{\bar{u}}^{\mathrm{b}})\right\|_{0,P_t} &\le C\big(t^{-1}\|g_3^0\|_{-1} + t^{1/2}\|g_3^0\|_0 + t\|g_3^0\|_1 + t^{-1}\|f_3^0\|_{-1} \\ &+ t^{1/2}\|f_3^2\|_0 + t\|f_3^2\|_1 + \|\underset{\sim}{g}^1\|_0 + t^{3/2}\|\operatorname{div}\underset{\sim}{g}^1\|_0 + t^2\|\operatorname{div}\underset{\sim}{g}^1\|_1 + \|\underset{\sim}{f}^1\|_0\big).\end{aligned}$$

REMARK. Again, the use of $\underline{\bar{u}}^{\mathrm{b}}$ instead of $\underline{u}^{\mathrm{b}}$ (which is necessary to satisfy the boundary conditions) results in the loss of half a power of t. Indeed one can check that

$$\begin{aligned}\left\|\underline{\underline{\underline{A}}}\underline{\underline{\sigma}}^{\mathrm{b}} - \underline{\underline{\varepsilon}}(\underline{u}^{\mathrm{b}})\right\|_{0,P_t} &\le Ct^{1/2}\big(t^{-1}\|g_3^0\|_{-1} + \|g_3^0\|_0 + t\|g_3^0\|_1 + t^{-1}\|f_3^0\|_{-1} \\ &+ \|f_3^2\|_0 + t\|f_3^2\|_1 + \|\underset{\sim}{g}^1\|_0 + t\|\operatorname{div}\underset{\sim}{g}^1\|_0 + t^2\|\operatorname{div}\underset{\sim}{g}^1\|_1 + \|\underset{\sim}{f}^1\|_0\big).\end{aligned}$$

We also obtain

$$\begin{aligned}(32)\quad \left\|\underline{\underline{\varepsilon}}(\underline{\bar{u}}^{\mathrm{b}} - \underline{u}^{\mathrm{b}})\right\|_{0,P_t} &\le Ct^{1/2}\Big[\|\underset{\sim}{\nabla}(\bar{\omega}_2 - \omega_2)\|_0 + t^{-1}\|\bar{\omega}_2 - \omega_2\|_0\Big] \le \|\omega_2\|_1 \\ &\le C\big(t^{-1}\|g_3^0\|_{-1} + t\|g_3^0\|_1 + t^{-1}\|f_3^0\|_{-1} + t\|f_3^2\|_1 \\ &\qquad + \|\underset{\sim}{g}^1\|_0 + t^2\|\operatorname{div}\underset{\sim}{g}^1\|_1 + \|\underset{\sim}{f}^1\|_0\big).\end{aligned}$$

Combining the bounds (28) and (31) and invoking the two energies principle, we obtain

$$\begin{aligned}\left\|\underline{\underline{\varepsilon}}(\underline{u}^* - \underline{\bar{u}})\right\|_{0,P_t} + \|\underline{\underline{\sigma}}^* - \underline{\underline{\sigma}}\|_{0,P_t} &\le C\big[N(\underset{\sim}{g}^0) + N(\underset{\sim}{f}^0) + tN(g_3^1) + t^{1/2}N(f_3^1) \\ &+ t^{-1}N(g_3^0) + t^{-1}N(f_3^0, f_3^2) + N(\underset{\sim}{g}^1) + N(\underset{\sim}{f}^1)\big],\end{aligned}$$

which is the estimate of Theorem 1 but with $\underline{u}$ replaced by $\underline{\bar{u}}$. Invoking (29) and (32) concludes the proof of the theorem.

6. Further Discussion

The application of the two energies principle to justify plate theories was initiated in the pioneering work of Morgenstern [**9**], where it was used to prove convergence of the biharmonic model of plate bending to the three-dimensional solution when the thickness tends to zero. The biharmonic model is not derived in that work and the expressions of the approximate stress and displacement in terms of the biharmonic solution are chosen in an ad hoc fashion, as needed for the convergence proof. Although the rate of convergence is not considered, the residual error is not unlike that treated here and one can obtain the same $O(t^{1/2})$ rate in relative energy norm.

Besides the variational approach used here, there are two other main approaches to the derivation of plate models, each with many variations. We do not attempt to review the vast literature. Surveys may be found in [**6**] and [**12**]. The most classical approach starts with a priori assumptions on the form of some components of the solution, or on constraints that the solution must satisfy. These are justified on physical or mechanical grounds as an attempt to capture the fact that the plate is thin. This approach does not seem to lend itself naturally to an error analysis. The second main approach is to use asymptotic analysis in which the plate thickness is viewed as a small parameter in the three-dimensional boundary-value problem and, with the help of assumptions on the data and scaling of the independent and/or dependent variables, a formal or rigorous passage to the limit of zero thickness is made. A very systematic treatment based on asymptotic analysis, can be found in the book [**6**] of Ciarlet. In the approach of Ciarlet and his coworkers, asymptotic analysis is not only used to derive a plate model, but also to rigorously justify convergence of the (appropriately scaled) 3-D solution to the (appropriately scaled) plate model solution. While no rate of convergence is found in that work, an $O(t^{1/2})$ estimate is proved in [**7**]. It is worth noting that the approach presented here, in contrast to the asymptotic approach, requires no assumptions on the scaling of the loads, in order to obtain the convergence result.

The asymptotic approach, as present in [**6**], leads to essentially the same model of plate stretching as was obtained here by the $\mathrm{HR}_1(1)$ method, namely the generalized plane stress problem in which the only loads are due to the even part of the in-plane surface load and the average over the plate thickness of the in-plane volume load. Recall that that equation, unlike the variant we obtained via $\mathrm{HR}_4'(1)$, ignores the effect of the odd part of the transverse loads on stretching.

For plate bending, the asymptotic approach leads to the Kirchhoff-Love or biharmonic plate equation, rather than to the Reissner-Mindlin equations. (It is easy to see that the biharmonic model is, after appropriate scaling, the zero thickness limit of the Reissner-Mindlin model.) To the best of our knowledge, there is no way to obtain Reissner-Mindlin type models of plate bending from the asymptotic approach. On the other hand, the variational approach presented here gives rise to many different methods, but the simplest model it yields is Reissner-Mindlin. We know of no way to obtain the biharmonic model from this approach without either making a priori mechanical assumptions (e.g., by enforcing the Kirchhoff hypothesis on the trial space), or by passing to a limit of zero thickness after obtaining the two-dimensional model. We also remark again that the minimum energy approaches ($\mathrm{HR}_2(p)$ and $\mathrm{HR}_3(p)$) do not yield either the plane stress system, nor the biharmonic equation, nor the Reissner-Mindlin system. It is sometimes suggested to obtain the Reissner-Mindlin system from the (divergent) $\mathrm{HR}_2(1)$ approach, by changing the compliance tensor in an ad hoc fashion. However it is difficult to justify this procedure except that it yields a model that is known from other derivations. In our opinion, a preferable procedure is to use the $\mathrm{HR}_1(1)$ model and eliminate the stress field in order to obtain, in a rational manner, a modification of the displacement energy form to minimize.

A striking feature of the error analysis is the slow rate of convergence obtained, namely $O(t^{1/2})$. Recall that in our analysis of the $\mathrm{HR}_4'(1)$ model, this rate arises as a consequence of the failure of the model to match the boundary conditions on the lateral boundary. Even for the formally higher-order methods, $\mathrm{HR}_4'(p)$, $p > 1$, the same difficulty will arise, and so these methods are *not* found to be of truly higher order. (This statement holds for most realistic plate boundary-value problems. For periodic plates, or plates subject to very special conditions on the lateral boundary, higher-order convergence is possible.) One might wonder whether the $O(t^{1/2})$ rate is realistic, or only a pessimistic consequence of the method of proof. Although we know of no definite results for the model treated here, it appears to be real. Combining the two energies approach with a more accurate choice of the boundary corrector ($\bar{\omega}_2 - \omega_2$ in our notation), and an asymptotic analysis of its behavior, Chen was able to prove that the $O(t^{1/2})$ rate is sharp for the biharmonic plate bending model in some situations [**5**]. He also investigated the question of whether higher-order error estimates hold for the restriction of the error to a subdomain which stays away from the lateral boundary, and, again for the biharmonic, answered this affirmatively for some boundary-value problems.

Appendix: Derivation of the $\mathrm{HR}_4'(1)$ model

In view of the definition of the spaces $\underset{\equiv}{\Sigma}{}_g^*$ and $\underline{V}^*$ in §3, the $\mathrm{HR}_4'(1)$ solution may be written in the form (12), (13) for some functions $\underset{\sim}{\eta}$, ρ, $\underset{\sim}{\phi}$, ω, ω_2, $\underset{\approx}{\sigma}{}^0$, σ_{33}^0, $\underset{\approx}{\sigma}{}^1$, $\underset{\sim}{\sigma}{}^0$, and σ_{33}^1 of $\underset{\sim}{x}$. We now verify that these functions satisfy (and so are determined by) equations (14)–(22).

To assist in the computations below we record the following integrals, all taken over the interval $(-t/2, t/2)$:

$$\int q(z)\,dz = t, \quad \int zq(z)\,dz = 0, \quad \int q(z)^2\,dz = 6t/5, \quad \int zq'(z)\,dz = -t,$$
$$\int r(z)\,dz = t/5, \quad \int s'(z)\,dz = 0, \quad \int s(z)z\,dz = t^2/12, \quad \int s(z)^2\,dz = 5t/42,$$
$$\int q(z)r(z)\,dz = 0, \quad \int s'(z)r(z)\,dz = -1, \quad \int z^2\,dz = t^3/12.$$

Turning first to the verification of the stretching problem, we note that

$$\underset{\equiv}{\underset{\equiv}{A}}\underset{=}{\sigma}{}^{\mathrm{s}}(\underset{\sim}{x})$$
$$= \begin{pmatrix} \underset{\approx}{\underset{\approx}{A}}\underset{\approx}{\sigma}{}^0(\underset{\sim}{x}) - \frac{\nu}{E}\big[g_3^1(\underset{\sim}{x}) + \sigma_{33}^0(\underset{\sim}{x})q(x_3)\big]\underset{\approx}{\delta} & \frac{1+\nu}{E}\frac{2x_3}{t}\underset{\sim}{g}{}^0(\underset{\sim}{x}) \\ \frac{1+\nu}{E}\frac{2x_3}{t}\underset{\sim}{g}{}^0(\underset{\sim}{x})^T & \frac{1}{E}\big[g_3^1(\underset{\sim}{x}) + \sigma_{33}^0(\underset{\sim}{x})q(x_3)\big] - \frac{\nu}{E}\,\mathrm{tr}\big[\underset{\approx}{\sigma}{}^0(\underset{\sim}{x})\big] \end{pmatrix},$$

and

$$\underline{\operatorname{div}}\,\underline{\underline{\sigma}}^{\mathrm{s}}(\underline{x}) = \begin{pmatrix} \underset{\sim}{\operatorname{div}}\,\underset{\approx}{\sigma}^0(\underset{\sim}{x}) + \frac{2}{t}\underset{\sim}{g}^0(\underset{\sim}{x}) \\ \frac{2x_3}{t}\operatorname{div}\underset{\sim}{g}^0(\underset{\sim}{x}) + \sigma_{33}^0(\underset{\sim}{x})q'(x_3) \end{pmatrix}.$$

We shall derive the equations for the stretching portion of the solution, (14)–(17), by taking a sequence of test functions in (9) and (10). First, let $v \in L^2(\Omega)$ be arbitrary and take $\underline{v}(\underline{x}) = \big(0, v(\underset{\sim}{x})x_3\big)^T$ in (10). This gives the expression (17b) for σ_{33}^0. Next take $\underline{v} = (\underset{\sim}{v}, 0)^T$ where $\underset{\sim}{v} \in \underset{\sim}{L}(\Omega)$ is arbitrary, obtaining

$$-t\,\underset{\sim}{\operatorname{div}}\,\underset{\approx}{\sigma}^0 = \underset{\sim}{l}_1. \tag{33}$$

Third, let $\underset{\approx}{\tau} \in \underset{\approx}{H}(\underset{\sim}{\operatorname{div}}, \Omega)$ be arbitrary and choose the test function

$$\underline{\underline{\tau}} = \begin{pmatrix} \underset{\approx}{\tau} & 0 \\ 0 & 0 \end{pmatrix},$$

in (9). Noting that $\underline{\operatorname{div}}\,\underline{\underline{\tau}} = (\underset{\sim}{\operatorname{div}}\,\underset{\approx}{\tau}, 0)^T$, we get

$$\int_\Omega \left[\underset{\approx\!\!\approx}{A}\underset{\approx}{\sigma}^0 : \underset{\approx}{\tau} - \frac{\nu}{E}(\sigma_{33}^0 + g_3^1)\operatorname{tr}(\underset{\approx}{\tau}) + \underset{\sim}{\eta} \cdot \underset{\sim}{\operatorname{div}}\,\underset{\approx}{\tau} \right] d\underset{\sim}{x} = 0 \quad \text{for all } \underset{\approx}{\tau} \in \underset{\approx}{H}(\underset{\sim}{\operatorname{div}}, \Omega).$$

This implies the boundary condition (15) and that

$$\underset{\approx\!\!\approx}{A}\underset{\approx}{\sigma}^0 = \underset{\approx}{\varepsilon}(\underset{\sim}{\eta}) + \frac{\nu}{E}(\sigma_{33}^0 + g_3^1)\underset{\approx}{\delta} = \underset{\approx}{\varepsilon}(\underset{\sim}{\eta}) + \frac{\nu}{E} l_2 \underset{\approx}{\delta},$$

where we have used (17b) and (16) to obtain the last equality. This immediately gives (17a) and, combining with (33), we obtain the differential equation (14). Thus we have verified that $\underset{\sim}{\eta}$ is determined by (14) and (15), $\underset{\approx}{\sigma}^0$ is determined by (17a), and σ_{33}^0 is determined by (17b). To obtain (17c), we choose

$$\underline{\underline{\tau}} = \begin{pmatrix} 0 & 0 \\ 0 & \tau q(x_3) \end{pmatrix},$$

in (9) with $\tau \in L^2(\Omega)$ arbitrary. Thus we have verified the stretching solution for $\mathrm{HR}_4'(1)$.

For the bending problem, we note that

$$\underline{\underline{\underline{\underline{A}}}}\,\underline{\underline{\sigma}}^{\mathrm{b}} = \begin{pmatrix} \underset{\approx\!\!\approx}{A}\underset{\approx}{\sigma}^1\frac{x_3}{t} - \frac{\nu}{E}\left[g_3^0\frac{2x_3}{t} + \sigma_{33}^1 s(x_3)\right]\underset{\approx}{\delta} & \frac{1+\nu}{E}\left[\underset{\sim}{g}^1 + \underset{\sim}{\sigma}^0 q(x_3)\right] \\ \frac{1+\nu}{E}\left[\underset{\sim}{g}^1 + \underset{\sim}{\sigma}^0 q(x_3)\right]^T & \frac{1}{E}\left[g_3^0\frac{2x_3}{t} + \sigma_{33}^1 s(x_3)\right] - \frac{\nu}{E}\operatorname{tr}[\underset{\approx}{\sigma}^1]\frac{x_3}{t} \end{pmatrix},$$

and

$$\underline{\operatorname{div}}\,\underline{\underline{\sigma}}^{\mathrm{b}}(\underline{x}) = \begin{pmatrix} \underset{\sim}{\operatorname{div}}\,\underset{\approx}{\sigma}^1(\underset{\sim}{x})\frac{x_3}{t} + \underset{\sim}{\sigma}^0(\underset{\sim}{x})q'(x_3) \\ \operatorname{div}\underset{\sim}{g}^1(\underset{\sim}{x}) + \operatorname{div}\underset{\sim}{\sigma}^0(\underset{\sim}{x})q(x_3) + 2g_3^0(\underset{\sim}{x})/t + \sigma_{33}^1(\underset{\sim}{x})s'(x_3) \end{pmatrix}.$$

Let $v \in L^2(\Omega)$ be arbitrary and take $\underline{v}(\underline{x}) = \big(0, v(\underset{\sim}{x})\big)^T$ in (10). This yields the equation

$$-t\operatorname{div}\underset{\sim}{\sigma}^0 = t\operatorname{div}\underset{\sim}{g}^1 + 2g_3^0 + f_3^0. \tag{34}$$

Next, take $\underline{v}(\underline{x}) = \big(0, v(\underset{\sim}{x})r(x_3)\big)^T$ where again $v \in L^2(\Omega)$ is arbitrary, to obtain (22c).

Third, let $\underline{v}(\underline{x}) = \big(\underset{\sim}{\psi}(\underset{\sim}{x})x_3, 0\big)^T$ with $\underset{\sim}{\psi} \in \underset{\sim}{L}^2(\Omega)$ arbitrary, to get

$$-\frac{t^2}{12}\,\underset{\sim}{\operatorname{div}}\,\underset{\approx}{\sigma}^1 + t\underset{\sim}{\sigma}^0 = t\underset{\sim}{f}^1. \tag{35}$$

Next we take various choices of $\underset{=}{\tau}$ in (9). First let $\underset{\approx}{\tau} \in \underset{\approx}{H}(\underset{\sim}{\operatorname{div}}, \Omega)$ be arbitrary and take

$$\underset{=}{\tau}(\underline{x}) = \begin{pmatrix} \underset{\approx}{\tau}(\underset{\sim}{x})x_3 & 0 \\ 0 & 0 \end{pmatrix}.$$

Noting that $\underline{\operatorname{div}}\,\underset{=}{\tau} = (\underset{\sim}{\operatorname{div}}\,\underset{\approx}{\tau}x_3, 0)^T$, we get, for all $\underset{\approx}{\tau} \in \underset{\approx}{H}(\underset{\sim}{\operatorname{div}}, \Omega)$, that

$$\int_\Omega \left[\frac{t^2}{12}\underset{\approxeq}{A}\underset{\approx}{\sigma}^1 : \underset{\approx}{\tau} - \frac{t^2}{6}\frac{\nu}{E}g_3^0 \operatorname{tr}(\underset{\approx}{\tau}) - \frac{\nu}{E}\frac{t^2}{12}\sigma_{33}^1 \operatorname{tr}(\underset{\approx}{\tau}) - \frac{t^3}{12}\underset{\sim}{\phi} \cdot \underset{\sim}{\operatorname{div}}\,\underset{\approx}{\tau}\right] d\underset{\sim}{x} = 0.$$

This implies that

$$\underset{\sim}{\phi} = 0 \text{ on } \partial\Omega,$$

and that

$$\underset{\approxeq}{A}\underset{\approx}{\sigma}^1 = -t\underset{\approx}{\varepsilon}(\underset{\sim}{\phi}) + \frac{\nu}{E}(\sigma_{33}^1 + 2g_3^0)\underset{\approx}{\delta},$$

or

$$\begin{aligned} \underset{\approx}{\sigma}^1 &= -t\underset{\approxeq}{A}^{-1}\underset{\approx}{\varepsilon}(\underset{\sim}{\phi}) + \frac{\nu}{1-\nu}(\sigma_{33}^1 + 2g_3^0)\underset{\approx}{\delta} \\ &= -t\underset{\approxeq}{A}^{-1}\underset{\approx}{\varepsilon}(\underset{\sim}{\phi}) + k_2\underset{\approx}{\delta}, \end{aligned} \tag{36}$$

where we have invoked (22c) at the last step. Thus (22a) is satisfied.

Next set

$$\underset{=}{\tau}(\underline{x}) = \begin{pmatrix} 0 & \underset{\sim}{\tau}(\underset{\sim}{x})q(x_3) \\ \underset{\sim}{\tau}(\underset{\sim}{x})^T q(x_3) & 0 \end{pmatrix},$$

with $\underset{\sim}{\tau} \in \underset{\sim}{H}(\operatorname{div}, \Omega)$. We find that

$$\int_\Omega \left\{t\left[\frac{2(1+\nu)}{E}\left(\underset{\sim}{g}^1 + \frac{6}{5}\underset{\sim}{\sigma}^0\right) + \underset{\sim}{\phi}\right] \cdot \underset{\sim}{\tau} + t\omega \operatorname{div}\underset{\sim}{\tau}\right\} d\underset{\sim}{x} = 0,$$

for all such $\underset{\sim}{\tau}$. We infer that

$$\omega = 0 \text{ on } \partial\Omega,$$

and (22b) holds.

Substituting (22b) into (34) leads to the equation (19). Similarly, inserting (36) and (22b) into (35) we obtain (18). We have thus verified (18)–(20) and the first three equations of (22). To obtain the final equation we choose the test function

$$\underset{=}{\tau}(\underline{x}) = \begin{pmatrix} 0 & \underset{\sim}{0} \\ 0 & \tau(\underset{\sim}{x})s(x_3) \end{pmatrix},$$

in (9).

References

1. S. M. Alessandrini, *Some two-dimensional plate models: Derivation, asymptotic properties, and numerical approximation*, Ph.D. thesis, Rutgers Univ., New Jersey, 1991.
2. D. N. Arnold and R. S. Falk, *A uniformly accurate finite element method for the Reissner-Mindlin plate model*, SIAM J. Numer. Anal. **26** (1989), 1276–1290.
3. ———, *Asymptotic analysis of the boundary layer for the Reissner-Mindlin plate model*, SIAM J. Math. Anal. **27** (1996), 486–514.
4. I. Babuška and L. Li, *The problem of plate modeling: Theoretical and computational results*, Comput. Methods Appl. Mech. Engrg. **100** (1992), 249–273.
5. C. Chen, *Asymptotic convergence rates for the Kirchhoff plate model*, Ph.D. thesis, Univ. Park, Pennsylvania, 1995.
6. P. G. Ciarlet, *Plates and junctions in elastic multi-structures, an asymptotic analysis*, Masson, Paris, 1990.
7. P. Destuynder, *Sur une justification des modèles de plaques et de coques par les méthodes asymptotiques*, Ph.D. thesis, Univ. P. et M. Curie, Paris, 1980.
8. K. H. Lo, R. M. Christensen, and E. M. Wu, *A high-order theory of plate deformations*, J. Appl. Mech. **46** (1977), 663–676.
9. D. Morgenstern, *Herleitung der Plattentheorie aus der dreidimensionalen Elastizitätstheorie*, Arch. Rational Mech. Anal. **4** (1959), 145–152.
10. W. Prager and J. L. Synge, *Approximations in elasticity based on the concept of the function space*, Quart. Appl. Math. **5** (1947), 241–269.
11. E. Reissner, *On bending of elastic plates*, Quart. Appl. Math. **5** (1947), 55-68.
12. ———, *Reflections on the theory of elastic plates*, Appl. Mech. Rev. **38** (1985), 1453–1464.
13. C. Schwab, *A-posteriori modeling error estimation for hierarchic plate models*, Numer. Math. **74** (1996), 1-259.

Lockheed Martin Corporation, Building 137-123, 199 Borton Landing Road, Moorestown, NJ 08057, USA

E-mail address: `stephen.m.alessandrini@lmco.com`

Department of Mathematics, Pennsylvania State University, University Park, PA 16802, USA

E-mail address: `dna@math.psu.edu`

Department of Mathematics, Rutgers University, Piscataway, NJ 08854, USA

E-mail address: `falk@math.rutgers.edu`

Department of Mathematics, Pennsylvania State University, University Park, PA 16802, USA

E-mail address: `alm@math.psu.edu`

Centre de Recherches Mathématiques
CRM Proceedings and Lecture Notes
Volume **21**, 1999

Simple Triangular Shell Elements for Large Strain Estimations of Sheet Metal Forming Parts

Jean-Louis Batoz, Ying Qiao Guo, and Frédéric Mercier

Abstract. An efficient algorithm to estimate the large elasto-plastic strains encountered in thin sheets obtained by deep drawing has been continuously developed by the authors since 1987. The algorithm is based on a finite element discretization of the known final shape. In this paper a new simple triangular shell element with constant membrane and bending strains is presented using Discrete Kirchhoff constraints. The expressions of the internal force vector and logarithmic strains through the thickness are derived. Two applications are considered to discuss the validity and efficiency of the numerical procedure.

Introduction. To avoid trial and error tryout procedures, the use of sheet metal forming simulation is in constant progress in the stamping industry to evaluate the deformation paths and the forming defects such as fracture and wrinkling. Many research groups are still developing and improving finite element codes for the analysis when an initial design of the blank and the tools is done and when the forming conditions are defined. Several analysis tools are described in some recent proceedings [**1, 9, 15, 17, 22, 30, 31, 33**]. They are based on membrane, shell or even solid elements considering static or dynamic, implicit or explicit approaches. These analysis tools can be very precise if used by well trained engineers but they are time consuming and they need expensive computer resources.

Computer simulation can also be very attractive and helpful for the process and tooling engineers to define the initial blank (thickness, contour and surface), some process parameter (boundary conditions, holding forces, lubrification conditions, draw bead types and positions, ...) and the material properties (thickness, yield stress, hardening, anisotropy, ...). This has been recognized by some industrial and academic research and development groups. As a result a number of simplified procedures have been developed in the last ten years [**1, 3–5, 9, 11–15, 17, 18, 22, 28, 30, 31, 33**]. They are mainly based on the fact that the surface of the desired part is known. A comparison is then made with the initial flat blank to estimate the deformation and formability of the final product taking into account simple constitutive equations and assumptions regarding the tool actions.

In this paper we present some theoretical aspects and numerical results taking membrane (or stretching) *and* bending effects into account using a simplified

1991 *Mathematics Subject Classification.* Primary: 73K15; Secondary: 73G05.
This is the final form of the paper.

procedure developed by the authors since 1987 and called the "inverse approach" [**3–5, 11–14, 18**]. For a large number of industrial applications the membrane effects are dominant but for a number of situations it is justified to consider bending effects in the model [**10, 16, 19, 23, 29, 32**]. The justifications are the same for the incremental (Analysis) approaches and for the simplified (Preliminary Design) approaches. In deep drawing and stretch forming operations the material undergoes bending and unbending when passing over the die radius. When sharp tools are used in stamping operations neglecting bending strains will result in gross underestimation of the maximum strains. Even with our simplified inverse approach we would like to estimate the wrinkling zones and the spring-back of the workpiece after removing the tools. We then need information on the variation of stresses and strains through the thickness. This is also necessary for the estimation of necking during drawing as found by Brunet and Sabourin [**7, 8**].

Several techniques have been proposed to take into account the bending effects when membrane finite elements are mainly used for the analysis. A strategy has been described by Onate *et al.* [**21**]. Bending effects can be included in triangular membrane elements without introduction of rotational degrees of freedom. This has successfully been developed and used by Rio *et al.* [**25**] and Brunet and Sabourin [**6–8**].

In the next section we describe our shell element formulation in the context of the inverse approach. Bending effects are taken into account (without additional degrees of freedom compared to the membrane model) using the simple plate bending element DKT6 developed in 1992 by the senior author [**2, 26, 27**] (the same element is described as DKTLL in [**20**]). In the last section numerical results are presented and discussed for some problems.

1. Formulation Aspects

1.1. Virtual displacements and rotations. The starting blank is assumed to be flat (although curved blanks can also be considered also [**14**]. The following displacement field is considered between the original flat blank C^0 and the final curved workpiece C obtained after drawing and stretching (Fig. 1):

$$\underset{\sim}{u}_q = \underset{\sim}{u}_p + z\underset{\sim}{\hat{n}} - z^0 n^0, \tag{1}$$

with

$$\underset{\sim}{u}_p = \underset{\sim}{x}_p - \underset{\sim}{x}_p^0 = U\underset{\sim}{i} + V\underset{\sim}{j} + W\underset{\sim}{k}, \tag{2}$$

$\underset{\sim}{n}^0$ and $\underset{\sim}{\hat{n}}$ are unit vectors normal to the mid-surfaces of C^0 and C respectively.

In the inverse approach, the final configuration is fully known except for the thickness h. The position vector of a point q in C is defined by:

$$\underset{\sim}{x}_q = \underset{\sim}{x}_p + z\underset{\sim}{\hat{n}}, \quad -\frac{h}{2} \leq z \leq \frac{h}{2}. \tag{3}$$

The other unknown quantities describing the deformation are the global displacement components U and V along the global axis X and Y since the vertical displacement component W is also known, i.e.:

$$W = \underset{\sim}{u}_p \cdot \underset{\sim}{k} = (\underset{\sim}{x}_p - \underset{\sim}{x}_p^0) \cdot \underset{\sim}{k} = Z - Z^0 = \overline{W}. \tag{4}$$

This means that, even if shell elements including bending effects and rotational dof are considered the virtual displacement field is simply given by:

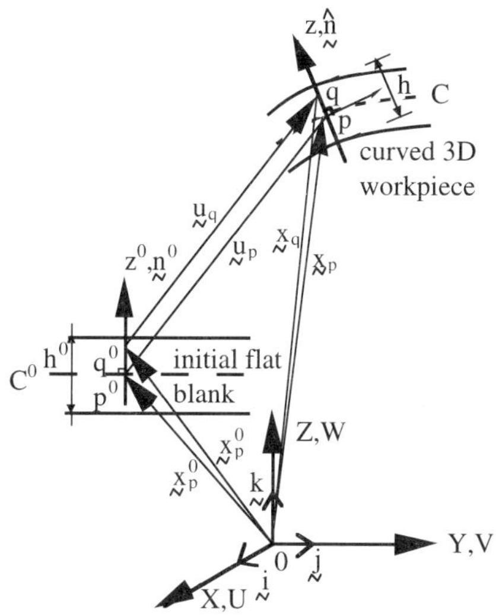

FIGURE 1. Shell geometry and kinematics.

(5) $$\underset{\sim}{u}_q^* = U^* \underset{\sim}{i} + V^* \underset{\sim}{j}; \quad W^* = 0; \quad \underset{\sim}{\theta}^* = \underset{\sim}{0}.$$

1.2. DKT12 shell element internal force vector. The known tridimensional workpiece of volume V is assumed to be discretized by flat triangular shell elements of constant thickness having 3 vertex nodes P_I, P_2, P_3 and 3 mid-side nodes P_4, P_5, P_6 (Fig. 2). The equilibrium of the workpiece is expressed in a weak form using the Principle of Virtual Work (PVW):

(6) $$W - \sum_e W_{\text{int}}^e - W_{\text{ext}} = 0,$$

for any kinematically admissible virtual displacements $\underset{\sim}{u}_q^*$ (Eq. (5)).

The external virtual work results from the tool actions.

Neglecting the transverse shear effects of the thin sheet, the element internal virtual work is given by the general expression:

(7) $$W_{\text{int}}^e = \int_{V^e} \langle \varepsilon^* \rangle \{\sigma\}\, dV; \quad dV = dz\, dA = dz\, dx\, dy,$$

$\langle \varepsilon^* \rangle = \langle \varepsilon_x^* \varepsilon_y^* \lambda_{xy}^* \rangle$ (virtual strains); $\langle \sigma \rangle = \langle \sigma_x \sigma_y \sigma_{xy} \rangle$ (Cauchy stresses) where x and y are orthogonal coordinates in the tangent plane of each element (Fig. 2). The thickness h is assumed to be constant per element. A linear variation of the virtual strains through the thickness is considered:

(8) $$\langle \varepsilon^* \rangle = \langle e^* \rangle + z \langle \chi^* \rangle.$$

For the present flat triangular element the virtual membrane strains are expressed in terms of the virtual displacement components u^*, v^* along the local coordinates x, y of the element (Fig. 2). Linear approximations are assumed for u^*

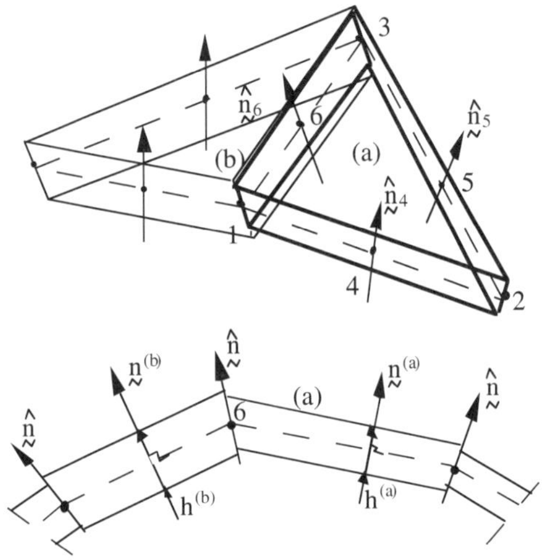

$$\hat{\underset{\sim}{n}}_6 = \frac{1}{2}\frac{\underset{\sim}{n}^{(a)} + \underset{\sim}{n}^{(b)}}{\left|\underset{\sim}{n}^{(a)} + \underset{\sim}{n}^{(b)}\right|}$$

FIGURE 2. Shell element in initial and deformed configurations.

and v^* (CST membrane element):

(9)
$$\langle e^* \rangle = \langle u^*_{,x}\, v^*_{,y} u^*_{,y} + v^*_{,x} \rangle; \quad \{e^*\} = [B_m]\{u^*_n\},$$
$$\langle u^*_n \rangle = \langle u^*_i v^*_i w^*_i i = 1,2,3 \rangle.$$

The virtual bending strains are expressed in a simple form considering the Discrete Kirchhoff Triangular plate bending element called DKT6 first presented by Batoz and Dhatt [**2**]. This bending element will lead to a $[B_f]$ constant matrix. The DKT6 bending stiffness matrix appears to be identical for the Hybrid Stress Model HSM6 and also for the Morley incompatible Kirchhoff element [**2**]. The DKT6 plate bending element was used in previous works (presented in 1991) for incremental analysis of large strain shell problems [**26, 27**]. The DKTLL element [**20**] is the same as the present DKT6 element. The DKT6 element has 6 dof (normal displacement w at the corner nodes and mid-side rotation θ_S along the element sides, Fig. 2). To define the virtual curvatures of the DKT6 element, the rotation components β^*_x and β^*_y (in the plane xz and yz) of the normal are linearly expressed with *half-*C^0 approximations in terms of the rotations at the mid-nodes.

(10)
$$\beta^*_x = \sum_{k=4,5,6} \underline{N}_k \beta^*_{x_k}; \quad \beta^*_y = \sum_{k=4,5,6} \underline{N}_k \beta^*_{y_k},$$

with $\underline{N}_4 = 1 - 2\eta$; $\underline{N}_5 = -1 + 2\xi + 2\eta$; $\underline{N}_6 = 1 - 2\xi$ with the area coordinates ξ and η.

Then a Kirchhoff constraint is introduced along the element sides to express the rotation β^*_{sk} in terms of the normal displacement w^*_i and w^*_j at the vertex nodes, i.e.:

(11)
$$\int_0^{L_k} \overset{*}{\gamma}_{sk}\, ds = \int_0^{L_k} (w,^*_s + \beta^*_s)\, ds = 0,$$

so that: $\beta^*_{sk} = (w^*_i - w^*_j)/L_k$

The virtual rotations $\beta^*_{nk} = \theta^*_{sk}$ are zero since the actual rotations at mid-nodes are known Eq. (5). We finally obtain the resulting $[B_f]$ matrix as (see [**2**] §6.4.3 for details):

$$\langle \chi^* \rangle = \langle \beta^*_{x,x} \beta^*_{y,y} \beta^*_{x,y} + \beta^*_{y,x} \rangle; \quad \{\chi^*\} = [B_f]\{u^*_n\}. \tag{12}$$

The $[B_m]$ and $[B_f]$ matrices are constant per element (and unchanged during the iteration solution process). Using Eqs. (8), (9), (11):

$$\{\varepsilon^*\} = \big([B_m] + z[B_f]\big)\{u^*_n\}. \tag{13}$$

We then transform $\{u^*_n\}$ to the global nodal virtual displacement $\{U^*_n\}$ using the direction cosines of $\underset{\sim}{t}_1$, $\underset{\sim}{t}_2$, $\underset{\sim}{n}$ and taking into account that $W^*_i = 0$ (Eq. (5)):

$$\begin{gathered} \{u^*_n\} = [T]\{U^*_n\}; \quad \langle U^*_n \rangle = \langle U^*_i V^*_i \quad i = 1, 2, 3 \rangle, \\ [T] \begin{bmatrix} [t] & & \\ & [t] & \\ & & [t] \end{bmatrix}; \quad [t] = \begin{bmatrix} t_{1X} & t_{1Y} \\ t_{2X} & t_{2Y} \\ n_X & n_Y \end{bmatrix}. \end{gathered} \tag{14}$$

The element internal force vector $\{f_{\text{int}}\}$ is defined using Eqs. (7), (12), (13):

$$W^e_{\text{int}} = \langle U^*_n \rangle \{f_{\text{int}}\}, \tag{15}$$

with

$$\{f_{\text{int}}\} = [T]^T \int_{V^e} \big([B_m]^T + z[B_f]^T\big)\{\sigma\}\, dz\, dA.$$

For a given value of z, the strains and stresses are constant on each element so that:

$$\{f_{\text{int}}\} = [T]^T\big([B_m]^T\{N\} + [B_f]^T\{M\}\big)A, \tag{16}$$

with the membrane forces and bending moments defined as:

$$\{N\} = \frac{h}{2}\int_{-1}^{1} \{\sigma\}\, d\zeta; \quad \{M\} = \frac{h^2}{4}\int_{-1}^{1} \{\sigma\}\zeta\, d\zeta; \quad -1 \le \zeta \le 1. \tag{17}$$

In Eqs. (16) and (17), only h and $\{\sigma(\zeta)\}$ are changing during the iteration process. The resultant forces can be obtained by numerical integration through the thickness at the centroïd of the element considering 3 (or 5) Lobatto integration points in order to have information on the strains and stresses on the upper and lower surfaces.

1.3. Constitutive equations for a planar isotropic elasto-plastic sheet. In the present inverse approach, only the initial and the final configurations C^0 and C are compared. Therefore the elasto-plastic strains are assumed to be independent of the loading path. As in our previous work ([**3**] to [**5, 11, 13, 14**]), the so-called Hencky-Ilyushin deformation theory of plasticity is considered. In that theory the total constitutive equation takes the simple form for transverse anisotropy:

$$\{\sigma\} = E_s[P]^{-1}\{\varepsilon\}, \tag{18}$$

where E_S is the secant modulus of the uniaxial stress-strain curve.

$$\begin{aligned} E_s &= \frac{\overline{\sigma}}{\overline{\varepsilon}}; \quad \overline{\varepsilon} = \left(\langle\varepsilon\rangle[P]^{-1}\{\varepsilon\}\right)^{1/2}, \\ \overline{\sigma}^2 &= \langle\sigma\rangle[P]\{\sigma\}; \quad \langle\sigma\rangle = \langle\sigma_x\sigma_y\sigma_{xy}\rangle, \\ [P] &= \begin{bmatrix} 1 & -v & 0 \\ -v & 1 & 0 \\ 0 & 0 & \frac{2(1+2r)}{1+r} \end{bmatrix}; \quad v = \frac{r}{1+r}, \end{aligned} \tag{19}$$

r: planar isotropic coefficient of the sheet.

With an estimation of the total strains $\{\varepsilon\}$, we can compute $\overline{\varepsilon}$ (Eq. (19)). Using the uniaxial stress strain curve of the material we estimate E_S to obtain the stresses using Eq. (18). This operation is performed for each numerical integration point through the thickness.

1.4. Logarithmic strains including curvature effects. The position vectors of a material point q are given by (Fig. 1):

$$\underset{\sim}{x} = \underset{\sim}{x}_p + z\underset{\sim}{\hat{n}}; \quad \underset{\sim}{x}_q^0 = \underset{\sim}{x}_p^0 + z^0\underset{\sim}{n}^0. \tag{20}$$

Since $d\underset{\sim}{n}^0 = \underset{\sim}{0}$ for a flat blank we obtain:

$$d\underset{\sim}{x}_q = d\underset{\sim}{x}_p + z\, d\underset{\sim}{\hat{n}} + \underset{\sim}{\hat{n}}\, dz; \quad d\underset{\sim}{x}_q^0 = d\underset{\sim}{x}_p^0 + \underset{\sim}{n}^0\, dz^0, \tag{21}$$

$\underset{\sim}{\hat{n}}$ represents the "true" normal of the deformed shell and $d\underset{\sim}{\hat{n}} \neq \underset{\sim}{0}$ can describe the curvatures.

Let us introduce two unit orthogonal tangent vectors $\underset{\sim}{\hat{t}}_1$ and $\underset{\sim}{\hat{t}}_2$ on the curved deformed middle surface and let x and y be the physical orthogonal curvilinear coordinates. Then:

$$\underset{\sim}{\hat{t}}_1 = \underset{\sim}{x}_{p,x}; \quad \underset{\sim}{\hat{t}}_2 = \underset{\sim}{x}_{p,y}; \quad [\underset{\sim}{\widehat{Q}}] = \left[\underset{\sim}{\hat{t}}_1 \ \vdots \ \underset{\sim}{\hat{t}}_2 \ \vdots \ \underset{\sim}{\hat{n}}\right], \tag{22}$$

and

$$d\underset{\sim}{x}_q = (\underset{\sim}{\hat{t}}_1 + z\underset{\sim}{\hat{n}}_{,x})\, dx + (\underset{\sim}{\hat{t}}_2 + z\underset{\sim}{\hat{n}}_{,y})\, dy + \underset{\sim}{\hat{n}}\, dz,$$

or in matrix form;

$$\{dx_q\} = [F_x]\{dx\}, \tag{23}$$

with

$$\langle dx\rangle = \langle dx\, dy\, dz\rangle,$$

and

$$[F_x] = \left[\underset{\sim}{\hat{t}}_1 + \underset{\sim}{\hat{n}}_{,x} \ \vdots \ \underset{\sim}{\hat{t}}_2 + z\underset{\sim}{\hat{n}}_{,y} \ \vdots \ \underset{\sim}{\hat{n}}\right].$$

If we assume that the thickness is constant per element then:

$$dz = d\zeta\frac{h}{2} \quad \text{and} \quad dz^0 = d\zeta\frac{h^0}{2} \quad \text{with } I \leq \zeta \leq 1, \tag{24}$$

leading to the definition of the constant (per element) thickness stretch:

$$\frac{dz}{dz^0} = \frac{h}{h^0} = \lambda_3. \tag{25}$$

This relation allows us to write:

$$d\underset{\sim}{x}_q^0 = (\underset{\sim}{\hat{t}}_1 - \underset{\sim}{u}_{p,x})\, dx + (\underset{\sim}{\hat{t}}_2 - \underset{\sim}{u}_{p,y})\, dy + \frac{1}{\lambda_3}\underset{\sim}{n}^0\, dz,$$

or in matrix form:

$$\{dx_q^0\} = [F_x^0]^{-1}\{dx\}, \tag{26}$$

with

$$[F_x^0]^{-1} = \left[\underset{\sim}{t}_1 - \underset{\sim}{u}_{p,x} \;\vdots\; \underset{\sim}{t}_2 - \underset{\sim}{u}_{p,y} \;\vdots\; \frac{1}{\lambda_3}\underset{\sim}{n}^0\right].$$

Using equations (23) and (26) we define the inverse deformation gradient tensor $[F]^{-I}$ such that:

$$\{dx_q^0\} = [F]^{-1}\{dx_q\} \text{ with } [F]^{-1} = [F_x^0]^{-1}[F_x]^{-1}. \tag{27}$$

The left Cauchy-Green (CG) tensor $[B]$ is defined as:

$$[B]^{-1} = [F]^{-T}[F]^{-1} \approx [\underset{\sim}{\widehat{Q}}][B_z]^{-1}[\underset{\sim}{\widehat{Q}}^T], \tag{28}$$

with

$$[B_z]^{-1} \approx \begin{bmatrix} a & b & 0 \\ b & c & 0 \\ 0 & 0 & (\lambda_3)^{-2} \end{bmatrix}, \tag{29}$$

a, b, c depend on z, on the curvatures given by $\hat{t}_1 \cdot \hat{n}_{,x}$, $\hat{\underset{\sim}{t}}_1 \cdot \hat{\underset{\sim}{n}}_{,y}$, $\hat{\underset{\sim}{t}}_2 \cdot \hat{\underset{\sim}{n}}_{,y}$ and on the first derivatives of the displacement vector $\underset{\sim}{u}_p$.

The eigenvalues of $[B_z]$ are the squares of the principal stretches λ_I and λ_2, i.e.,

$$\begin{Bmatrix} \lambda_1 \\ \lambda_2 \end{Bmatrix} = \left(\frac{1}{2}(a+c) \pm \frac{1}{2}\sqrt{(a-c)^2 + 4b^2}\right)^{1/2}. \tag{30}$$

These quantities are varying through the thickness. Full incompressibility is assumed and the element thickness stretch is obtained from the values of λ_1 and λ_2 of the mid-surface:

$$\lambda_3 = (\lambda_1\lambda_2)^{-1} \text{ at } z = 0; \quad h = \lambda_3 h^0. \tag{31}$$

The eigenvectors of $[B]^{-I}$ define the principal directions of strains:

$$[M] = \left[\underset{\sim}{m}_1 \;\vdots\; \underset{\sim}{m}_2 \;\vdots\; \underset{\sim}{m}_3\right], \tag{32}$$

with

$$\underset{\sim}{m}_1 = \hat{\underset{\sim}{t}}_1 \cos\alpha + \hat{\underset{\sim}{t}}_2 \sin\alpha; \quad \underset{\sim}{m}_2 = -\hat{\underset{\sim}{t}}_1 \sin\alpha + \hat{\underset{\sim}{t}}_2 \cos\alpha,$$

$$\underset{\sim}{m}_3 = \hat{\underset{\sim}{n}}; \quad \alpha = (\hat{\underset{\sim}{t}}_1, \underset{\sim}{m}_1) = \frac{1}{2}\,\mathrm{Arctg}\,\frac{2b}{a-c}.$$

The left CG tensor is given in terms of $[M]$ and λ_i $(i = 1, 2, 3)$ as:

$$[B] = [M][\lambda^2][M]^T. \tag{33}$$

The large logarithmic strain components at a point through the shell thickness are given by ε_x, ε_y, γ_{xy} such that:

$$\langle\varepsilon\rangle = \langle\varepsilon_x \varepsilon_y \gamma_{xy}\rangle,$$

$$[\varepsilon] = \begin{bmatrix} \varepsilon_x & \frac{1}{2}\gamma_{xy} & 0 \\ \frac{1}{2}\gamma_{xy} & \varepsilon_y & 0 \\ 0 & 0 & \ell n\lambda_3 \end{bmatrix} = [M][\ell n\lambda][M]^T. \tag{34}$$

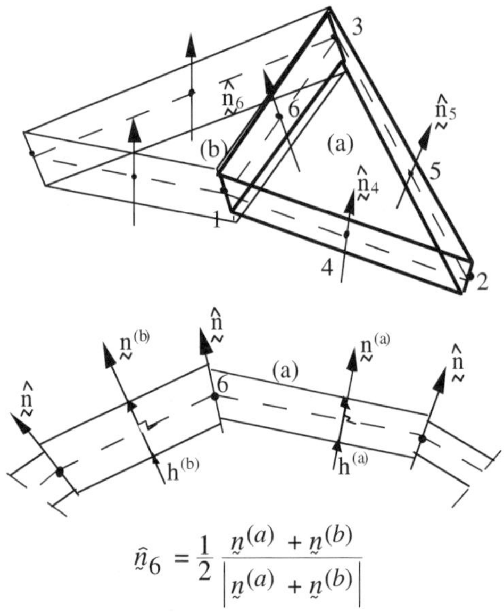

FIGURE 3. Flat facet elements and normals at mid-nodes.

In the present study we consider that the shell is discretized by triangular elements with C^0 continuity of the middle surface and half C^0 continuity of the normals $\hat{\underset{\sim}{n}}$ (Fig. 3) such that:

$$\hat{\underset{\sim}{n}} = \sum_{k=4,5,6} \underline{\mathrm{N}}_k \hat{\underset{\sim}{n}}_k, \tag{35}$$

where $\underline{\mathrm{N}}_k(\xi, \eta)$ are given by Eq. (10). $\hat{\underset{\sim}{n}}_k$ are unique at the mid-side nodes. They can be obtained prior to calculations from the average of $\underset{\sim}{n}^{(a)}$ and $\underset{\sim}{n}^{(b)}$ where $\underset{\sim}{n}^{(a)}$ and $\underset{\sim}{n}^{(b)}$ are the unit normals of the facet elements (a) and (b). The curvatures are constant per element. They are estimated using

$$r \approx -\underset{\sim}{t}_1 \cdot \hat{\underset{\sim}{n}}_{,x}; \quad t \approx -\underset{\sim}{t}_2 \cdot \hat{\underset{\sim}{n}}_{,y}; \quad s \approx -\underset{\sim}{t}_1 \cdot \hat{\underset{\sim}{n}}_{,y} = -\underset{\sim}{t}_2 \cdot \hat{\underset{\sim}{n}}_{,x}. \tag{36}$$

Then we compute the principal stretches (Eq. (30)) and directions (Eq. (31)) at the centroid of the triangular element using Eqs. (28) and (29) with $\hat{\underset{\sim}{t}}_1 \approx \underset{\sim}{t}_1$ and $\hat{\underset{\sim}{t}}_2 \approx \underset{\sim}{t}_2$. The values are first evaluated at the mid-surface ($z = 0$), in order to estimate the constant thickness stretch λ_3 (Eq. (31)). Then the logarithmic strains and directions at the other numerical integration points are computed using Eqs. (29), (30), (32) and (34).

1.5. External virtual work. A number of assumptions can be made regarding the effect of the punch and die and the effect of blank holder forces with friction. Some details are given in previous papers [**4, 12**].

If nodal concentrated "vertical" forces are assumed, then the reaction associated to the imposed "vertical" displacement $\overline{W}_j$ at node j is obtained from the assembly of the element internal force vector $\{f_{\text{int}}\}$ (Eq. (16)) with a full 9 by 9$[T]$ matrix (Eq. (14)), i.e., without elimination of the components corresponding to W_i^* $i = 1$,

2, 3). The nodal force/reaction vector at node j is then defined as:

$$\langle F\rangle = \langle 00R_z\rangle, \tag{37}$$

with

$$R_z = \underset{\sim}{F}{}^j_{\text{int}} \cdot \underset{\sim}{k} = (F^j_{\text{int}})_z.$$

If "normal" nodal forces are assumed then the nodal force vector is given by:

$$\underset{\sim}{F} = F\underset{\sim}{n} \text{ with } \langle n\rangle = \langle n_x n_y n_z\rangle,$$

and

$$F = R_z/n_z, \tag{38}$$

where $\underset{\sim}{n}$ is a unit vector normal to the sheet at node j with $n_z > 0$.

Friction between the tools and the sheet can be considered in a simplified manner if we introduce a knowledge of the relative movement between the tool (punch or die) and the sheet at the end of the forming process [**12**].

An equivalent external load vector can also be defined due to the friction under the blankholder. For the flat blank the tangential force per unit area is defined as:

$$\underset{\sim}{f} = -2\mu q_n \frac{\underset{\sim}{u}}{\|\underset{\sim}{u}\|}, \tag{39}$$

where μ is the friction coefficient, q_n is the normal blank holder pressure acting on the two faces of the sheet and $\underset{\sim}{u}$ is the estimated tangential displacement (varying during the iteration process).

The effect of drawbead can be replaced by equivalent restraining forces at the nodes situated along the drawbead line or by an artificial increase of the friction coefficient using equation (39).

1.6. Resolution of the non linear equations. From equations (6) and (15) we can define the global residual vector $\{R\}$ such that:

$$W = \langle U^*\rangle\{R\} = 0 \quad \forall\{U\}^* \text{ with } \{R\} = \{F_{\text{int}}\} - \{F_{\text{ext}}\}. \tag{40}$$

The internal force vector $\{F_{\text{int}}\}$ is obtained by assembling of the element vectors $\{f_{\text{int}}\}$ (Eq. (15)). The $\{F_{\text{ext}}\}$ vector results from the assumption of the tool actions.

The system of nonlinear equations $\{R(U)\} = \{0\}$ can be solved using various algorithmic approaches such as the classical implicit static (Newton-Raphson) method with equilibrium check. Other approaches are presently studied such as the explicit static method (no equilibrium check) or the dynamic relaxation method. The results presented in the next section have been obtained using the Newton-Raphson method (in several intermediate steps for convergence reasons). The tangent stiffness matrices taking the bending effects into account have been obtained numerically by finite differences (perturbation technique). Since $W = \overline{W}$ has been introduced at all nodes a minimum of three additional boundary conditions on U, V, have to be introduced to eliminate the three rigid body modes (translations along X and Y and rotations around Z).

To start the computation it is necessary to know the whole deformed structure (the useful 3D workpiece and the trimming zone). Part of the result is the estimated contour of the initial blank. However it is possible to obtain in an iterative manner the shape of the trimming contour for an imposed shape of the blank. This is an attractive aspect of the simplified inverse approach. The numerical procedure is described in [**12**] and [**11**].

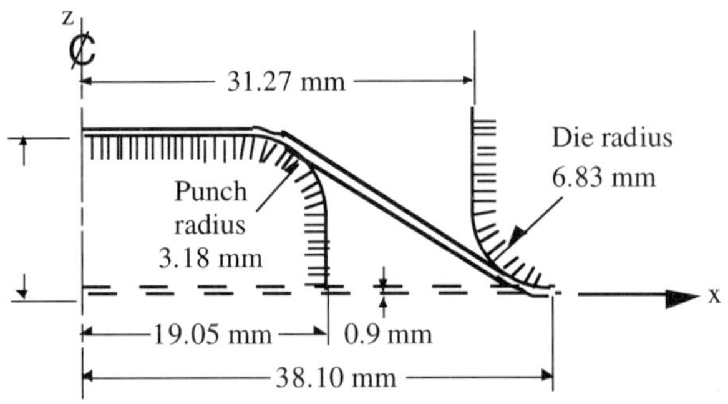

$$E = 2.1 \times 10^5\ MPa\ ;\ \nu = 0.3\ ;\ \mu = 0.25\ ;\ \bar{\sigma} = 562.1\left(\varepsilon^p\right)^{0.248} MPa$$

FIGURE 4. Plane strain stretching by a square punch.

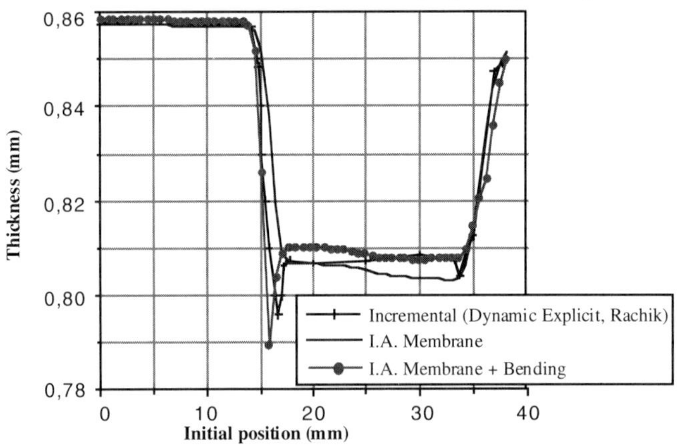

FIGURE 5. Thickness distribution for the plane strain stretching problem with friction effects.

2. Numerical Results

2.1. Plane strain stretching of a thin sheet by a square punch. The problem presented in Fig. 4 has been studied by Stoughton [**29**] (experimental and numerical results are reported). It was also considered to evaluate the bending effects by Wang and Tang [**32**], Peric *et al.* [**23**], Rachik [**24**] using incremental static or dynamic formulations and taking into account the J_2 flow theory and the contact conditions with friction between the sheet and the tools. In the present inverse approach the final shape is defined from simple geometrical considerations for a punch travel of 12 mm. A deformed strip of 2.8 mm width is discretized with 60 elements along X and 4 elements along Y (total of 480 elements). The friction effects can be considered in a simplified manner considering the normal nodal forces (Eq. (38)) and the relative movement between the sheet and the tools. The thickness distribution considering friction effects is presented in Fig. 5. The results obtained with the simplified inverse approach considering membrane and membrane + bending are compared with those obtained by Rachik [**24**] using an incremental explicit dynamic shell approach. All results are in good agreement. It

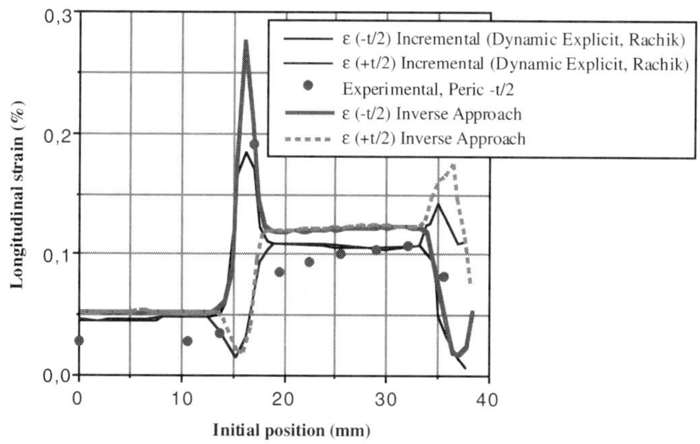

FIGURE 6. Longitudinal strain distribution on the external surfaces for the plane strain stretching problem with friction.

is observed that the membrane + bending model can estimate the local thickness variations due to the tool radius. The longitudinal strain values on the outside surfaces are reported on Fig. 6. The results obtained using the inverse approach (for membrane + bending), and using the explicit dynamic incremental approach [**24**] are compared with some experimental values given in [**23**] for the outside surface. It appears that our simplified approach overestimates the strain predictions. However the overall distribution is rather satisfactory for the simplicity of our model.

2.2. Drawing of a circular blank with a flat headed punch. A cylindrical cup drawing is described in Figs. 7 and 8 (1250 elements). The material is an anisotropic XD340 steel. The same problem has been studied extensively both numerically and experimentally by Arrieux, Brunet, Boivin *et al.* [**6–8**] in relation with forming limit stress state diagram criterion to predict necking. If we do not consider the friction effects the draw in values along X and Y are 6.33 and 7.04 respectively, using our simplified membrane + bending (DKT12) shell model. The values of the positive principal stresses σ_1 and σ_2 on the outside and inside surfaces are represented in Fig. 9. The bending effects are found quite significant. However when friction effects are taken into account these bending effects vanish: the stresses are found almost constant through the thickness. The results obtained for σ_1 and σ_2 considering friction and a membrane model are reported in Fig. 10. In this case some σ_1, σ_2 values are close to the forming limit stress curves. In fact the experimental and numerical results using an incremental approach [**6–8**] did predict necking at 22 mm punch travel near the limit of the flat head and punch radius (and near the X rolling direction). The draw-in values considering friction effects are estimated as $DX = 4.41\,\text{mm}$ and $DY = 5.60\,\text{mm}$. The difference in draw-in values along X and Y is attributed to the anisotropy of the sheet. The results are obtained in only 8 mn of CPU on a DEC ALPHA 3000/900 workstation with 320 MB of RAM for the membrane + bending model without friction and in 4 mn of CPU for the membrane model with friction.

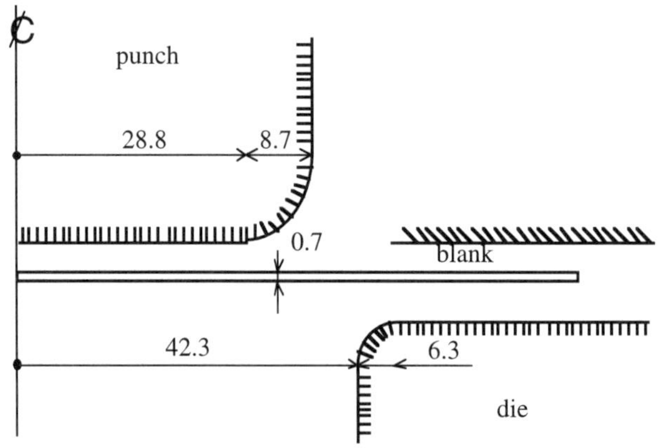

blank radius = 75 mm

$\overline{\sigma} = 729.09\ (0.01345 + \overline{\varepsilon})^{0.1657}$ MPa

$r_0 = 0.831$; $r_{45} = 0.949$; $r_{90} = 1.07$

Holder pressure : 6.5 MPa

friction (holder and die) : 0.13 ; (punch) : 0.03

FIGURE 7. Drawing with cylindrical punch (data).

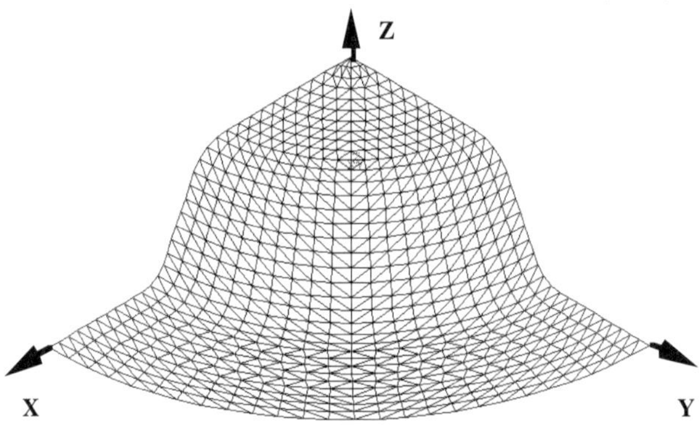

FIGURE 8. Finite element mesh.

Conclusion

A simple triangular shell element obtained by superposition of the CST membrane element and a low order Discrete Kirchhoff bending element has been derived to estimate the large elastoplastic strains resulting from deep drawing of flat metallic sheets. Satisfactory results have been obtained for a small amount of CPU time despite a number of questionable assumptions regarding the simplified constitutive equations and the action of the tools. The bending effects can be significant in the zones with ratio over thickness less than 10. They can be estimated without

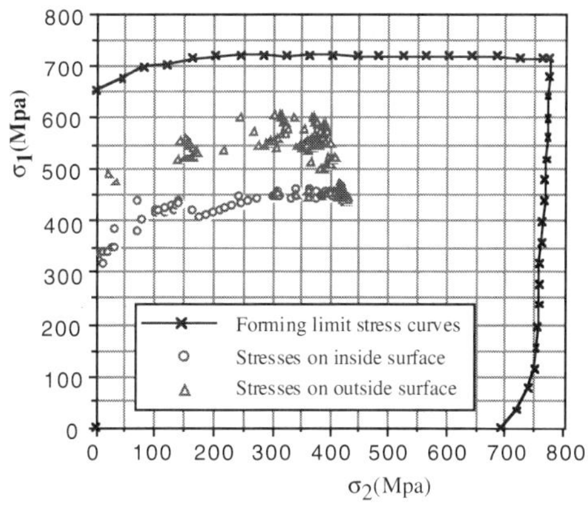

FIGURE 9. Drawing of a circular blank. Positive principal stresses at the external surfaces for the membrane + bending model without friction.

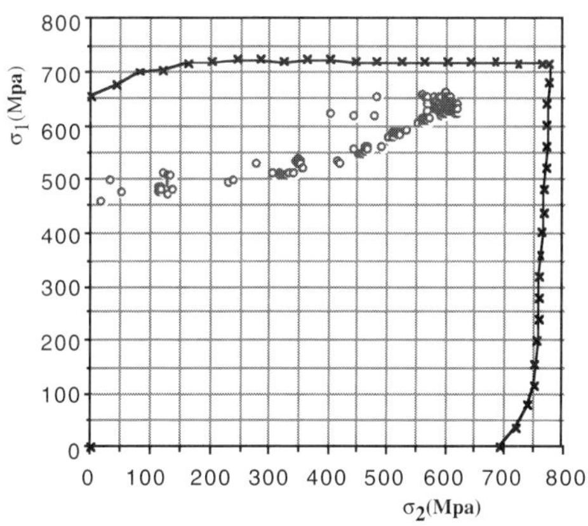

FIGURE 10. Drawing of a circular blank. Positive principal stresses for the membrane model with friction.

additional degrees of freedom compared to the membrane model. The shape optimization of the initial blank is presently undertaken using the nonlinear finite element model described in this paper.

References

1. T. Altan (ed.), *Sheet metal forming technology*, J. Mat. Proc. Tech. **46**, no. 3-4, Elsevier, 1994.
2. J. L. Batoz and G. Dhatt, *Modélisation des structures des éléments finis*, "Coques", vol. 3, Hermès, Paris, 1992.
3. J. L. Batoz, P. Duroux, Y. Q. Guo, and J. M. Detraux, *An efficient algorithm to estimate the large strains in deep drawing*, Numiform 89 (Thompson *et al.*, eds.), A. A. Balkema, Rotterdam-Brookfield, 1989, pp. 383–388.
4. J. L. Batoz, Y. Q. Guo, and J. M. Detraux, *An inverse finite element procedure to estimate the large plastic strain in sheet metal forming*, 3rd Intern. Conf. Technology of Plasticity (Kyoto, 1990), 1990, pp. 1403–1408.

5. J. L. Batoz, R. Narainen, P. Duroux, and Y. Q. Guo, *Comparison of the implicit, explicit and inverse approaches for the estimation of the strain distribution in axisymmetrical thin sheets*, 4th Conf. Techn. Plasticity (China, 1993), 1993, pp. 1695–1700.
6. M. Brunet, R. Arrieux, and T. Nguyen Nhat, *Necking prediction using forming limit stress surfaces in* 3D *sheet metal forming*, Numiform 95 (Shen and Dawson, eds.), A. A. Balkema, Rotterdam-Brookfield, 1995, pp. 669–674.
7. M. Brunet and F. Sabourin, *A simplified triangular shell element with necking criterion for* 3D *sheet forming analysis*, [**17**], pp. 229–238.
8. ______, *Prediction of necking and wrinkles with a simplified triangular shell element in sheet forming*, Proc. Intern. Conf. Metal Forming Process Simulation in Industry (Baden-Baden, Germany, 1994), 1994, pp. 75–93.
9. J. L. Chenot *et al.* (eds.), *Numerical methods in industrial forming processes*, Numiform 92, A. A. Balkema, Rotterdam-Brookfield, 1992.
10. S. Choudhury and J. K. Lee, *Numerical simulation of thin sheet metal forming processes including bending effects*, Microstructural Development and Control in Material Processing (D. R. Duraham and A. Saigal, eds.), MD-vol. 14, ASME, 1989.
11. P. Duroux, *Évaluation numérique des déformations dans les tôles embouties*, Ph.D. thesis, Univ. de Technologie de Compiègne, France, 1992.
12. Y. Q. Guo, J. L. Batoz, J. M. Detraux, and P. Duroux, *Finite element procedures for strain estimations of sheet metal forming parts*, Internat. J. Numer. Methods Engrg. **30** (1990), 1385–1401.
13. Y. Q. Guo, J. L. Batoz, M. El Mouatassim, and J. M. Detraux, *On the estimation of thickness strains in thin car panels by the inverse approach*, [**9**], pp. 1403–1408.
14. ______, *Determination of the trimming contour under the curved blankholders with the inverse approach*, Conf. Intern. ATMMF 92 (Wang *el al.*, eds.) (Wuham, China, 1992), 1992, pp. 279–289.
15. B. Kröplin and E. Luckey (eds.) *Metal forming process simulation in industry*, Intern. Conf. Workshop, (Baden-Baden, Germany, 1994), Fastenrath, Wuppertal, Germany, 1994.
16. D. Lee, *Recent innovations in sheet material forming*, J. Materials Proc. Techn. **46** (1994) vol. 3–4, 333–349.
17. A. Makinouchi *et al.* (eds), *Numerical simulation of* 3D *sheet metal forming processes. Verification of simulation with experiments*, Numisheet 93 (Isehara, Japan, 1993), 1993.
18. R. Narainen, *Comparaison de trois méthodes de calcul pour la simulation de l'emboutissage de tôles axisymétriques*, Ph.D. thesis, Univ. Technologie de Compiègne, France, 1993.
19. Q. Q. Nie and D. Lee, *Comparison of different analytical and experimental results obtained under simple sheet metal forming processes*, Proc. 4th ICTP (Beijing, China, 1993), 1993, pp. 1630–1634.
20. E. Onate, *A review of some finite element families for thick and thin plate and shell analysis*, Recents Developments in Finite Element Analysis (T. J. R. Hughes *et al.*, eds.) (Barcelona, 1994), CIMNE, 1994, pp. 98–111.
21. E. Onate, C. Agelet de Saracibar, and J. B. Dalin, *Finite element analysis of sheet metal forming problems using a selective voided viscous shell membrane formulation*, Numiform 89 (Thompson *et al.*, eds.) A. A. Balkema, Rotterdam-Brookfield, 1989, pp. 23–29.
22. D. R. J. Owen *et al.*, *Computational plasticity—Fundamentals and applications*, COMPLAS IV, Pineridge Press, Swansea, U.K., 1995.
23. D. Peric, D. R. J. Owen, and M. E. Honnor, *Simulation of thin sheet metal forming processes employing a thin shell element*, F. E. Simulation of 3D Sheet Metal Forming Processes in Automotive Industry (Zürich, 1991), 1991, pp. 569–600.
24. M. Rachik, *Simulation de la mise en forme des structures minces* (*soufflage des plastiques et emboutissage des tôles*), Ph.D. thesis, Univ. Technologie de Compiègne, France, 1993.
25. G. Rio, B. Tathi, and F. Horkay, *Introducing bending rigidity in a finite element membrane sheet metal forming model*, [**30**], 1993, pp. 449–456.
26. J. M. Roelandt and J. L. Batoz, *Shell finite element for deep drawing problems*: *Computational aspects and results in Finite Inelastic Deformations—Theory and Applications*, IUTAM Symposium (D. Besdo and E. Stein, eds.), Springer-Verlag, Berlin, Heidelberg, 1992, pp. 423–430.
27. J. M. Roelandt, X. J. Liu, J. L. Batoz, and J. P. Jameux, *Axisymmetric and general shell elements for large transformations (formulation and applications)*, [**30**], 1993, pp. 457–463.

28. S. F. Shen and P. Dawson (eds.), *Simulation of material processing: theory, methods and applications*, Numiform 95, A. A. Balkema, Rotterdam-Brookfield, 1995.
29. T. B. Stoughton, *Finite element modeling of* 1008 AK *sheet steel stretched over a rectangular punch with bending effects*, Computer Modeling of Sheet Metal Forming Process (Wang and Tang, eds.), The Metallurgical Society, 1985, pp. 143–159.
30. C. Teodosiu *et al.* (eds), *Large plastic deformations, fundamental aspects and applications to metal forming* A. A. Balkema, Rotterdam, Brookfield, 1993.
31. Z. R. Wang and Y. He (eds), *Advanced technology of plasticity* 1993, Proc. 4th Intern. Conf. Technology of Plasticity (Beijing, China, 1993), Intern. Academic Publ., 1993.
32. N. M. Wang and S. C. Tang, *Analysis of bending effects in sheet forming operations*, Int. J. Numer. Methods Engrg. **25** (1988), no. 1, 253–267.
33. Z. R. Wang *et al.* (eds.), *Advanced technology and machinery in metal forming*, Proc. Intern. Conf. ATMMF'92, Huazhong Univ. Science and Technology Press, Wuhan, China, 1992.

Membres du Pôle Modélisation de la Région Picardie, Université de Technologie de Compiègne, Division MNM, Laboratoire LG2mS, UPRESA 6066, B.P. 20529, 60205 Compiègne Cedex, France

E-mail address, Jean-Louis Batoz: `Batoz@utc.fr`

E-mail address, Ying Qiao Guo: `Guo@utc.fr`

Centre de Recherches Mathématiques
CRM Proceedings and Lecture Notes
Volume **21**, 1999

Some Approximation Methods for Linear Thin Shell Problems

Michel Bernadou

Scope

This is a review paper of some approximation methods for linear thin shell problems. Further details can be found in [**B2**] as well as some other results for linear buckling and shape optimization.

Firstly we recall the two basic models of thin shells: the thin shell model by Koiter (see [**K, KO**]) which belongs to the Kirchhoff [**KI**] and Love [**L**] family of models, and the thin and semi-thick shell model of Naghdi [**N, NA**] which uses the ideas of Cosserat brothers [**CC**] and takes into account the effects of transverse shear. In both cases, we give the associated existence and uniqueness results.

Next, in paragraph 2, we present some approximation methods for the displacement model of Koiter. They include conforming and nonconforming methods and the effects of numerical integration techniques and possible curved boundary approximations. For these methods, we have proved their convergence, obtained *a priori* error estimates and checked their efficiency for practical applications in the real world of industrial problems.

In paragraph 3, we report two approximation methods which assume that we only know the geometry of the shell at a finite number of points. The first one uses a flat triangular facet approximation of the middle surface, the second one a B-spline approximation method.

Finally in paragraph 4, we consider the problem of structures made up of an assemblage of elementary shells, in other words, we study junction problems between thin shells.

1. Basic Thin Shell Models; Existence and Uniqueness

In this paragraph, we recall briefly the geometrical definition of a thin shell and the two models that we subsequently use.

1991 *Mathematics Subject Classification.* Primary: 73K15; Secondary: 65N30.
This is the final form of the paper.

1.1. Geometrical definition of a thin shell. Let $\mathcal{E}^3$ be the usual euclidean space referred to an orthonormal fixed system $(0, \vec{e}_1, \vec{e}_2, \vec{e}_3)$ and let Ω be a bounded open subset in a plane $\mathcal{E}^2$, with a boundary Γ. Then, the *middle surface* $\overline{S}$ of the shell is the image in $\mathcal{E}^3$ of the set $\overline{\Omega}$ by a sufficiently smooth mapping $\vec{\phi}$: $(\xi^1, \xi^2) \in \overline{\Omega} \subset \mathcal{E}^2 \to \vec{\phi}(\xi^1, \xi^2) \in \mathcal{E}^3$.

Then we introduce two local basis in the tangent plane to the middle surface: the *covariant basis* which is defined by $\vec{a}_\alpha = \vec{\phi}_{,\alpha}, \alpha = 1, 2$, and the *contravariant basis* $\{\vec{a}^\alpha, \alpha = 1, 2\}$ which is such that $\vec{a}^\alpha \cdot \vec{a}_\beta = \delta^\alpha_\beta$, and we complete these bases in $\mathcal{E}^3$ by adding $\vec{a}^3 = \vec{a}_3 = \vec{a}_1 \times \vec{a}_2 / \mid \vec{a}_1 \times \vec{a}_2 \mid$.

The *thickness* e of the shell is defined through a mapping

$$e\colon (\xi^1, \xi^2) \in \overline{\Omega} \to \{x \in \mathbb{R}; x > 0\}.$$

Then the *shell* $\mathcal{C}$ is the closed subset of $\mathcal{E}^3$ defined by

$$\mathcal{C} = \left\{ M \in \mathcal{E}^3; \overrightarrow{OM} = \vec{\phi}(\xi^1, \xi^2) + \xi^3 \vec{a}_3, (\xi^1, \xi^2) \in \overline{\Omega}, -\frac{1}{2} e(\xi^1, \xi^2) \leq \xi^3 \leq \frac{1}{2} e(\xi^1, \xi^2) \right\}.$$

Deformation of the shell. There are essentially two different families of thin shell models depending of the consideration of transverse shear effects or not.

Mindlin-Reissner models. This family of equations includes transverse shear effects: the particles located along the normal to the middle surface before deformation, remain aligned after the deformation but this normal is no longer orthogonal to the deformed middle surface. In these models we get five unknowns u_i, $i = 1, 2, 3$ and $\delta_\alpha, \alpha = 1, 2$:

$$\vec{u} = u_i \vec{a}^i; \quad \underline{\delta} = \delta_\alpha \vec{a}^\alpha.$$

Corresponding set of equations is a system of order two with respect to the five unknowns.

Kirchhoff-Love models. This family of equations neglects transverse shear effects, e.g., the normals remain normal during the deformation. In this model, we only get three unknowns u_i, $i = 1, 2, 3$. Corresponding set of equations is a system of order two with respect to (u_1, u_2) and of order four with respect to u_3.

Link between both families. Naghdi's model give back Koiter's model by imposing relation

$$\delta_\alpha + u_{3,\alpha} + b^\lambda_\alpha u_\lambda = 0$$

where (b^λ_α) denotes the second fundamental form of the middle surface $\overline{S}$.

Strain-displacement relations (linear theory). For clarity, we display in Table 1 the different characteristics concerned with strain-displacement relations for both families of models.

Behaviour law. For simplicity we assume that the constitutive material of the shell is elastic, homogeneous and isotropic.

1.2. Variational formulation of Naghdi's model. The variational formulation can be stated as follows:

$$(1) \qquad \begin{cases} \text{For any } \vec{p} \in \left(L^2(\Omega)\right)^3, \vec{N} \in \left(L^2(\Gamma_1)\right)^3 \text{ and } \underline{M} \in \left(L^2(\Gamma)\right)^2, \\ \text{find } (\vec{u}, \underline{\delta}) \in \vec{W} \text{such that} \\ b\big[(\vec{u}, \underline{\delta}); (\vec{v}, \underline{\beta})\big] + c\big[(\vec{u}, \underline{\delta}); (\vec{v}, \underline{\beta})\big] = \ell(\vec{v}, \underline{\beta}), \quad \forall (\vec{v}, \underline{\beta}) \in \vec{W}, \end{cases}$$

TABLE 1. Strain displacement relations (linear theory).

COSSERAT, NAGHDI models	KIRCHHOFF, LOVE, KOITER models
• *Approximation of the* 3d *strain tensor*	
$\varepsilon_{\alpha\beta}(\vec{U}) \simeq \gamma_{\alpha\beta}(\vec{u}) + \xi^3\chi_{\alpha\beta}(\vec{u},\underline{\delta})$	$\varepsilon_{\alpha\beta}(\vec{U}) \simeq \gamma_{\alpha\beta}(\vec{u}) + \xi^3\bar{\rho}_{\alpha\beta}(\vec{u})$
$\varepsilon_{3\alpha}(\vec{U}) \simeq \frac{1}{2}(u_{3,\alpha} + b^\lambda_\alpha u_\lambda + \delta_\alpha)$	$\varepsilon_{3\alpha}(\vec{U}) \simeq 0$
$\varepsilon_{33}(\vec{U})$ determined by a plane stress hypothesis	
• *Strain and change of curvature tensor of the middle surface*	
$\gamma_{\alpha\beta}(\vec{u}) = \frac{1}{2}(u_{\alpha\|\beta} + u_{\beta\|\alpha}) - b_{\alpha\beta}u_3$	$\gamma_{\alpha\beta}(\vec{u}) = \frac{1}{2}(u_{\alpha\|\beta} + u_{\beta\|\alpha}) - b_{\alpha\beta}u_3$
$\chi_{\alpha\beta}(\vec{u},\underline{\delta}) = \frac{1}{2}(\delta_{\alpha\|\beta} + \delta_{\beta\|\alpha}) - \frac{1}{2}(b^\lambda_\alpha u_{\lambda\|\beta} + b^\lambda_\beta u_{\lambda\|\alpha}) + b^\lambda_\alpha b_{\lambda\beta}u_3$	$\bar{\rho}_{\alpha\beta}(\vec{u}) = u_{3\|\alpha\beta} - b^\lambda_\alpha b_{\lambda\beta}u_3 + b^\lambda_\alpha u_{\lambda\|\beta} + b^\lambda_\beta u_{\lambda\|\alpha} + b^\lambda_\alpha\|_\beta u_\lambda$

where

$$b\big[(\vec{u},\underline{\delta});(\vec{v},\underline{\beta})\big] = \int_\Omega eE^{\alpha\beta\lambda\mu}\left[\gamma_{\alpha\beta}(\vec{u})\gamma_{\lambda\mu}(\vec{v}) + \frac{e^2}{12}\chi_{\alpha\beta}(\vec{u},\underline{\delta})\chi_{\lambda\mu}(\vec{v},\underline{\beta})\right]\sqrt{a}\,d\xi^1\,d\xi^2,$$

$$c\big[(\vec{u},\underline{\delta});(\vec{v},\underline{\beta})\big] = \int_\Omega \frac{eEa^{\alpha\beta}}{2(1+\nu)}[u_{3,\alpha} + b^\lambda_\alpha u_\lambda + \delta_\alpha][v_{3,\beta} + b^\mu_\beta v_\mu + \beta_\beta]\sqrt{a}\,d\xi^1\,d\xi^2,$$

$$\ell(\vec{v},\underline{\beta}) = \int_\Omega p^i v_i\sqrt{a}\,d\xi^1\,d\xi^2 + \int_{\Gamma_1}(N^i v_i - M^\lambda\beta_\lambda)\,d\gamma,$$

$$\vec{W} = \big(\{w \in H^1(\Omega); w|_{\Gamma_1} = 0\}\big)^5,$$

$$E^{\alpha\beta\lambda\mu} = \frac{E}{2(1+\nu)}\left(a^{\alpha\lambda}a^{\beta\mu} + a^{\alpha\mu}a^{\beta\lambda} + \frac{2\nu}{1-\nu}a^{\alpha\beta}a^{\lambda\mu}\right),$$

$E =$ Young's modulus; $\nu =$ Poisson's coefficient,

$$a^{\alpha\beta} = \vec{a}^\alpha \cdot \vec{a}^\beta.$$

This formulation corresponds to the deformation of a shell which is clamped along a part $\vec{\phi}(\Gamma_0)$ of its boundary (with meas$(\Gamma_0) > 0$), loaded by a distribution of volumes forces with a resultant $\vec{p}$ upon the middle surface S of the shell and loaded by a distribution of surface forces applied to its lateral surface $\vec{\phi}(\Gamma_1)\times]-\frac{e}{2},\frac{e}{2}[$, $\Gamma_1 = \Gamma - \Gamma_0$, with resultant $\vec{N}$ and resultant moment $\underline{M} = \varepsilon_{\alpha\beta}M^\beta\vec{a}^\alpha$ along $\vec{\phi}(\Gamma_1)$.
Then, we have:

THEOREM 1 (see [**BCM**]). *Problem* (1) *has one and only one solution.*

1.3. Variational formulation of the Koiter model. The variational formulation can be stated as follows:

$$(2)\quad \begin{cases} \text{For any } \vec{p} \in \big(L^2(\Omega)\big)^3, \vec{N} \in \big(L^2(\Gamma_1)\big)^3 \text{ and } \underline{M} \in \big(L^2(\Gamma_1)\big)^2, \text{ find } \vec{u}^* \in \vec{V} \\ \text{such that } a(\vec{u}^*,\vec{v}) = f(\vec{v}), \quad \forall\vec{v} \in \vec{V}, \end{cases}$$

where

$$a(\vec{u},\vec{v}) = \int_\Omega eE^{\alpha\beta\lambda\mu}\Big[\gamma_{\alpha\beta}(\vec{u})\gamma_{\lambda\mu}(\vec{v}) + \frac{e^2}{12}\bar{\rho}_{\alpha\beta}(\vec{u})\bar{\rho}_{\lambda\mu}(\vec{v})\Big]\sqrt{a}\,d\xi^1\,d\xi^2,$$

$$f(\vec{v}) = \int_\Omega p^i v_i \sqrt{a}\,d\xi^1\,d\xi^2 + \int_{\Gamma_1}\big[N^i v_i + M^\lambda(v_{3,\lambda} + b^\mu_\lambda v_\mu)\big]\,d\gamma,$$

$$\vec{V} = \big\{\vec{v}\in\big(H^1(\Omega)\big)^2\times H^2(\Omega); \vec{v}\,|_{\gamma_0} = \vec{0}; \partial_n v_{3|\Gamma_0} = 0\big\}.$$

Then, we obtain

THEOREM 2 (see [**BCM, BC**]). *Problem* (2) *has one and only one solution.*

2. Approximation of the Displacement

The definition of the middle surface S through the application $\vec{\phi}$ is very interesting with respect to the approximation by finite element methods. Indeed, according to (1) or (2), the problem is defined on *the plane reference domain* Ω instead to be defined on the middle surface S so that the geometry of the shell appears only *through variable coefficients.* More precisely, the variational formulation can be rewritten as

$$\int_\Omega {}^tU[A_{IJ}]V\,d\xi^1\,d\xi^2 = \int_\Omega {}^tFV\,d\xi^1\,d\xi^2, \tag{3}$$

where the column matrix V (and similarly for the matrix U) is given by

$${}^tV = [v_1 v_{1,1} v_{1,2} v_2 v_{2,1} v_{2,2} v_3 v_{3,1} v_{3,2} v_{3,11} v_{3,12} v_{3,22}], \tag{4}$$

and where the square matrix $[A_{IJ}]_{12\times 12}$ is only dependent on e, E, ν and on the first, second and third partial derivatives of the mapping $\vec{\phi}$.

Subsequently, we assume for simplicity that the reference domain Ω is polygonal so that it can be exactly covered by a triangulation $\mathcal{T}_h$ made of nondegenerated triangles. Then, we consider three different families of approximation methods.

2.1. Conforming finite element methods. In Sections 2.1 and 2.2, we restrict our attention to the approximation of Koiter's model (2). In order to approximate the solution of this model:

(i) we construct a finite element subspace $\vec{V}_h$ of space $\vec{V}$:

$$\begin{cases} V_{h1}\subset V_1 = \big\{v\in H^1(\Omega), v|_{\Gamma_0} = 0\big\},\\ V_{h2}\subset V_2 = \big\{v\in H^2(\Omega), v|_{\Gamma_0} = \frac{\partial v}{\partial n}|_{\Gamma_0} = 0\big\},\\ \vec{V}_h = V_{h1}\times V_{h1}\times V_{h2}. \end{cases} \tag{5}$$

(ii) we introduce a *numerical integration scheme*, i.e., for any element K of the triangular $\mathcal{T}_h$:

$$\int_K \phi(\xi)\,d\xi \sim \sum_{\ell=1}^{L}\omega_{\ell,K}\phi(b_{\ell,K}). \tag{6}$$

V_{h1} (u_1 and u_2) $V_{h2}(u_3)$	Argyris 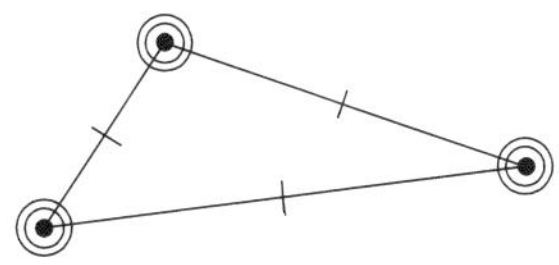	Ganev 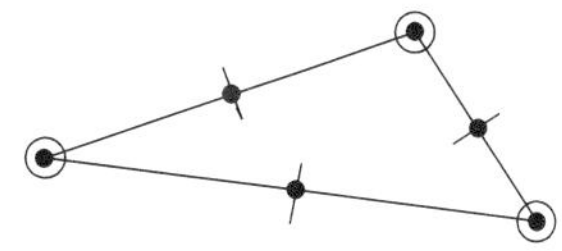
Argyris 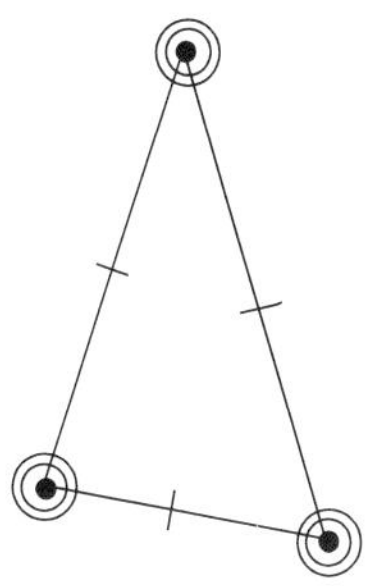	$\|\vec{u} - \vec{u}_h\| = 0(h^4)$ $\vec{u} \in \left(H^5(\Omega)\right)^2 \times H^6(\Omega)$, $A_{IJ} \in W^{4,\infty}(\Omega)$ $p^i \in W^{4,q}(\Omega), q \geq 2$ Integration scheme exact for polynomials of degree 8 (16 nodes)	$\|\vec{u} - \vec{u}_h\| = 0(h^4)$ $\vec{u} \in \left(H^5(\Omega)\right)^2 \times H^6(\Omega)$, $A_{IJ} \in W^{4,\infty}(\Omega)$ $p^i \in W^{4,q}(\Omega), q \geq 2$ Integration scheme exact for polynomials of degree 6 (12 nodes)

FIGURE 1. Examples of approximations of high degree of precision.

Then, from (3) and (4), we obtain

$$(7)\qquad \begin{aligned} a_h(\vec{u}_h, \vec{v}_h) &= \sum_{K \in \mathcal{T}_h} \sum_{\ell=1}^{L} \omega_{\ell,K} \left({}^t U_h [A_{IJ}] V_h\right)(b_{\ell,K}), \\ f_h(\vec{v}_h) &= \sum_{K \in \mathcal{T}_h} \sum_{\ell=1}^{L} \omega_{\ell,K} ({}^t F V_h)(b_{\ell,K}). \end{aligned}$$

Thus, the approximate problem can be stated as follows:

$$\text{Find } \vec{u}_h \in \vec{V}_h \text{ such that } a_h(\vec{u}_h, \vec{v}_h) = f(\vec{v}_h), \quad \forall \vec{v}_h \in \vec{V}_h.$$

The mathematical analysis of such an approximation is complicated owing to the presence of variable coefficients and to the coupling between tangential components and normal component of the displacement (see [**B1**]). Fig. 1 gives some illustrations of these results in case of an approximation of high degree of precision.

The space V_{h2} is constructed from Argyris finite elements (see [**A**]). For the space V_{h1}, we can take $V_{h1} = V_{h2}$, but this is expensive and useless. Indeed, in this case, the mathematical analysis shows that the optimal choice to construct V_{h1} is to use the Ganev's element (see [**G**]). For the same error estimate, for the same regularity on the data, that means less degrees of freedom and less integration nodes. This is illustrated by the example considered in Fig. 2. It consists in the approximation of the classical benchmark of a clamped hyperbolic paraboloid roof by methods of Fig. 1.

The mesh in use is described in Fig. 3 and takes into account the symmetry conditions. We also indicate the number of degrees of freedom, the number of unknowns, the number of integration nodes and the computing time corresponding to both methods: Argyris/Argyris or Argyris/Ganev.

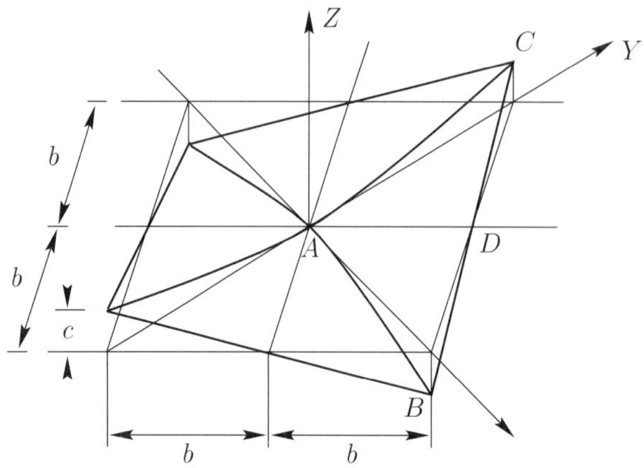

Equation:
$Z = c(Y^2 - X^2)/2a^2$
where
$a = 50\text{cm}$
$c = 10\text{cm}$
$h = 0.8\text{cm}$
$E = 2.85.10^4\text{kp/cm}^2$
$\nu = 0.4$
$q = 0.1\text{kp/cm}^2$

FIGURE 2. The hyperbolic paraboloid roof.

$V_{h1}(u_1 \text{ and } u_2)$	Argyris	Ganev
$V_{h2}(u_3)$	Argyris	Argyris
Number of d.o.f.	693	651
Number of unknowns	483	471
Number of integration nodes	576	432
Computing time (on BULL DPS8/70)		
· Rigidity matrix	693s	412s
· Second member	14s	9s
· Resolution	153s	107s
Total	860s	528s

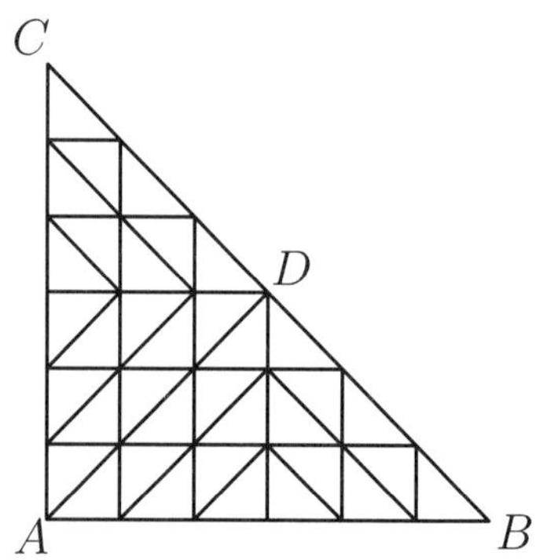

FIGURE 3. The mesh and some characteristic values of the implementation.

Thus the Argyris/Ganev method allows us to win about 40% of CPU time: 25% on the number of integration nodes plus about 20% on the obtention of elementary rigidity matrices (51 degrees of freedom for Argyris/Ganev instead of 63 in case of Argyris/Argyris).

Example of an engineering application. We have used such an approximation to compute stress distribution in an arch dam (see [**BB**]).

2.2. Nonconforming finite element methods. The inclusion $\vec{V}_h \subset \vec{V}$ requires to use C^1-finite elements the implementation of which is not really easy. Thus it could be interesting to use nonconforming elements: the associated finite element spaces are close to $\vec{V}$ but no longer subspaces of $\vec{V}$.

A review of such elements can be found in [**C**] for plate problems and in [**T**] for thin shell problems. Figures 4 and 5 exhibit "semiloof" [**M, FV**] and "delinquent" [**FV, S**] elements with corresponding a priori error estimates.

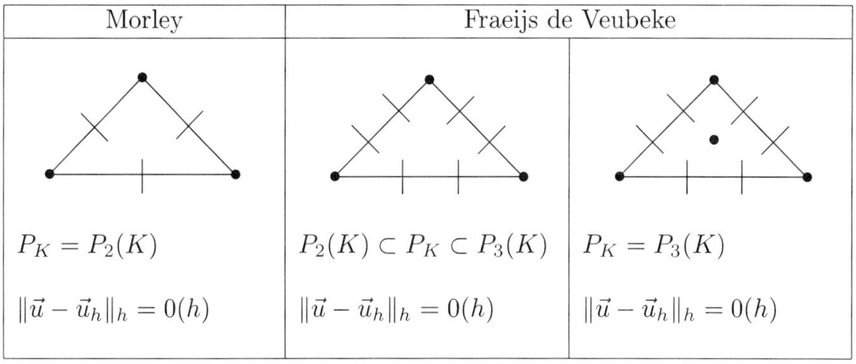

FIGURE 4. The semiloof finite elements.

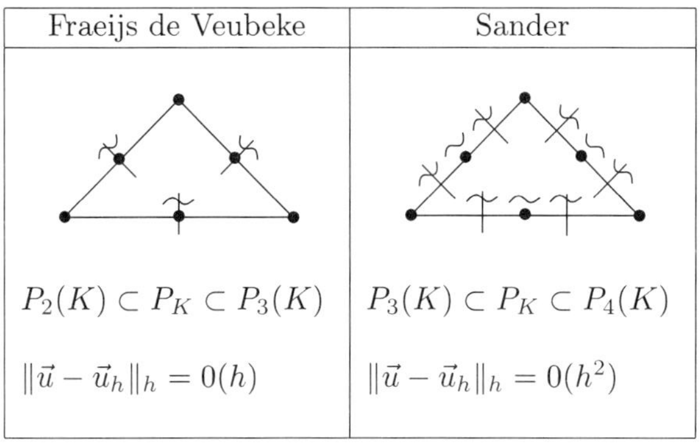

FIGURE 5. The delinquent elements.

In order to decrease the level of required regularity, the definition of delinquent finite elements uses mean values of the function (note $\tilde{\bullet}$) and mean values of its normal derivative (noted $\tilde{|}$) along the sides of the triangles. These delinquent finite elements are very interesting since they:

(i) are easy to implement;
(ii) are very well adapted to junction problems (see paragraph 4);
(iii) have good degree of accuracy, i.e., $0(h^2)$ for Sander element. Theoretical studies will be developed in [**BP**].

2.3. The DKT (Discrete Kirchhoff Triangle) methods. These methods are somewhat different from the previous ones since the starting model is that by Naghdi, i.e., the variational formulation (1). The main features of these methods are as follows

(i) The finite element spaces to approximate the displacement $\vec{v}$ and the rotation $\underset{\sim}{\delta}$: each component v_i, $i = 1, 2, 3$ and δ_α, $\alpha = 1, 2$ is approximated in a finite dimensional space associated with one of the finite elements shown in Fig. 6.

Other possibilities exist for choosing finite element spaces to approximate the components v_i and δ_α: keeping the same finite elements for the components v_3, δ_1 and δ_2, we can use linear triangles or reduced cubic Hermite triangles to approximate the components u_1 and u_2. In the same way,

Components	Type of finite element	Representation of this finite element	Associated space of finite elements
v_1, v_2, δ_1, δ_2	Quadratic triangle	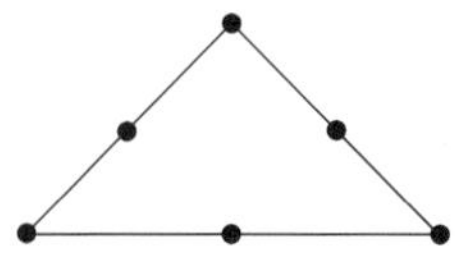	X_{h1}
v_3	Reduced cubic Hermite triangle	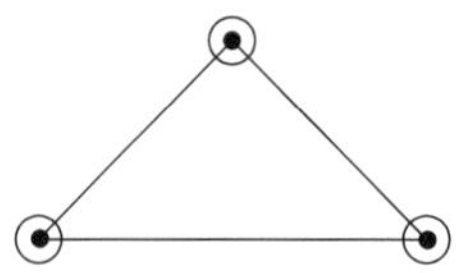	X_{h2}

FIGURE 6. The finite elements used.

instead of using triangles, we can use quadrilaterals, which leads to D.K.Q. methods (see [**BA1, BA2**]).

(ii) *Kirchhoff-Love type constraints imposed on the degrees of freedom*: for a given triangle, we have a total of 33 degrees of freedom. We impose 12 degrees, thus bringing the total down to 21 effective degrees of freedom:

$$v_{h\alpha}(a_i), v_{h\alpha}(b_i), v_{h3}(a_i), \delta_{h\alpha}(a_i), \quad \alpha = 1,2; i = 1,2,3,$$

where a_i, b_i denote respectively the vertices and midsides of the triangle. The aim of these constraints is to obtain a well-posed discrete problem: in the definition of the discrete problem, we neglect the transverse shear energy $c\big[(\vec{v}, \underline{\delta}), (\vec{v}, \underline{\delta})\big]$ (see (1)). These constraints then enable us to satisfactorily link the displacement and rotation degrees of freedom. The twelve *Kirchhoff-Love type constraints* are:

$$\text{(8)} \qquad \begin{cases} 6 \text{ constraints at the vertices } a_i, i = 1,2,3; \\ 3 \text{ “tangential” constraints at the midsides } b_i, i = 1,2,3; \\ 3 \text{ “normal” averaged constraints on the sides } a_{i+1}a_{i-1}. \end{cases}$$

Then, the approximated DKT problem can be defined as follows:

$$\text{Find } (\vec{u}_h, \underline{\beta}_h) \in \vec{Z}_h \text{ such that } b\big[(\vec{u}_h, \underline{\beta}_h), (\vec{v}_h, \underline{\delta}_h)\big] = f(\vec{v}_h, \underline{\delta}_h), \quad \forall (\vec{v}_h, \underline{\delta}_h) \in \vec{Z}_h,$$

where the space $\vec{Z}_h$ is defined as follows:

$$\vec{Z}_h = \big\{(\vec{v}_h, \underline{\delta}_h) \text{ such that } \vec{v}_h \in (V_{h1})^2 \times V_{h2}, \underline{\delta}_h \in (V_{h1})^2 \text{ and } (\vec{v}_h, \underline{\delta}_h)|_K \text{ satisfies relations (8), for all } K \in \mathcal{T}_h\big\},$$

and where the spaces $V_{h\alpha} = \{v_h \in X_{h\alpha}; v_{h|\Gamma_0} = 0\}$, the spaces $X_{h\alpha}$ being defined in Fig. 6.

In [**BM**], we have proved the convergence of this DKT method and obtained the following a priori error estimate:

$$\|\vec{u} - \vec{u}_h\|_h \leq Ch\big(\|u_1\|^2_{2,\Omega} + \|u_2\|^2_{2,\Omega} + \|u_3\|^2_{3,\Omega}\big)^{1/2}$$

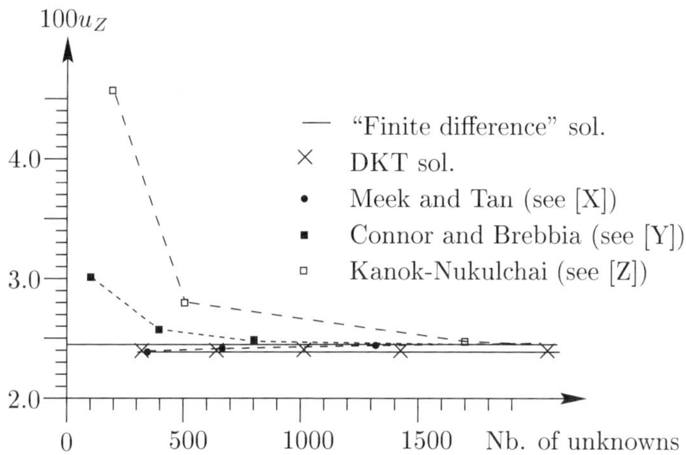

FIGURE 7. Comparison between the values $u_Z(A)$ given by various authors (see [X]).

where $\vec{u}$ is the unique solution of Koiter's model and where the norm $\|\vec{v}_h\|_h = \left\{\|\vec{v}_h\|_{1,\Omega} + \sum_{K\in T_h} |v_{h3}|^2_{2,T}\right\}^{1/2}$ takes into account the nonconformity of the space $\vec{V}_h$.

The efficiency of this method has been checked on several benchmarks. For instance for the example described in Fig. 2 and with the mesh indicated in Fig. 3, we have obtained results displayed in Fig. 7

3. Joint Approximation of the Geometry and of the Displacement

In the engineering world, the shell structures we encounter are seldom as geometrically well defined as those we have considered in the first two paragraphs. For this reason, we often need an extra approximation for the shell geometry when analysing the shells used in industry; this is the case, for example,

(i) *in shells where we only have a point by point description of the middle surface*: the explicit analytical expression of the mapping $\vec{\phi}$ is not available. Here, we consider two methods for the approximation of the geometry:
  - *approximation by flat facets.* This leads to methods of low accuracy;
  - *approximation by B-spline methods.* Using this method, we were able to perform a highly accurate approximation of a turbine blade for which we had no access to the mapping $\vec{\phi}$.

(ii) *in shells which are actually an assemblage of various elementary shells.* Here, we have to take the shell junctions into account (see paragraph 4).

3.1. Flat facet element methods. Hereunder we describe one of the most representative flat plate element methods due to Clough-Johnson (see [**CJ1**,**CJ2**]). The main steps of this method are (see Fig. 8):

STEP 1. The middle surface $\overline{\mathcal{S}}$ is replaced by a *continuous faceted surface* $\overline{\mathcal{S}}_h$ whose vertices are on $\overline{\mathcal{S}}$.

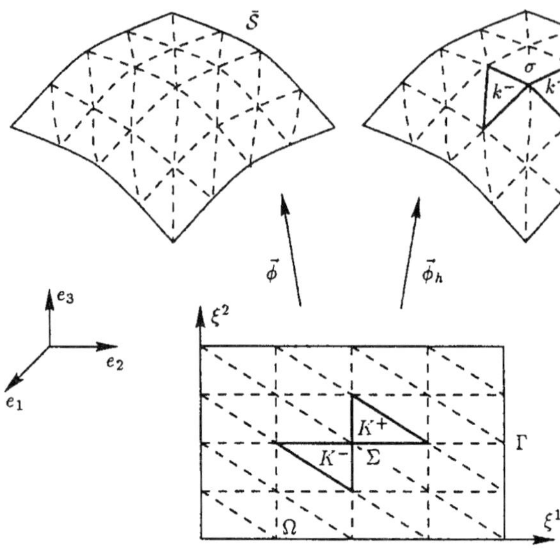

FIGURE 8. The initial middle surface $\bar{\mathcal{S}}$ and the flat facet middle surface $\bar{\mathcal{S}}_h$.

STEP 2. On each plane facet, we define the corresponding *plate energy* which is approximated, *facet by facet*, through conforming finite element methods: P_1-triangles for membrane components of the displacement and reduced HCT-triangle for the deflection.

STEP 3. In order to get an approximate energy which is consistent (i.e. sufficiently closed) to the exact energy of the shell, we introduce the following *compatibility conditions* on the discrete space:

(i) the displacement is continuous at the vertices of the surface $\mathcal{S}_h$;
(ii) the tangential components to the initial middle surface $\mathcal{S}$ of the rotation vector, are continuous at the vertices of the surface $\mathcal{S}_h$;

By using such a method, computational mechanicians have obtained satisfactory or unsatisfactory numerical results according to the considered problems. For our part, we have tried to justify such methods from mathematical analysis point of view. The conclusions are the following:

(i) The approximation of thin shell problems by flat facets is not always convergent for general thin shells. In this way, we have clearly identified the term which prevents such an unconditionally convergence.
(ii) To circumvent this lack of convergence, we have shown
  - that for sufficiently shallow shells, the previous term remains sufficiently small and leads to a good convergence result;
  - that by imposing a slight perturbation to the bending term, it is possible to obtain a new method which is unconditionally convergent even for shells of general shape.
(iii) Along the same lines, the addition of drilling degrees of freedom does not change significantly the problem;

iv) On the contrary, the replacement of flat facet elements ($P1$-approximation of the geometry) by curved facet elements ($P2$-approximation of the geometry) leads to an unconditionally convergent method.

Anyway, all these kinds of improvements lead in the best to $0(h)$ convergent methods only. In the next section, we consider a $0(h^4)$ high precision approximation of the shell geometry.

3.2. Approximation by B-spline functions of the shell geometry. The geometrical data of the shell are still the same: coordinates of a finite number of points located on the middle surface and values of the thickness at these points. Then we assume that the mappings $\vec{\phi}$ and e exist, are sufficiently regular but not available. Then, we obtain a high precision approximation method (geometry and displacement) by combining:

(i) The definition of an approximation $\vec{\phi}_{h_S}$ of the mapping $\vec{\phi}$ such that $\vec{\phi}_{h_S} : \Omega \to \mathcal{S}_{h_S} = \vec{\phi}_{h_S}(\Omega)$.
To do this, we use the tensor products of B-spline functions of order k (integer ≥ 1) which are piecewise polynomial of degree $k-1$ and which effect $\mathcal{C}^{k-2}$-class junctions. The order k will be chosen in relation to the shell theory being used (general or shallow) and the accuracy of other approximations carried out, especially displacement approximations. For a detailed account of *B-spline functions*, we refer the reader to [**BO**].

(ii) The corresponding approximation result, i.e., for any element K of the triangulation

$$|\vec{\phi} - \vec{\phi}_{h_S}|_{r,\infty,K} \leq C h_{\mathcal{S}}^{t-r} |\vec{\phi}|_{t,\infty,K}, \quad t = 1, \ldots, k \text{ and } r = 0, \ldots, t.$$

(iii) The approximation e_{h_S} of the thickness function e. Here again, we use products of B-spline functions of order $\ell \geq 1$, which leads to the following approximation result: for any element K in the mesh

$$|e - e_{h_S}|_{r,\infty,K} \leq C h_{\mathcal{S}}^{t-r} |e|_{t,\infty,K}, \quad t = 1, \ldots, \ell \text{ and } r = 0, \ldots, t.$$

(iv) The two approximation results for the shell geometry (respectively, middle surface and thickness), are combined with those obtained previously for the approximation of the displacement field, in order to deduce the convergence properties of approximated solutions $\vec{u}_{\underset{\sim}{h}}$ towards the solution $\vec{u}$ of the continuous problem, together with asymptotic estimations of the corresponding error (the index $h_{\underset{\sim}{h}} = (h_{\mathcal{S}}, h)$ is a reminder that the approximated solutions depend on the geometry approximation ($h_{\mathcal{S}}$) and the displacement approximation (h)). This study is undertaken for the case of general shells; it can readily be transposed to shallow shells.

(v) These various results with examples of error estimates obtained for various given choices of B-spline and finite element methods. The combinations preferred are those which ensure the best possible coherence between the various components of the error.

Engineering application. This method was developed jointly with Turboméca to perform very precise calculations on the vibration eigenmodes of turbine and compressor blades (see [**BL**]).

4. Junction Problems

In industrial constructions, the assemblages between thin structures are very common, in particular those between thin shells. Subsequently, we propose a modelization of such a junction between thin shells. For more details, we refer to [**BF**] for junctions between plates, and to [**BC1, BC2**] for junctions between shells.

4.1. Variational formulation. Let us consider two thin shells whose middle surfaces S and $\widetilde{S}$ join along a common boundary Γ. We assume that these middle surfaces S and $\widetilde{S}$ are the images of plane domains Ω and $\widetilde{\Omega}$ through bijections $\vec{\phi}$ and $\vec{\widetilde{\phi}}$. Moreover, we assume that:

(i) The shell S is clamped along a part ∂S_0 of its boundary, with meas $(\partial S_0) > 0$ and is loaded along ∂S_1 so that $\partial S = \partial S_0 \cup \partial S_1 \cup \Gamma$ and $\partial\Omega = \partial\Omega_0 \cup \partial\Omega_1 \cup \gamma$ where $\gamma = (\vec{\phi})^{-1}(\Gamma)$.

(ii) The shell $\widetilde{S}$ is loaded along $\partial\widetilde{S}_1$ so that $\partial\widetilde{S} = \partial\widetilde{S}_1 \cup \Gamma, \partial\widetilde{\Omega} = \partial\widetilde{\Omega}_1 \cup \widetilde{\gamma}$ where $\widetilde{\gamma} = (\vec{\widetilde{\phi}})^{-1}(\Gamma)$.

The application of the action-reaction principle at any point of the hinge Γ implies the transmission of the efforts, i.e. $\vec{N}(P) = \vec{\widetilde{N}}(P)$ and $\vec{M}(P) = \vec{\widetilde{M}}(P)$, $\forall P \in \Gamma$. Subsequently, we examine two types of hinge behaviours :

(i) a rigid behaviour for which the continuity of the displacements and of the tangential rotations along the hinge is insured at any points P of Γ, i.e. $\vec{u}(P) = \vec{\widetilde{u}}(P), (\vec{\psi}.\vec{t})(P) = (\vec{\widetilde{\psi}}.\vec{t})(P) = \big[(\vec{\widetilde{t}}.\vec{t})(\vec{\widetilde{\psi}}.\vec{\widetilde{t}})\big](P)$;

(ii) an elastic behaviour, i.e. $\vec{u}(P) = \vec{\widetilde{u}}(P), M_n(P) = k\big[(\vec{\psi} - \vec{\widetilde{\psi}}).\vec{t}\big](P)$, where $\vec{\psi}$ (resp. $\vec{\widetilde{\psi}}$) is the infinitesimal rotation vector and where the coefficient k of elastic stiffness of the hinge is a given positive constant attached to the problem into consideration; it should be determined experimentally. The extension to the case of a variable coefficient k along the hinge is always possible.

The equations of the junction problem. The equilibrium of shells S and $\widetilde{S}$ lead to (see [**K**]):

$$\begin{cases} \big[n^{\alpha\beta} + \frac{1}{2}b^\alpha_\nu m^{\beta\nu} - \frac{1}{2}b^\beta_\nu m^{\alpha\nu}\big]\big|_\beta + b^\alpha_\nu m^{\beta\nu}|_\beta + p^\alpha = 0, \\ m^{\alpha\beta}|_{\alpha\beta} - b_{\alpha\beta}n^{\alpha\beta} - p^3 = 0, \text{ in } \Omega, \end{cases}$$

$$\begin{cases} \big[\widetilde{n}^{\alpha\beta} + \frac{1}{2}\widetilde{b}^\alpha_\nu \widetilde{m}^{\beta\nu} - \frac{1}{2}\widetilde{b}^\beta_\nu \widetilde{m}^{\alpha\nu}\big]\big|_\beta + \widetilde{b}^\alpha_\nu \widetilde{m}^{\beta\nu}|_\beta + \widetilde{p}^\alpha = 0, \\ \widetilde{m}^{\alpha\beta}|_{\alpha\beta} - \widetilde{b}_{\alpha\beta}\widetilde{n}^{\alpha\beta} - \widetilde{p}^3 = 0, \text{ in } \widetilde{\Omega}, \end{cases}$$

while the boundary conditions are respectively given by $\vec{u} = \vec{0}; u_{3,n} = 0$, on $\partial\Omega_0$,

$$\begin{cases} N^\alpha + b^{\alpha\beta}(M_n n_\beta + M_t t_\beta) = \big[n^{\alpha\beta} + \frac{3}{2}b^\alpha_\nu m^{\beta\nu} - \frac{1}{2}b^\beta_\nu m^{\alpha\nu}\big]n_\beta, \\ N^3 - M_{t,s} = -m^{\alpha\beta}|_\alpha n_\beta - (m^{\alpha\beta}n_\alpha t_\beta)_{,s}, \quad M_n = m^{\alpha\beta}n_\alpha n_\beta, \text{ on } \partial\Omega_1, \end{cases}$$

$$\begin{cases} \widetilde{N}^\alpha + \widetilde{b}^{\alpha\beta}(\widetilde{M}_n \widetilde{n}_\beta + \widetilde{M}_t \widetilde{t}_\beta) = \big[\widetilde{n}^{\alpha\beta} + \frac{3}{2}\widetilde{b}^\alpha_\nu \widetilde{m}^{\beta\nu} - \frac{1}{2}\widetilde{b}^\beta_\nu \widetilde{m}^{\alpha\nu}\big]\widetilde{n}_\beta, \\ \widetilde{N}^3 - \widetilde{M}_{\widetilde{t},\widetilde{s}} = -\widetilde{m}^{\alpha\beta}|_\alpha \widetilde{n}_\beta - (\widetilde{m}^{\alpha\beta}\widetilde{n}_\alpha \widetilde{t}_\beta)_{,\widetilde{s}}, \quad \widetilde{M}_{\widetilde{n}} = \widetilde{m}^{\alpha\beta}\widetilde{n}_\alpha \widetilde{n}_\beta, \text{ on } \partial\widetilde{\Omega}_1. \end{cases}$$

Moreover, the conditions of junction upon Γ are depending on the type of hinge ; for a rigid or an elastic hinge, we get respectively:

$$\vec{u} = \vec{\tilde{u}}; \quad -n^\beta(u_{3,\beta} + b_\beta^\alpha u_\alpha) = -(\vec{t}.\vec{\tilde{t}})\tilde{n}^\beta(\tilde{u}_{3,\beta} + \tilde{b}_\beta^\alpha \tilde{u}_\alpha),$$

$$\vec{u} = \vec{\tilde{u}}; \quad k\big[(\vec{t}.\vec{\tilde{t}})\tilde{n}^\beta(\tilde{u}_{3,\beta} + \tilde{b}_\beta^\alpha \tilde{u}_\alpha) - n^\beta(u_{3,\beta} + b_\beta^\alpha u_\alpha)\big] = M_n.$$

Then, these elastic and rigid junctions problems can be variationaly formulated as follows:

Case of an elastic hinge. Find $(\vec{u}^k, \vec{\tilde{u}}^k) \in W_{e\ell}$ *such that*

(9) $$a\big[(\vec{u}^k; \vec{\tilde{u}}^k), (\vec{v}; \vec{\tilde{v}})\big] + kb\big[(\vec{u}^k; \vec{\tilde{u}}^k), (\vec{v}; \vec{\tilde{v}})\big] = \ell(\vec{v}; \vec{\tilde{v}}), \quad \forall(\vec{v}; \vec{\tilde{v}}) \in W_{e\ell},$$

where

$$\begin{cases} a\big[(\vec{u}; \vec{\tilde{u}}), (\vec{v}; \vec{\tilde{v}})\big] = \int_\Omega eE^{\alpha\beta\lambda\mu}\big[\gamma_{\lambda\mu}(\vec{u})\gamma_{\alpha\beta}(\vec{v}) + \frac{e^2}{12}\rho_{\lambda\mu}(\vec{u})\rho_{\alpha\beta}(\vec{v})\big]\sqrt{a}\, d\xi^1\, d\xi^2 \\ \qquad + \int_{\tilde{\Omega}} \tilde{e}\tilde{E}^{\alpha\beta\lambda\mu}\big[\tilde{\gamma}_{\lambda\mu}(\vec{\tilde{u}})\tilde{\gamma}_{\alpha\beta}(\vec{\tilde{v}}) + \frac{\tilde{e}^2}{12}\tilde{\rho}_{\lambda\mu}(\vec{\tilde{u}})\tilde{\rho}_{\alpha\beta}(\vec{\tilde{v}})\big]\sqrt{\tilde{a}}\, d\tilde{\xi}^1\, d\tilde{\xi}^2, \end{cases}$$

$$b\big[(\vec{u}; \vec{\tilde{u}}), (\vec{v}; \vec{\tilde{v}})\big] = \int_\gamma \big[n^\beta(u_{3,\beta} + b_\beta^\alpha u_\alpha) - \tilde{n}^\beta(\vec{t}.\vec{\tilde{t}})(\tilde{u}_{3,\beta} + \tilde{b}_\beta^\alpha \tilde{u}_\alpha)\big]$$
$$\times\big[n^\lambda(v_{3,\lambda} + b_\lambda^\mu v_\mu) - \tilde{n}^\lambda(\vec{t}.\vec{\tilde{t}})(\tilde{v}_{3,\lambda} + \tilde{b}_\lambda^\mu \tilde{v}_\mu)\big]\, d\gamma,$$

$$\ell(\vec{v}; \vec{\tilde{v}}) = \int_\Omega \vec{p}.\vec{v}\sqrt{a}\, d\xi^1\, d\xi^2 + \int_{\tilde{\Omega}} \vec{\tilde{p}}.\vec{\tilde{v}}\sqrt{\tilde{a}}\, d\tilde{\xi}^1\, d\tilde{\xi}^2 + \int_{\partial\Omega_1} \big[\vec{N}.\vec{v} + \vec{M}.\vec{\psi}(\vec{v})\big]\, d\omega$$
$$+ \int_{\partial\tilde{\Omega}_1} \big[\vec{\tilde{N}}.\vec{\tilde{v}} + \vec{\tilde{M}}.\vec{\tilde{\psi}}(\vec{\tilde{v}})\big]\, d\tilde{\omega},$$

$$W_{e\ell} = \big\{(\vec{w}; \vec{\tilde{w}}) \in \vec{V} \times \vec{\tilde{V}}; \vec{w} = \vec{\tilde{w}} \text{ on } \gamma\big\}, \quad \vec{V} = V_1 \times V_1 \times V_2,$$
$$\vec{\tilde{V}} = \big(H^1(\tilde{\Omega})\big)^2 \times H^2(\tilde{\Omega}),$$

(10) $$V_1 = \big\{v \in H^1(\Omega); v = 0 \text{ on } \partial\Omega_0\big\}; \; V_2 = \big\{v \in H^2(\Omega), v = v_{,n} = 0 \text{ on } \partial\Omega_0\big\}.$$

Then, we obtain :

THEOREM 3 (see [**BC1**]). *Assume that the data are sufficiently smooth. Then, problem* (9) *has one and only one solution.*

Case of a rigid hinge. With previous notations, corresponding variational formulation can be written as : Find $(\vec{u}; \vec{\tilde{u}}) \in W_{\text{rig}}$ such that

(11) $$a\big[(\vec{u}; \vec{\tilde{u}}), (\vec{v}; \vec{\tilde{v}})\big] = \ell(\vec{v}; \vec{\tilde{v}}), \quad \forall(\vec{v}; \vec{\tilde{v}}) \in W_{\text{rig}},$$

where

$$W_{\text{rig}} = \big\{(\vec{w}; \vec{\tilde{w}}) \in \vec{V} \times \vec{\tilde{V}}; \vec{w} = \vec{\tilde{w}} \text{ on } \gamma;$$
$$n^\beta(w_{3,\beta} + b_\beta^\alpha w_\alpha) - (\vec{t}.\vec{\tilde{t}})\tilde{n}^\beta(\tilde{w}_{3,\beta} + \tilde{b}_\beta^\alpha \tilde{w}_\alpha) = 0 \text{ on } \gamma\big\}.$$

Then, one can prove that:

THEOREM 4. *Under the assumptions of Theorem* 3, *the problem* (11) *has one and only one solution.*

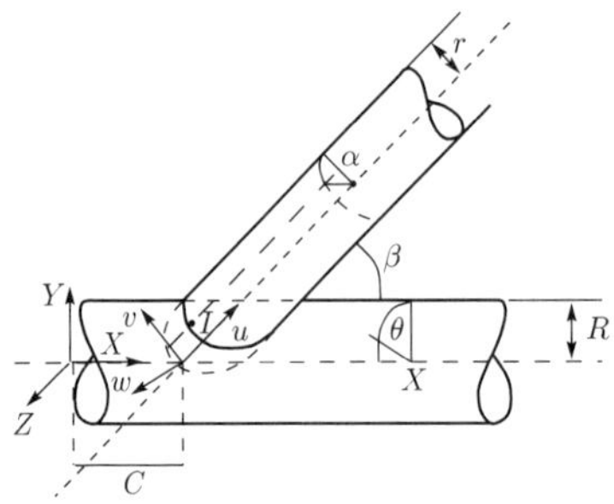

Main tube

$$Y^2+Z^2=R^2 \iff \vec{\phi}=\begin{cases} X=X \\ Y=R\cos\theta \\ Z=R\sin\theta \end{cases}$$

Secondary tube

$$\vec{\tilde{\phi}}=\begin{cases} X=C+u\cos\beta-r\sin\beta\cos\alpha \\ Y=u\sin\beta+r\cos\beta\cos\alpha \\ Z=r\sin\alpha \end{cases}$$

FIGURE 9. Junction between two circular cylinders.

Main tube

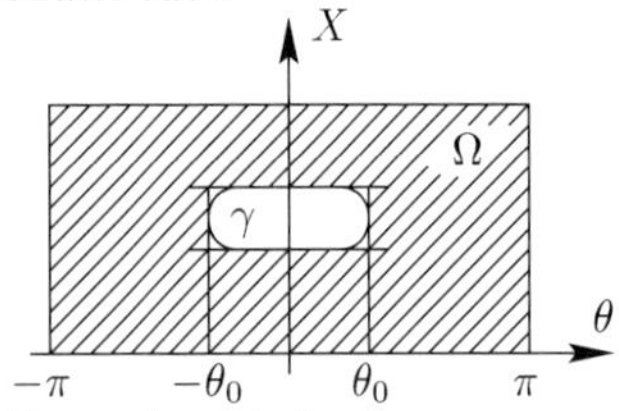

$$\sin\theta_0=\tfrac{r}{R},\ 0<\theta_0<\tfrac{\pi}{2}$$

$$(\gamma)\begin{cases} X=C+R\frac{\cos\theta}{tg\beta}\pm\frac{r}{\sin\beta}\sqrt{1-\frac{R^2}{r^2}\sin^2\theta} \\ \theta\in[-\theta_0,\theta_0] \end{cases}$$

Secondary tube

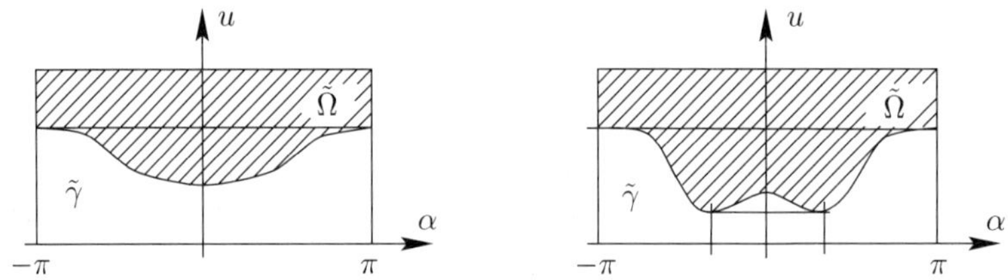

Case $r\le R\cos\beta$ Case $R\cos\beta<r$

Reference domains $\left((\tilde{\gamma}):u=\left\{\sqrt{R^2-r^2\sin^2\alpha}-r\cos\alpha\cos\beta\right\}/\sin\beta\right)$

FIGURE 10. Corresponding reference domains and parameterization of intersection line.

THEOREM 5. *Let* $(\vec{u}^k;\vec{\tilde{u}}^k)$ *and* $(\vec{u};\vec{\tilde{u}})$ *be the solutions of the elastic and rigid hinge problem* (9) *and* (11). *Then,* $\lim_{k\to+\infty}\left\|(\vec{u}^k;\vec{\tilde{u}}^k)-(\vec{u};\vec{\tilde{u}})\right\|_{W_{e\ell}}=0$.

4.2. Engineering example: junction between two circular cylinders. Such a junction is very common in oil platforms. Although the geometry is simple, the determination of the intersection line is not easy. The mappings $\vec{\phi}$ and $\vec{\tilde{\phi}}$, and the reference domains Ω and $\tilde{\Omega}$ are given in Fig.s 9 and 10. These domains are no longer polygonal and we have to consider the additional approximation of curved boundaries.

4.3. Some finite element approximation.

Conforming finite element approximation. Concerning the junction problem, let us refer to [**BF**] for a junction between plates and to [**BC1,BC2**] for a junction between shells. This last case is much more complicated and it remains many open problems, even if we consider junctions as simple as the one presented in Fig. 9.

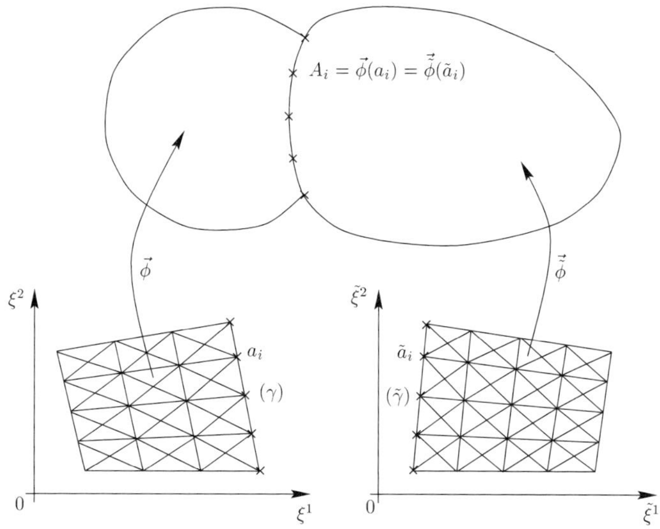

FIGURE 11. Triangulations associated to the junction problem (case of polygonal domains Ω and $\widetilde{\Omega}$.

Let us point out these difficulties in case of an elastic hinge. In that case, the variational formulation is given by relation (9). For simplicity, we assume that the domains Ω and $\widetilde{\Omega}$ are polygonal so that we can construct finite element spaces $\vec{V}_h$ and $\vec{\tilde{V}}_h$ such that (see (10)):

$$\vec{V}_h \subset \vec{V} \text{ and } \vec{\tilde{V}}_h \subset \vec{\tilde{V}}.$$

To fix the ideas, assume that these spaces are constructed by means of $\mathcal{C}^1$-quintic elements (see [**A**]). Then, thanks to the spaces $\vec{V}_h$ and $\vec{\tilde{V}}_h$, we can approximate the solution $(\vec{u}^k, \vec{\tilde{u}}^k)$ of problem (9) in the finite dimensional space

$$\begin{aligned}(12)\quad W_{e\ell h} = \big\{&(\vec{w}_h; \vec{\tilde{w}}_h) \in \vec{V}_h \times \vec{\tilde{V}}_h; \vec{w}_h(a_i) = \vec{\tilde{w}}_h(\widetilde{a}_i); t^\alpha(a_i)\vec{w}_{h,\alpha}(a_i) \\ &= \tilde{t}^\alpha(\widetilde{a}_i)\vec{\tilde{w}}_{h,\alpha}(\widetilde{a}_i); t^\alpha(a_i)t^\beta(a_i)\vec{w}_{h,\alpha\beta}(a_i) = \tilde{t}^\alpha(\widetilde{a}_i)\tilde{t}^\beta(\widetilde{a}_i)\vec{\tilde{w}}_{h,\alpha\beta}(\widetilde{a}_i)\big\},\end{aligned}$$

where a_i and $\widetilde{a}_i$ are the vertices of the triangulations $\mathcal{T}_h$ and $\widetilde{\mathcal{T}}_h$ of the domains Ω and $\widetilde{\Omega}$ whose images through the mappings $\vec{\phi}$ and $\vec{\tilde{\phi}}$ are located on the hinge (Γ) and such that $A_i = \vec{\phi}(a_i) = \vec{\tilde{\phi}}(\widetilde{a}_i)$ (see Fig. 11).

Let us observe that :

(i) the constraints used in the definition (12) come from the study of nonhomogeneous boundary conditions;

(ii) these constraints are not sufficient to imply $\vec{w}_h = \vec{\tilde{w}}_h$ along $\Gamma = \vec{\phi}(\gamma) = \vec{\tilde{\phi}}(\widetilde{\gamma})$ so that, in general, we obtain $W_{e\ell h} \not\subset W_{e\ell}$.

The approximated problem. By similarity with problem (9), the associate approximated problem can be formulated as:

$$
(13)\qquad \begin{cases} \textit{Find } (\vec{u}_h^k;\vec{\tilde{u}}_h^k) \in W_{e\ell h} \quad \textit{such that} \\ a\big[(\vec{u}_h^k;\vec{\tilde{u}}_h^k),(\vec{v}_h;\vec{\tilde{v}}_h)\big] + kb_h\big[(\vec{u}_h^k;\vec{\tilde{u}}_h^k),(\vec{v}_h;\vec{\tilde{v}}_h)\big] \\ \qquad\qquad = \ell(\vec{v}_h;\vec{\tilde{v}}_h), \forall(\vec{v}_h;\vec{\tilde{v}}_h) \in W_{e\ell h}, \end{cases}
$$

where $b_h[.,.]$ is a suitable approximation of $b[.,.]$ associated for example to the use of numerical integration schemes. This approximation takes into account that generally the bilinear form $b[.,.]$ cannot be computed exactly along Γ. Of course, we can—or, generally, we must!—also use some numerical integration techniques to approximate $a[.,.]$ and $\ell(.)$. In [**BC2**], we prove the convergence of this approximation method and we obtain an a priori error estimate which is at the best possible level:

$$\big\|(u^k,\tilde{u}^k) - (u_h^k,\tilde{u}_h^k)\big\| = 0(h^4).$$

Nonconforming finite element methods. Two sets of nonconforming finite element methods are very well adapted to the study of junctions

(i) The P_1-triangle, used to approximate the tangential components of the displacement, when combined with the Morley element [**M**] used to approximate the normal component. The degrees of freedom (see Fig. 4) are very well adapted since, along the junction, we have to impose the continuity of displacement at the vertices, and, in case of a rigid hinge, the continuity of rotations at midsides. In studying the error estimate for the approximation of the junction problem, we should find back the results of [**T**] obtained in case of one shell.

(ii) The P_2-triangle (membrane components) combined with the "delinquent" triangle introduced by Sander [**S**] (bending components). Here again the choice of degrees of freedom is very well adapted to junction problem (see Fig. 5 and [**BP**]).

4.4. Some remarks. The modelization of junctions assumes that the shell is thin and that the junctions are "perfect". Many problems remain open like:

- precise determination of the junction;
- case of big increase of the thickness around the junction. Comparison with 3D approaches of the junction should be fruitful as well as mathematical justifications by the asymptotic development method;
- extensions to Reissner-Mindlin modelizations (which include transverse shear effects) and study of the associated boundary layers.

Acknowledgments

This work is a part of the Human Capital and Mobility Program "Shells: Mathematical Modeling and Analysis, Scientific Computing" of the Commission of the European Communities (Contract no. ERBCHRXCT 940536).

References

[A] J. H. Argyris, I. Fried, and D. W. Scharpf, *The* TUBA *family of plate elements for the matrix displacement method*, J. Royal Aero. Soc. **72** (1968), 701-709.

[BA1] J. L. Batoz and M. Ben Tahar, *Evaluation of a new thin plate quadrilateral element*, Internat. J. Numer. Methods Engrg. **18** (1982), 1655–1677.

[BA2] ———, *Modélisation des structures par éléments finis. II. Poutres et plaques*; III. *coques*, Hermes, Paris, 1990 and 1992.

[B1] M. Bernadou, *Convergence of conforming finite element methods for general shells problems*, Internat. J. Engrg. Sci. **18** (1980), 249–276.

[B2] ———, *Finite element methods for thin shell problems*, J. Wiley & Sons, Chichester, 1996.

[BB] M. Bernadou and J. M. Boisserie, *The finite element method in thin shell theory; Application to arch dam simulations*, Birkhaüser, Boston, 1982.

[BC] M. Bernadou and P. G. Ciarlet, *Sur l'ellipticité du modèle linéaire de coques de W. T. Koiter*, Computing Methods in Applied Sciences and Engineering (R. Glowinski and J. L. Lions, eds.), Lectures Notes in Economics and Mathematical Systems, vol. 134, Springer-Verlag, Berlin, pp. 89–136.

[BCM] M. Bernadou, P. G. Ciarlet, and B. Miara, *Existence theorems for two-dimensional linear shell theories*, J. Elasticity **34** (1994), 111–138.

[BC1] M. Bernadou and A. Cubier, *Numerical analysis of junctions between thin shells.* I. *Continuous problems*, Comput. Methods Appl. Mech. Engrg. **161** (1998), 349–363.

[BC2] ———, *Numerical analysis of junctions between thin shells.*II. *Continuous problems*, Comput. Methods Appl. Mech. Engrg. **161** (1998), 365–387.

[BF] M. Bernadou, S. Fayolle, and F. Léné, *On the approximation of junctions of beams and plates*, Comput. Methods Appl. Mech. Engrg. **74** (1989), 307–326.

[BL] M. Bernadou and B. Lalanne, *On the approximation of thin shells by* B-*spline and finite element methods*, Innovative Numerical Methods in Engineering (R. P. Shaw and J. Périaux, eds.) (Atlanta, 1986), 1986, pp. 585–592.

[BM] M. Bernadou, P. Mato Eiroa, and P. Trouvé, *On the convergence of a* D.K.T. *method valid for shells or arbitrary shape*, Comput. Methods Appl. Mech. Engrg. **118** (1994), 373–391.

[BP] M. Bernadou, F. Palma, P. Trouvé, and P. Vilariño, *Approximation of junctions between plates or between shells by "delinquent" finite element* (to appear).

[BO] C. de Boor, *A practical guide to splines*, Springer-Verlag, New-York, 1978.

[C] P. G. Ciarlet, *The finite element method for elliptic problems*, Amsterdam, North-Holland, 1978.

[CJ1] R. W. Clough and C. P. Johnson, *A finite element approximation for the analysis of thin shells*, Internat. J. Solids Structures **4** (1968), 43–60.

[CJ2] ———, *Finite element analysis of arbitrary thin shells*, Proc. ACI Symposium on Concrete Thin Shells, New York, 1970, pp. 333–363.

[CB] J. J. Connor and C. Brebbia, *Stiffness matrix for shallow rectangular shell elements*, Proc. ASCE **93** (1967), no. EM5, 43–65.

[CC] E. Cosserat and F. Cosserat, *Théorie des corps déformables*, Hermann, Paris, 1909.

[FV] F. Fraeijs de Veubeke, *Bending and stretching of plates*, Proc. Matrix Methods in Structural Mechanics, Wright Patterson, AFB Ohio, 1965.

[G] H. G. Ganev and Tch. T. Dimitrov, *Calculation of arch dams as a shell using* IBM-370 *computer and curved finite elements*, Theory of Shells (Proc. Third IUTAM Sympos., Tbilisi, 1978), North-Holland, Amsterdam-New York, 1980, pp. 691–696.

[KA] W. Kanok Nukulchai, *A simple and efficient finite element for general shell analysis*, Internat. J. Numer. Methods Engrg. **14** (1979), 179–200.

[KI] G. Kirchhoff, Vorlesungen über Mathematische Physik, Mechanik, Leipzig, 1876.

[K] W. T. Koiter, *On the nonlinear theory of thin elastic shells*, Proc. Konink. Nederl. Akad. Wetensch. B **69** (1966), 1–54.

[KO] ———, *On the foundations of the linear theory of thin elastic shells*, Proc. Konink. Nederl. Akad. Wetensch. B **73** (1970), 169–195.

[L] A. E. H. Love, *The mathematical theory of elasticity*, Cambridge University Press, 1934.

[MT] J. L. Meek and H. S. Tan, *A faceted shell elements with loof nodes*, Internat. J. Numer. Methods Engrg. **23** (1986), 49–67.

[M] L. S. D. Morley, *The constant moment plate bending element*, J. Strain Anal. **6** (1971), 20–24.

[N] P. M. Naghdi, *Foundations of elastic shell theory*, Progress in Solid Mechanics, vol. 4, North-Holland, Amsterdam, 1963, pp. 1–90.

[NA] P. M. Naghdi, *The theory of shells and plates*, Handbuch der Physik, vol. VI a.2, Springer-Verlag, Berlin, 1972, pp. 425–640.

[S] G. Sander, *Modèles d'éléments finis délinquants*, Méthodes Numériques pour les Problèmes de Coques (Janvier 1985), Clamart, 1985.

[T] P. Trouvé, *Sur la convergence des méthodes d'éléments finis nonconformes pour des problèmes linéaires de coques minces*, Numer. Math. **57** (1990), 481–524.

Pôle Universitaire Léonard de Vinci, 92016 Paris La Défense Cedex, and, Institut National de Recherche en Informatique et en Automatique, Domaine de Voluceau, Rocquencourt, 78153 Le Chesnay Cedex, France

E-mail address: `Michel.Bernadou@inria.fr`

Centre de Recherches Mathématiques
CRM Proceedings and Lecture Notes
Volume **21**, 1999

Numerical Analysis of Piezoelectric Shells

Michel Bernadou and Christophe Haenel

Scope

Piezoelectric materials have interesting properties used to control the acoustic or vibration characteristics of a structure. In this paper we study basic equations of piezoelectric shells: their possible derivation from three dimensional equations, the existence of solutions and their approximation by finite element methods.

1. Piezoelectricity

The piezoelectric phenomena were discovered in 1880 by Jacques and Pierre Curie; they can be viewed as an interaction between mechanical and electrical phenomena (see [**I**]):

- a mechanical deformation of the material generates an electric field in the material; this property is used as the direct piezoelectric effect;
- the application to the material of an electric field, or of a potential difference, generates a deformation; this effect is used as the inverse piezoelectric effect.

These material properties are more and more extensively used in industry for the last ten to fifteen years, in particular to make an active control of elastic structures: in this way, by using a distributed system of sensors and actuators displayed inside or upon the surface of the material, one can decrease or stop the vibrations of the structure. Their increased usages is mainly linked to the fact that

- these effects can be obtained without applying mechanical loading;
- sensors and actuators are very light: they can be sticked to the material and they do not change significantly the weight or the dynamical properties of the structure.

Beyond vibration control, such materials can be used to realize shape control for plane propellers, plane wings, telescope mirrors as well as control of metal fatigue, artificial organs in biomechanics and much more. An extended presentation of such possibilities can be found for instance in [**I, RR, T**].

1991 *Mathematics Subject Classification.* Primary: 73K15; Secondary: 65N30; 73R05.
This is the final form of the paper.

2. The Three-Dimensional Piezoelectric Problem

Whether the considered thin shell has a pure mechanical behaviour or a piezoelectric behaviour we take into account the smallness of the shell thickness to obtain approximated equations just set upon the middle surface. In this way, it is fruitful to introduce a geometrical representation of the 3d shell as a tensor product between a mapping of the middle surface and a mapping of the thickness.

Thus, we introduce successively the following concepts.

2.1. Local bases. Let $\mathcal{E}^3$ be the euclidean space and let $(0, \vec{e}_1, \vec{e}_2, \vec{e}_3)$ be a reference system of $\mathcal{E}^3$. For simplicity, this system is assumed to be orthonormal. Any point M of $\mathcal{E}^3$ is then referred as

$$\overrightarrow{OM} = x^i \vec{e}_i,$$

where we use, here and subsequently, the summation convention for a repeated index, occurring once as a subscript and once as a superscript. The range of latin (resp. greek) indices is $\{1,2,3\}$ (resp. $\{1,2\}$). This point M may also be referred to a curvilinear coordinate system (ξ^1, ξ^2, ξ^3) such that $x^i = x^i(\xi^1, \xi^2, \xi^3)$, with $\det(\partial x^i/\partial \xi^j) \neq 0$.

Then, with any point M, we associate two different local basis, i.e.,

$$(1) \qquad \begin{aligned} \vec{g}_i &= \frac{\partial \overrightarrow{OM}}{\partial \xi^i} = \frac{\partial x^j}{\partial \xi^i} \vec{e}_j = \text{covariant basis}, \\ \vec{g}^j &= \text{contravariant basis, such that } \vec{g}^j . \vec{g}_i = \delta_i^j, \end{aligned}$$

where $\delta_i^j = 0$ if $j \neq i$, and 1 if $j = i$.

With these covariant and contravariant bases, we associate the contravariant and covariant components of a tensor, for example $\vec{v} = v_i \vec{g}^i = v^i \vec{g}_i$, or $T = T_{ij} \vec{g}^i \otimes \vec{g}^j = T_i^{\cdot j} \vec{g}^i \otimes \vec{g}_j = T^i_{\cdot j} \vec{g}_i \otimes \vec{g}^j = T^{ij} \vec{g}_i \otimes \vec{g}_j$. The passage between covariant, mixed and contravariant components is insured by the metric tensors (g_{ij}) or (g^{ij}), i.e., $v^i = g^{ij} v_j$, $T_i^{\cdot j} = g_{ik} T^{kj} = g^{jk} T_{ik}$.

Since these local bases $(\vec{g}_i)$ or $(\vec{g}^i)$ are not orthonormal, the computation of tensor derivatives is made easier by introducing Christoffel symbols $\overset{*}{\Gamma}{}^k_{ij}$. Indeed, consider for instance $\vec{v} = v^i \vec{g}_i$; then, $\vec{v}_{,j} = v_{i\|j} \vec{g}^i = v^i\|_j \vec{g}_i$, where the covariant derivatives of v_i and v^i are given by

$$(2) \quad v_{i\|j} = v_{i,j} - \overset{*}{\Gamma}{}^k_{ij} v_k, \quad v^i\|_j = v^i_{,j} + \overset{*}{\Gamma}{}^i_{kj} v^k, \text{ with } \overset{*}{\Gamma}{}^k_{ij} = \vec{g}^k \cdot \vec{g}_{i,j} = g^{k\ell} \vec{g}_\ell \cdot \vec{g}_{i,j}.$$

Finally, we introduce the volume element

$$(3) \qquad d\mathcal{V} = \sqrt{g}\, d\xi^1\, d\xi^2\, d\xi^3, \quad \sqrt{g} = (\vec{g}_1 \times \vec{g}_2) . \vec{g}_3.$$

For more details, the reader is referred to chapter one of [**G**].

2.2. Equations of 3d piezoelectricity. Let $\mathcal{V}$ be the three-dimensional domain occupied by the unloaded piezoelectric body. We note $\partial\mathcal{V}$ its boundary. We assume that this body is submitted to the following loads:

(i) Mechanical loads: A distributed loading of density $\vec{p}$ in $\mathcal{V}$ and a distributed surface loading of density $\vec{q}$ on a part $\partial\mathcal{V}_1^M$ of the boundary $\partial\mathcal{V}$. Moreover, the body is assumed to be clamped on the complementary part $\partial\mathcal{V}_0^M$ of its boundary, with measure $(\partial\mathcal{V}_0^M) > 0$.

(ii) Electrical loads: No electrical loads in the body; fixed potential, say $\varphi = \varphi_0$, on a part $\partial\mathcal{V}_0^E$ of the boundary (applied through electrodes), with measure $(\partial\mathcal{V}_0^E) > 0$, and no electrical loading on the complementary part $\partial\mathcal{V}_1^E = \partial\mathcal{V} - \partial\mathcal{V}_0^E$, i.e., $\vec{D}.\vec{n} = 0$, where $\vec{D}$ denotes the electrical displacement vector and $\vec{n}$ denotes the external unit normal to the boundary.

The piezoelectric equations. These equations are detailed for instance in [**I, E, M**] when an orthonormal reference system is used and in [**L**] for a representation by a general curvilinear coordinate system. It is fruitful to use these references in conjunction with [**G**] for the elasticity equations expressed in a general curvilinear coordinate system. We denote

σ^{ij}: components of the stress tensor $\big((\sigma) = \sigma^{ij}\vec{g}_i \otimes \vec{g}_j\big)$;
D^i: components of the electrical displacement $(\vec{D} = D^i\vec{g}_i)$;
n_i: components of the external unit normal to $\partial\mathcal{V}(\vec{n} = n_i\vec{g}^i)$;
$E_i = -\varphi\|_i = -\varphi_{,i}$: components of the electrical field $(\vec{E} = E_i\vec{g}^i)$;
φ: electric potential;
$\vec{\overset{*}{u}}$: displacement vector $(\vec{\overset{*}{u}} = \overset{*}{u}_i\vec{g}^i)$;
$\overset{*}{\gamma}_{ij} = \frac{1}{2}(\overset{*}{u}_{i\|j} + \overset{*}{u}_{j\|i})$: components of the strain tensor.

Then, the equations can be written:

(i) Equilibrium equations:

$$-\sigma^{ij}\|_j = p^i \text{ in } \mathcal{V}, \tag{4}$$

$$\vec{\overset{*}{u}} = \vec{0} \text{ on } \partial\mathcal{V}_0^M, \quad \sigma^{ij}n_j = q^i \text{ on } \partial\mathcal{V}_1^M.$$

(ii) Maxwell-Gauss equations:

$$D^i\|_i = 0 \text{ in } \mathcal{V}, \tag{5}$$

$$\varphi = \varphi_0 \text{ on } \partial\mathcal{V}_0^E, \quad D^in_i = 0 \text{ on } \partial\mathcal{V}_1^E. \tag{6}$$

It remains to give the constitutive laws to express all these equations as functions of the main unknowns of the problem, i.e., $\overset{*}{u}_i$ and φ.

Constitutive laws.

$$\sigma^{ij}(\vec{\overset{*}{u}}, \varphi) = C^{ijk\ell}\overset{*}{\gamma}_{k\ell}(\vec{\overset{*}{u}}) - e^{mij}E_m(\varphi), \tag{7}$$

$$D^i(\vec{\overset{*}{u}}, \varphi) = e^{ik\ell}\overset{*}{\gamma}_{k\ell}(\vec{\overset{*}{u}}) + e^{ij}E_j(\varphi), \tag{8}$$

where the coefficients $C^{ijk\ell}, e^{mij}, e^{ij}$ satisfy the properties

$$C^{ijk\ell} = C^{jik\ell} = C^{k\ell ij},$$

$$C^{ijk\ell}\varepsilon_{ij}\varepsilon_{k\ell} \geq \alpha \sum_{i,j=1}^{3}(\varepsilon_{ij})^2, \quad \forall\varepsilon_{ij} = \varepsilon_{ji} \in \mathbb{R}, \ \alpha = \text{constant} > 0,$$

$$e^{mij} = e^{mji}, \quad e^{ij} = e^{ji},$$

$$e^{ij}w_iw_j \geq \beta\sum_{i=1}^{3}(w_i)^2, \quad \forall w_i \in \mathbb{R}, \ \beta = \text{constant} > 0. \tag{9}$$

Main unknowns. The substitution of equations (7) (8) into the equations (4) and (5) show that these equations are only dependent on $\vec{\overset{*}{u}}$ and φ which are the main unknowns, i.e.,

$$\overset{*}{u}_i = \vec{\overset{*}{u}}.\vec{g}_i = \text{components of the displacement field}, \quad \varphi = \text{potential}.$$

Before giving the variational formulation of the problem, it is convenient to substitute to the nonhomogeneous boundary condition (6) an homogeneous ones. In this way, we assume the existence of a function $\widehat{\varphi}$ defined on $\mathcal{V}$ such that $\widehat{\varphi}|_{\partial\mathcal{V}_0^E} = \varphi_0$ and we set $\overline{\varphi} = \varphi - \widehat{\varphi}$. Then, $\overline{\varphi} = 0$ on $\partial\mathcal{V}_0^E$ and the new set of main unknowns becomes $(\overset{*}{u}_i, \overline{\varphi})$ and is solution of

$$-\sigma^{ij}(\vec{\overset{*}{u}}, \varphi)\|_j = p^i - \left(e^{mij}\, E_m(\overline{\varphi})\right)\big\|_j, \tag{10}$$

$$D^i(\vec{\overset{*}{u}}, \overline{\varphi})\|_i = -\left(e^{ij} E_j(\overline{\varphi})\right)\big\|_i, \tag{11}$$

with associated suitable boundary conditions.

2.3. Variational formulation. From equations (10) (11) and from the associated boundary conditions, we obtain the following variational formulation. Find:

$$(\vec{\overset{*}{u}}, \overline{\varphi}) \in \vec{V} \times \Theta \text{ such that } \overset{*}{a}((\vec{\overset{*}{u}}, \overline{\varphi}), (\vec{\overset{*}{v}}, \psi)) = \overset{*}{f}(\vec{\overset{*}{v}}, \psi), \quad \forall (\vec{\overset{*}{v}}, \psi) \in \vec{V} \times \Theta, \tag{12}$$

where

$$\overset{*}{a}((\vec{\overset{*}{u}}, \overline{\varphi}), (\vec{\overset{*}{v}}, \psi)) = \int_{\mathcal{V}} [\{C^{ijk\ell}\overset{*}{\gamma}_{k\ell}(\vec{\overset{*}{u}}) - e^{mij} E_m(\overline{\varphi})\}\overset{*}{\gamma}_{ij}(\vec{\overset{*}{v}})$$
$$+ \{e^{ik\ell}\overset{*}{\gamma}_{k\ell}(\vec{\overset{*}{u}}) + e^{ij} E_j(\overline{\varphi})\} E_i(\psi)]\, d\mathcal{V}.$$

$$\overset{*}{f}(\vec{\overset{*}{v}}, \psi) = \int_{\mathcal{V}} [p^i \overset{*}{v}_i + e^{mij} E_m(\widehat{\varphi})\overset{*}{\gamma}_{ij}(\vec{\overset{*}{v}}) - e^{ji} E_i(\widehat{\varphi}) E_j(\psi)]\, d\mathcal{V} + \int_{\partial\mathcal{V}_1^M} q^i \overset{*}{v}_i\, dS.$$

$$\vec{V} = V \times V \times V, \quad V = \{v \in H^1(\mathcal{V}); v|_{\partial\mathcal{V}_0^M} = 0\}, \quad \Theta = \{\psi \in H^1(\mathcal{V}); \psi|_{\partial\mathcal{V}_0^E} = 0\}.$$

Then, we prove the existence and uniqueness of a solution for problem (12):

THEOREM 1. *Problem* (12) *has one and only one solution.*

PROOF. We use Lax-Milgram lemma [**C**]:

- $\vec{V} \times \Theta$ is a Hilbert space.
- $\overset{*}{a}(.,.)$ and $\overset{*}{f}(.)$ are bilinear and linear $\vec{V} \times \Theta$-continuous forms.
- It remains to prove that $\overset{*}{a}(.,.)$ is $\vec{V} \times \Theta$-elliptic, i.e., there exists a constant $\gamma > 0$ such that

$$\overset{*}{a}((\vec{\overset{*}{v}}, \psi), (\vec{\overset{*}{v}}, \psi)) \geq \gamma\left\{\sum_{i=1}^{3} \|\overset{*}{v}_i\|_{1,\mathcal{V}}^2 + \|\psi\|_{1,\mathcal{V}}^2\right\}, \quad \forall (\vec{\overset{*}{v}}, \psi) \in \vec{V} \times \Theta.$$

Let us observe that

$$\overset{*}{a}((\vec{\overset{*}{v}}, \psi), (\vec{\overset{*}{v}}, \psi)) = \int_{\mathcal{V}} \{C^{ijk\ell}\overset{*}{\gamma}_{ij}(\vec{\overset{*}{v}})\overset{*}{\gamma}_{k\ell}(\vec{\overset{*}{v}}) + e^{ij} E_i(\psi) E_j(\psi)\}\, d\mathcal{V}.$$

The first part of this result comes from the $\vec{V}$-ellipticity property of the three-dimensional elasticity problem when a representation by a curvilinear coordinates system is used [**CI**]. The second part of the result comes from $E_i = -\varphi_{,i}$, (9), the assumption measure $(\partial\mathcal{V}_0^E) > 0$ and Poincaré's inequality [**NE**]; then, there exists a constant $\delta_2 > 0$ such that

$$\int_{\mathcal{V}} e^{ij}\psi_{,i}\psi_{,j}\, d\mathcal{V} \geq \delta_2 \|\psi\|^2_{1,\mathcal{V}}, \quad \forall\psi \in \Theta. \quad \square$$

3. Piezoelectric Thin Shells

3.1. Geometrical representation of a thin shell. Subsequently we use a geometric representation of a general thin shell as the tensor product of its middle surface S by its thickness e (see [**B, BE**]). In other words, we specialize the geometrical representation introduced in paragraph 2 so that:

- the curvilinear coordinates (ξ^1, ξ^2) describe the middle surface S;
- ξ^3 is the thickness coordinate.

This system (ξ^1, ξ^2, ξ^3) is, at least locally, a system of curvilinear coordinates of $\mathcal{E}^3$. The thickness e of the shell is defined by the application $e : (\xi^1, \xi^2) \in \overline{\Omega} \to \{x \in \mathbb{R}; x > 0\}$. Then, the shell $\mathcal{C}$ is the closed subset of $\mathcal{E}^3$ defined by

$$\mathcal{C} = \left\{ M \in \mathcal{E}^3; \overrightarrow{OM} = \vec{\phi}(\xi^1, \xi^2) + \xi^3\vec{a}_3; (\xi^1, \xi^2) \in \overline{\Omega}; -\frac{1}{2}e(\xi^1, \xi^2) \leq \xi^3 \leq \frac{1}{2}e(\xi^1, \xi^2) \right\}.$$

According to relations (1) and $\vec{a}_{3,\alpha} = -b_\alpha^\lambda \vec{a}_\lambda$, the derivatives of the vector $\overrightarrow{OM} = \vec{\phi}(\xi^1, \xi^2) + \xi^3\vec{a}_3$ give

$$\vec{g}_\alpha = (\delta_\alpha^\nu - \xi^3 b_\alpha^\nu)\vec{a}_\nu; \quad \vec{g}_3 = \vec{a}_3. \tag{13}$$

The vectors $\vec{g}_1$ and $\vec{g}_2$ are parallel to the tangent plane to the middle surface at point P while $\vec{g}_3$ is normal to that plane. It is proved in [**BC**] that $(M, \vec{g}_1, \vec{g}_2, \vec{g}_3)$ is a local reference system at any point of the shell $\mathcal{C}$. Finally, according to [**G**], the volume element (3) can be written

$$d\mathcal{C} = \sqrt{g}\, d\xi^1\, d\xi^2\, d\xi^3 = \sqrt{\frac{g}{a}}\sqrt{a}\, d\xi^1\, d\xi^2\, d\xi^3 = \sqrt{\frac{g}{a}}\, d\mathcal{S}\, d\xi^3, \tag{14}$$

where

$$\sqrt{\frac{g}{a}} = 1 - 2H\xi^3 + K(\xi^3)^2, \quad H = \text{mean curvature} = \frac{1}{2}b_\alpha^\alpha, \tag{15}$$
$$K = \text{total curvature} = b_1^1 b_2^2 - b_2^1 b_1^2.$$

It is interesting to note that, for a thin shell, we have [**BC**]:

$$|H| \leq \frac{1}{10e}, \quad |K| \leq \frac{1}{100e^2}, \text{ so that } 1 - 2H\xi^3 + K(\xi^3)^2 \geq \frac{89}{100}. \tag{16}$$

3.2. Two-dimensional piezoelectric thin shell equations. The explicit obtention of two-dimensional modelizations of piezoelectric thin shells is beyond the aim of this paper. In this way, let us mention the works by [**T**] and [**R1,R2,R3,R4**] who restrict the representation of the middle surface to curvilinear coordinates giving the principal lines of curvatures, and [**L**] who uses general curvilinear coordinates. Let us add that the mathematical justifications of such equations is still open, even for purely mechanical deformations like those given by [**N, NA, K**].

Paradoxically, without giving the explicit variational formulations of piezoelectric thin shells, we give hereunder a general existence and uniqueness theorem valid for modelizations which take, or not, into account the effect of transverse shear strains. For clarity, this theorem allows to take into account any piezoelectric shell modelization whose associated purely mechanical modelization enters in the framework of the existence studies developed in [**B, BC, BCM**].

The most popular theories of two-dimensional thin shells are based on two hypotheses:

(i) Theories including transverse shear effects [**N, NA**]:
- the particles located along the normals before deformation remain aligned after deformation along a line which is generally no longer normal to the deformed middle surface; then

$$\vec{\overset{*}{u}}(\xi^1, \xi^2, \xi^3) \simeq \vec{u}(\xi^1, \xi^2) + \xi^3 \beta_\alpha(\xi^1, \xi^2)\vec{a}^\alpha, \tag{17}$$

$\beta_\alpha =$ components of the rotation of the normal.

- the stresses are approximatively plane and parallel to the middle plane, i.e.,

$$\sigma^{3\alpha} \simeq \sigma^{\alpha 3} \simeq \sigma^{33} \simeq 0. \tag{18}$$

In these theories the main unknowns are $u_i(\xi^1, \xi^2) = \vec{u}(\xi^1, \xi^2).\vec{a}_i(\xi^1, \xi^2)$ and $\beta_\alpha(\xi^1, \xi^2)$.

(ii) Theories without transverse shear effects [**K**]: the hypotheses are similar except that the normals to the middle surface remain normal during the deformation. In this case, equations (18) remain valid while equation (17) is replaced by

$$\vec{\overset{*}{u}}(\xi^1, \xi^2, \xi^3) \simeq \vec{u}(\xi^1, \xi^2) - \xi^3\big(u_{3,\alpha}(\xi^1, \xi^2) + b_\alpha^\lambda u_\lambda(\xi^1, \xi^2)\big)\vec{a}^\alpha. \tag{19}$$

In these modelizations, the main unknowns are the three components $u_i(\xi^1, \xi^2) = \vec{u}(\xi^1, \xi^2).\vec{a}_i(\xi^1, \xi^2)$ of the displacement.

Besides these assumptions concerning the behaviour of the displacement vector $\vec{\overset{*}{u}}(\xi^1, \xi^2, \xi^3)$ through the thickness, we have to indicate how the potential $\overline{\varphi}$ behaves across the thickness. Let us assume that $\overline{\varphi}$ can be developed as follows

$$\overline{\varphi}(\xi^1, \xi^2, \xi^3) = \sum_{n=0}^{+\infty} \frac{(\xi^3)^n}{n!} \overline{\varphi}_n(\xi^1, \xi^2). \tag{20}$$

Then by substitution of expressions like (2) (13) (14) (15) (17) (or (19)) (18) (20) ... into the variational formulation (12) and by integration through the thickness, we could obtain the associated two-dimensional theories. These theories are dependent on the basic hypotheses (17) to (20) but also on various simplifications more or less justified ... See [**H**] for a discussion on these aspects

We just want to conclude by the following result which is really interesting with respect to possible applications based to direct or associated inverse problems.

THEOREM 2. *Any two-dimensional linear piezoelectric thin shell variational formulation*:

(a) *which is obtained from the three-dimensional formulation* (12) *by integration through the thickness as described above*;

(b) *which gives back to a purely mechanical variational formulation which enters in the framework of the existence and uniqueness theorems of* [**BC**][**BCM**];

(c) *which is obtained by restricting the potential to satisfy*

$$\overline{\varphi}(\xi^1,\xi^2,\xi^3) = \overline{\varphi}_0(\xi^1,\xi^2) + \xi^3\overline{\varphi}_1(\xi^1,\xi^2) + \frac{1}{2}(\xi^3)^2\overline{\varphi}_2(\xi^1,\xi^2); \tag{21}$$

(d) *which is obtained for an applied potential φ_0 along a part $\partial\mathcal{C}_0$ of the lateral boundary of the shell $\mathcal{C}$;*

has one and only one solution.

PROOF. These two-dimensional linear piezoelectric thin shell variational formulations are clearly of the same type than (12) but now, the remaining integrations are performed on $\mathcal{S}$ or, equivalently, on Ω. In particular, the associated functional spaces are of Hilbert type, the associated forms $a(.,.)$ and $f(.)$ are bilinear and linear and they are continuous. Then, to apply the Lax-Milgram lemma [**C**], and to conclude, it remains to prove that $a(.,.)$ is elliptic. It this way, we note that relation (12) involves for any $(\vec{\overset{*}{v}},\psi) \in \vec{V}\times\Theta$

$$\overset{*}{a}\big((\vec{\overset{*}{v}},\psi),(\vec{\overset{*}{v}},\psi)\big) = \overset{*}{a}_M(\vec{\overset{*}{v}},\vec{\overset{*}{v}}) + \overset{*}{a}_E(\psi,\psi), \tag{22}$$

with

$$\overset{*}{a}_M(\vec{\overset{*}{v}},\vec{\overset{*}{v}}) = \int_{\mathcal{C}} C^{ijk\ell}\overset{*}{\gamma}_{ij}(\vec{\overset{*}{v}})\overset{*}{\gamma}_{k\ell}(\vec{\overset{*}{v}})\,d\mathcal{C}, \tag{23a}$$

$$\overset{*}{a}_E(\psi,\psi) = \int_{\mathcal{C}} e^{ij}\psi_{,i}\psi_{,j}\,d\mathcal{C}, \tag{23b}$$

so that the global ellipticity is dependent on

(i) the ellipticity of the two-dimensional approximated bilinear form $a_M(.,.)$ associated to (23a); this is secured by assumption (b) and results form [**BC, BCM**];

(ii) the ellipticity of the two-dimensional approximated bilinear form associated to (23b) with assumption (21), we obtain

$$\begin{aligned} I &= a_E\big((\psi_0,\psi_1,\psi_2),(\psi_0,\psi_1,\psi_2)\big) \\ &= \int_{\mathcal{C}} e^{ij}\Big\{\psi_0+\xi^3\psi_1+\frac{1}{2}(\xi^3)^2\psi_2\Big\}_{,i}\Big\{\psi_0+\xi^3\psi_1+\frac{1}{2}(\xi^3)^2\psi_2\Big\}_{,j}\,d\mathcal{C}, \end{aligned}$$

and with (9) and (14)–(16):

$$I \geq \frac{89}{100}\beta\int_\Omega\Big[\int_{-e/2}^{e/2}\sum_{i=1}^{3}\Big(\Big(\psi_0+\xi^3\psi_1+\frac{1}{2}(\xi^3)^2\psi_2\Big)_{,i}\Big)^2 d\xi^3\Big]\sqrt{a}\,d\xi^1\,d\xi^2.$$

Since ψ_j, $j=0,1,2$, is only dependent on ξ^1 and ξ^2, we obtain

$$I \geq \frac{89}{100}\beta\int_\Omega\Big\{\int_{-e/2}^{e/2}\Big[\sum_{\alpha=1}^{2}(\psi_{0,\alpha}+\xi^3\psi_{1,\alpha}+\frac{1}{2}(\xi^3)^2\psi_{2,\alpha})^2 + (\psi_1+\xi^3\psi_2)^2\Big]d\xi^3\Big\}\sqrt{a}\,d\xi^1\,d\xi^2,$$

so that (odd exponents of ξ^3 give zero):

$$\begin{aligned} I \geq \frac{89}{100}\beta\int_\Omega\Big\{\int_{-e/2}^{e/2}\Big[&(\psi_{0,1})^2+(\psi_{0,2})^2+(\psi_1)^2 \\ &+(\xi^3)^2\Big\{(\psi_2)^2+(\psi_{1,1})^2+(\psi_{1,2})^2+\psi_{0,1}\psi_{2,1}+\psi_{0,2}\psi_{2,2}\Big\} \\ &+\frac{1}{4}(\xi^3)^4\{(\psi_{2,1})^2+(\psi_{2,2})^2\}\Big]d\xi^3\Big\}\sqrt{a}\,d\xi^1\,d\xi^2. \end{aligned}$$

Finally, by denoting $\bar{e} = \min_{\xi \in \Omega} e(\xi) > 0$,

$$I \geq \frac{89}{100} \beta \bar{e} \left\{ \frac{1}{4} \left(|\psi_{0,1}|^2_{0,\Omega} + |\psi_{0,2}|^2_{0,\Omega} + 4|\psi_1|^2_{0,\Omega} \right) \right.$$
$$\left. + \frac{\bar{e}^2}{12} \left(|\psi_{1,1}|^2_{0,\Omega} + |\psi_{1,2}|^2_{0,\Omega} + |\psi_2|^2_{0,\Omega} \right) + \frac{7\bar{e}^4}{8640} \left(|\psi_{2,1}|^2_{0,\Omega} + |\psi_{2,2}|^2_{0,\Omega} \right) \right\},$$

so that, with assumption (d) and with Poincaré's inequality [**NE**], we obtain the existence of a constant $\gamma > 0$ such that

$$a_E\big((\psi_0, \psi_1, \psi_2), (\psi_0, \psi_1, \psi_2)\big) \geq \gamma \sum_{i=0}^{2} \|\psi_i\|^2_{1,\Omega}. \quad \square$$

REMARK 1. In statement of Theorem 2, assumptions (c) and (d) are made for simplicity. More general situations could be considered.

4. Approximation by Conforming Finite Element Methods

By taking into account the ellipticity of the bilinear form $a(.,.)$, we can infer that the conforming finite element approximation of piezoelectric thin shell equations gives results entirely similar to those obtained for the conforming approximation of the associated purely mechanical thin shell problem [**B**]. For instance, to approximate the piezoelectric extension of Koiter's equations [**K**] we can combine:

(24) reduced-HCT (resp. P_1-triangles) elements to approximate

$$u_3 \text{ (resp. } u_1, u_2, \psi_o, \psi_1, \psi_2),$$

or

(25) Argyris (resp. Ganev) finite element to approximate

$$u_3 \text{ (resp. } u_1, u_2, \psi_o, \psi_1, \psi_2).$$

By using methods developed in [**B**], it is possible to prove the convergence of these approximations and to obtain a priori energy norm estimates in $0(h)$ for (24) and in $0(h^4)$ for (25), where h denotes the higher diameter of the elements of the triangulation

Of course, like for the purely mechanical problem, we find again the difficulties coming from possible membrane and/or shear locking. Another additional difficulty is due to the fact that the associated matrix is no longer symmetric since the bilinear form $a(.,.)$ itself is no longer symmetric.

These problems will be more extensively developed in [**H**]. Other approximation methods like nonconforming finite element methods or DKT methods can also be considered.

Acknowledgments

This work is a part of the Human Capital and Mobility Program "Shells: Mathematical Modeling and Analysis, Scientific Computing" of the Commission of the European Communities (Contract no. ERBCHRXCT 940536).

References

[B] M. Bernadou, *Finite element methods for thin shell problems*, J. Wiley & Sons, Chichester, 1996.

[BE] ———, *Some approximation methods for linear thin shell problems*, (this volume).

[BC] M. Bernadou and P. G. Ciarlet, *Sur l'ellipticité du modèle linéaire de coques de W. T. Koiter*, Computing Methods in Applied Sciences and Engineering (Second Internat. Sympos., Versailles, 1975) (R. Glowinski and J. L. Lions, eds.), Lectures Notes in Economics and Mathematical Systems, Part I, vol. 134, Springer, Berlin, 1976, pp. 89–136.

[BCM] M. Bernadou, P. G. Ciarlet, and B. Miara, *Existence theorems for two-dimensional linear shell theories*, J. Elasticity **34** (1994), 111–138.

[C] P. G. Ciarlet, *Basic error estimates for elliptic problems, Handbook of Numerical Analysis. II. Finite element methods*, Part 1 (P. G. Ciarlet and J. L. Lions, eds.), North-Holland, Amsterdam, 1991, pp. 17–351.

[CI] ———, *Théorie mathématique des coques, une introduction linéairement elastique*, École de Printemps 1994, INRIA, Rocquencourt, 1994.

[E] A. C. Eringen and G. A. Maugin, *Electrodynamics of continua. I. Foundations and solid media*, Springer-Verlag, New-York, 1990.

[G] A. E. Green and W. Zerna, *Theoretical elasticity*, 2nd ed., Oxford Univ. Press, Oxford, 1968.

[H] C. Haenel, *Modélisation, analyse et simulation numérique de coques piézoélectriques*, Doctoral dissertation (to appear).

[I] T. Ikeda, *Fundamental of piezoelectricity*, Oxford Univ. Press, Oxford, 1990.

[K] W. T. Koitier, *On the nonlinear theory of thin elastic shells*, Proc. Konink Nederl. Akad. Wetensch. B **69** (1966), 1–54.

[L] N. N. Le Khan'chau, *The theory of piezoelectric shells*, Prikl. Mat. Mekh. USSR **50** (1986), no. 1, 98–105.

[M] G. A. Maugin, *Continuum mechanics of electromagnetic solids*, North-Holland, Amsterdam, 1988.

[N] P. M. Naghdi, *Foundations of elastic shell theory*, 1963 Progress in Solid Mechanics, vol. IV, North-Holland, Amsterdam, 1963, pp. 1–90.

[NA] ———, *The theory of shells and plates*, Handbuch der Physik, vol. VI a.2, Springer-Verlag, Berlin, 1972, pp. 425–640.

[NE] J. Nečas, *Les méthodes directes en théorie des équations elliptiques*, Masson, Paris, 1967.

[RR] C. A. Rogers and R. C. Rogers, *Recent advances in adaptive and sensory materials and their applications*, Tecnomic Publ. Co., Lancaster, 1992.

[R1] N. N. Rogacheva, *Equations of state of piezoceramic shells*, Prikl. Mat. Mekh. USSR **45** (1981), 677–684.

[R2] ———, *On boundary conditions in the theory of piezoceramic shells polarized along coordinate lines*, Prikl. Mat. Mekh. USSR **47** (1984), no. 2, 220–226.

[R3] ———, *On Saint-Venant type conditions in the theory of piezoelastic shells*, Prikl. Mat. Mekh. USSR **48** (1984), no. 2, 213–216.

[R4] ———, *Classification of free piezoceramic shell vibrations*, Prikl. Mat. Mekh. USSR **50** (1986), no. 1, 106–111.

[T] H. S. Tzou, *Piezoelectric shells: Distributed sensing and control of continua*, Kluwer Academic Publ., Dordrecht, The Netherlands, 1993.

Pôle Universitaire Léonard de Vinci, 92016 Paris La Défense Cedex, and Institut National de Recherche en Informatique et en Automatique, Domaine de Voluceau, Rocquencourt, 78153 Le Chesnay Cedex, France

E-mail address: Michel.Bernadou@inria.fr

Institut National de Recherche en Informatique et en Automatique, Domaine de Voluceau, Rocquencourt, 78153 Le Chesnay Cedex, France

E-mail address: Christophe.Haenef@inria.fr

Centre de Recherches Mathématiques
CRM Proceedings and Lecture Notes
Volume **21**, 1999

Étude intrinsèque d'un problème de coque en grandes déformations

M. Carrive et P. Le Tallec

Ce papier est dédicacé à J. C. Simo

Résumé. Dans ce papier nous établissons une formulation intrinsèque du modèle « géométriquement exact »de J. C Simo que nous analysons ensuite. Cette analyse permet de justifier les fondements d'une telle théorie à la fois du point de vue mécanique et mathématique (formelle). L'approche est basée sur une paramétrisation originale de la coque qui consiste à introduire un feuilletage de manière à travailler directement sur la surface moyenne (ce qui permet en particulier de traiter les ovoides). Nous montrons un résultat global de caractérisation d'une surface de Cosserat et proposons une justification de la loi de comportement hyperélastique pour les coques.

Description géométrique de la coque

1. Paramétrisation de la coque au repos

Nous supposons que la coque au repos occupe un ouvert régulier feuilleté, c'est-à-dire : il existe une surface gauche régulière $\omega_0 \subset \mathcal{B}$, telle que $\mathcal{B}$ soit l'image d'un sous-ensemble

$$\Omega_0 = \big\{(m_0, \xi) \in \omega_0 \times \mathbb{R}, \xi \in I(m_0)\big\} \subset \omega_0 \times \mathbb{R}$$

par l'application injective continue

$$\begin{aligned} \phi^0 \colon \Omega_0 &\to \mathbb{R}^3 \\ (m_0, \xi) &\mapsto m_0 + \xi\, t^0(m_0), \end{aligned}$$

où l'intervalle $I(m) = \big[h^-(m), h^+(m)\big] \subset \mathbb{R}$ et le vecteur unitaire $t^0(m) \in \mathbb{R}^3$ sont des fonctions données du point $m_0 \in \omega_0$. Les grandeurs ω_0, $t^0(.)$ et $I(.)$ définissent une paramétrisation de la coque au repos. Pour que la théorie soit mécaniquement raisonnable, il faut que la coque soit d'épaisseur faible suivant la direction t^0 et

1991 *Mathematics Subject Classification.* Primary : 54C40, 14E20 ; Secondary : 46E25, 20C20.

This is the final form of the paper.

de section $I(m_0)$ lentement variable. La surface ω_0 est prise quasi-parallèle aux faces supérieures et inférieures du corps $\mathcal{B}$ et le vecteur $t^0(m_0)$ est transverse à ω_0 et tel que $\|t^0\| = 1$ (soit $t^0 \in S^2$, la sphère unité de $\mathbb{R}^3$). Le vecteur t^0 est généralement choisi perpendiculaire à ω_0. L'application ϕ^0 définit un **feuilletage** $\big(\phi^0(\omega_0 \times \{\xi\}), \xi = \text{cste}\big)$ et **une fibration** $\big(\phi^0(\{m_0\} \times I(m_0)), m_0 = \text{cste}\big)$ de la coque. La feuille $\omega_0 = \phi^0(\omega_0 \times \{0\})$ est appelée **surface moyenne** . On peut déformer arbitrairement ω_0 de manière à construire une surface régulière $\omega = (\varphi^0)^{-1}(\omega_0)$ qui servira de **configuration de référence**. C'est sur cette surface que seront mesurées les quantités cinétiques et cinématiques que nous introduirons. Notre modèle sera cependant construit indépendamment du choix de cette surface de référence qui ne sera qu'un auxiliaire de calcul. À partir de cette configuration ω, on peut définir un **domaine de référence abstrait**

$$\Omega = \big\{(m, \xi) \in \omega \times \mathbb{R}, \xi \in I(m)\big\} \subset \omega \times \mathbb{R}.$$

Il suffit pour cela de remplacer $m_0 \in \omega_0$ par son image $m = (\varphi^0)^{-1}(m_0) \in \omega$. Par abus de notation ϕ^0 est maintenant défini sur Ω.

2. Hypothèse cinématique de Cosserat

Nous introduisons à présent une hypothèse cinématique à *un directeur inextensible* de type Cosserat [**COS**], permettant d'exprimer le mouvement de la coque en termes de sa surface moyenne déformée. Pour cela on fait l'hypothèse de comportement physique suivante :

(a) Tout point initialement défini le long d'une fibre ($m = \overline{m}$) reste après déformation le long de cette même fibre.

(b) Il n'y a pas de changement de forme de la fibre dans l'épaisseur.

Moyennant ces hypothèses, tout point $X = \phi^0(m, \xi) \in \mathcal{B}$ se déforme de la façon suivante :

$$x(X) = \phi(m, \xi) = \varphi(m) + \xi t(m), \tag{2.1}$$

où φ est l'application injective qui transforme la surface de référence ω en surface moyenne déformée $\omega_\varphi = \phi\big(\omega \times \{0\}\big)$ et l'application t définit un vecteur transverse à ω_φ tel que $t \in S^2$. L'hypothèse cinématique (2.1) est purement géométrique : elle est indépendante de toute notion de base locale ou de calcul différentiel. Elle s'appuie uniquement sur la construction du couple d'applications (φ, t).

L'ensemble des configurations cinématiquement admissibles de la coque est alors décrit par la variété

$$\mathcal{C} = \big\{(\varphi, t) : \omega \to \mathbb{R}^3 \times \mathcal{S}^2 / \varphi \text{ est injective et } t \text{ transverse à } \varphi(\omega)\big\}. \tag{2.2}$$

3. Cadre algébrique

Pour cette étude nous nous plaçons dans le cadre de la géométrie différentielle intrinsèque et adoptons une écriture tensorielle.

Tout d'abord afin d'éviter toute confusion nous choisissons de ne pas identifier l'espace physique $\mathbb{R}^3$ à son dual. Nous le noterons E et E^* désignera son dual. Le tenseur $G \in E^* \otimes E^*$ désignera la métrique euclidienne associée à l'espace E. On notera $T_m\omega$ (ou $T\omega$ lorsqu'il n'y aura pas d'ambiguité), le plan tangent au point m à la surface ω. On désigne par i l'injection canonique de $T\omega$ dans l'espace tangent à Ω, $T_\Omega = T\omega \times \mathbb{R}$, et π la projection associée. On note $\widehat{N} \in T_m\omega \times \mathbb{R}$ le vecteur de composante $\widehat{N} = (0,1)$ et $\widehat{N}^*$ son dual (qui est la projection de $T_m\omega \times \mathbb{R}$ sur $\mathbb{R}$).

4. Tenseurs surfaciques de déformation

Afin de mesurer les déformations subies par la coque entre la configuration initiale (φ^0, t^0) et la configuration actuelle (φ, t),nous introduisons des tenseurs surfaciques fondamentaux, calculés sur la configuration de référence ω, qui décriront les comportements mécaniques et interviendront dans la formulation variationnelle.

Pour mesurer les dilatations de la surface moyenne actuelle on introduit le tenseur métrique $\mathbf{C} \in T\omega^* \otimes T\omega^*$ défini par :

$$\mathbf{C} = \nabla\varphi^t \cdot G \cdot \nabla\varphi.$$

Ses composantes $C_{\alpha\beta}$ en base locale correspondent à la première forme fondamentale de la surface moyenne, introduite en théorie classique (Koiter [**KOI**], Naghdi [**NAG**]). Les variations d'orientations de la fibre t quand on parcourt la surface moyenne actuelle sont mesurées par un tenseur $\mathbf{K} \in T\omega^* \otimes T\omega^*$ défini par :

$$\mathbf{K} = \nabla\varphi^t \cdot G \cdot \nabla t.$$

Si t est le vecteur unitaire normal au plan tangent, ses composantes $K_{\alpha\beta}$ dans la base locale correspondent à la deuxième forme fondamentale. Enfin, pour mesurer les écarts entre la fibre t et la normale à la surface moyenne on introduit le tenseur $\mathbf{R} \in T\omega^*$ défini par :

$$\mathbf{R} = t \cdot G \cdot \nabla\varphi.$$

Les déformations de la coque entre la position initiale (φ^0, t^0) et la position finale (φ, t) sont alors représentées par les trois tenseurs $\boldsymbol{\varepsilon}$, $\boldsymbol{\rho}$, $\boldsymbol{\gamma}$ suivants :

- Déformation de membrane : $\boldsymbol{\varepsilon} = \frac{1}{2}(\mathbf{C} - \mathbf{C}^0) \in T\omega^* \otimes T\omega^*$.
- Déformation de flexion : $\boldsymbol{\rho} = \mathbf{K} - \mathbf{K}^0 \in T\omega^* \otimes T\omega^*$.
- Déformation de cisaillement : $\boldsymbol{\gamma} = \mathbf{R} - \mathbf{R}^0 \in T\omega^*$.

LEMME 4.1. *Si $\boldsymbol{\gamma} = 0$ en tout point $m \in \omega$ (pas de cisaillement) alors le tenseur des déformations de courbure $\boldsymbol{\rho}$ est symétrique.*

DÉMONSTRATION. Il suffit de montrer que $K - K^t = \nabla\mathbf{R} - \nabla\mathbf{R}^t$, d'où l'on déduit $\boldsymbol{\rho} - \boldsymbol{\rho}^t = \nabla\boldsymbol{\gamma} - \nabla\boldsymbol{\gamma}^t = 0$. □

ÉQUATIONS D'ÉQUILIBRE DE COQUES

Dans ce paragraphe nous construisons les équations d'équilibre régissant le comportement de la coque à partir des équations des milieux continus tridimensionnels. Pour cela on se place en configuration physique, c'est-à-dire que l'on choisi comme surface de référence la surface moyenne déformée ω_φ. Afin de ne pas alourdir les notations et lorsqu'il n'y a pas d'ambiguïté possible nous laisserons sans indice les grandeurs calculées sur la surface ω_φ.

5. Équations d'équilibre réduites

5.1. Équilibre d'une feuille. Soit $S \subset \omega_\varphi$ un élément de surface et soit une cote $\bar{\xi} \in I(m_\varphi)$, $\forall m_\varphi \in S$. On définit le volume élémentaire tridimensionnel

$$\mathcal{V}_{\bar{\xi}} = \left\{\varphi(m_\varphi) + \xi t(m_\varphi), m_\varphi \in S \subset \omega_\varphi / \xi \in I_{\bar{\xi}} = [\bar{\xi}, \bar{\xi} + \delta\xi]\right\} = \phi(S \times I_{\bar{\xi}}).$$

Par construction, $\mathcal{V}_{\bar{\xi}}$ est un élément de volume dont les bords latéraux sont parallèles aux fibres t et les bords horizontaux sont les feuilles de hauteur $\bar{\xi}$ et $\bar{\xi} + \delta\xi$. La frontière $\partial\mathcal{V}_{\bar{\xi}}$ s'exprime alors par

$$\partial\mathcal{V}_{\bar{\xi}} = \partial\mathcal{V}^l_{\bar{\xi}} \cup \partial\mathcal{V}^h_{\bar{\xi}} = \phi(\partial S \times I_{\bar{\xi}}) \cup \left[\phi(S \times \{\bar{\xi}\}) \cup \phi(S \times \{\bar{\xi} + \delta\xi\})\right].$$

On note $\boldsymbol{\sigma}$ le tenseur des contraintes de Cauchy, dn^l et dn^h la normale élémentaire à l'élément de surface $\partial\mathcal{V}^l_{\bar{\xi}}$, respectivement $\partial\mathcal{V}^h_{\bar{\xi}}$, et b la force volumique appliquée mesurée par unité de volume en configuration déformée. Avec ces notations, l'équilibre des forces exercées sur $\mathcal{V}_{\bar{\xi}}$ se décompose comme suit :

$$\iint_{\partial\mathcal{V}^l_{\bar{\xi}}} \boldsymbol{\sigma} \cdot dn^l + \iint_{\partial\mathcal{V}^h_{\bar{\xi}}} \boldsymbol{\sigma} \cdot dn^h + \iiint_{\mathcal{V}_{\bar{\xi}}} b \, d\mathcal{V} = 0. \tag{5.1}$$

En effectuant un changement de variables afin de se ramener sur le domaine feuilleté $S \times I_{\bar{\xi}}$, l'équation (5.1) devient

$$\begin{aligned} \iint_{\partial S \times I_{\bar{\xi}}} \boldsymbol{\sigma} \cdot (\nabla\phi^{-t} \cdot i^*) \cdot \nu^* J \, d\Gamma \, d\xi + \iint_{S \times \{\bar{\xi} + \delta\xi\}} \boldsymbol{\sigma} \cdot t^* J \, dS \\ - \iint_{S \times \{\bar{\xi}\}} \boldsymbol{\sigma} \cdot t^* J \, dS + \iiint_{\mathcal{V}_{\bar{\xi}}} b \, J \, dS \, d\xi = 0. \end{aligned} \tag{5.2}$$

Ci-dessus, dS et $d\Gamma$ désigne la mesure sur la surface ω_φ, respectivement sur le bord ∂S, J est le jacobien de la transformation ϕ et ν^* est la normale à ∂S au point m. Les intégrales sur la surface et suivant l'épaisseur étant alors découplées, les équations d'équilibre en configuration ω_φ peuvent s'écrire sous la forme

$$\int_{I_{\bar{\xi}}} \int_{\partial S} J\boldsymbol{\sigma} \cdot (\nabla\phi^{-t} \cdot i^*) \cdot \nu^* d\Gamma \, d\xi + \iint_S [J\boldsymbol{\sigma} \cdot t^*]_{\bar{\xi}}^{\bar{\xi} + \delta\xi} \, dS + \int_{I_{\bar{\xi}}} \iint_S J \, b \, dS \, d\xi = 0,$$

soit encore

$$\int_{I_{\bar{\xi}}} \left[\int_{\partial S} J\boldsymbol{\sigma} \cdot (\nabla\phi^{-t} \cdot i^*) \cdot \nu^* d\Gamma + \iint_S \frac{\partial}{\partial\xi}(J\boldsymbol{\sigma} \cdot t^*) \, dS + \iint_S J \, b \, dS\right] d\xi = 0.$$

Comme ceci est vrai pour tout intervalle $I_{\bar{\xi}}$, $\bar{\xi} \in I$, on en déduit la condition d'équilibre équivalente de toute feuille élémentaire :

$$\int_{\partial S} J\boldsymbol{\sigma} \cdot (\nabla\phi^{-t} \cdot i^*) \cdot \nu^* d\Gamma + \iint_S \frac{\partial}{\partial\xi}(J\boldsymbol{\sigma} \cdot t^*) \, dS + \iint_S J \, b \, dS = 0, \quad \forall S \subset \omega_\varphi. \tag{5.3}$$

5.2. Équilibre des forces et des efforts résultants. En intégrant l'équation d'équilibre d'une feuille (5.3) sur l'intervalle $[h^-, h^+]$, on obtient l'équilibre d'un élément de volume traversant toute l'épaisseur, soit

$$\int_{\partial S} N \cdot \nu^* d\Gamma + \iint_S \overline{n} \, dS = 0, \tag{5.4}$$

où $N = \int_{h^-}^{h^+} J\boldsymbol{\sigma} \cdot (\nabla\phi^{-t} \cdot i^*)\, d\xi \in E \otimes T\omega_\varphi$ est le tenseur mesurant la *contrainte résultante de traction* et où

$$\overline{n} = \left(\int_{h^-}^{h^+} J\, b\, d\xi + [J\, \boldsymbol{\sigma} \cdot t^*]_{h^-}^{h^+}\right) \in E,$$

est un vecteur qui mesure les forces extérieures agissant sur la coque. Il prend en compte la résultante des forces volumiques et les efforts de traction appliqués sur les bords horizontaux extérieurs de la coque.

5.3. Équilibre des moments résultants autour de la surface moyenne. En multipliant l'équation d'équilibre (5.3) par ξ et en l'intégrant à travers l'épaisseur, on obtient l'équation d'équilibre des premiers moments autour de la surface moyenne pour un élément de volume traversant toute l'épaisseur, soit

$$\text{(5.5)}\quad \int_{\partial S}\left(\int_{h^-}^{h^+} J\, \xi\boldsymbol{\sigma} \cdot (\nabla\phi^{-t} \cdot i^*)\, d\xi\right) \cdot \nu^* d\Gamma + \iint_S \left(\int_{h^-}^{h^+} \xi \frac{\partial}{\partial \xi}(J\boldsymbol{\sigma} \cdot t^*)\, d\xi\right) + \left(\int_{h^-}^{h^+} J\, \xi\, b\, d\xi\right) dS = 0.$$

En intégrant par parties le second terme de l'équation, il vient

$$\int_{h^-}^{h^+} \xi \frac{\partial}{\partial \xi}(J\, \boldsymbol{\sigma} \cdot t^*)\, d\xi = [J\, \xi\boldsymbol{\sigma} \cdot t^*]_{h^-}^{h^+} - \int_{h^-}^{h^+} J\, \boldsymbol{\sigma} \cdot t^*\, d\xi,$$

et donc l'équation d'équilibre des moments (5.5) devient

$$\text{(5.6)}\qquad \int_{\partial S} M \cdot \nu^* d\Gamma - \iint_S l\, dS + \iint_S \overline{m}\, dS = 0,$$

où $M = \int_{h^-}^{h^+} J\, \xi\boldsymbol{\sigma}(\nabla\phi^{-t} \cdot i^*)\, d\xi \in E \otimes T\omega_\varphi$, est le tenseur mesurant le *moment de contrainte résultant de flexion*,

$$\overline{m} = [J\, \xi\boldsymbol{\sigma} \cdot t^*]_{h^-}^{h^+} + \int_{h^-}^{h^+} J\, \xi b\, d\xi,$$

le moment d'ordre un par rapport à la surface moyenne des densités de forces extérieures exercées sur la tranche de coque considérée et où

$$l = \int_{h^-}^{h^+} J\, \boldsymbol{\sigma} \cdot t^* d\xi,$$

mesure la « moyenne » des efforts de traction normaux, agissant d'une feuille à l'autre de la coque et intervient comme un contrainte interne.

5.4. Équilibre des couples résultant—Condition de symétrie. Les conditions d'équilibre écrites jusqu'à présent intègrent à travers l'épaisseur la condition d'équilibre des forces et des moments exercés sur chaque feuille de la coque. À l'équilibre, la somme des couples exercés sur un élément de volume traversant toute la coque est aussi nulle ce qui se traduit, en élasticité tridimensionnelle, par une condition de symétrie du tenseur des contraintes $\boldsymbol{\sigma}$. Il nous faut donc maintenant intégrer cette condition de symétrie à travers l'épaisseur, soit

$$0 = \int_{h^-}^{h^+} \operatorname{skew}(J\boldsymbol{\sigma}) = \operatorname{skew}\left(\int_{h^-}^{h^+} J\boldsymbol{\sigma}\right) = \operatorname{skew}\left(\int_{h^-}^{h^+} (J\boldsymbol{\sigma} \cdot \nabla\phi^{-t} \cdot \nabla\phi^t)\right).$$

En décomposant la fonction ϕ donnée par (2.1) il vient

$$\int_{h^-}^{h^+} J\boldsymbol{\sigma} = \left(\int_{h^-}^{h^+} J\boldsymbol{\sigma}\cdot\nabla\phi^{-t}\right)\cdot(\pi^t\cdot\nabla\varphi^t+\widehat{N}^*\otimes t)+\left(\int_{h^-}^{h^+}\xi J\boldsymbol{\sigma}\cdot\nabla\phi^{-t}\right)\cdot\pi^t\cdot\nabla t^t.$$

En utilisant la décomposition de $\nabla\phi^{-t}$ suivant surface et épaisseur, $\nabla\phi^{-t}=(\nabla\phi^{-t}\cdot i^*)\cdot i^t+t^*\otimes\widehat{N}$ et la définition de M, N et l on obtient finalement

$$\text{skew}(N\cdot\nabla\varphi^t+l\otimes t+M\cdot\nabla t^t)=0. \tag{5.7}$$

Cette dernière équation exprime une condition de symétrie d'un certain tenseur

$$\widetilde{\boldsymbol{N}}=N\cdot\nabla\varphi^t+l\otimes t+M\cdot\nabla t^t \tag{5.8}$$

qui est l'analogue du tenseur de Cauchy de la théorie tridimensionnelle.

5.5. Équations d'équilibre réduites. En appliquant la formule de Stokes pour les équations (5.4) et (5.6), on obtient un système de trois équations bidimensionnelles sur ω_φ

$$\begin{cases}\text{(i)} & \text{DIV}(N)+\overline{n}=0,\\ \text{(ii)} & \text{DIV}(M)-l+\overline{m}=0,\\ \text{(iii)} & \text{skew}(N\cdot\nabla\varphi^t+l\otimes t+M\cdot\nabla t^t)=0.\end{cases} \tag{5.9}$$

Si l'on exprime ces équations en coordonnées locales on retrouve les équations d'équilibre de Simo [**SIM**].

6. Formulation variationnelle et tenseurs effectifs

L'espace des variations admissibles $\mathcal{V}$ est l'espace tangent $T_\phi\mathcal{C}$ en chaque configuration $\phi=(\varphi,t)$, à la variété $\mathcal{C}$ définie par (2.2). Comme $\mathcal{C}$ est un ouvert de la variété $\mathcal{M}=\{(\varphi,t)\colon\omega_\varphi\to\mathbb{R}^3\times S^2\}$, leur espace coïncide et

$$\mathcal{V}=\{\delta\phi=(\delta\varphi,\delta t)\colon\omega_\varphi\to\mathbb{R}^3\times\mathbb{R}^3 \text{ tel que } \delta t\cdot G\cdot t=0\}.$$

Compte tenu des hypothèses cinématiques (a) et (b) on s'attend à trouver une formulation variationnelle des équations d'équilibre telle que le travail des efforts internes se décompose en trois parties : énergie de membrane, de courbure et de cisaillement ; les contraintes internes apparaissant alors comme les variables duales des trois tenseurs de déformation élémentaires (ou linéarisés) $\delta\boldsymbol{\varepsilon}$, $\delta\boldsymbol{\rho}$ et $\delta\boldsymbol{\gamma}$.

Pour obtenir cette formulation, on multiplie les équations d'équilibre des efforts et des moments résultant (5.9) (i), (5.9) (ii) par le déplacement admissible $\delta\varphi$ respectivement δt et on intègre sur toute l'épaisseur de la coque. En utilisant la condition d'inextensibilité $(\|t\|=1)$, la condition de symétrie (5.9) (iii) ainsi que les propriétés classiques du calcul tensoriel, on obtient effectivement (après quelques calculs un peu techniques cf. [**CAR**]) la formulation variationnelle suivante :

$$\begin{cases}\mathcal{G}(\phi,\delta\phi)=\displaystyle\int_\omega(\widetilde{\boldsymbol{n}}\colon\delta\boldsymbol{\varepsilon}+\widetilde{\boldsymbol{q}}\cdot\delta\boldsymbol{\gamma}+\mathbf{m}\colon\delta\boldsymbol{\rho})\,dS\\ \qquad\qquad -\displaystyle\int_\omega(\overline{n}\cdot G\cdot\delta\varphi+\overline{m}\cdot G\cdot\delta t)\,dS\\ \qquad\qquad -\displaystyle\int_{\partial\omega}(N\cdot\nu^*\cdot G\cdot\delta\varphi+M\cdot\nu^*\cdot G\cdot\delta t)\,d\Gamma=0,\quad\forall\delta\phi\in\mathcal{V}\\ \widetilde{\boldsymbol{n}}^t=\widetilde{\boldsymbol{n}},\end{cases} \tag{6.1}$$

où

$$(6.2)\qquad \begin{cases} \widetilde{\boldsymbol{n}} = \pi \cdot \mathcal{N} \cdot \pi^t \in T\omega_\varphi \otimes T\omega_\varphi, \\ \widetilde{\boldsymbol{q}} = \widehat{N}^* \cdot \mathcal{N} \cdot \pi^t = \pi \cdot \mathcal{N} \cdot \widehat{N}^* \in T\omega_\varphi, \\ \mathbf{m} = \pi \cdot f^{-1} \cdot M \in T\omega_\varphi \otimes T\omega_\varphi, \end{cases}$$

sous la notation : $f = \nabla\phi/_{\xi=o} = \nabla\varphi \cdot \pi + t \otimes \widehat{N}^*$ et $\mathcal{N} = f^{-t} \cdot \widetilde{\boldsymbol{N}} \cdot f^{-1}$. Les tenseurs $\widetilde{\boldsymbol{n}}$, $\mathbf{m}$ et $\widetilde{\boldsymbol{q}}$ sont qualifiés de tenseurs effectifs car ce sont eux qui travaillent dans un déplacement virtuel $(\delta\varphi, \delta t)$. On montre **[CAR]** que les tenseurs de contraintes bidimensionnels N, M et l s'expriment en fontion de ces trois tenseurs effectifs par la relation.

$$(6.3)\qquad \begin{cases} N = \nabla\varphi \cdot \widetilde{\boldsymbol{n}} + t \otimes \widetilde{\boldsymbol{q}} + \nabla t \cdot \mathbf{m}^t, \\ M = \nabla\varphi \cdot \mathbf{m} + t \otimes (t \cdot M), \\ l = \nabla\varphi \cdot \widetilde{\boldsymbol{q}} + \nabla t \cdot (t^* \cdot M) + \lambda t \end{cases}$$

où λ est la composante d'épaississement $(\widehat{N}^* \cdot \mathcal{N} \cdot \widehat{N}^*)$.

Analyse formelle des équations d'équilibre

Dans ce paragraphe nous étudions les propriétés du modèle proposé dans le but d'une part de valider le modèle du point de vue mathématique et mécanique et d'autre part de montrer la pertinence de l'approche intrinsèque. Nous supposerons que les variables (φ, t) et $(\delta\varphi, \delta t)$ sont suffisamment régulières.

7. Transport des équations

Pour construire les équations d'équilibre, nous avons toujours travaillé sur la configuration réelle et utilisé une notation tensorielle de sorte que nous ne privilégions aucune base et aucun référentiel. En ce sens la formulation obtenue est bien intrinsèque. En particulier cette formulation est valable pour des surfaces fermées (ovoides). Nous pouvons à présent choisir une surface régulière ω quelconque (par exemple la surface moyenne au repos, ou dans un état précontraint) comme configuration de référence et y formuler le problème ci-dessus par simple transport. Il suffit pour cela de multiplier les tenseurs de déformation par le gradient de la transformée $T\colon \omega \to \omega_\varphi$ et les tenseurs de contraintes par l'inverse du gradient de T et son jacobien $\jmath$ (exemple : $\boldsymbol{\varepsilon}_\omega = \nabla T^t \cdot \boldsymbol{\varepsilon} \cdot \nabla T$; $\widetilde{\boldsymbol{n}}_\omega = \jmath \nabla T^{-1} \cdot \widetilde{\boldsymbol{n}} \circ T \cdot \nabla T^{-t}$). Les équations d'équilibre conservent alors la même forme (6.1) mais sont maintenant exprimées sur la surface choisie ω.

8. Interprétation de la formulation variationnelle

Dans cette section nous montrons que la formulation variationnelle (6.1) est formellement équivalente aux équations d'équilibre réduites (5.9) moyennant l'introduction d'une hypothèse de comportement. Pour cela on se débarrasse de la condition d'inextensibilité incluse dans l'espace des déplacements admissibles en introduisant l'application $L\colon (\eta, h) \to t \cdot G \cdot h$, $(\eta, h) \in \mathcal{D}^6(\omega)$, $\mathcal{D}$ désignant l'espace des fontions à support compact indéfiniment dérivables. L'équation (6.1) s'écrit alors

$$\mathcal{G}(\varphi, t) \in (\mathrm{Ker} L)^\perp \subset \mathcal{D}'(\omega)^6.$$

L'application L étant surjective on peut appliquer le théorème de l'image fermée **[BRE]** et l'on déduit qu'il existe $\bar{\lambda} \in \mathcal{D}'(\omega)$ tel que

$$\mathcal{G}\big((\varphi, t), (\eta, h)\big) - \int_{\omega} \bar{\lambda} t \cdot G \cdot h \, dS = 0.$$

En utilisant la condition de symétrie sur $\widetilde{\boldsymbol{n}}$, les propriétés de l'opérateur trace et comme (η, h) est à support compact, on obtient finalement

$$\begin{cases} \mathrm{DIV}(\nabla\varphi \cdot \widetilde{\boldsymbol{n}} + t \otimes \widetilde{\boldsymbol{q}} + \nabla t \cdot \mathbf{m}^t) + \jmath\overline{n} = 0, \\ \mathrm{DIV}(\nabla\varphi \cdot \mathbf{m}) - \nabla\varphi \cdot \widetilde{\boldsymbol{q}} + \bar{\lambda} t + \jmath\overline{m} = 0. \end{cases} \tag{8.1}$$

La réciproque s'obtient naturellement, en utilisant la condition d'inextensibilité $(t \cdot G \cdot \delta t = 0)$ et en appliquant la formule de Stokes. Si l'on exprime les tenseurs N, M et l en fonction des tenseurs effectifs en utilisant la relation (6.3), on peut alors énoncer la proposition suivante :

PROPOSITION 8.1. *Sous l'hypothèse de comportement $t^* \cdot M = 0$, la formulation variationnelle* (6.1) *est équivalente aux équations d'équilibre réduites* (5.9).

9. Caractérisation globale du mouvement d'une surface de Cosserat

Nous établissons ici un théorème qui permet de caractériser les positions de la coque à l'aide des tenseurs géométriques $(\mathbf{C}, \mathbf{K}, \mathbf{R})$. Du point de vue mécanique ceci signifie que la donnée des tenseurs $(\mathbf{C}, \mathbf{K}, \mathbf{R})$ suffit à contrÔler, à un mouvement rigide près, la position (φ, t) de la coque. Ce résultat est une généralisation du théorème de Bonnet sur les surfaces **[SPI]** au cas des surfaces fibrées.

THÉORÈME 9.1. *Soit (φ, t), (φ', t') deux positions admissibles de la coque, $(\mathbf{C}, \mathbf{K}, \mathbf{R})$, respectivement $(\mathbf{C}', \mathbf{K}', \mathbf{R}')$, leurs tenseurs de déformations associés. Si $(\mathbf{C}, \mathbf{K}, \mathbf{R}) = (\mathbf{C}', \mathbf{K}', \mathbf{R}')$ alors $\exists Q \in \mathcal{SO}^3$, et $v \in \mathbb{R}^3$ tels que*

$$\begin{cases} \varphi' = Q \cdot \varphi + v, \\ t' = Q \cdot t. \end{cases}$$

DÉMONSTRATION. Montrons tout d'abord l'égalité suivante :

$$\mathbf{K}_n = \frac{\mathbf{K} - \mathbf{C} \cdot \nabla\tau}{\sqrt{1 - \tau \cdot \mathbf{C} \cdot \tau}}. \tag{9.1}$$

Ici, $K_n = \nabla\varphi^t \cdot G \cdot \nabla n$ est le tenseur de courbure géométrique (n étant la normale en chaque point au plan tangent à la surface moyenne), $\tau = \mathbf{C}^{-1} \cdot \mathbf{R} \in T\omega$ et $\nabla\tau \in T\omega \otimes T\omega^*$ désigne la dérivée covariante de τ pour la métrique $\mathbf{C}$.

Soit $p\colon E \to T\omega_\varphi \subset E$, la projection orthogonale pour la métrique G. Comme $\|t\| = 1$ on a,

$$t = p(t) + \alpha n, \quad \alpha = \sqrt{1 - \|p(t)\|^2}. \tag{9.2}$$

Le vecteur $p(t)$ étant le «pusforward»**[SPI]** d'un vecteur τ du plan tangent $T\omega$, c.-à-d.

$$p(t) = \varphi_* \tau = \nabla\varphi \cdot \tau,$$

on en déduit

$$\|p(t)\|^2 = (\nabla\varphi \cdot \tau)^t \cdot G \cdot \nabla\varphi \cdot \tau = \tau \cdot \mathbf{C} \cdot \tau \quad \text{et} \quad \alpha = \sqrt{1 - \tau \cdot \mathbf{C} \cdot \tau}.$$

Par ailleurs, comme n est orthogonal à $T\omega_\varphi$ on obtient

$$\mathbf{R} = t \cdot G \cdot \nabla\varphi = p(t) \cdot G \cdot \nabla\varphi = (\nabla\varphi^t \cdot G \cdot \nabla\varphi) \cdot \tau = \mathbf{C} \cdot \tau,$$

d'où $\tau = \mathbf{C}^{-1} \cdot \mathbf{R}$.

Calculons à présent $\mathbf{K} = \nabla\varphi^t \cdot G \cdot \nabla t$ en fonction de τ et de n. En dérivant (9.2), en utilisant l'orthogonalité de n à $T\omega_\varphi$ et la définition de $\mathbf{K}_n$ il vient

$$\mathbf{K} = \nabla\varphi^t \cdot G \cdot \nabla(\varphi_*\tau) + \alpha \mathbf{K}_n.$$

De plus, toujours par orthogonalité, on a $\forall v \in T\omega$,

$$\nabla\varphi^t \cdot G \cdot \nabla(\varphi_*\tau) \cdot v = \nabla\varphi^t \cdot G \cdot p\big(\nabla(\varphi_*\tau) \cdot v\big).$$

Or Spivak [**SPI**, vol. III, pp. 1–3], par construction de la dérivée covariante, on a

$$p\big(\nabla(\varphi_*\tau) \cdot v\big) = (\varphi_*\nabla\tau) \cdot v = \nabla\varphi \cdot \nabla\tau \cdot v.$$

On obtient donc finalement pour $\mathbf{K}$ l'expression suivante :

$$\mathbf{K} = \mathbf{C} \cdot \nabla\tau + \sqrt{1 - \tau \cdot \mathbf{C} \cdot \tau}\mathbf{K}_n \text{ avec } \tau = \mathbf{C}^{-1} \cdot \mathbf{R},$$

qui donne bien la relation annoncée (9.1).

D'après l'hypothèse du théorème, on a $\mathbf{C} = \mathbf{C}'$, $\mathbf{R} = \mathbf{R}'$ d'où $\tau' = \mathbf{C}'^{-1} \cdot \mathbf{R}' = \mathbf{C}^{-1} \cdot \mathbf{R} = \tau$, $\alpha = \alpha'$ et par (9.1) on déduit $\mathbf{K}_n = \mathbf{K}'_n$. D'après le théorème de Bonnet il existe une rotation $Q \in E \otimes E^*$ et une translation v telle que

$$\varphi' = Q \cdot \varphi + v.$$

Il nous reste donc à montrer $t' = Q \cdot t$. D'après la relation précédente, on a

$$p(t') = \nabla\varphi' \cdot \tau' = Q \cdot \nabla\varphi \cdot \tau = Q \cdot p(t).$$

Comme Q est un tenseur orthogonal on a d'une part

$$(\nabla\varphi')^t \cdot G \cdot Q \cdot n = \nabla\varphi^t \cdot Q^t \cdot G \cdot Q \cdot n = \nabla\varphi^t \cdot G \cdot n = 0,$$

c'est-à-dire $(Q \cdot n) \perp T\omega_{\varphi'}$ et d'autre part $\|Q \cdot n\| = 1$. Par conséquent, $n' = \varepsilon Q \cdot n$ avec $\varepsilon = \pm 1$. Mais alors

$$\mathbf{K}'_n = \varepsilon \nabla\varphi^t \cdot Q^t \cdot G \cdot Q \cdot \nabla n = \varepsilon \nabla\varphi^t \cdot G \cdot \nabla n = \mathbf{K}_n,$$

et par identification on déduit $\varepsilon = 1$. Finalement, on trouve bien

$$t' = p(t') + \alpha n' = Q \cdot p(t) + \alpha Q \cdot n = Q \cdot t. \quad \square$$

10. Existence local du mouvement d'une surface de Cosserat et conditions de compatibilité

Dans ce paragraphe nous montrons un résultat d'existence local sous certaines conditions de compatibilité. Ce résultat traduit une propriété d'indépendance des tenseurs géométriques **C**, **K** et **R**. Soit $\mathcal{E}$ un espace vectoriel, on note $\mathrm{Sym}(\mathcal{E})$ et $\mathrm{Sym}^+(\mathcal{E})$, le sous-espace des tenseurs symétriques respectivement symétriques définis positifs de $\mathcal{E} \otimes \mathcal{E}$. On introduit l'ensemble

$$M^+_{m,\omega} = \begin{cases} (C, K, R) \in \mathrm{Sym}(T_m\omega^*) \times (T_m\omega^* \otimes T_m\omega^*) \times T_m\omega^*, \text{ tel que} \\ c_f = \pi^t \cdot C \cdot \pi + \pi^t \cdot R \otimes \widehat{N}^* + \widehat{N}^* \otimes R \cdot \pi \\ \qquad + \widehat{N}^* \otimes \widehat{N}^* \in \mathrm{Sym}^+(T_m\omega \times \mathbb{R})^*. \end{cases}$$

THÉORÈME 10.1. *À tout champ* $(C, K, R) \in M^+_{m,\omega}$ *correspond une position admissible* (φ, t) *définie sur un voisinage de* $m \in T_m\omega$ *telle que*

$$\begin{cases} \mathbf{C}(m) = \nabla_m\varphi^t \cdot G \cdot \nabla_m\varphi = C, \\ \mathbf{K}(m) = \nabla_m\varphi^t \cdot G \cdot \nabla_m t = K, \\ \mathbf{R}(m) = t(m) \cdot G \cdot \nabla_m\varphi = R. \end{cases}$$

DÉMONSTRATION. La preuve s'effectue en trois étapes. La première et la deuxième étape étant obtenues de manière très naturelle nous ne les détaillerons pas ici.

1^RE^ÉTAPE. Le résultat annoncé étant local nous pouvons nous ramener localement à un ouvert $\widehat{\mathcal{O}} \subset \mathbb{R}^2$ par un changement de variable $\hat\theta \colon \widehat{\mathcal{O}} \to \mathcal{O} \subset \omega$. Les tenseurs géométriques $(C, K, R) \in T_m\omega$ sont alors transportés sur $\mathbb{R}^2$ par composition avec l'application $\nabla\hat\theta$. Les tenseurs images sont notés $(\widehat{C}, \widehat{K}, \widehat{R})$. On montre que le tenseur $\hat c_f$ correspondant, obtenue par composition avec l'application surjective $T_{\hat\theta} = i \cdot \nabla\hat\theta \cdot \widehat{\pi} + \widehat{N} \otimes \widehat{N}^*$, est symétrique défini positif et donc que $(\widehat{C}, \widehat{K}, \widehat{R}) \in M^+_{\mathbb{R}^2}$. On peut alors écrire

$$\hat c_f = \hat c_f^{1/2} \cdot G^{-1} \cdot \hat c_f^{1/2},$$

où G^{-1} est le tenseur métrique associé à E^*.

2^E^ÉTAPE. Construction d'une position admissible $(\widehat{\varphi}, \hat t\,)$. Soit $\mathcal{O}^\varepsilon$ une boule de rayon ε de centre m. $\hat\theta^{-1}(\mathcal{O}^\varepsilon)$ est un voisinage de $x = \hat\theta^{-1}(m)$ et contient une boule de centre m de rayon $k\varepsilon$ notée $\widehat{\mathcal{O}}^\varepsilon$. Soit $\widehat{\varphi}$ et $\hat t$ les applications linéaires définies sur $\widehat{\mathcal{O}}^\varepsilon$ par

$$\begin{aligned} \widehat{\varphi}(x^\varepsilon) &= (G^{-1} \cdot \hat c_f^{1/2} \cdot \hat\imath) \cdot (x^\varepsilon - x), \\ \hat t(x^\varepsilon) &= \frac{a + AX^\varepsilon}{\|a + AX^\varepsilon\|}, \end{aligned}$$

sous la notation

$$a = G^{-1} \cdot \hat c_f^{1/2} \cdot \widehat{N}, \quad A = \hat c_f^{-1/2} \cdot \hat\imath^* \cdot \widehat{K}, \quad X^\varepsilon = x^\varepsilon - x.$$

Par construction, a est un vecteur de E et A est un tenseur de $E \otimes \mathbb{R}^{2*}$, donc $\widehat{\varphi}$ et $\hat t$ sont bien à valeurs dans E et $\widehat{\varphi}$ est injective comme composée de deux applications bijectives et d'une injection. Pour que le couple d'applications $(\widehat{\varphi}, \hat t\,)$ ainsi construit définisse une position admissible, il faut que

$$\forall x^\varepsilon \in \widehat{\mathcal{O}}^\varepsilon \begin{cases} \text{(i)} & \|\hat t\,\| = 1, \\ \text{(ii)} & \hat t(x) \notin \operatorname{Im}\big(\nabla(\widehat{\varphi})\big) \ (\hat t \text{ est transverse}). \end{cases}$$

Par construction (i) est vérifié. Pour montrer (ii) il suffit de montrer que $\{\widehat{\varphi}_{,1}, \widehat{\varphi}_{,2}, \hat t\,\}$ forme une base locale en chaque point $\widehat{\varphi}(x^\varepsilon) \in \mathcal{O}^\varepsilon$, autrement dit $\hat f = \nabla\widehat{\varphi} \cdot \widehat{\pi} + \hat t \otimes \widehat{N}^*$ est bijective où encore $\det(\hat f^t \cdot G \cdot \hat f) > 0$. On sait que

$$\hat f^t \cdot G \cdot \hat f = \widehat{\pi}^t \cdot \widehat{\mathbf{C}} \cdot \widehat{\pi} + \widehat{\pi}^t \cdot \widehat{\mathbf{R}} \otimes \widehat{N}^* + \widehat{N}^* \otimes \widehat{\mathbf{R}} \cdot \widehat{\pi} + \widehat{N}^* \otimes \widehat{N}^*,$$

où $\widehat{\mathbf{C}} = \nabla\widehat{\varphi}^t \cdot G \cdot \nabla\widehat{\varphi}$ et $\widehat{\mathbf{R}} = \hat t \cdot G \cdot \nabla\widehat{\varphi}$. Par ailleurs on a

$$\|a + AX^\varepsilon\|^2 = a \cdot G \cdot a + 2a \cdot G \cdot (A \cdot X^\varepsilon) + \|A \cdot X^\varepsilon\|^2.$$

Or, après substitution le second terme se réduit à

$$a^t \cdot G \cdot A \cdot X^\varepsilon = \widehat{N} \cdot \hat c_f^{1/2} \cdot G^{-1} \cdot G \cdot \hat c_f^{-1/2} \cdot \hat\imath^* \cdot \widehat{K} \cdot X^\varepsilon = (\widehat{N} \cdot \hat\imath^*) \cdot \widehat{K} \cdot X^\varepsilon = 0,$$

et le premier terme à

$$a^t \cdot G \cdot a = \widehat{N} \cdot \hat{c}_f^{1/2} \cdot G^{-1} \cdot G \cdot G^{-1} \cdot \hat{c}_f^{1/2} \cdot \widehat{N} = \widehat{N} \cdot \hat{c}_f \cdot \widehat{N} = \widehat{N} \cdot \widehat{N}^* \otimes \widehat{N}^* \cdot \widehat{N} = 1.$$

Il s'en déduit

$$\hat{t} = \frac{a + A \cdot X^\varepsilon}{\|a + A \cdot X^\varepsilon\|} = \frac{a + A \cdot X^\varepsilon}{(1 + \|A \cdot X^\varepsilon\|^2)^{1/2}} = a + A \cdot X^\varepsilon + \mathcal{O}(\|X^\varepsilon\|^2).$$

Par construction de $\widehat{\varphi}$ et en utilisant l'expression de $\hat{t}$ sous la forme ci-dessus, on obtient

$$\begin{aligned}
(10.1) \qquad \widehat{\mathbf{C}}(x^\varepsilon) &= \hat{\imath}^t \cdot \hat{c}_f^{1/2} \cdot G^{-1} \cdot G \cdot G^{-1} \cdot \hat{c}_f^{1/2} \cdot \hat{\imath} = \hat{\imath}^t \cdot \hat{c}_f \cdot \hat{\imath} = \widehat{C} \\
\widehat{\mathbf{R}}(x^\varepsilon) &= (\hat{\imath}^t \cdot \hat{c}_f^{1/2} \cdot G^{-1} \cdot G \cdot G^{-1} \cdot \hat{c}_f^{1/2} \cdot \widehat{N}) + \mathcal{O}(\|x^\varepsilon - x\|) \\
&= \hat{\imath}^t \cdot \hat{c}_f \cdot \widehat{N} + \mathcal{O}(\|x^\varepsilon - x\|) \\
&= \widehat{R} + \mathcal{O}(\|x^\varepsilon - x\|), \quad \forall x^\varepsilon \in \widehat{\mathcal{O}}^\varepsilon.
\end{aligned}$$

Par addition, on obtient

$$\hat{f}^t \cdot G \cdot \hat{f} = \hat{c}_f + \mathcal{O}(\|x^\varepsilon - x\|).$$

Or $\hat{c}_f$ est défini positif donc $\det(\hat{c}_f)$ est positif strictement. En conséquence, pour ε suffisamment petit on a

$$\det(\hat{f}^t \cdot G \cdot \hat{f}) > 0.$$

Nous avons ainsi construit une position $(\widehat{\varphi}, \hat{t}\,)$ admissible définie sur une boule suffisamment petite. Il nous reste à montrer que les tenseurs $\widehat{\mathbf{C}}$, $\widehat{\mathbf{K}}$ et $\widehat{\mathbf{R}}$ ont les bonnes valeurs au point m. D'après (10.1) on a immédiatement $\widehat{\mathbf{C}}(x) = \widehat{C}$ et $\widehat{\mathbf{R}}(x) = \widehat{R}$. Quant au tenseur de courbure $\widehat{\mathbf{K}}$, un calcul direct donne

$$\widehat{\mathbf{K}}(x) = \nabla\widehat{\varphi}^t \cdot G \cdot \nabla\hat{t} = (\hat{\imath}^t \cdot \hat{c}_f^{1/2} \cdot G^{-1}) \cdot G \cdot (\hat{c}_f^{-1/2} \cdot \hat{\imath}^* \cdot \widehat{K}) = (\hat{\imath}^t \cdot \hat{\imath}^* \cdot \widehat{K}) = \widehat{K}.$$

3^E^ÉTAPE. Retour à la configuration ω.

Pour tout $m^\varepsilon \in \mathcal{O}^\varepsilon$, on pose

$$\begin{cases} \varphi(m^\varepsilon) = \widehat{\varphi} \circ \hat{\theta}^{-1}(m^\varepsilon), \\ t(m^\varepsilon) = \hat{t} \circ \hat{\theta}^{-1}(m^\varepsilon). \end{cases}$$

Le couple (φ, t) est obtenu comme composition d'une position admissible avec une injection. Pour ε suffisamment petit, (φ, t) définit donc bien une position admissible sur la boule $\mathcal{O}^\varepsilon$. De plus, au point m on a bien par composition

$$\mathbf{C}(m, (\varphi, t)) = C, \mathbf{K}(m, (\varphi, t)) = K, \mathbf{R}(m, (\varphi, t)) = R. \quad \square$$

JUSTIFICATION DE LA LOI HYPERÉLASTIQUE POUR LES COQUES

Pour calculer la position d'équilibre (φ, t) de la coque déformée, nous devons compléter les équations d'équilibre régissant le comportement de la coque en introduisant des lois de comportement qui relient les contraintes aux variables cinématiques. Pour écrire ces lois de comportement on se place dans la configuration au repos $\omega = \omega_0$ qui est connue et donc sur laquelle sont définies les propriétés du matériau.

11. Hypothèse d'élasticité pour une coque

Comme l'application de feuilletage $\phi^0(m_0,\xi) = m_0 + \xi\, t^0(m_0)$ définit un difféomorphisme de Ω_0 sur $\mathcal{B}$, nous pouvons identifier ces deux espaces. Avec cette identification la déformation réelle de la coque est donnée par l'application ϕ_0 définie sur Ω_0 par

$$\phi_0(m_0,\xi) = \varphi_0(m_0) + \xi\, t_0(m_0).$$

Nous avons vu à la section (6) que dans une déformation cinématiquement admissible $(\delta\varphi_0, \delta t_0)$, ce sont les contraintes effectives $(\widetilde{\boldsymbol{n}}_0, \mathbf{m}_0, \widetilde{\boldsymbol{q}}_0)$ qui travaillent. Ces contraintes correspondent à des moments d'ordre 0 et 1 du tenseur de Piola $J\boldsymbol{\sigma}\cdot\nabla\phi_0^{-t}$. Ainsi, nous dirons qu'une coque est élastique si et seulement si le champ de tenseurs $(\widetilde{\boldsymbol{n}}_0, \mathbf{m}_0, \widetilde{\boldsymbol{q}}_0)$ dépend uniquement des moments du gradient de déformation $\nabla\phi_0$, ou de ses puissances. D'après le théorème de décomposition polaire, on a

$$\nabla\phi_0 = Q\cdot c_0^{1/2},$$

où $c_0 = \nabla\phi_0^t \cdot G \cdot \nabla\phi_0$ est le tenseur de Cauchy-Green à gauche. D'après l'axiome d'indifférence matérielle, ceci implique que $(\widetilde{\boldsymbol{n}}_0, \mathbf{m}_0, \widetilde{\boldsymbol{q}}_0)$ dépend uniquement de c_0. En utilisant l'expression de $\nabla\phi_0$ en fonction des applications φ_0 et t_0 et la définition de $\mathbf{C}_0$, $\mathbf{K}_0$ et $\mathbf{R}_0$, on obtient l'égalité suivante :

$$\begin{aligned} c_0 = \pi_0^t\cdot\big(\mathbf{C}_0 + \xi(\mathbf{K}_0+\mathbf{K}_0^t) + \xi^2(\nabla t_0^t\cdot G\cdot\nabla t_0)\big)\cdot\pi_0 \\ + \pi_0^t\cdot\mathbf{R}_0\otimes\widehat{N}^* + \widehat{N}^*\otimes\mathbf{R}_0\cdot\pi_0 + \widehat{N}^*\otimes\widehat{N}^*. \end{aligned}$$

Or, on vérifie aisément que $\nabla t_0^t\cdot G\cdot\nabla t_0 = \mathbf{K}_0^t\cdot\mathbf{C}_0^{-1}\cdot\mathbf{K}_0$. Nous avons ainsi montré que le tenseur c_0 (plus exactement ses moments) ne dépend que de $\mathbf{C}_0$, $\mathbf{K}_0$ et $\mathbf{R}_0$. L'hypothèse d'élasticité s'écrit donc finalement

$$(11.1)\qquad \begin{cases} \widetilde{\boldsymbol{n}}_0(m_0) = \widetilde{\boldsymbol{n}}_0(m_0,\mathbf{C}_0,\mathbf{K}_0,\mathbf{R}_0), \\ \mathbf{m}_0(m_0) = \mathbf{m}_0(m_0,\mathbf{C}_0,\mathbf{K}_0,\mathbf{R}_0), \\ \widetilde{\boldsymbol{q}}_0(m_0) = \widetilde{\boldsymbol{q}}_0(m_0,\mathbf{C}_0,\mathbf{K}_0,\mathbf{R}_0). \end{cases}$$

12. Loi hyperélastique

Dans cette section nous donnons une justification de la loi hyperélastique pour les coques en utilisant uniquement les équations bidimensionnelles. La définition de l'hyperélasticité que nous utilisons ici est celle de Gurtin [**GUR**].

DÉFINITION 12.1. Une coque est hyperélastique si et seulement si elle est élastique et si le travail des efforts intérieurs est nul sur un processus clos.

Définissons tout d'abord la notion de processus clos dans le cadre d'une théorie bidimensionnelle de coque développée ici. Soit $\mathcal{C}_0[0,\tau]$ l'ensemble des chemins admissibles définis par

$$\phi_{0t} = (\varphi_0, t_0)_t \colon [0,\tau]\to\mathcal{C}_0.$$

(Rappelons que $\mathcal{C}_0$ désigne la variété décrivant l'ensemble des configurations admissibles (φ_0, t_0) définis sur ω_0). La vitesse définie pour toute particule $m_0\in\omega_0$ est alors donnée par

$$\big(\dot{\boldsymbol{\varphi}}_0(m_0), \dot{t}_0(m_0)\big) = \left(\left.\frac{d\varphi_{0t}}{dt}\right|_{t=0}(m_0), \left.\frac{dt_{0t}}{dt}\right|_{t=0}(m_0)\right),$$

où $t\to(\varphi_{0\varepsilon}, t_{0\varepsilon})$ est un chemin de configuration admissible. Par construction $(\dot{\boldsymbol{\varphi}}_0, \dot{\boldsymbol{t}}_0)$ appartient au plan tangent $T_{\phi_0}\mathcal{C}_0$. C'est donc un champ admissible et il définit donc un mouvement virtuel. De plus il induit un champ de vitesse de déformations virtuelles $(\dot{\mathbf{C}}_0, \dot{\mathbf{K}}_0, \dot{\mathbf{R}}_0)$. On appelle processus une paire $\big(\phi_{0t}, (\widetilde{\boldsymbol{n}}_0, \mathbf{m}_0, \widetilde{\boldsymbol{q}}_0)\big)$ où $(\widetilde{\boldsymbol{n}}_0, \mathbf{m}_0, \widetilde{\boldsymbol{q}}_0)$ est le champ de contraintes effectives associé au chemin ϕ_{0t} et on dit que ce processus est clos sur l'intervalle $[0, \tau]$ si

$$\phi_{0t}(0) = \phi_{0t}(\tau) \quad \text{et} \quad \dot{\boldsymbol{\phi}}_{0t}(0) = \dot{\boldsymbol{\phi}}_{0t}(\tau).$$

Le travail des efforts intérieurs $W_{\text{int}}(S_0)$, $S_0 \subset \omega_0$, développé au cours d'un processus $\big(\phi_{0t}, (\widetilde{\boldsymbol{n}}_0, \mathbf{m}_0, \widetilde{\boldsymbol{q}}_0)\big)$, $t \in [0, \tau]$, est obtenu en intégrant la puissance des efforts intérieurs sur l'intervalle $[0, \tau]$, soit

$$W_{\text{int}}(S_0) = -\int_0^\tau \int_{S_0} (\widetilde{\boldsymbol{n}}_0 \colon \dot{\mathbf{C}}_0 + \widetilde{\boldsymbol{q}}_0 \cdot \dot{\mathbf{R}}_0 + \mathbf{m}_0 : \dot{\mathbf{K}}_0)\, dS\, dt.$$

DÉFINITION 12.2. On dit que le travail des efforts intérieurs est nul sur un processus clos si pour toute partie $S_0 \subset \omega_0$ et pour tout $\tau \in \mathbb{R}$, $W_{\text{int}}(S_0) = 0$ dans tout processus clos sur $[0, \tau]$.

On démontre alors immédiatement le résultat suivant :

PROPOSITION 12.3. *Le travail des efforts intérieurs est nul sur un processus clos si et seulement si étant donné $m_0 \in \omega_0$ et étant donné un intervalle $\tau \in \mathbb{R}$*

$$\lambda(m_0) = \int_0^\tau (\widetilde{\boldsymbol{n}}_0 : \dot{\mathbf{C}}_0 + \widetilde{\boldsymbol{q}}_0 \cdot \dot{\mathbf{R}}_0 + \mathbf{m}_0 \colon \dot{\mathbf{K}}_0)(m_0)\, dt = 0 \tag{12.1}$$

dans tout processus qui est clos sur $[0, \tau]$.

Nous pouvons maintenant introduire la loi de comportement pour les coques.

THÉORÈME 12.4. *Une coque est hyperélastique si et seulement si en tout point $m_0 \in \omega_0$ il existe une fonction scalaire $\psi_{m_0} \colon M^+_{m_0,\omega_0} \to \mathbb{R}$ dont dérivent tous les efforts intérieurs :*

$$\begin{cases} \widetilde{\boldsymbol{n}}_0(m_0, \mathbf{C}_0, \mathbf{K}_0, \mathbf{R}_0) = 2\frac{\partial \psi_{m_0}}{\partial C_0}(\mathbf{C}_0, \mathbf{K}_0, \mathbf{R}_0), \\ \mathbf{m}_0(m_0, \mathbf{C}_0, \mathbf{K}_0, \mathbf{R}_0) = \frac{\partial \psi_{m_0}}{\partial K_0}(\mathbf{C}_0, \mathbf{K}_0, \mathbf{R}_0), \\ \widetilde{\boldsymbol{q}}_0(m_0, \mathbf{C}_0, \mathbf{K}_0, \mathbf{R}_0) = \frac{\partial \psi_{m_0}}{\partial R_0}(\mathbf{C}_0, \mathbf{K}_0, \mathbf{R}_0). \end{cases} \tag{12.2}$$

Cette fonction ψ est la **densité d'énergie libre** du système.

DÉMONSTRATION. On note par $\mathcal{F}\big([0,\tau], M^+_{m_0,\omega_0}\big)$, l'ensemble des chemins admissibles

$$(C_0, K_0, R_0)_t \colon [0, \tau] \to M^+_{m_0,\omega_0}.$$

Soit $(C_0, K_0, R_0)_t$ un chemin fermé donné de $M^+_{m_0,\omega_0}$ c'est-à-dire vérifiant

$$(C_0, K_0, R_0)_t(0) = (C_0, K_0, R_0)_t(\tau), \quad (\dot{C}_0, \dot{K}_0, \dot{R}_0)_t(0) = (\dot{C}_0, \dot{K}_0, \dot{R}_0)_t(\tau).$$

D'après le théorème d'existence (10.1), pour chaque $t \in [0, \tau]$, il existe une configuration $(\varphi_0, t_0)_t$ définie sur un voisinage $\mathcal{O}_0^\varepsilon$ de m_0 telle que

$$(\mathbf{C}_0, \mathbf{K}_0, \mathbf{R}_0)\big(m_0, (\varphi_0, t_0)\big)(t) = (C_0, K_0, R_0)(t), \quad \forall t \in [0, \tau].$$

Pour montrer que le chemin $(\varphi_0, t_0)_t$ est fermé, nous reprenons la démarche du théorème. Comme $(C_0, K_0, R_0)_t$ est fermé, il s'ensuit que le chemin $(\widehat{C}, \widehat{K}, \widehat{R})_t \in \mathcal{F}\big([0,\tau], M^+_{\mathbb{R}^2}\big)$ obtenu par transport est fermé. On en déduit immédiatement que

$$\hat{c}_{f_t} \in \mathcal{F}\big([0, \tau], \mathrm{Sym}^+(\mathbb{R}^3)\big),$$

est aussi fermé. Par construction on obtient que $(\widehat{\varphi}, \hat{t})_t$ est un chemin fermé. Le chemin $(\varphi_0, t_0)_t$ est donc fermé comme composition d'un champ admissible fermé avec une injection. Soit $\big(\phi_{0t}, (\widetilde{\boldsymbol{n}}_0, \mathbf{m}_0, \widetilde{\boldsymbol{q}}_0)\big)$ le processus clos sur $[0, \tau]$ associé à $\phi_{0t} = (\varphi_0, t_0)_t$ défini sur l'ouvert $\mathcal{O}_0^\varepsilon$. D'après la proposition (12.3), on a

$$\forall m_0^\varepsilon \in \mathcal{O}_0^\varepsilon, \int_0^\tau (\widetilde{\boldsymbol{n}}_0 \colon \dot{\mathbf{C}}_0 + \widetilde{\boldsymbol{q}}_0 \cdot \dot{\mathbf{R}}_0 + \mathbf{m}_0 \colon \dot{\mathbf{K}}_0)(m_0^\varepsilon)\, dt = 0.$$

En particulier au point m_0, on obtient

$$\int_0^\tau (\widetilde{\boldsymbol{n}}_0 \colon \dot{C}_0 + \widetilde{\boldsymbol{q}}_0 \cdot \dot{R}_0 + \mathbf{m}_0 \colon \dot{K}_0)\, dt = 0. \tag{12.3}$$

La coque considérée étant élastique, les efforts intérieurs $(\widetilde{\boldsymbol{n}}_0, \mathbf{m}_0, \widetilde{\boldsymbol{q}}_0)(m_0, t)$ sont une fonction donnée du point m_0 et des tenseurs C_0, K_0 et R_0 définis dans $M^+_{m_0,\omega_0}$. Par ailleurs, $M^+_{m_0,\omega_0}$ est un ouvert connexe de l'espace vectoriel $\mathrm{Sym}(T\omega_0^*)$ $\times\, (T\omega_0^* \otimes T\omega_0^*) \times T\omega_0^*$ (voir annexe). L'égalité (12.3) étant vraie pour tout chemin fermé $(C_0, K_0, R_0)_t \in \mathcal{F}\big([0, \tau], M^+_{m_0,\omega_0}\big)$ et ceci quelque soit $m_0 \in \omega_0$, d'après un résultat classique d'analyse vectorielle **[NIC]**, la proposition (12.2) est vérifiée. □

Remarque 12.5. L'allure de ψ doit être obtenue par des considérations mécaniques. Le plus souvent on identifie ψ à partir d'un modèle tridimensionnel et d'une hypothèse de contraintes planes.

Annexe

Lemme 12.6. *L'ensemble $M^+_{m,\omega}$ est un ouvert connexe de l'espace vectoriel* $\mathrm{Sym}(T\omega^*) \times (T\omega^* \otimes T\omega^*) \times T\omega^*$.

Démonstration. Soit F et F^+ les ensembles définis par

$$F = \big\{\mathcal{T} \in \mathrm{Sym}(T\omega \times \mathbb{R})^* / \widehat{N} \cdot \mathcal{T} \cdot \widehat{N} = 1\big\},$$
$$F^+ = \big\{\mathcal{T} \in \mathrm{Sym}^+(T\omega \times \mathbb{R})^* / \widehat{N} \cdot \mathcal{T} \cdot \widehat{N} = 1\big\}.$$

Par construction F^+ est un ouvert de F comme image réciproque de $(\mathbb{R}^+)^7$ par l'application qui a tout tenseur $\mathcal{T} \in F$ fait correspondre le déterminant des mineurs de $\mathcal{T}$. Montrons que F^+ est connexe par arc.

Soit $\mathcal{T}_0$, $\mathcal{T}_1 \in F^+$. Par hypothèse $\mathcal{T}_0$ et $\mathcal{T}_1$ sont des tenseurs symétriques définis positifs, donc il existe L_0, $L_1 \in (T\omega \times \mathbb{R})^* \otimes (T\omega \times \mathbb{R})^*$ tel que

$$\mathcal{T}_0 = L_0^t \cdot L_0 \text{ et } \mathcal{T}_1 = L_1^t \cdot L_1.$$

Soit $L(t) = tL_1 + (1-t)L_0$, $t \in [0,1]$ un arc entre L_0 et L_1. Pour tout $t \in [0,1]$, posons

$$\mathcal{T}(t) = \frac{L^t(t) \cdot L(t)}{\widehat{N} \cdot [L^t(t) \cdot L(t)] \cdot \widehat{N}}.$$

Par construction $\mathcal{T}(t)$ est symétrique définie positif et $\widehat{N} \cdot \mathcal{T}(t) \cdot \widehat{N} = 1$. De plus en vérifie aisément que $\mathcal{T}(0) = \mathcal{T}_0$ et $\mathcal{T}(1) = \mathcal{T}_1$. L'ensemble F^+ est donc connexe par arc. Soit Θ l'application

$$\begin{aligned}\Theta \colon \mathrm{Sym}(T\omega^*) \times (T\omega^* \otimes T\omega^*) \times T\omega^* &\to F \times (T\omega^* \otimes T\omega^*)\\ (C, K, R) &\mapsto (\pi^t \cdot C \cdot \pi + \pi^t \cdot R \otimes \widehat{N}^* + \widehat{N}^* \otimes R \cdot \pi + \widehat{N}^* \otimes \widehat{N}^*, K).\end{aligned}$$

Par construction Θ est une application bicontinue et $M^+_{m,\omega}$ est l'image réciproque de F^+. Comme Θ est continue on en déduit que $M^+_{m,\omega}$ est un ouvert de $\mathrm{Sym}(T\omega^*) \times$

$(T\omega^* \otimes T\omega^*) \times T\omega^*$. De plus comme Θ^{-1} est aussi continue et F^+ est connexe par arc, on en déduit que $M^+_{m,\omega}$ est connexe. □

Conclusion

Nous avons montré que le modèle intrinsèque bidimensionnel proposé ici est capable de modéliser le comportement de coques (y compris les ovoides) en grandes déformations (mais à variations d'épaisseurs négligeable), à condition d'introduire une loi de comportement pour les matériaux à réponse non linéaire. Ce modèle a abouti à une méthode de résolution originale en base cartésienne fixe pour le cas d'une coque mince en grands déplacements **[CTM]**. La méthode a en particulier été utilisée pour résoudre des problèmes d'interaction fluide/structure en grands déplacements (Le Tallec, Mouro, à paraître dans Comp. Meth. Appl. Mech. Engin.)

Références

[BRE] H. Brezis, *Analyse fonctionnelle*, Masson, Paris, 1987.

[CAR] M. Carrive, *Modélisation intrinsèque et analyse numérique d'un problème de coque mince en grands déplacement*, thèse de l'Univ. Paris-Dauphine, 1995.

[CTM] M. Carrive, P. Le Tallec, and J. Mouro, *Finite element approximation of a geometrically exact shell model*, Rev. Européenne Élém. Finis, vol 4, 1995.

[COS] E. Cosserat and F. Cosserat, *Théorie des corps déformables*, Hermann, Paris, 1909.

[GUR] M. E. Gurtin, *Introduction to continuum mechanics*, Academic Press, New York, 1981.

[KOI] W. T. Koiter, *On the nonlinear theory of thin elastic shells.* I, II, III, Nederl. Akad. Wetensch. Proc. Ser. B **69** (1966), 1–54.

[NAG] P. M. Naghdi, *The theory of shells and plates*, Handbuch der Physik A **6** (1972), no. 2, Springer-Verlag, Berlin, 425–640.

[NIC] H. K. Nickerson, D. C Spencer and N. E Steenrod, *Advanced calculus*, Van Nostrand-Reinhold, New York, 1959.

[SIM] J. C. Simo and D. D. Fox, *On a stress resultant geometrically exact shell model.*I. *Formulation and optimal parametrization*, Comput. Methods Appl. Mech. Eng. (1989), North Holland, 267–304.

[SPI] M. Spivak, *A comprehensive introduction to differential geometry*, Publish or Perish Inc., 1979.

Laboratoire de Modélisation Mécanique et de Mathématiques Appliquées, Université de Poitiers et E.N.S.M.A., SP2MI, BP 179, 86960 Futuroscope Cedex, France

CEREMADE, Université Paris Dauphine, 75775 Paris Cedex 16, France
E-mail address: letallec@ceremade.dauphine.fr

Centre de Recherches Mathématiques
CRM Proceedings and Lecture Notes
Volume **21**, 1999

Locking-Free Mixed Stabilized Finite Element Methods for Bending-Dominated Shells

Dominique Chapelle and Rolf Stenberg

ABSTRACT. We consider the design of finite element methods for the Naghdi shell model in the case when the deformation is bending-dominated. Two formulations based on stabilizing techniques are introduced and it is proved that they are stable, hence free from locking.

1. Introduction

The problem of numerical locking for shells has been a matter of concern to engineers and applied mathematicians for years (see, e.g., [**BD86**]). It is now better apprehended, and considerable hindsight has been gained in particular from theoretical works on the asymptotic behaviour of (two- or three-dimensional) shell models when the thickness tends to zero, the mid-surface geometry being preserved [**SP89a, SP89b, PP93, PP95, CL94a, CLM94, CL94b**]. Out of the two possible causes of locking, namely the shear and membrane parts of the strain energy, it is now clear that the second one is the commanding factor. The mechanical behaviour of a thin shell indeed dramatically depends on whether or not there exist nontrivial displacements with vanishing membrane strains, which is an essentially geometric property. In the affirmative the model is said to be bending-dominated because, for a thin shell, minimizing the total energy will yield solutions with finite bending, but vanishing shear and membrane strains. In this case, locking can arise at the approximation stage, since the finite element spaces have very little chance to contain any field of this very special type. If the shell is not bending-dominated, locking in its accepted sense is impossible, but other difficulties can occur (cf., e.g., [**LSP95**]).

Many ad hoc shell finite element procedures, using various techniques and tricks, have been proposed with a view to avoiding or alleviating numerical locking. To our knowledge, however, they either apply to specific geometries such as cylinders [**Pit92**] or spheres [**Kir93**], or else they remain short of a complete mathematical analysis. An important step towards this goal is accomplished in

1991 *Mathematics Subject Classification.* Primary: 65N30; Secondary: 73K15, 73V05.

The first author was supported in part by the Human Capital and Mobility Program "Shells: Mathematical Modeling and Analysis, Scientific Computing" of the Commission of the European Communities (Contract # ERBCHRXCT940536).

This is the final form of the paper.

[**AB97**], where a mixed method is introduced and proved to satisfy the inf-sup condition. Yet, the analysis there crucially relies on a drastic geometric assumption, which is in general not fulfilled in practice. In the present article we summarize some results, developed in the full-length paper [**CS**], pertaining to some shell finite element methods based on mixed stabilized formulations and valid for the bending-dominated case. Stabilization techniques have earlier been shown to give methods free from locking for several related problems in continuum mechanics such as incompressible elasticity [**HF87, FHLM88, FS91**], beams [**LHFM87**], arches [**LFHM87**] and Reissner-Mindlin plates [**Pit88, HF88, Ste95, LS94**]. For the finite element methods presented here, we are able to establish the stability, thereby proving that locking is ruled out, without resorting to any restrictive assumptions on the geometry. Shear and membrane stresses are used as auxiliary unknowns for the mixed formulation, but discretized by discontinuous functions so that they can be eliminated at the element level to obtain a purely displacement-based matrix problem.

An outline of the paper is as follows. In the next section we specify our notation regarding the differential geometry of shell surfaces. Then, in Section 3 we recall the Naghdi shell model [**Nag63, Ber94**] and give the equations in the bending-dominated case that we focus on. The two proposed finite element formulations, relative to different choices of the stabilization weights, are eventually introduced in Sections 4 and 5, where we also sketch the corresponding mathematical analyses fully detailed in [**CS**].

2. Shell Geometry and Notation

We use the classical representation of the shell geometry, as described in [**GZ68**]. Thus, the shell mid-surface is characterized by a map $\vec{r}$ which is a one-to-one mapping from Ω, an open domain of $\mathbb{R}^2$, into $\mathbb{R}^3$. $\vec{r}$ will be assumed to be as smooth as required (in practice $\mathcal{C}^3(\overline{\Omega})$). The actual surface we denote by $\mathcal{S}$, i.e., $\mathcal{S} = \vec{r}(\overline{\Omega})$. We consider a shell of uniform thickness t.

Regarding the concepts and quantities of the differential geometry of surfaces that we use in this study, we again refer to [**GZ68**] for both definitions and notation. However, with a view to making equations more compact, we also make use of the alternate notation by which surface tensors are represented by letters with a number of underbars equal to their order. In particular, a scalar is denoted by a simple letter. With this notation, a tensor is considered as independent of the type (covariant or contravariant) of its representations in the curvilinear coordinate system, and transforming one set of components into another is simply done by using the transformation formulas. Thus we will write

$$\underline{\underline{T}} = T^{\alpha\beta}\, \vec{a}_\alpha \otimes \vec{a}_\beta = T_{\alpha\beta}\, \vec{a}^\alpha \otimes \vec{a}^\beta = T_\alpha^{\cdot\beta}\, \vec{a}^\alpha \otimes \vec{a}_\beta = T^\alpha_{\cdot\beta}\, \vec{a}_\alpha \otimes \vec{a}^\beta.$$

Here $\vec{a}_\alpha = \partial\vec{r}/\partial\xi^\alpha$, $\alpha = 1, 2$, are the covariant surface base vectors and $\vec{a}^\alpha$, $\alpha = 1, 2$, is the contravariant base. A single dot will denote the simple contraction operator:

$$\underline{u} \cdot \underline{v} \stackrel{\text{def}}{=} u^\alpha v_\alpha = u_\alpha v^\alpha,$$

$$\underline{\underline{T}} \cdot \underline{v} \stackrel{\text{def}}{=} T^{\alpha\beta} v_\beta\, \vec{a}_\alpha = T_{\alpha\beta} v^\beta\, \vec{a}^\alpha = \ldots,$$

$$\underline{\underline{T}} \cdot \underline{\underline{X}} \stackrel{\text{def}}{=} T^{\alpha\beta} X_{\beta\gamma}\, \vec{a}_\alpha \otimes \vec{a}^\gamma = T_{\alpha\beta} X^{\beta\gamma}\, \vec{a}^\alpha \otimes \vec{a}_\gamma = \ldots,$$

whereas a colon denotes a double contraction:

$$\underline{\underline{T}} : \underline{\underline{X}} \stackrel{\text{def}}{=} T^{\alpha\beta} X_{\beta\alpha}.$$

The metric tensor for the surface we denote by $\underline{\underline{a}}$ and the second and third fundamental forms are denoted by $\underline{\underline{b}}$ and $\underline{\underline{c}}$ $(= \underline{\underline{b}} : \underline{\underline{b}})$, respectively.

For a second-order tensor, we define the transposition and symmetry operators as follows

$${}^t\underline{\underline{T}} \stackrel{\text{def}}{=} T^{\beta\alpha}\, \vec{a}_\alpha \otimes \vec{a}_\beta, \quad \underline{\underline{\Sigma}}(\underline{\underline{T}}) \stackrel{\text{def}}{=} \frac{1}{2}(\underline{\underline{T}} + {}^t\underline{\underline{T}}).$$

We next consider differential operators. The gradient of a tensor is obtained by taking the covariant derivative and adding one index

$$\underline{\nabla} w \stackrel{\text{def}}{=} w_{|\alpha}\, \vec{a}^\alpha = w_{,\alpha}\, \vec{a}^\alpha, \quad \underline{\underline{\nabla}}\, \underline{u} \stackrel{\text{def}}{=} u_{\alpha|\beta}\, \vec{a}^\alpha \otimes \vec{a}^\beta,$$

and the divergence by contracting the gradient on the last two indices

$$\operatorname{div} \underline{u} \stackrel{\text{def}}{=} u^\alpha{}_{|\alpha}, \quad \underline{\operatorname{div}}\, \underline{\underline{T}} \stackrel{\text{def}}{=} T^{\alpha\beta}{}_{|\beta}\, \vec{a}_\alpha.$$

For a surface integral over ω, a subdomain of Ω, we use the following compact notation

$$\int_\omega w\, dS \stackrel{\text{def}}{=} \int_\omega w(\xi_1, \xi_2) \sqrt{a}\, d\xi_1\, d\xi_2.$$

We also write

$$\int_C w\, ds \stackrel{\text{def}}{=} \int_C w(\xi_1, \xi_2)\, ds,$$

for an integral along a curve C, with $(ds)^2 = a_{\alpha\beta}\, d\xi_\alpha\, d\xi_\beta$. With this notation, Gauss theorem reads

$$\int_\omega \operatorname{div} \underline{u}\, dS = \int_{\partial\omega} \underline{u} \cdot \underline{\nu}\, ds, \tag{2.1}$$

where $\underline{\nu}$ is the unit outward normal, in the tangential plane, to the boundary $\vec{r}(\partial\omega)$. From this the following Green formulas are derived

$$\int_\omega \underline{u} \cdot \underline{\nabla} w\, dS = \int_{\partial\omega} (\underline{u} \cdot \underline{\nu}) w\, ds - \int_\omega (\operatorname{div} \underline{u}) w\, dS, \tag{2.2}$$

$$\int_\omega {}^t\underline{\underline{X}} : \underline{\underline{\nabla}}\, \underline{u}\, dS = \int_{\partial\omega} (\underline{\underline{X}} \cdot \underline{\nu}) \cdot \underline{u}\, ds - \int_\omega (\underline{\operatorname{div}}\, \underline{\underline{X}}) \cdot \underline{u}\, dS. \tag{2.3}$$

For our analysis it will be convenient to use special norms for Sobolev spaces of tensors. If two zero-order tensors v and w are in $L^2(\omega)$ we define the "intrinsic" inner-product and norm

$$\langle v, w \rangle_\omega \stackrel{\text{def}}{=} \int_\omega v w\, dS, \quad \|w\|_{0,\omega} \stackrel{\text{def}}{=} \langle w, w \rangle_\omega^{1/2}.$$

This new norm is equivalent to the regular $L^2(\omega)$-norm, cf. [**Rou69**]. We also extend this definition to higher order tensors and write

$$\langle \underline{u}, \underline{v} \rangle_\omega \stackrel{\text{def}}{=} \int_\omega \underline{u} \cdot \underline{v}\, dS, \quad \langle \underline{\underline{T}}, \underline{\underline{X}} \rangle_\omega \stackrel{\text{def}}{=} \int_\omega {}^t\underline{\underline{T}} : \underline{\underline{X}}\, dS.$$

We also denote the corresponding norms by $\|\cdot\|_{0,\omega}$ and these can be shown to be equivalent to the regular $L^2(\omega)$-norm of any of the representations of these tensors in covariant or contravariant components [**Rou69**]. Finally, we introduce new $H^1(\omega)$-norms through

$$\|w\|_{1,\omega}^2 \stackrel{\text{def}}{=} \|w\|_{0,\omega}^2 + \|\underline{\nabla} w\|_{0,\omega}^2, \quad \|\underline{u}\|_{1,\omega}^2 \stackrel{\text{def}}{=} \|\underline{u}\|_{0,\omega}^2 + \|\underline{\underline{\nabla}}\, \underline{u}\|_{0,\omega}^2.$$

Recalling the relationship between covariant and regular derivatives, it is clear that this new norm is again equivalent to the standard $H^1(\omega)$-norm of the tensors in any representation. Using higher-order covariant differentiation, there is no difficulty in similarly defining intrinsic H^k-norms ($k \geq 2$) and establishing the corresponding equivalence properties. For all norm and inner-product signs, we will omit the domain subscript when the domain considered is Ω itself.

3. The Variational Shell Model

The shell model we consider, the so-called Naghdi model, is of the Reissner-Mindlin type, i.e., it includes the effect of shear deformation, cf. [**Nag63, Ber94, Val95, Wem81, Des90**] and the references therein. The unknowns are $\vec{u} = (\underline{u}, u_3)$, the displacement at each point of the midsurface decomposed into tangential and transverse parts (respectively a first-order and zero-order surface tensor), and $\underline{\theta}$ as the first-order tensor representing the rotation of a fiber normal to the mid-surface $\mathcal{S}$ in the undeformed configuration. We let $\partial\Omega = \Gamma_0 \cup \Gamma_1$ and we assume that the boundary conditions on $\partial\mathcal{S} = \vec{r}(\Gamma_0) \cup \vec{r}(\Gamma_1)$ are given so that $\vec{r}(\Gamma_0)$ is the part of the boundary along which the shell is fully clamped (i.e., $\vec{u} = \vec{0}$, $\underline{\theta} = \underline{0}$) and $\vec{r}(\Gamma_1)$ the part where all displacements are left free. The problem is posed in the domain Ω and we define the displacement space $\mathcal{U}$ by

$$\mathcal{U} \stackrel{\text{def}}{=} \left\{ (\vec{v}, \underline{\eta}) \in \left[H^1(\Omega)\right]^3 \times \left[H^1(\Omega)\right]^2 \mid \vec{v} = \vec{0} \text{ and } \underline{\eta} = \underline{0} \text{ on } \Gamma_0 \right\}.$$

The variational formulation then reads [**Ber94**]: find $(\vec{u}, \underline{\theta}) \in \mathcal{U}$ such

$$\text{(3.1)} \quad \frac{t^3}{12} \int_\Omega \underline{\underline{\kappa}}(\vec{u}, \underline{\theta}) : \underline{\underline{\underline{\underline{E}}}} : \underline{\underline{\kappa}}(\vec{v}, \underline{\eta})\, dS + t \int_\Omega \underline{\underline{\varepsilon}}(\vec{u}) : \underline{\underline{\underline{\underline{E}}}} : \underline{\underline{\varepsilon}}(\vec{v})\, dS$$
$$+ t \int_\Omega \underline{\gamma}(\vec{u}, \underline{\theta}) \cdot \underline{\underline{G}} \cdot \underline{\gamma}(\vec{v}, \underline{\eta})\, dS = \int_\Omega \vec{g} \cdot \vec{v}\, dS \quad \forall (\vec{v}, \underline{\eta}) \in \mathcal{U}.$$

Here the material properties are given by the two tensors

$$E^{\alpha\beta\lambda\mu} = \frac{E}{2(1+\nu)} \left[a^{\alpha\lambda} a^{\beta\mu} + a^{\alpha\mu} a^{\beta\lambda} + \frac{2\nu}{1-\nu} a^{\alpha\beta} a^{\lambda\mu} \right], \quad G^{\alpha\beta} = \frac{E}{2(1+\nu)} a^{\alpha\beta},$$

and $\underline{\underline{\kappa}}$, $\underline{\underline{\varepsilon}}$, $\underline{\gamma}$ denote the bending, membrane and shear strain tensors, respectively:

$$\underline{\underline{\kappa}}(\vec{u}, \underline{\theta}) \stackrel{\text{def}}{=} \underline{\underline{\Sigma}}(\underline{\nabla}\, \underline{\theta}) - \underline{\underline{\Sigma}}(\underline{\underline{b}} \cdot \underline{\nabla}\, \underline{u}) + \underline{\underline{c}}\, u_3, \quad \underline{\underline{\varepsilon}}(\vec{u}) \stackrel{\text{def}}{=} \underline{\underline{\Sigma}}(\underline{\nabla}\, \underline{u}) - \underline{\underline{b}}\, u_3,$$
$$\underline{\gamma}(\vec{u}, \underline{\theta}) \stackrel{\text{def}}{=} \underline{\nabla} u_3 + \underline{\underline{b}} \cdot \underline{u} + \underline{\theta}.$$

We also recall that $\underline{\underline{b}}$ and $\underline{\underline{c}}$ are the symmetric tensors corresponding to the second and third fundamental forms of the surface, respectively.

We also write equation (3.1) in the shorter form:

$$\text{(3.2)} \quad t^3 A(\vec{u}, \underline{\theta}; \vec{v}, \underline{\eta}) + t D(\vec{u}, \underline{\theta}; \vec{v}, \underline{\eta}) = (\vec{g}, \vec{v}),$$

with

$$\text{(3.3)} \quad A(\vec{u}, \underline{\theta}; \vec{v}, \underline{\eta}) \stackrel{\text{def}}{=} \int_\Omega \underline{\underline{\kappa}}(\vec{u}, \underline{\theta}) : \underline{\underline{\underline{\underline{E}}}} : \underline{\underline{\kappa}}(\vec{v}, \underline{\eta})\, dS,$$

and

$$\text{(3.4)} \quad D(\vec{u}, \underline{\theta}; \vec{v}, \underline{\eta}) \stackrel{\text{def}}{=} \int_\Omega \underline{\underline{\varepsilon}}(\vec{u}) : \underline{\underline{\underline{\underline{E}}}} : \underline{\underline{\varepsilon}}(\vec{v})\, dS + \int_\Omega \underline{\gamma}(\vec{u}, \underline{\theta}) \cdot \underline{\underline{G}} \cdot \underline{\gamma}(\vec{v}, \underline{\eta})\, dS.$$

Implicit in the use of a shell model is the assumption that the shell is "thin". Hence, it is essential to study how the solution behaves in the limit when the thickness $t \to 0$. This limit behavior is very different depending on the geometry of the shell and the boundary conditions and depends in a crucial way (cf. [**SP89a, SP89b, CL94a, CLM94, PP93, PP95**]) on the subspace

$$\mathcal{U}^0 \stackrel{\text{def}}{=} \left\{(\vec{v},\underline{\eta}) \in \mathcal{U} | D(\vec{v},\underline{\eta};\vec{v},\underline{\eta}) = 0\right\},$$

consisting of the displacement fields with vanishing shear and membrane strains.

In this paper we only consider the case when the shell is said to be in a *bending-dominated state of deformation.* This is the case when (cf. [**SP89a, CLM94, PP95**])

$$\mathcal{U}^0 \neq \{(\vec{0},\underline{0})\}.$$

Then the limit problem is obtained by assuming the loading to be given by $\vec{g} = t^3\vec{f}$ with $\vec{f}$ independent of the thickness t. In the sequel we therefore study the following problem.

$\mathcal{P}_t$: Find $(\vec{u},\underline{\theta}) \in \mathcal{U}$ such that

$$A(\vec{u},\underline{\theta};\vec{v},\underline{\eta}) + t^{-2}D(\vec{u},\underline{\theta};\vec{v},\underline{\eta}) = F(\vec{v}) \quad \forall(\vec{v},\underline{\eta}) \in \mathcal{U}. \tag{3.5}$$

Here we introduced the notation $F(\cdot) \stackrel{\text{def}}{=} (\vec{f},\cdot)$.

From [**BC75**] and [**Cou78**] we know that, for any $t > 0$, $\mathcal{P}_t$ has a unique solution, provided that the distributed force field $\vec{f}$ is in an appropriate space, say $\left[L^2(\Omega)\right]^3$. Furthermore, as t tends to zero, $\mathcal{P}_t$ can be seen as a standard penalty problem. The limit problem $\mathcal{P}_0$ is defined as:

$\mathcal{P}_0$: Find $(\vec{u}_0,\underline{\theta}_0)$ in $\mathcal{U}^0$ such that

$$A(\vec{u}_0,\underline{\theta}_0;\vec{v},\underline{\eta}) = F(\vec{v}) \quad \forall(\vec{v},\underline{\eta}) \in \mathcal{U}^0. \tag{3.6}$$

The term "limit problem" is justified since, as $t \to 0$, it can be shown that the solution $(\vec{u},\underline{\theta})$ of $\mathcal{P}_t$ converges strongly in $\mathcal{U}$ to the solution $(\vec{u}_0,\underline{\theta}_0)$ of $\mathcal{P}_0$ (cf. [**CP94, Pit92**]).

In the approximation of plate problems it has often turned out to be advantageous to use a mixed formulation, cf. [**BF91, BFS91, DN88**]. This is the approach that we use here as well. Hence, we define as new unknowns stress variables dual to the strains in the variational formulation (3.5). By $\underline{\underline{n}}$ we denote the symmetric membrane force tensor, Here we depart from the usual notation, cf. Remark 3.1 below. The shear force we denote by $\underline{q}$. These are connected to the membrane and shear strains through the constitutive equations

$$\underline{\underline{n}} = \frac{1}{t^2}\underline{\underline{\underline{\underline{E}}}}:\underline{\underline{\varepsilon}}(\vec{u}), \tag{3.7}$$

$$\underline{q} = \frac{1}{t^2}\underline{\underline{G}}\cdot\underline{\gamma}(\vec{u},\underline{\theta}). \tag{3.8}$$

The symmetric bending moment tensor is not taken as an independent unknown, but we use the following abbreviation based on the bending constitutive equation:

$$\underline{\underline{m}}(\vec{u},\underline{\theta}) \stackrel{\text{def}}{=} \frac{1}{12}\underline{\underline{\underline{\underline{E}}}}:\underline{\underline{\kappa}}(\vec{u},\underline{\theta}). \tag{3.9}$$

By introducing

$$(\underline{\underline{n}}, \underline{q}) \in \mathcal{Q} \stackrel{\text{def}}{=} \left\{\underline{\underline{p}} \in \left[L^2(\Omega)\right]^{2\times 2} \mid {}^t\underline{\underline{p}} = \underline{\underline{p}}\right\} \times \left[L^2(\Omega)\right]^2,$$

as new unknowns and by writing the constitutive equations (3.8) and (3.9) in a weak form, we obtain an equivalent "mixed" formulation of the problem:

$\mathcal{M}_t$: Find $(\vec{u}, \underline{\theta}, \underline{\underline{n}}, \underline{q}) \in \mathcal{U} \times \mathcal{Q}$ such that

$$B(\vec{u}, \underline{\theta}, \underline{\underline{n}}, \underline{q}; \vec{v}, \underline{\eta}, \underline{\underline{p}}, \underline{r}) = F(\vec{v}) \quad \forall(\vec{v}, \underline{\eta}, \underline{\underline{p}}, \underline{r}) \in \mathcal{U} \times \mathcal{Q}, \tag{3.10}$$

with the new bilinear forms defined through

$$B(\vec{u}, \underline{\theta}, \underline{\underline{n}}, \underline{q}; \vec{v}, \underline{\eta}, \underline{\underline{p}}, \underline{r}) \stackrel{\text{def}}{=} A(\vec{u}, \underline{\theta}; \vec{v}, \underline{\eta}) + M(\vec{v}, \underline{\eta}; \underline{\underline{n}}, \underline{q}) + M(\vec{u}, \underline{\theta}; \underline{\underline{p}}, \underline{r}) - N(\underline{\underline{n}}, \underline{q}; \underline{\underline{p}}, \underline{r}),$$

$$M(\vec{v}, \underline{\eta}; \underline{\underline{p}}, \underline{r}) \stackrel{\text{def}}{=} \int_\Omega \left(\underline{\underline{\varepsilon}}(\vec{v}) : \underline{\underline{p}} + \underline{\gamma}(\vec{v}, \underline{\eta}) \cdot \underline{r}\right) dS,$$

and

$$N(\underline{\underline{n}}, \underline{q}; \underline{\underline{p}}, \underline{r}) \stackrel{\text{def}}{=} t^2 \int_\Omega \left(\underline{\underline{n}} : \underline{\underline{\underline{\underline{\breve{E}}}}} : \underline{\underline{p}} + \underline{q} \cdot \underline{\underline{\breve{G}}} \cdot \underline{r}\right) dS.$$

Here we use the notation

$$\breve{E}_{\alpha\beta\lambda\mu} = \frac{1+\nu}{2E}\left[a_{\alpha\lambda}a_{\beta\mu} + a_{\alpha\mu}a_{\beta\lambda} - \frac{2\nu}{1+\nu}a_{\alpha\beta}a_{\lambda\mu}\right], \quad \breve{G}_{\alpha\beta} = \frac{2(1+\nu)}{E}a_{\alpha\beta}.$$

The second order tensor $\underline{\underline{\breve{G}}}$ is the inverse of $\underline{\underline{G}}$ (i.e., $\underline{\underline{G}} \cdot \underline{\underline{\breve{G}}} = \underline{\underline{\breve{G}}} \cdot \underline{\underline{G}} = \underline{\underline{a}}$), whereas $\underline{\underline{\underline{\underline{\breve{E}}}}}$ is such that, for any two symmetric tensors $\underline{\underline{T}}$ and $\underline{\underline{X}}$:

$$\underline{\underline{T}} = \underline{\underline{\underline{\underline{E}}}} : \underline{\underline{X}} \iff \underline{\underline{X}} = \underline{\underline{\underline{\underline{\breve{E}}}}} : \underline{\underline{T}}. \tag{3.11}$$

Finally, we recall that using the symmetry of the second order tensors, and the Green formulas (2.2) and (2.3), the following integration by parts formulas are obtained

$$\begin{aligned} \int_\omega \underline{\underline{m}} : \underline{\underline{\kappa}}(\vec{v}, \underline{\eta})\, dS = &- \int_\omega \underline{\operatorname{div}}\, \underline{\underline{m}} \cdot \underline{\eta}\, dS + \int_\omega \underline{\operatorname{div}}\, (\underline{\underline{b}} \cdot \underline{\underline{m}}) \cdot \underline{v}\, dS \\ &+ \int_\omega (\underline{\underline{b}} : (\underline{\underline{b}} \cdot \underline{\underline{m}}))\, v_3\, dS + \int_{\partial\omega} (\underline{\underline{m}} \cdot \underline{\nu}) \cdot \underline{\eta}\, ds \\ &- \int_{\partial\omega} ((\underline{\underline{b}} \cdot \underline{\underline{m}}) \cdot \underline{\nu}) \cdot \underline{v}\, ds, \end{aligned} \tag{3.12}$$

$$\int_\omega \underline{\underline{n}} : \underline{\underline{\varepsilon}}(\vec{v})\, dS = - \int_\omega \underline{\operatorname{div}}\, \underline{\underline{n}} \cdot \underline{v}\, dS - \int_\omega (\underline{\underline{b}} : \underline{\underline{n}})\, v_3\, dS + \int_{\partial\omega} (\underline{\underline{n}} \cdot \underline{\nu}) \cdot \underline{v}\, ds, \tag{3.13}$$

and

$$\begin{aligned} \int_\omega \underline{q} : \underline{\gamma}(\vec{v}, \underline{\eta})\, dS = &\int_\omega \underline{q} \cdot \underline{\eta}\, dS + \int_\omega (\underline{\underline{b}} \cdot \underline{q}) \cdot \underline{v}\, dS - \int_\omega \operatorname{div} \underline{q}\, v_3\, dS \\ &+ \int_{\partial\omega} (\underline{q} \cdot \underline{\nu})\, v_3\, ds. \end{aligned} \tag{3.14}$$

Collecting these formulas gives

$$
\begin{aligned}
(3.15)\quad \int_\omega \underline{\underline{m}} : \underline{\underline{\kappa}}(\vec{v}, \underline{\eta})\, dS + \int_\omega \underline{\underline{n}} : \underline{\underline{\varepsilon}}(\vec{v})\, dS + \int_\omega \underline{q} : \underline{\gamma}(\vec{v}, \underline{\eta})\, dS \\
= - \int_\omega (\underline{\operatorname{div}}\, \underline{\underline{m}} - \underline{q}) \cdot \underline{\eta}\, dS - \int_\omega \big(\underline{\operatorname{div}}\, (\underline{\underline{n}} - \underline{\underline{b}} \cdot \underline{\underline{m}}) - \underline{\underline{b}} \cdot \underline{q}\big) \cdot \underline{v}\, dS \\
- \int_\omega \big(\operatorname{div} \underline{q} + \underline{\underline{b}} : (\underline{\underline{n}} - \underline{\underline{b}} \cdot \underline{\underline{m}})\big)\, v_3\, dS + \int_{\partial\omega} (\underline{\underline{m}} \cdot \underline{\nu}) \cdot \underline{\eta}\, ds \\
+ \int_{\partial\omega} \big((\underline{\underline{n}} - \underline{\underline{b}} \cdot \underline{\underline{m}}) \cdot \underline{\nu}\big) \cdot \underline{v}\, ds + \int_{\partial\omega} (\underline{q} \cdot \underline{\nu})\, v_3\, ds.
\end{aligned}
$$

Hence, recalling the constitutive equations (3.7)–(3.9), problem $\mathcal{P}_t$ gives the following equations of equilibrium

$$\underline{\operatorname{div}}\, \underline{\underline{m}} - \underline{q} = 0, \tag{3.16}$$

$$\underline{\operatorname{div}}\, (\underline{\underline{n}} - \underline{\underline{b}} \cdot \underline{\underline{m}}) - \underline{\underline{b}} \cdot \underline{q} + \underline{f} = 0, \tag{3.17}$$

$$\operatorname{div} \underline{q} + \underline{\underline{b}} : (\underline{\underline{n}} - \underline{\underline{b}} \cdot \underline{\underline{m}}) + f_3 = 0, \tag{3.18}$$

in the domain Ω, and the natural boundary conditions

$$\underline{\underline{m}} \cdot \underline{\nu} = 0, \tag{3.19}$$

$$(\underline{\underline{n}} - \underline{\underline{b}} \cdot \underline{\underline{m}}) \cdot \underline{\nu} = 0, \tag{3.20}$$

$$\underline{q} \cdot \underline{\nu} = 0, \tag{3.21}$$

on the free boundary Γ_1.

REMARK 3.1. In the above equations the quantity

$$\tilde{\underline{\underline{n}}} = \underline{\underline{n}} - \underline{\underline{b}} \cdot \underline{\underline{m}}, \tag{3.22}$$

is the tensor obtained by integrating the three-dimensional membrane stresses over the shell thickness, cf., e.g., [**Koi59, GZ68, Nag63, Wem81**]. Usually this is called the membrane force. For designing a stable method it seems more practical to introduce the variable $\underline{\underline{n}}$, and not $\tilde{\underline{\underline{n}}}$, as an new unknown. For simplicity we refer to $\underline{\underline{n}}$ as the membrane force.

4. A First Approximation Scheme

We use a finite element partitioning $\mathcal{C}_h$ of $\overline{\Omega}$ into straight-sided triangles or quadrilaterals. Naturally, the partitioning is assumed to satisfy the usual regularity and compatibility conditions, cf. [**Cia78, Bra92, RT88**]. The diameter of an element $K \in \mathcal{C}_h$ we denote by h_K, and we let $h = \max_{K \in \mathcal{C}_h} h_K$. By Γ_h we denote the collection of edges in the mesh, and by h_E the length of an edge $E \in \Gamma_h$. The letter C is used to denote a generic strictly positive constant, independent of both t and h, which is allowed to take different values at different occurrences except when appearing with indices.

Let $\mathcal{U}^n$ and $\mathcal{Q}^n$ be finite element subspaces of $\mathcal{U}$ and $\mathcal{Q}$, respectively. A mixed finite element method of the "classical" saddle point type would be based on the formulation $\mathcal{M}_t$, viz.

$\mathcal{M}_t^h$**:** Find $(\vec{u}^h, \underline{\theta}^h, \underline{\underline{n}}^h, \underline{q}^h) \in \mathcal{U}^n \times \mathcal{Q}^n$ such that

$$B(\vec{u}^h, \underline{\theta}^h, \underline{\underline{n}}^h, \underline{q}^h; \vec{v}, \underline{\eta}, \underline{\underline{p}}, \underline{r}) = F(\vec{v}, \underline{\eta}, \underline{\underline{p}}, \underline{r}) \quad \forall (\vec{v}, \underline{\eta}, \underline{\underline{p}}, \underline{r}) \in \mathcal{U}^n \times \mathcal{Q}^n. \tag{4.1}$$

It appears difficult to design a stable finite element method directly based on this formulation. Hence, we follow an approach that has turned out to be fruitful for the related beam, arch and plate problems, cf. [**FHLM88, FS91, Ste95**]. The approximation scheme we propose is derived from (4.1) by adding weighted least-squares-type terms of the equilibrium equations (3.16))–(3.18) and the constitutive relations (3.7))–(3.8). In order to simplify the presentation of our method, we introduce the weighted inner-product

$$\langle v, w\rangle_h \stackrel{\text{def}}{=} \sum_{K\in\mathcal{C}_h} h_K^2 \langle v, w\rangle_K, \tag{4.2}$$

with analogous definitions for higher order tensors. Further, in accordance with (3.9) and (3.22) we introduce the notation

$$\underline{\underline{\tilde{n}}}(\vec{v}, \underline{\eta}, \underline{\underline{p}}) \stackrel{\text{def}}{=} \underline{\underline{p}} - \underline{\underline{b}} \cdot \underline{\underline{m}}(\vec{v}, \underline{\eta}). \tag{4.3}$$

Our first stabilized method will now be defined as follows:

$\mathcal{S}_t^h$: Find $(\vec{u}^h, \underline{\theta}^h, \underline{\underline{n}}^h, \underline{q}^h) \in \mathcal{U}^n \times \mathcal{Q}^n$ such that

$$B_h(\vec{u}^h, \underline{\theta}^h, \underline{\underline{n}}^h, \underline{q}^h; \vec{v}, \underline{\eta}, \underline{\underline{p}}, \underline{r}) = F_h(\vec{v}, \underline{\eta}, \underline{\underline{p}}, \underline{r}) \quad \forall(\vec{v}, \underline{\eta}, \underline{\underline{p}}, \underline{r}) \in \mathcal{U}^n \times \mathcal{Q}^n, \tag{4.4}$$

where

$$\begin{aligned} B_h(\vec{w}, \underline{\tau}, \underline{\underline{k}}, \underline{s}; \vec{v}, \underline{\eta}, \underline{\underline{p}}, \underline{r}) \stackrel{\text{def}}{=} B(\vec{w}, \underline{\tau}, \underline{\underline{k}}, \underline{s}; \vec{v}, \underline{\eta}, \underline{\underline{p}}, \underline{r}) &+ S_h^e(\vec{w}, \underline{\tau}, \underline{\underline{k}}, \underline{s}; \vec{v}, \underline{\eta}, \underline{\underline{p}}, \underline{r}) \\ &+ S_h^c(\vec{w}, \underline{\tau}, \underline{\underline{k}}, \underline{s}; \vec{v}, \underline{\eta}, \underline{\underline{p}}, \underline{r}), \end{aligned} \tag{4.5}$$

with a stabilizing term originating from the equilibrium equations

$$\begin{aligned} S_h^e(\vec{w}, \underline{\tau}, \underline{\underline{k}}, \underline{s}; \vec{v}, \underline{\eta}, \underline{\underline{p}}, \underline{r}) \stackrel{\text{def}}{=} &-\alpha_1 \langle \underline{\operatorname{div}}\, \underline{\underline{m}}(\vec{w}, \underline{\tau}) - \underline{s}, \underline{\operatorname{div}}\, \underline{\underline{m}}(\vec{v}, \underline{\eta}) - \underline{r}\rangle_h \\ &- \alpha_2 \langle \underline{\operatorname{div}}\, \underline{\underline{\tilde{n}}}(\vec{w}, \underline{\tau}, \underline{\underline{k}}) - \underline{\underline{b}} \cdot \underline{s}, \underline{\operatorname{div}}\, \underline{\underline{\tilde{n}}}(\vec{v}, \underline{\eta}, \underline{\underline{p}}) - \underline{\underline{b}} \cdot \underline{r}\rangle_h \\ &- \alpha_3 \langle \operatorname{div} \underline{s} + \underline{\underline{b}} : \underline{\underline{\tilde{n}}}(\vec{w}, \underline{\tau}, \underline{\underline{k}}), \operatorname{div} \underline{r} + \underline{\underline{b}} : \underline{\underline{\tilde{n}}}(\vec{v}, \underline{\eta}, \underline{\underline{p}})\rangle_h, \end{aligned} \tag{4.6}$$

and one from the constitutive equations

$$\begin{aligned} S_h^c(\vec{w}, \underline{\tau}, \underline{\underline{k}}, \underline{s}; \vec{v}, \underline{\eta}, \underline{\underline{p}}, \underline{r}) \stackrel{\text{def}}{=} \alpha_4 &\int_\Omega \big[\underline{\underline{\varepsilon}}(\vec{w}) - t^2 \underline{\underline{\underline{\underline{\breve{E}}}}} : \underline{\underline{k}}\big] : \underline{\underline{\underline{\underline{E}}}} : \big[\underline{\underline{\varepsilon}}(\vec{v}) - t^2 \underline{\underline{\underline{\underline{\breve{E}}}}} : \underline{\underline{p}}\big]\, dS \\ + \alpha_5 &\int_\Omega \big[\underline{\gamma}(\vec{w}, \underline{\tau}) - t^2 \underline{\underline{\breve{G}}} \cdot \underline{s}\big] \cdot \underline{\underline{G}} \cdot \big[\underline{\gamma}(\vec{v}, \underline{\eta}) - t^2 \underline{\underline{\breve{G}}} \cdot \underline{r}\big]\, dS. \end{aligned} \tag{4.7}$$

The right hand side is defined as

$$\begin{aligned} F_h(\vec{v}, \underline{\eta}, \underline{\underline{p}}, \underline{r}) \stackrel{\text{def}}{=} F(\vec{v}) &+ \alpha_2 \langle \underline{f}, \underline{\operatorname{div}}\, \underline{\underline{\tilde{n}}}(\vec{v}, \underline{\eta}, \underline{\underline{p}}) - \underline{\underline{b}} \cdot \underline{r}\rangle_h \\ &+ \alpha_3 \langle f_3, \operatorname{div} \underline{r} + \underline{\underline{b}} : \underline{\underline{\tilde{n}}}(\vec{v}, \underline{\eta}, \underline{\underline{p}})\rangle_h. \end{aligned} \tag{4.8}$$

The norm used for the displacement is the modified $H^1(\Omega)$-norm that we previously defined, i.e.,

$$\|\vec{v}, \underline{\eta}\|_1 \stackrel{\text{def}}{=} \left(\|\underline{v}\|_1^2 + \|v_3\|_1^2 + \|\underline{\theta}\|_1\right)^{1/2}. \tag{4.9}$$

For the stresses, we use a discrete norm defined by

$$\|\underline{\underline{p}}, \underline{r}\|_{t,h} \stackrel{\text{def}}{=} \left(t^2 \|\underline{\underline{p}}, \underline{r}\|_0^2 + |\underline{\underline{p}}, \underline{r}|_I^2 + |\underline{\underline{p}}, \underline{r}|_J^2\right)^{1/2}, \tag{4.10}$$

with "interior" terms

$$(4.11)\quad |\underline{\underline{p}}, \underline{r}|_I \stackrel{\text{def}}{=} \Big(\sum_{K\in\mathcal{C}_h} h_K^2 \|\underline{r}\|_{0,K}^2 + \sum_{K\in\mathcal{C}_h} h_K^2 \|\underline{\operatorname{div}}\, \underline{\underline{p}}\|_{0,K}^2 + \sum_{K\in\mathcal{C}_h} h_K^2 \|\operatorname{div} \underline{r} + \underline{\underline{b}} : \underline{\underline{p}}\|_{0,K}^2 \Big)^{1/2},$$

and "jump" terms

$$(4.12)\quad |\underline{\underline{p}}, \underline{r}|_J \stackrel{\text{def}}{=} \Big(\sum_{E\in\Gamma_h\setminus\Gamma_0} h_E \int_E \big[[\![\underline{\underline{p}} \cdot \underline{\nu}]\!]^2 + [\![\underline{r} \cdot \underline{\nu}]\!]^2\big]\, ds \Big)^{1/2},$$

where $[\![\cdot]\!]$ stands for the jump of the quantity if the edge is in the interior, and for the quantity itself if it is on Γ_1. Finally, we let

$$(4.13)\quad \|\vec{v}, \underline{\eta}, \underline{\underline{p}}, \underline{r}\|_{t,h} \stackrel{\text{def}}{=} \big(\|\vec{v}, \underline{\eta}\|_1^2 + \|\underline{\underline{p}}, \underline{r}\|_{t,h}^2\big)^{1/2}.$$

We begin our analysis by noting the consistency of formulation $\mathcal{S}_t^h$ with respect to the continuous problem $\mathcal{M}_t$.

LEMMA 4.1. *Suppose that $\vec{f} \in \big[L^2(\Omega)\big]^3$. Then the solution $(\vec{u}, \underline{\theta}, \underline{\underline{n}}, \underline{q}) \in \mathcal{U} \times \mathcal{Q}$ of $\mathcal{M}_t$ satisfies*

$$(4.14)\quad B_h(\vec{u}, \underline{\theta}, \underline{\underline{n}}, \underline{q}; \vec{v}, \underline{\eta}, \underline{\underline{p}}, \underline{r}) = F_h(\vec{v}, \underline{\eta}, \underline{\underline{p}}, \underline{r}) \quad \forall(\vec{v}, \underline{\eta}, \underline{\underline{p}}, \underline{r}) \in \mathcal{U}^n \times \mathcal{Q}^n.$$

In order to obtain a stable method we have to specify the finite element spaces. To this end we let, for $l \geq 0$,

$$(4.15)\quad R_l(K) \stackrel{\text{def}}{=} \begin{cases} P_l(K) & \text{if } K \text{ is a triangle,} \\ Q_l(K) & \text{if } K \text{ is a quadrilateral,} \end{cases}$$

and for $k \geq 1$ we now define

$$(4.16)\quad \mathcal{U}^n \stackrel{\text{def}}{=} \big\{(\vec{v}, \underline{\eta}) \in \mathcal{U} \bigm| v_i|_K \in R_{k+1}(K),\ i = 1,2,3, \text{ and } \eta_\alpha|_K \in R_k(K),\ \alpha = 1,2,\ \forall K \in \mathcal{C}_h\big\},$$

$$(4.17)\quad \mathcal{Q}^n \stackrel{\text{def}}{=} \big\{(\underline{\underline{p}}, \underline{r}) \in \mathcal{Q} \bigm| p^{\alpha\beta}|_K \in R_{k-1}(K),\ \alpha,\beta = 1,2, \text{ and } r^\alpha|_K \in R_{k-1}(K),\ \alpha = 1,2,\ \forall K \in \mathcal{C}_h\big\}.$$

REMARK 4.2. With the choice of finite elements spaces that we made, the stresses are interpolated discontinuously between elements. This implies that they can be eliminated at the element level to obtain a purely displacement-based numerical scheme.

We are now ready to state the key result, i.e., the stability of our scheme. We assume that the shell thickness is in the range $0 < t \leq t_0$, with t_0 fixed.

LEMMA 4.3. *Assume that:*

(S1) $0 < \alpha_i < C_i^I$, *for $i = 1, 2$, where the positive constants C_i^I are fixed, derived from inverse inequalities.*
(S2) $\alpha_3 > 0$.
(S3) $0 < \alpha_i < t_0^{-2}$, *for $i = 4, 5$.*
(S4) *h is sufficiently small.*

Then the stabilized formulation $\mathcal{S}_t^h$ is stable, i.e., there is a positive constant C such that, for all $(\vec{w}, \underline{\tau}, \underline{\underline{k}}, \underline{s}) \in \mathcal{U}^n \times \mathcal{Q}^n$, there exist $(\vec{z}, \underline{\eta}, \underline{\underline{p}}, \underline{r}) \in \mathcal{U}^n \times \mathcal{Q}^n$ with

$$B_h(\vec{w}, \underline{\tau}, \underline{\underline{k}}, \underline{s}; \vec{z}, \underline{\eta}, \underline{\underline{p}}, \underline{r}) \geq C \|\vec{w}, \underline{\tau}, \underline{\underline{k}}, \underline{s}\|_{t,h}, \tag{4.18}$$

and

$$\|\vec{z}, \underline{\eta}, \underline{\underline{p}}, \underline{r}\|_{t,h} \leq 1. \tag{4.19}$$

PROOF. We only give an outline of the proof, decomposed into three steps. Let $(\vec{w}, \underline{\tau}, \underline{\underline{k}}, \underline{s}) \in \mathcal{U}^n \times \mathcal{Q}^n$ be arbitrary.

STEP 1. Under conditions (S1)–(S4) and using ad hoc inverse inequalities, it can be shown that

$$B_h(\vec{w}, \underline{\tau}, \underline{\underline{k}}, \underline{s}; \vec{w}, \underline{\tau}, -\underline{\underline{k}}, -\underline{s}) \geq C_1 \big(\|\vec{w}, \underline{\tau}\|_1^2 + t^2 \|\underline{\underline{k}}, \underline{s}\|_0^2 + |\underline{\underline{k}}, \underline{s}|_I^2 \big). \tag{4.20}$$

STEP 2. We obtain a stability inequality involving the jump-terms of the norm by showing that there exists $(\vec{v}, \underline{0}) \in \mathcal{U}^n$ such that

$$\sum_{E \in \Gamma_h \setminus \Gamma_0} \int_E \big([\![\underline{\underline{k}} \cdot \underline{\nu}]\!] \cdot \underline{v} + [\![\underline{s} \cdot \underline{\nu}]\!]\, v_3 \big)\, ds \geq C |\underline{\underline{k}}, \underline{s}|_J^2, \tag{4.21}$$

and

$$\sum_{K \in \mathcal{C}_h} h_K^{-2} \|\vec{v}\|_{0,K}^2 + \|\vec{v}\|_1^2 \leq |\underline{\underline{k}}, \underline{s}|_J^2. \tag{4.22}$$

This result crucially relies on a geometric property of the normal vector $\underline{\nu}$ which we establish in [**CS**, Appendix]. With this $\vec{v}$, we get

$$B_h(\vec{w}, \underline{\tau}, \underline{\underline{k}}, \underline{s}; \vec{v}, \underline{0}, \underline{\underline{0}}, \underline{0}) \geq C_2 |\underline{\underline{k}}, \underline{s}|_J^2 - C_3 \big(\|\vec{w}, \underline{\tau}\|_1^2 + |\underline{\underline{k}}, \underline{s}|_I^2 \big). \tag{4.23}$$

STEP 3. We finally consider $(\vec{z}, \underline{\eta}, \underline{\underline{p}}, \underline{r}) = (\vec{w} + \delta \vec{v}, \underline{\tau}, -\underline{\underline{k}}, -\underline{s})$, with $\delta > 0$. The relations (4.20) and (4.23) proved at the previous steps then give

$$\begin{aligned} &B_h(\vec{w}, \underline{\tau}, \underline{\underline{k}}, \underline{s}; \vec{z}, \underline{\eta}, \underline{\underline{p}}, \underline{r}) \\ &\qquad \geq (C_1 - \delta C_3) \big(\|\vec{w}, \underline{\tau}\|_1^2 + |\underline{\underline{k}}, \underline{s}|_I^2 \big) + C_1 t^2 \|\underline{\underline{k}}, \underline{s}\|_0^2 + \delta C_2 |\underline{\underline{k}}, \underline{s}|_J^2 \\ &\qquad \geq C_4 \|\vec{w}, \underline{\tau}, \underline{\underline{k}}, \underline{s}\|_{t,h}^2, \end{aligned}$$

when choosing $\delta < C_1/C_3$. Using (4.22) we get

$$\begin{aligned} \|\vec{z}, \underline{\eta}, \underline{\underline{p}}, \underline{r}\|_{t,h} &\leq \|\vec{w}, \underline{\tau}, \underline{\underline{k}}, \underline{s}\|_{t,h} + \delta \|\vec{v}, \underline{0}, \underline{\underline{0}}, \underline{0}\|_{t,h} = \|\vec{w}, \underline{\tau}, \underline{\underline{k}}, \underline{s}\|_{t,h} + \delta \|\vec{v}\|_1 \\ &\leq \|\vec{w}, \underline{\tau}, \underline{\underline{k}}, \underline{s}\|_{t,h} + \delta |\underline{\underline{k}}, \underline{s}|_J \leq C_5 \|\vec{w}, \underline{\tau}, \underline{\underline{k}}, \underline{s}\|_{t,h}, \end{aligned} \tag{4.24}$$

and the assertion is thus proved.

□

REMARK 4.4. The inverse inequalities which the constants C_i^I, $i = 1, 2$ are deduced from are local, i.e., they can be written within each element. This implies that these constants can be *numerically* determined, for each element, using a simple local eigenvalue problem, cf. [**HH92**] where this idea is introduced for several stabilized methods.

Consistency and stability being ascertained, to obtain a priori error estimates it essentially remains to establish some interpolation estimates. This being a rather standard procedure, we directly state our final result.

THEOREM 4.5. *Assume that conditions* (S1)–(S4) *of Lemma* 4.3 *hold and that the solution of* $\mathcal{M}_t$ *is such that* $\vec{u} \in [H^{k+1}(\Omega)]^3$, $\underline{\theta} \in [H^{k+1}(\Omega)]^2$, $\underline{\underline{n}} \in [H^k(\Omega)]^{2\times 2}$ *and* $\underline{q} \in [H^k(\Omega)]^2$. *Then the finite element formulation* $\mathcal{S}_t^h$ *has a unique solution satisfying*

$$\|\vec{u} - \vec{u}^h, \underline{\theta} - \underline{\theta}^h\|_1 + \|\underline{\underline{n}} - \underline{\underline{n}}^h, \underline{q} - \underline{q}^h\|_{t,h} \le Ch^k\big(\|\vec{u}, \underline{\theta}\|_{k+1} + \|\underline{\underline{n}}, \underline{q}\|_k\big). \tag{4.25}$$

5. A Second Stabilized Method

In this section, we propose an alternative stabilized scheme, obtained from the previous formulation by changing the weights in front of the stabilizing terms that originate from the constitutive equations, in the spirit of what is considered for plates in [**Pit88, HF88, Ste95, LSV93, LS94, LS99**]. This second stabilized method reads:

$\widetilde{\mathcal{S}}_t^h$: Find $(\vec{u}^h, \underline{\theta}^h, \underline{\underline{n}}^h, \underline{q}^h) \in \mathcal{U}^n \times \mathcal{Q}^n$ such that

$$\widetilde{B}_h(\vec{u}^h, \underline{\theta}^h, \underline{\underline{n}}^h, \underline{q}^h; \vec{v}, \underline{\eta}, \underline{\underline{p}}, \underline{r}) = F_h(\vec{v}, \underline{\eta}, \underline{\underline{p}}, \underline{r}) \quad \forall (\vec{v}, \underline{\eta}, \underline{\underline{p}}, \underline{r}) \in \mathcal{U}^h \times \mathcal{Q}^h, \tag{5.1}$$

where

$$\begin{aligned}\widetilde{B}_h(\vec{w}, \underline{\tau}, \underline{\underline{k}}, \underline{s}; \vec{v}, \underline{\eta}, \underline{\underline{p}}, \underline{r}) \stackrel{\text{def}}{=} B(\vec{w}, \underline{\tau}, \underline{\underline{k}}, \underline{s}; \vec{v}, \underline{\eta}, \underline{\underline{p}}, \underline{r}) &+ S_h^e(\vec{w}, \underline{\tau}, \underline{\underline{k}}, \underline{s}; \vec{v}, \underline{\eta}, \underline{\underline{p}}, \underline{r})\\ &+ \widetilde{S}_h^c(\vec{w}, \underline{\tau}, \underline{\underline{k}}, \underline{s}; \vec{v}, \underline{\eta}, \underline{\underline{p}}, \underline{r}),\end{aligned} \tag{5.2}$$

with the stabilizing term S_h^e originating from the equilibrium equations defined as before in (4.6), and the term from the constitutive equations redefined by

$$\begin{aligned}&\widetilde{S}_h^c(\vec{w}, \underline{\tau}, \underline{\underline{k}}, \underline{s}; \vec{v}, \underline{\eta}, \underline{\underline{p}}, \underline{r})\\ &\quad\stackrel{\text{def}}{=} \frac{1}{2}\sum_{K\in\mathcal{C}_h} \frac{1}{t^2 + \alpha_4 h_K^2}\int_K \big[\underline{\underline{\varepsilon}}(\vec{w}) - t^2 \breve{\underline{\underline{\underline{\underline{E}}}}} : \underline{\underline{k}}\big] : \underline{\underline{\underline{\underline{E}}}} : \big[\underline{\underline{\varepsilon}}(\vec{v}) - t^2 \breve{\underline{\underline{\underline{\underline{E}}}}} : \underline{\underline{p}}\big]\, dS\\ &\quad+ \frac{1}{2}\sum_{K\in\mathcal{C}_h} \frac{1}{t^2 + \alpha_5 h_K^2}\int_K \big[\underline{\gamma}(\vec{w}, \underline{\tau}) - t^2 \breve{\underline{\underline{G}}} \cdot \underline{s}\big] \cdot \underline{\underline{G}} \cdot \big[\underline{\gamma}(\vec{v}, \underline{\eta}) - t^2 \breve{\underline{\underline{G}}} \cdot \underline{r}\big]\, dS.\end{aligned} \tag{5.3}$$

The right hand side F_h remains unchanged as defined by (4.8).

For the displacement variables we will use the following norm

$$[\![\vec{v}, \underline{\eta}]\!]_{t,h} \stackrel{\text{def}}{=} \left(\|\vec{v}, \underline{\eta}\|_1^2 + \sum_{K\in\mathcal{C}_h} \frac{1}{t^2 + h_K^2}\big(\|\underline{\gamma}(\vec{v}, \underline{\eta})\|_{0,K}^2 + \|\underline{\underline{\varepsilon}}(\vec{v})\|_{0,K}^2\big)\right)^{1/2},$$

which is obviously stronger than the H^1-norm used for our first method. The norm for the stresses we now have to weaken by first redefining the "jump" term

$$|\underline{\underline{p}}, \underline{r}|_J \stackrel{\text{def}}{=} \left(\sum_{E\in\Gamma_h\setminus\Gamma_0} h_E^3 \int_E \big[[\![\underline{\underline{p}} \cdot \underline{\nu}]\!]^2 + [\![\underline{r} \cdot \underline{\nu}]\!]^2\big]\, ds\right)^{1/2},$$

and then defining

$$[\![\underline{\underline{p}}, \underline{r}]\!]_{t,h} \stackrel{\text{def}}{=} \big(t^2\|\underline{\underline{p}}, \underline{r}\|_0^2 + |\underline{\underline{p}}, \underline{r}|_I^2 + |\underline{\underline{p}}, \underline{r}|_J^2\big)^{1/2},$$

by keeping the interior part as defined in (4.11). We will also use the combined notation

$$[\![\vec{v}, \underline{\eta}, \underline{\underline{p}}, \underline{r}]\!]_{t,h} \stackrel{\text{def}}{=} \big([\![\vec{v}, \underline{\eta}]\!]_{t,h}^2 + [\![\underline{\underline{p}}, \underline{r}]\!]_{t,h}^2\big)^{1/2}.$$

The analysis of this new scheme will now essentially follow the same lines as in Section 4. First, we again note that the formulation is consistent.

LEMMA 5.1. *Suppose that* $\vec{f} \in \left[L^2(\Omega)\right]^3$. *Then the solution* $(\vec{u}, \underline{\theta}, \underline{\underline{n}}, \underline{q}) \in \mathcal{U} \times \mathcal{Q}$ *of* $\mathcal{M}_t$ *satisfies*

$$\widetilde{B}_h(\vec{u}, \underline{\theta}, \underline{\underline{n}}, \underline{q}; \vec{v}, \underline{\eta}, \underline{\underline{p}}, \underline{r}) = F_h(\vec{v}, \underline{\eta}, \underline{\underline{p}}, \underline{r}) \quad \forall (\vec{v}, \underline{\eta}, \underline{\underline{p}}, \underline{r}) \in \mathcal{U}^n \times \mathcal{Q}^n.$$

The finite element subspace $\mathcal{U}^n \times \mathcal{Q}^n$ we keep as defined by (4.16) and (4.17) in Section 4. We then state the new stability result, which can be established by a strategy similar to that used for Lemma 4.3.

LEMMA 5.2. *Assume that:*

($\widetilde{S}1$) $0 < \alpha_i < C_i^I$, *for* $i = 1, 2$, *where the positive constants* C_i^I *are fixed, derived from inverse inequalities.*
($\widetilde{S}2$) $\alpha_i > 0$, *for* $i = 3, 4, 5$.
($\widetilde{S}3$) h *is sufficiently small.*

Then the stabilized formulation $\tilde{\mathcal{S}}_t^h$ *is stable, i.e., there is a positive constant* C *such that, for all* $(\vec{w}, \underline{\tau}, \underline{\underline{k}}, \underline{s}) \in \mathcal{U}^n \times \mathcal{Q}^n$, *there exist* $(\vec{z}, \underline{\eta}, \underline{\underline{p}}, \underline{r}) \in \mathcal{U}^n \times \mathcal{Q}^n$ *with*

$$\widetilde{B}_h(\vec{w}, \underline{\tau}, \underline{\underline{k}}, \underline{s}; \vec{z}, \underline{\eta}, \underline{\underline{p}}, \underline{r}) \geq C [\![\vec{w}, \underline{\tau}, \underline{\underline{k}}, \underline{s}]\!]_{t,h}, \tag{5.4}$$

and

$$[\![\vec{z}, \underline{\eta}, \underline{\underline{p}}, \underline{r}]\!]_{t,h} \leq 1. \tag{5.5}$$

We can now state our final approximation result, again obtained by using the consistency and stability properties with specific interpolation estimates.

THEOREM 5.3. *Assume that conditions* ($\widetilde{S}1$)–($\widetilde{S}3$) *of Lemma* 5.2 *hold and that the solution of* $\mathcal{M}_t$ *is such that* $\vec{u} \in \left[H^{k+2}(\Omega)\right]^3$, $\underline{\theta} \in \left[H^{k+1}(\Omega)\right]^2$, $\underline{\underline{n}} \in \left[H^k(\Omega)\right]^{2\times 2}$ *and* $\underline{q} \in \left[H^k(\Omega)\right]^2$. *Then the finite element formulation* $\tilde{\mathcal{S}}_t^h$ *has a unique solution satisfying*

$$[\![\vec{u} - \vec{u}^h, \underline{\theta} - \underline{\theta}^h, \underline{\underline{n}} - \underline{\underline{n}}^h, \underline{q} - \underline{q}^h]\!]_{t,h} \leq C h^k \left(\|\vec{u}\|_{k+2} + \|\underline{\theta}\|_{k+1} + \|\underline{\underline{n}}, \underline{q}\|_k \right). \tag{5.6}$$

Acknowledgement

Most of this work was performed while R. Stenberg was invited at INRIA by Professor M. Bernadou, under whose supervision D. Chapelle was pursuing his doctorate. Both authors are grateful to Professor Bernadou for giving them this opportunity, for several helpful discussions, and for his continuous support during this project.

References

[AB97] D. N. Arnold and F. Brezzi, *Locking free finite elements for shells*, Math. Comp. **66** (1997), 405–421.

[BD86] K. J. Bathe and E. N. Dvorkin, *A formulation of general shell elements—the use of mixed interpolation of tensorial components*, Internat. J. Numer. Methods Engrg. **22** (1986), 697–722.

[Ber94] M. Bernadou, *Méthodes d'éléments finis pour les problèmes de coques minces*, Masson, 1994.

[BC75] M. Bernadou and P. G. Ciarlet, *Sur l'ellipticité du modèle linéaire de coques de W. T. Koiter*, Computing Methods in Applied Sciences and Engineering (R. Glowinski and J. L. Lions, eds.) (Second Internat. Sympos., Versailles, 1975), Part 1, pp. 89–136; Lecture Notes in Econom. and Math. Systems, vol. 134, Springer, Berlin, 1976.

[Bra92] D. Braess, *Finite Elemente*, Springer-Verlag, Heidelberg, 1992.

[BF91] F. Brezzi and M. Fortin, *Mixed and hybrid finite element methods*, Springer Ser. Comput. Math., vol. 15, Springer-Verlag, New York, 1991.

[BFS91] F. Brezzi, M. Fortin, and R. Stenberg, *Error analysis of mixed-interpolated elements for Reissner-Mindlin plates*, Math. Models Methods Appl. Sci. **1** (1991), no. 2, 125–151.

[CS] D. Chapelle and R. Stenberg, *Stabilized finite element formulations for shells in a bending dominated state*, SIAM J. Numer. Anal. **36** (1999), no. 1, 32–73.

[CP94] D. Chenais and J.-C. Paumier, *On the locking phenomenon for a class of elliptic problems*, Numer. Math. **67** (1994), 427–440.

[Cia78] P. G. Ciarlet, *The finite element method for elliptic problems*, North-Holland, 1978.

[CL94a] P. G. Ciarlet and V. Lods, *Analyse asymptotique des coques linéairement élastiques. I. Coques "membranaires"*, C. R. Acad. Sci. Paris Sér. I Math. **318** (1994), no. 9, 863–868.

[CL94b] ______, *Analyse asymptotique des coques linéairement élastiques.* III. *Une justification du modèle de W. T. Koiter*, C. R. Acad. Sci. Paris Sér. I Math. **319** (1994), no. 3, 299–304.

[CLM94] P. G. Ciarlet, V. Lods, and B. Miara, *Analyse asymptotique des coques linéairement élastiques.* II. *Coques "en flexion"*, C. R. Acad. Sci. Paris Sér. I Math. **319** (1994), no. 1, 95–100.

[Cou78] N. Coutris, *Théorème d'existence et d'unicité pour un problème de coque élastique dans le cas d'un modèle linéaire de P. M. Naghdi*, RAIRO Anal. Numér. **12** (1978), no. 1, 51–57.

[Des90] P. Destuynder, *Modélisation des coques minces élastiques*, Masson, 1990.

[DN88] P. Destuynder and T. Nevers, *A new finite element scheme for bending plates*, Comput. Methods Appl. Mech. Engrg. **68** (1988), 127–139.

[FHLM88] L. P. Franca, T. J. R. Hughes, A. F. D. Loula, and I. Miranda, *A new family of stable elements for nearly incompressible elasticity based on a mixed Petrov-Galerkin finite element formulation*, Numer. Math. **53** (1988), 123–141.

[FS91] L. Franca and R. Stenberg, *Error analysis of some Galerkin-least-squares methods for the elasticity equations*, SIAM J. Numer. Anal. **28** (1991), 1680–1697.

[GZ68] A. E. Green and W. Zerna, *Theoretical elasticity*, 2nd ed., Oxford University Press, Oxford, 1968.

[HH92] I. Harari and T. J. R. Hughes, *What is C and h: Inequalities for the analysis and design of finite element methods*, Comput. Methods Appl. Mech. Engrg. **97** (1992), 157–192.

[HF87] T. J. R. Hughes and L. P. Franca, *A new finite element formulation for computational fluid dynamics:* VII. *The Stokes problem with various well-posed boundary conditions: Symmetric formulations that converge for all velocity/pressure spaces*, Comput. Methods Appl. Mech. Engrg. **65** (1987), 85–96.

[HF88] ______, *A mixed finite element formulation for Reissner-Mindlin plate theory: Uniform convergence of all higher-order spaces*, Comput. Methods Appl. Mech. Engrg. **67** (1988), 223–240.

[Kir93] A. Kirmse, *Bending-dominated deformation of thin spherical shells: analysis and finite element approximation*, SIAM J. Numer. Anal. **30** (1993), no. 4, 1015–1040.

[Koi59] W. T. Koiter, *A consistent first approximation in the general theory of thin elastic shells.* I. *Foundations and linear theory*, Research report 713, Delft Technological University, 1959.

[LSP95] D. Leguillon and E. Sanchez-Palencia, *Expériences de non-calculabilité dans les problèmes sensitifs. Coques minces chancelantes*, Deuxième Colloque National en Calcul des Structures, Hermes, 1995, pp. 275–280.

[LFHM87] A. F. D. Loula, L. P. Franca, T. J. R. Hughes, and I. Miranda, *Stability, convergence and accuracy of a new finite element method for the circular arch problem*, Comput. Methods Appl. Mech. Engrg. **63** (1987), 281–303.

[LHFM87] A. F. D. Loula, T. J. R. Hughes, L. P. Franca, and I. Miranda, *Mixed Petrov-Galerkin methods for the Timoshenko beam*, Comput. Methods Appl. Mech. Engrg. **63** (1987), 133–154.

[LS99] M. Lyly and R. Stenberg, *Stabilized finite element methods for Reissner-Mindlin plates*, Inst. für Mathematik und Geometrie, Univ. Inssbruck, preprint, 1999.

[LS94] ———, *Stabilized* MITC *plate bending elements*, Advances in Finite Element Techniques (M. Papadrakakis and B. H. V. Topping, eds.), CIVIL-COMP Ltd, 1994, pp. 11–16.

[LSV93] M. Lyly, R. Stenberg, and T. Vihinen, *A stable bilinear element for Reissner-Mindlin plates*, Comput. Methods Appl. Mech. Engrg. **110** (1993), no. 3, 343–357.

[Nag63] P. M. Naghdi, *Foundations of elastic shell theory*, Progress in Solid Mechanics, vol. IV, North-Holland, 1963, pp. 1–90.

[PP93] J. Piila and J. Pitkäranta, *Energy estimates relating different linear elastic models of a thin cylindrical shell.* I. *The membrane-dominated case*, SIAM J. Math. Anal. **24** (1993), no. 1, 1–22.

[PP95] ———, *Energy estimates relating different linear elastic models of a thin cylindrical shell.* II. *The case of free boundary*, SIAM J. Math. Anal. **26** (1995), no. 4, 820–849.

[Pit88] J. Pitkäranta, *Analysis of some low-order finite element schemes for Mindlin-Reissner and Kirchhoff plates*, Numer. Math. **53** (1988), 237–254.

[Pit92] ———, *The problem of membrane locking in finite element analysis of cylindrical shells*, Numer. Math. **61** (1992), 523–542.

[Rou69] P. Rougée, *Équilibre des coques élastiques minces inhomogènes en théorie non linéaire*, Ph.D. thesis, Univ. de Paris, 1969.

[RT88] P. A. Raviart and J.-M. Thomas, *Introduction à l'analyse numérique des équations aux dérivées partielles*, Masson, 1988.

[SP89a] E. Sanchez-Palencia, *Statique et dynamique des coques minces.* I. *Cas de flexion pure non inhibée*, C. R. Acad. Sci. Paris Sér. I **309** (1989), 411–417.

[SP89b] ———, *Statique et dynamique des coques minces.* II. *Cas de flexion pure inhibée - Approximation membranaire*, C. R. Acad. Sci. Paris Sér. I **309** (1989), 531–537.

[Ste95] R. Stenberg, *A new finite element formulation for the plate bending problem*, Asymptotic Methods for Elastic Structures (Lisbon, 1993) (Ciarlet, Trabucho, and Viano, eds.), Walter de Gruyter & Co., 1995, pp. 209–221.

[Val95] R. Valid, *The nonlinear theory of shells through variational principles*, John Wiley & Sons, 1995.

[Wem81] G. R. Wempner, *Mechanics of solids with applications to thin bodies*, Sijthoff & Nordhoff, 1981.

INRIA-Rocquencourt, BP 105, 78153 Le Chesnay Cedex, France
E-mail address: `Dominique.Chapelle@inria.fr`

Institut für Mathematik und Geometrie, Universität Innsbruck. Technikerstrasse 13, A-6020 Innsbruck, Austria
E-mail address: `Rolf.Stenberg@uibk.ac.at`

Centre de Recherches Mathématiques
CRM Proceedings and Lecture Notes
Volume **21**, 1999

Un élément de coque simple à trois nœuds en grandes rotations et élastoplasticité

F. Dammak, S. Chamlal, A. Gakwaya et G. Dhatt

Résumé. Nous proposons un élément de coque simple et efficace pour traîter des non-linéarités géométriques et matérielles avec grandes rotations. L'hypothèse cinématique de Reissner-Mindlin a été retenue avec intégration explicite sur l'épaisseur. Les mouvements de déplacements et de rotations sont actualisés d'une façon précise. Afin d'éviter le problème de blocage en cisaillement dans le cas des coques minces, un champ de déformation de cisaillement est convenablement choisi, constant sur l'élément. La matrice tangente consistante est symétrique et donnée sous forme explicite. Deux variables de rotation par nœud ont été utilisées ce qui conduit à un élément de coque triangulaire à 15 degrés de liberté. De bons résultats ont été obtenus dans les cas linéaires et non-linéaires. La bonne performance de cet élément simple est démontré à travers quelques tests sélectionnés de la littérature.

Abstract. We present a very simple and efficient shell element for the analysis of geometrical and material nonlinearities with finite rotations. The Reissner-Mindlin kinematics is considered with an explicit integration through the thickness. Displacements and rotations are exactly updated. A constant assumed shear strain field is used to avoid the shear locking in the thin shell limit. The consistent tangent stiffness matrix is symmetric and given explicitly in this paper. Two rotational variables are used at each node which lead to 15 degrees of freedom triangular shell element. It is shown that this type of element is very simple and gives good responses for linear and nonlinear applications. Its excellent performance is demonstrated with a number of discriminating examples selected from the literature.

1. Introduction

Dans l'analyse des structures plaques ou coques avec non-linéarités géométriques et matérielles, plusieurs opérations répétitives, comme le calcul de la matrice tangente, se font au niveau de l'élément. Pour réduire le coût des ces opérations répétitives, il est avantageux d'avoir des éléments simples et efficaces d'ordre le plus bas : éléments à trois ou quatre nœuds. Contrairement à la théorie de Kirchhoff-Love qui nécessite une continuité C^1, seule une continuité C^0 est à satisfaire pour la théorie de Reissner-Mindlin. Ceci entraîne une réduction de l'ordre des fonctions d'interpolation. Cependant, les éléments basés sur la théorie de Reissner-Mindlin

1991 *Mathematics Subject Classification.* Primary : 73G20 ; Secondary : 65C20.
This is the final form of the paper.

avec un modèle déplacement et une intégration exacte présentent un blocage en cisaillement dans le cas des coques minces.

Pour résoudre le problème de blocage en cisaillement, différentes techniques ont été employées. L'intégration réduite ou sélective a été l'une des premières méthodes qui a considérablement amélioré les performances des éléments. Différents éléments triangulaires à trois nœuds ont été examinés par Batoz *et al.* [**B2**]. Selon ces auteurs, l'élément obtenu avec intégration réduite/sélective SRI n'est pas acceptable. D'autres méthodes ont été proposées pour éviter le problème de blocage en cisaillement. Pour l'élément triangulaire à trois noeuds, une méthode plus efficace que le SRI, à un seul point d'intégration a été développé dans Hughes [**H1**]. Cet élément est obtenu par chevauchement de deux nœuds d'un élément quadrilatère à quatre nœuds. Belytschko *et al.* [**B3**] ont proposé un autre élément triangulaire à trois nœuds et un seul point d'intégration basé sur une décomposition des déplacements en modes de flexion et de cisaillement. Selon Belytschko *et al.* [**B3**], l'élément dans [**H1**] peut être équivalent au leur en un certain point d'intégration. Un élément de coque dégénéré triangulaire linéaire à trois nœuds simple et efficace a été développé par Boisse *et al.* [**B4**]. Cet élément est basé sur une formulation de substitution en cisaillement de type ANS (Assumed Natural Strain). Cette formulation, dans laquelle le champ de cisaillement est interpolé indépendamment, est similaire à celle proposé par Bathe et Dvorkin [**B1**] pour l'élément quadrilatère à quatre nœuds. Ce dernier élément à quatre nœuds est utilisé dans la formulation des éléments de coques non-linéaires vu sa simplicité et son efficacité [**D1, P1, S2**].

Les éléments quadrilatéraux à quatre nœuds sont de plus en plus populaires en analyse non-linéaire des coques. Cependant, les éléments triangulaires à trois nœuds ont été souvent utilisés comme éléments de coques minces à facette plane. Dans cet article, nous présentons un élément triangulaire à trois nœuds simple et efficace basé sur la théorie de Reissner-Mindlin. Cet élément initialement à facette plane devient courbe après déformation. La partie cisaillement de cet élément sera développée selon la formulation ANS. En se basant sur la formulation de Boisse *et al.* [**B4**], nous avons construit une mesure de déformation en cisaillement. L'élément obtenu ayant des parties membrane, flexion et cisaillement constantes, ne nécessite pas de boucle d'intégration. Suite à cette brève introduction, nous présentons dans la Section 2, un rappel de quelques notions sur les rotations qui seront utilisées dans la formulation de l'élément développé. Dans la section suivante, nous présentons le modèle de coque non-linéaire avec grands déplacements et grandes rotations. Les aspects géométrique, cinématique, déformations, forme faible et linéarisation sont traités dans cette même section. La Section 4 est relative à la loi de comportement. La mise en œuvre numérique de l'élément de coque à trois nœuds est donnée dans la Section 5. Cet élément est évalué à travers quelques tests standards dans la Section 6 et une conclusion générale est donnée dans la Section 7.

2. Description des rotations

Dans cette section, nous présentons un bref rappel de quelques aspects concernant les rotations finies. Les détails peuvent être trouvés dans [**A1, C1, D1, S1, S2**]. On définit une rotation $\mathbf{Q}$ comme un endomorphisme de R^3 dont la matrice associée est orthogonale. Cette rotation sera paramétrisée par le vecteur rotation défini tel que :

$$(1) \qquad \boldsymbol{\Theta} = \Theta \mathbf{e}, \quad \Theta = \|\boldsymbol{\Theta}\| \quad \text{et} \quad \|\mathbf{e}\| = 1,$$

où Θ est l'angle de rotation autour de l'axe défini par le vecteur unitaire $\mathbf{e}$. À chaque vecteur $\boldsymbol{\Theta}$ on associe une rotation définie par l'application exponentielle [**C1, S2**] :

$$\mathbf{Q} = \exp(\bar{\boldsymbol{\Theta}}) = I + \frac{\sin(\Theta)}{\Theta}\bar{\boldsymbol{\Theta}} + \frac{1-\cos(\Theta)}{\Theta^2}\bar{\boldsymbol{\Theta}}^2, \tag{2}$$

$$\mathbf{Q}^T = \exp(-\bar{\boldsymbol{\Theta}}); \exp(\bar{\boldsymbol{\Theta}}) \cdot \boldsymbol{\Theta} = \boldsymbol{\Theta}, \tag{3}$$

où $\overline{\boldsymbol{\Theta}}$ est la matrice anti-symétrique construite à partir du vecteur axial $\boldsymbol{\Theta}$ telle que

$$\bar{\boldsymbol{\Theta}} \cdot \mathbf{x} = \boldsymbol{\Theta} \wedge \mathbf{x} \quad \forall \mathbf{x} \in R^3. \tag{4}$$

La variation d'une matrice rotation est donnée par [**C1**] :

$$\delta\mathbf{Q} = \mathbf{Q}\delta\bar{\boldsymbol{\Theta}} = \delta\bar{\theta}\mathbf{Q}. \tag{5}$$

Les matrices anti-symétriques $\delta\bar{\boldsymbol{\Theta}}$ et $\delta\bar{\theta}$ représentent respectivement les variations angulaires matérielles et spatiales. La notion de matérielle et spatiale est introduite ici pour distinguer entre la multiplication à gauche ou à droite de la rotation $\mathbf{Q}$ [**C1**].

3. Relations de base de la théorie des coques

3.1. Géométrie, cinématique et déformations. Dans cette section, nous présentons les aspects géométrique et cinématique de la théorie des coques. Nous considérons deux configurations, une initiale notée C_0, et l'autre actuelle notée C_t, définies dans un repère orthonormé global $\mathbf{E}_i$, $i = 1, 2, 3$. La configuration initiale est supposée lisse, continue et différentiable. La notation suivante est utilisée : les vecteurs, tenseurs et matrices sont notés en lettres grasses, majuscules dans la configuration initiale, et minuscules dans la configuration actuelle. Les indices grecs prennent les valeurs 1, 2 alors que les indices latins prennent les valeurs 1, 2, 3. La description géométrique d'un point quelconque (q) du domaine de la coque dans les configurations, initiale et déformée, est donnée par :

$$\mathbf{X}_q(\xi^1, \xi^2, z) = \mathbf{X}_p(\xi^1, \xi^2) + z\mathbf{D}(\xi^1, \xi^2), \tag{6}$$

$$\mathbf{x}_q(\xi^1, \xi^2, z) = \mathbf{x}_p(\xi^1, \xi^2) + z\mathbf{d}(\xi^1, \xi^2), \tag{7}$$

où (p) est un point de la surface moyenne qui est paramétrisée par les deux coordonnées curvilignes ξ^1 et ξ^2. Les vecteurs $\mathbf{D}$ et $\mathbf{d}$ représentent les vecteurs directeurs unitaires dans les configurations respectives, initiale et déformée. Le choix le plus naturel du vecteur $\mathbf{D}$ est la normale ($\mathbf{N}$) à la surface moyenne initiale au point (p). Dans chaque configuration, la base covariante, associée aux coordonnées paramétriques ξ^α, est donnée par :

$$\begin{aligned} \mathbf{G}_\alpha &= \mathbf{A}_\alpha + z\mathbf{D}_{,\alpha}\,; \quad \mathbf{G}_3 = \mathbf{D}, \\ \mathbf{g}_\alpha &= \mathbf{a}_\alpha + z\mathbf{d}_{,\alpha}\,; \quad \mathbf{g}_3 = \mathbf{d}, \end{aligned} \tag{8}$$

où $(\ )_{,\alpha}$ désigne une dérivation par rapport aux coordonnées paramétriques ξ^α. Les vecteurs $\mathbf{A}_\alpha$ et $\mathbf{a}_\alpha$, représentent les vecteurs de base, tangents à la surface moyenne, dans les configurations initiale et actuelle :

$$\mathbf{A}_\alpha = \mathbf{X}_{p,\alpha}; \quad \mathbf{a}_\alpha = \mathbf{x}_{p,\alpha}. \tag{9}$$

L'élément de surface dans la configuration initiale est donnée par :

$$\mathbf{A} = \sqrt{\mathbf{A}}\, d\mathbf{A}_\xi; \quad \sqrt{\mathbf{A}} = \|\mathbf{A}_1 \wedge \mathbf{A}_2\|; \quad d\mathbf{A}_\xi = d\xi^1 d\xi^2. \tag{10}$$

Les composantes covariantes du tenseur métrique au point (q) dans chaque configuration sont définies par :

$$G_{ij} = \mathbf{G}_i \cdot \mathbf{G}_j; \quad g_{ij} = \mathbf{g}_i \cdot \mathbf{g}_j, \tag{11}$$

où les composantes covariantes g_{ij} du tenseur métrique $\mathbf{g}$ sont données par :

$$\begin{cases} g_{\alpha\beta} = a_{\alpha\beta} + z b_{\alpha\beta} + z^2 c_{\alpha\beta}, \\ g_{\alpha 3} = \varpi_\alpha, \\ g_{33} = 1, \end{cases} \tag{12}$$

où les variables cinématiques $a_{\alpha\beta}$, $b_{\alpha\beta}$, $c_{\alpha\beta}$, $\varpi_{\alpha\beta}$ sont définies par :

$$\begin{cases} a_{\alpha\beta} = \mathbf{a}_\alpha \cdot \mathbf{a}_\beta; & b_{\alpha\beta} = \mathbf{a}_\alpha \cdot \mathbf{d}_{,\beta} + \mathbf{a}_\beta \cdot \mathbf{d}_{,\alpha}, \\ c_{\alpha\beta} = \mathbf{d}_{,\alpha} \cdot \mathbf{d}_{,\beta}; & \varpi_\alpha = \mathbf{a}_\alpha \cdot \mathbf{d}. \end{cases} \tag{13}$$

La mesure de déformation adoptée est la mesure lagrangienne donnée par :

$$\varepsilon_{ij} = \frac{1}{2}(g_{ij} - G_{ij}). \tag{14}$$

En termes de composantes, cette mesure peut s'écrire :

$$\begin{cases} \varepsilon_{\alpha\beta} = e_{\alpha\beta} + z\chi_{\alpha\beta} + z^2 \kappa_{\alpha\beta}, \\ 2\varepsilon_{\alpha 3} = \gamma_\alpha, \\ \varepsilon_{33} = 0, \end{cases} \tag{15}$$

avec

$$\begin{cases} e_{\alpha\beta} = (a_{\alpha\beta} - A_{\alpha\beta})/2, \\ \boldsymbol{\chi}_{\alpha\beta} = (b_{\alpha\beta} - B_{\alpha\beta})/2, \\ \boldsymbol{\gamma}_{\alpha} = \varpi_\alpha - \varpi_\alpha^0, \end{cases} \tag{16}$$

où $e_{\alpha\beta}$ représente les déformations de membrane, $\boldsymbol{\chi}_{\alpha\beta}$ les déformations de flexion, $\boldsymbol{\kappa}_{\alpha\beta}$ les déformations de flexion d'ordre supérieur qui sont négligées et le terme $\boldsymbol{\gamma}_\alpha$ représente les déformations de cisaillement transversal.

3.2. Forme faible. La forme faible des équations d'équilibre tridimensionnelles, dans la configuration initiale, est donnée par :

$$G = \int_V S^{ij} \delta\varepsilon_{ij} dV - G_{\text{ext}} = 0, \tag{17}$$

où ($dV = \sqrt{G}\, d\xi^1\, d\xi^2\, dz$) est l'élément de volume de la coque dans la configuration initiale, $\delta\varepsilon_{ij}$ sont les composantes covariantes du tenseur de déformation de Green Lagrange et S^{ij} sont les composantes contravariantes du deuxième tenseur des contraintes de Piola Kirchhoff. En utilisant la relation (15) qui définit la mesure de déformation et après intégration sur l'épaisseur de la coque, l'équation (17) peut s'écrire :

$$G = \int_A (\delta e \cdot \mathbf{N} + \delta\chi \cdot \mathbf{M} + \delta\gamma \mathbf{T})\, dA - G_{\text{ext}} = 0, \tag{18}$$

où on'a introduit les vecteurs contraintes résultantes de Piola Kirchhoff de membrane $\mathbf{N}$, de flexion $\mathbf{M}$ et de cisaillement $\mathbf{T}$ qui s'écrivent sous la forme matricielle suivante :

$$\mathbf{N} = \begin{Bmatrix} N^{11} \\ N^{22} \\ N^{12} \end{Bmatrix}, \quad \mathbf{M} = \begin{Bmatrix} M^{11} \\ M^{22} \\ M^{12} \end{Bmatrix}, \quad \mathbf{T} = \begin{Bmatrix} T^{1} \\ T^{2} \end{Bmatrix}, \tag{19}$$

et dont les composantes sont données par :

$$N^{\alpha\beta} = \int_{-h/2}^{+h/2} S^{\alpha\beta}\sqrt{\frac{G}{A}}\,dz, \tag{20}$$

$$M^{\alpha\beta} = \int_{-h/2}^{+h/2} zS^{\alpha\beta}\sqrt{\frac{G}{A}}\,dz, \tag{21}$$

$$T^{\alpha} = \int_{-h/2}^{+h/2} S^{\alpha 3}\sqrt{\frac{G}{A}}\,dz, \tag{22}$$

et les vecteurs colonne des déformations de membrane $\mathbf{e}$, de flexion $\boldsymbol{\chi}$ et de cisaillement $\boldsymbol{\gamma}$ sont définis par :

$$\mathbf{e} = \begin{Bmatrix} e_{11} \\ e_{22} \\ 2e_{12} \end{Bmatrix}, \quad \boldsymbol{\chi} = \begin{Bmatrix} \chi_{11} \\ \chi_{22} \\ 2\chi_{12} \end{Bmatrix}, \quad \boldsymbol{\gamma} = \begin{Bmatrix} \gamma_{1} \\ \gamma_{2} \end{Bmatrix}. \tag{23}$$

Les variations virtuelles de membrane $\delta e_{\alpha\beta}$, de flexion $\delta\chi_{\alpha\beta}$ et de cisaillement $\delta\gamma_\alpha$ peuvent être déduites des relations (13) et (16) et s'écrivent :

$$\delta e_{\alpha\beta} = \frac{1}{2}(\mathbf{a}_\alpha \cdot \delta\mathbf{x}_{,\beta} + \mathbf{a}_\beta \cdot \delta\mathbf{x}_{,\alpha}), \tag{24}$$

$$\boldsymbol{\delta\chi}_{\alpha\beta} = \frac{1}{2}(\mathbf{a}_\alpha \cdot \boldsymbol{\delta d}_{,\beta} + \mathbf{a}_\beta \cdot \boldsymbol{\delta d}_{,\alpha} + \boldsymbol{\delta x}_{,\alpha} \cdot \mathbf{d}_{,\beta} + \boldsymbol{\delta x}_{,\beta} \cdot \mathbf{d}_{,\alpha}), \tag{25}$$

$$\boldsymbol{\delta\gamma}_\alpha = (\mathbf{a}_\alpha \cdot \boldsymbol{\delta d} + \boldsymbol{\delta x}_{,\alpha} \cdot \mathbf{d}). \tag{26}$$

Dans les expressions (25) et (26) des déformations virtuelles de flexion et de cisaillement, apparaît le vecteur variation $\boldsymbol{\delta d}$. Ce vecteur variation peut être exprimé en terme des variations angulaires spatiale et matérielle selon la relation (5) :

$$\delta\mathbf{d} = \delta\theta \wedge \mathbf{d} = \widetilde{\mathbf{Q}}\delta\theta; \qquad \widetilde{\mathbf{Q}} = -\bar{\mathbf{d}}, \tag{27}$$

$$\delta\mathbf{d} = \mathbf{Q}\delta\bar{\boldsymbol{\Theta}}\mathbf{E}_3 = \widetilde{\mathbf{Q}}\delta\boldsymbol{\Theta}; \quad \widetilde{\mathbf{Q}} = -\mathbf{Q}\overline{\mathbf{E}}_3, \tag{28}$$

où nous avons supposé la relation $\mathbf{d} = \mathbf{Q}E_3$ reliant le vecteur directeur $\mathbf{d}$ de la configuration actuelle avec le vecteur $\mathbf{E}_3$. Le vecteur $\delta\Theta$ est appelé variation virtuelle angulaire matérielle alors que $\delta\theta$ est appelé variation virtuelle angulaire spatiale. La transformation $\widetilde{\mathbf{Q}} = -\bar{\mathbf{d}}$, reliant le vecteur variation $\delta\mathbf{d}$ à la variation angulaire $\delta\theta$, nécessite l'introduction d'une rigidité additionnelle définie par une rotation fictive autour d'un vecteur directeur $\mathbf{d}$. Par contre, la transformation $\widetilde{\mathbf{Q}} = -\mathbf{Q}\overline{\mathbf{E}}_3$ est optimale et permet la réduction de trois à deux les degrés de liberté en rotation **[S1, S2]** :

$$\mathbf{Q} = [\mathbf{d}_1\mathbf{d}_2\mathbf{d}]; \quad -\mathbf{Q}\overline{E}_3 = [-\mathbf{d}_2\mathbf{d}_1\mathbf{0}]. \tag{29}$$

3.3. Linéarisation. Si nous définissons l'opérateur différentiel $\mathbf{B}$ des déformations comme suit :

$$\mathbf{B} = \begin{bmatrix} \mathbf{B}_m & \mathbf{0} \\ \mathbf{B}_{fu} & \mathbf{B}_{fr} \\ \mathbf{B}_{cu} & \mathbf{B}_{cr} \end{bmatrix}, \tag{30}$$

où les différents opérateurs $\mathbf{B}_m, \mathbf{B}_{fu}, \mathbf{B}_{fr}, \mathbf{B}_{cu}$ et $\mathbf{B}_{cr}$ correspondent respectivement aux effets de membrane, flexion et cisaillement sont donnés par :

$$\mathbf{B}_m = \begin{bmatrix} \mathbf{a}_1^T \frac{\partial}{\partial \xi^1} \\ \mathbf{a}_2^T \frac{\partial}{\partial \xi^2} \\ \mathbf{a}_1^T \frac{\partial}{\partial \xi^2} + \mathbf{a}_2^T \frac{\partial}{\partial \xi^1} \end{bmatrix}; \quad \mathbf{B}_{cu} = \begin{bmatrix} \mathbf{d}^T \frac{\partial}{\partial \xi^1} \\ \mathbf{d}^T \frac{\partial}{\partial \xi^2} \end{bmatrix}; \quad \mathbf{B}_{cr} = \begin{bmatrix} \mathbf{a}_1^T \\ \mathbf{a}_2^T \end{bmatrix} \widetilde{\mathbf{Q}}, \tag{31}$$

$$\mathbf{B}_{fu} = \begin{bmatrix} \mathbf{d}_{,1}^T \frac{\partial}{\partial \xi^1} \\ \mathbf{d}_{,2}^T \frac{\partial}{\partial \xi^2} \\ \mathbf{d}_{,1}^T \frac{\partial}{\partial \xi^2} + \mathbf{d}_{,2}^T \frac{\partial}{\partial \xi^1} \end{bmatrix}; \quad \mathbf{B}_{fr} = \begin{bmatrix} \mathbf{a}_1^T \frac{\partial}{\partial \xi^1} \\ \mathbf{a}_2^T \frac{\partial}{\partial \xi^2} \\ \mathbf{a}_1^T \frac{\partial}{\partial \xi^2} + \mathbf{a}_2^T \frac{\partial}{\partial \xi^1} \end{bmatrix} \widetilde{\mathbf{Q}}, \tag{32}$$

et si nous définissons le vecteur colonne des contraintes résultantes totales $\mathbf{\Sigma}$ comme suit :

$$\mathbf{\Sigma} = \begin{Bmatrix} \mathbf{N} \\ \mathbf{M} \\ \mathbf{T} \end{Bmatrix}, \tag{33}$$

alors la forme faible (17) devient :

$$G(\mathbf{\Phi}, \boldsymbol{\delta\Phi}) = \int_A \mathbf{B}\boldsymbol{\delta\Phi} \cdot \mathbf{\Sigma}\, dA - G_{\text{ext}}(\mathbf{\Phi}, \boldsymbol{\delta\Phi}) = 0, \tag{34}$$

où le vecteur colonne $\delta\mathbf{\Phi}$ peut avoir six composantes si nous utilisons la transformation (27) ($\delta\mathbf{\Phi} = (\delta x_1, \delta x_2, \delta x_3, \delta\theta_1, \delta\theta_2, \delta\theta_3)$: trois déplacements et trois rotations dans le repère global) ou cinq composantes si nous utilisons la transformation (28) ($\delta\mathbf{\Phi} = (\delta x_1, \delta x_2, \delta x_3, \delta\Theta_1, \delta\Theta_2)$: trois déplacements et deux rotations dans le repère local). La linéarisation de la forme faible (34) s'écrit :

$$DG = D_G G \cdot \Delta\mathbf{\Phi} + D_M G \cdot \Delta\mathbf{\Phi}. \tag{35}$$

La partie géométrique $D_G G \cdot \Delta\mathbf{\Phi}$ est obtenue par linéarisation des déformations virtuelles et s'écrit :

$$D_G G \cdot \Delta\mathbf{\Phi} = \int_A (\mathbf{N} \cdot \Delta\delta\mathbf{e} + \mathbf{M} \cdot \Delta\delta\boldsymbol{\chi} + \mathbf{T} \cdot \Delta\delta\boldsymbol{\gamma})\, dA, \tag{36}$$

où la deuxième variation des déformations de membrane, de flexion et de cisaillement est donnée par :

$$\Delta\delta e_{\alpha\beta} = \frac{1}{2}(\Delta\mathbf{x}_{,\alpha} \cdot \boldsymbol{\delta x}_{,\beta} + \Delta\mathbf{x}_{,\beta} \cdot \boldsymbol{\delta x}_{,\alpha}), \tag{37}$$

$$\begin{aligned} \Delta\boldsymbol{\delta\chi}_{\alpha\beta} = \frac{1}{2}(\Delta\mathbf{x}_{,\alpha} \cdot \boldsymbol{\delta d}_{,\beta} + \Delta\mathbf{x}_{,\beta} \cdot \boldsymbol{\delta d}_{,\alpha} + \boldsymbol{\delta x}_{,\alpha} \cdot \Delta\mathbf{d}_{,\beta} + \boldsymbol{\delta x}_{,\beta} \cdot \Delta\mathbf{d}_{,\alpha}) \\ + \frac{1}{2}(\mathbf{a}_\alpha \cdot \Delta\delta\mathbf{d}_{,\beta} + \mathbf{a}_\beta \cdot \Delta\delta\mathbf{d}_{,\alpha}), \end{aligned} \tag{38}$$

$$\Delta\boldsymbol{\delta}\gamma_\alpha = (\Delta\mathbf{x}_{,\alpha} \cdot \boldsymbol{\delta d} + \boldsymbol{\delta x}_{,\alpha} \cdot \Delta\mathbf{d} + \mathbf{a}_\alpha \cdot \Delta\delta\mathbf{d}), \tag{39}$$

où la deuxième variation du vecteur directeur $\mathbf{d}$ peut être déduite de (5), (27) et (28) :

$$\Delta\delta\mathbf{d} = -(\Delta\theta \cdot \delta\theta)\mathbf{d} = -(\Delta\boldsymbol{\Theta} \cdot \delta\boldsymbol{\Theta})\mathbf{d}. \tag{40}$$

La partie matérielle $D_M G \cdot \Delta\boldsymbol{\Phi}$ est obtenue par linéarisation des efforts résultants et s'écrit :

$$D_M G \cdot \Delta\boldsymbol{\Phi} = \int_A \mathbf{B}\delta\boldsymbol{\Phi} \cdot \Delta\boldsymbol{\Sigma}\, dA = \int_A \mathbf{B}\delta\boldsymbol{\Phi} \cdot \mathbf{H}_T^{ep} \cdot \mathbf{B}\Delta\boldsymbol{\Phi}\, dA, \tag{41}$$

où $\mathbf{H}_T^{ep}$ est le module tangent élastoplastique obtenu par intégration sur l'épaisseur du module tangent élastoplastique local cohérent avec l'intégration implicite de la loi de comportement.

4. Loi de comportement

Avec l'hypothèse des contraintes normales nulles, les relations entre les contraintes et les déformations peuvent être condensées et s'écrivent sous la forme matricielle suivante :

$$\mathbf{S} = \mathbf{H}\varepsilon^e = \mathbf{H}(\varepsilon - \varepsilon^p), \tag{42}$$

où $\mathbf{S}$ est le vecteur colonne des cinq composantes des contraintes non-nulles, ε et ε^p sont respectivement les déformations élastique et plastique et $\mathbf{H}$ est la loi élastique donnée par :

$$\mathbf{H} = \begin{bmatrix} \mathbf{H}_{11} & \mathbf{H}_{12} \\ \text{sym} & \mathbf{H}_{22} \end{bmatrix}, \tag{43}$$

avec

$$\mathbf{H}_{11} = \frac{E}{1-\nu^2}\begin{bmatrix} G^{11}G^{11} & \nu G^{11}G^{22} + (1-\nu)G^{12}G^{12} & G^{11}G^{12} \\ & G^{22}G^{22} & G^{22}G^{12} \\ \text{sym} & & \frac{1-\nu}{2}G^{11}G^{22} + \frac{1+\nu}{2}G^{12}G^{12} \end{bmatrix} \tag{44}$$

$$\mathbf{H}_{22} = \frac{E}{2(1+\nu)}\begin{bmatrix} G^{11} & G^{12} \\ G^{12} & G^{22} \end{bmatrix}; \quad \mathbf{H}_{12} = \mathbf{0}_{3\times 2}, \tag{45}$$

où E et ν sont respectivement le module de Young et le coefficient de Poisson.

Pour formuler le problème élastoplastique incrémentale, il est plus commode d'écrire les équations de la plasticité dans un repère local. Ces équations se résument au critère de Von Mises, à un écrouissage isotrope non linéaire et à un écrouissage cinématique linéaire et s'écrivent :

$$\begin{aligned} \varphi &= \frac{1}{2}\boldsymbol{\eta} \cdot \mathbf{P}\,\eta - \frac{1}{3}\kappa^2(\bar{\varepsilon}^p); \quad \eta = \mathbf{S} - \mathbf{k}, \\ \dot{\varepsilon}^p &= \dot{\lambda}\frac{\partial\varphi}{\partial\mathbf{S}} = \dot{\lambda}\mathbf{P}\eta, \\ \dot{\eta} &= \frac{2}{3}H\dot{\lambda}\eta, \\ \dot{\bar{\varepsilon}}^p &= \sqrt{\frac{2}{3}\,\dot{\varepsilon}^p_{ij}\,\dot{\varepsilon}^p_{ij}}, \end{aligned} \tag{46}$$

où $\mathbf{P}$ est une matrice de projection 5×5 donnée dans **[D1]**, η est les vecteur des contraintes effectives, H est le paramètre d'écrouissage cinématique et κ est la fonction non-linéaire qui définie l'écrouissage isotrope. Ces équations sont intégrées numériquement selon un algorithme de prédiction élastique/retour radial (return

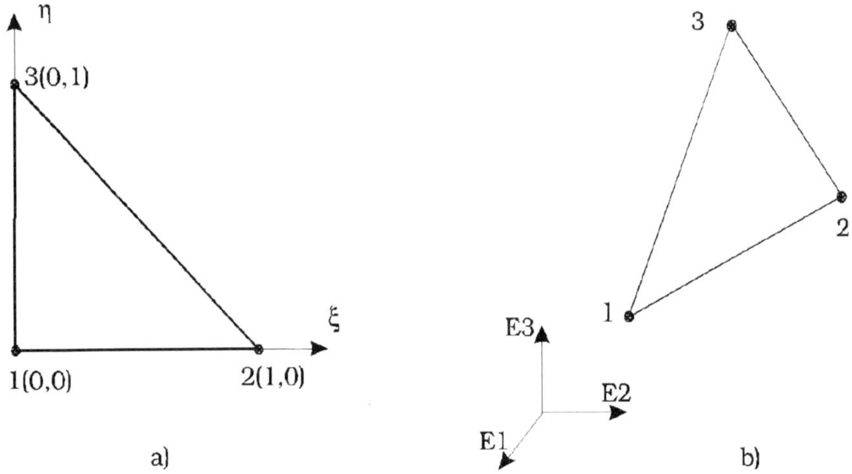

FIGURE 1. Élément triangulaire à trois nœuds.

mapping algorithm) analogue à celui proposé dans Simo and Taylor [**S3**] pour le cas des contraintes planes. Le module tangent élastoplastique est obtenu alors par intégration sur l'épaisseur du module tangent élastoplastique consistent local.

$$\mathbf{H}^{ep} = \begin{bmatrix} \mathbf{H}_{11}^{ep} & \mathbf{H}_{12}^{ep} \\ \text{sym} & \mathbf{H}_{22}^{ep} \end{bmatrix}; \quad \mathbf{H}_T^{ep} = \int_{-h/2}^{+h/2} \begin{bmatrix} \mathbf{H}_{11}^{ep} & z\mathbf{H}_{11}^{ep} & \mathbf{H}_{12}^{ep} \\ & z^2\mathbf{H}_{11}^{ep} & z\mathbf{H}_{12}^{ep} \\ \text{sym} & & \mathbf{H}_{22}^{ep} \end{bmatrix} dz. \tag{47}$$

5. Formulation éléments finis

Dans cette section, nous présentons la mise en œuvre numérique du modèle de coque non-linéaire en se basant sur l'élément triangulaire à trois nœuds.

Les vecteurs $\mathbf{x}$ et $\mathbf{d}$ sont interpolés indépendamment. Nous utilisons une même approximation pour le vecteur position, le champ de déplacement, ainsi que le vecteur directeur

$$\mathbf{x} = \sum_{i=1}^{3} N_i \cdot \mathbf{x}_i; \quad \mathbf{d} = \sum_{i=1}^{3} N_i \cdot \mathbf{d}_i, \tag{48}$$

$$N_1 = 1 - \xi - \eta; \quad N_2 = \xi; \quad N_3 = \eta. \tag{49}$$

Si nous désignons par $\mathbf{B}$ la matrice de l'opérateur de déformations défini pour un nœud (i) comme suit :

$$\mathbf{B}^i = \begin{bmatrix} \mathbf{B}_m^i & \mathbf{0} \\ \mathbf{B}_{fu}^i & \mathbf{B}_{fr}^i \\ \mathbf{B}_{cu}^i & \mathbf{B}_{cr}^i \end{bmatrix}, \tag{50}$$

alors la matrice tangente matérielle totale incluant tous les effets de membrane, flexion et cisaillement s'écrit

$$\mathbf{M}K_T = \int_A \mathbf{B}^T \mathbf{H}_T^{ep} \mathbf{B}\, dA = \frac{\sqrt{A}}{2} \mathbf{B}^T \mathbf{H}_T^{ep} \mathbf{B}. \tag{51}$$

Dans les sous sections suivantes, nous définissons les sous matrices appropriées de $\mathbf{B}^i$, relativement aux effets de membrane, flexion et cisaillement, ainsi que le résidu et la matrice tangente géométrique associée.

5.1. Membrane. La matrice $\mathbf{B}_m$ relative à la partie membrane s'écrit pour un nœud (i) :

$$\mathbf{B}_m^i = \begin{bmatrix} \mathbf{a}_1^T N_{i,1} \\ \mathbf{a}_2^T N_{i,2} \\ \mathbf{a}_1^T N_{i,2} + \mathbf{a}_2^T N_{i,1} \end{bmatrix}. \tag{52}$$

La contribution de la partie membrane au résidu total s'écrit :

$$\mathbf{R}_m = \int_A \mathbf{B}_m^T \mathbf{N}\, dA = \frac{\sqrt{A}}{2} \mathbf{B}_m^T \mathbf{N}, \tag{53}$$

alors que la matrice tangente géométrique provenant de la partie membrane relativement aux couples de nœuds (i, j) s'écrit selon (37) :

$$\mathbf{G}K_m^{ij} = \frac{\sqrt{A}}{2} \begin{bmatrix} u_{ij}\mathbf{I} & \mathbf{0} \\ \mathbf{0} & \mathbf{0} \end{bmatrix}; \tag{54}$$
$$u_{ij} = N_{i,1}(\mathbf{N}^{11} N_{j,1} + \mathbf{N}^{12} N_{j,2}) + N_{i,2}(\mathbf{N}^{12} N_{j,1} + \mathbf{N}^{22} N_{j,2}).$$

5.2. Flexion. Les matrices $\mathbf{B}_{fu}$ et $\mathbf{B}_{fr}$ relative à la partie flexion sont données pour un nœud (i) par

$$\mathbf{B}_{fu}^i = \begin{bmatrix} \mathbf{d}_{,1}^T N_{i,1} \\ \mathbf{d}_{,2}^T N_{i,2} \\ \mathbf{d}_{,1}^T N_{i,2} + \mathbf{d}_{,2}^T N_{i,1} \end{bmatrix}; \quad \mathbf{B}_{fr}^i = \begin{bmatrix} \mathbf{a}_1^T N_{i,1} \\ \mathbf{a}_2^T N_{i,2} \\ \mathbf{a}_1^T N_{i,2} + \mathbf{a}_2^T N_{i,1} \end{bmatrix} \widetilde{\mathbf{Q}}_i. \tag{55}$$

La contribution de la partie flexion au résidu total s'écrit :

$$\mathbf{R}_f = \int_A \mathbf{B}_f^T \mathbf{M}\, dA = \frac{\sqrt{A}}{2} \mathbf{B}_f^T \mathbf{M}. \tag{56}$$

La matrice tangente géométrique en flexion est défini pour un couple de nœuds (i, j) par :

$$\mathbf{G}K_f^{ij} = \frac{\sqrt{A}}{2} \begin{bmatrix} \mathbf{0} & fu_{ij}\widetilde{\mathbf{Q}}_j \\ fu_{ji}\widetilde{\mathbf{Q}}_i^T & fr_{ij}\mathbf{I} \end{bmatrix}, \tag{57}$$

où les termes fu_{ij} et fr_{ij} sont donnés par les expressions suivantes :

$$fu_{ij} = N_{i,1}(\mathbf{M}^{11} N_{j,1} + \mathbf{M}^{12} N_{j,2}) + N_{i,2}(\mathbf{M}^{12} N_{j,1} + \mathbf{M}^{22} N_{j,2}), \tag{58}$$

$$fr_{ii} = -N_{i,1}\{(\mathbf{a}_1 \cdot \mathbf{d}_i)\mathbf{M}^{11} + (\mathbf{a}_2 \cdot \mathbf{d}_i)\mathbf{M}^{12}\} - N_{i,2}\{(\mathbf{a}_1 \cdot \mathbf{d}_i)\mathbf{M}^{12} + (\mathbf{a}_2 \cdot \mathbf{d}_i)\mathbf{M}^{22}\}, \tag{59}$$

$$fr_{ij} = 0 \text{ pour } i \neq j. \tag{60}$$

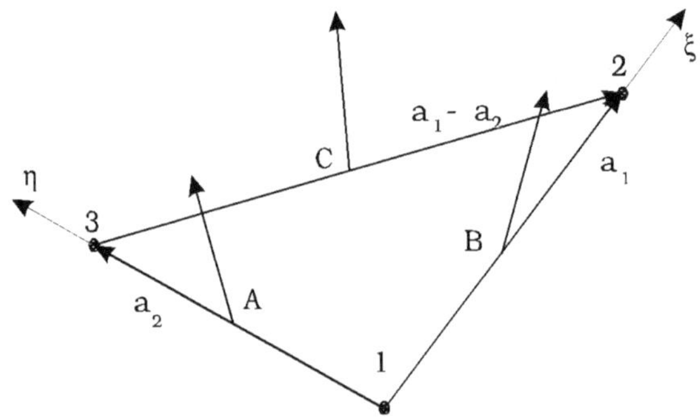

FIGURE 2. Élément TCSS.

5.3. Cisaillement. Pour éviter le blocage en cisaillement, nous avons utilisé la méthode des déformations de substitution ANS (Assumed Natural Strain). La mesure de déformation en cisaillement est obtenue en se basant sur la formulation de Boisse *et al.* [**B4**] et en introduisant l'intégration réduite en un seul point. L'élément obtenu passe les patch-tests mais présente un mode parasite qui n'a pas d'importance puisqu'il disparaît lors d'un assemblage de deux ou plusieurs éléments. Cette mesure de déformation de cisaillement transverse est définie de la façon suivante :

$$(61)\qquad \begin{cases} \overline{\gamma}_1 = \frac{1}{3}\{\mathbf{a}_1(2\mathbf{d}_B + \mathbf{d}_C) + \mathbf{a}_2(\mathbf{d}_A - \mathbf{d}_C) - \mathbf{A}_1(2\mathbf{D}_B + \mathbf{D}_C) \\ \qquad\qquad -\mathbf{A}_2(\mathbf{D}_A - \mathbf{D}_C)\}, \\ \overline{\gamma}_2 = \frac{1}{3}\{\mathbf{a}_1(\mathbf{d}_B - \mathbf{d}_C) + \mathbf{a}_2(2\mathbf{d}_A + \mathbf{d}_C) - \mathbf{A}_1(\mathbf{D}_B - \mathbf{D}_C) \\ \qquad\qquad -\mathbf{A}_2(2\mathbf{D}_A + \mathbf{D}_C)\}, \end{cases}$$

où les indices A, B et C représentent les milieux des arêtes de la frontière de l'élément présenté à la figure 2. Selon la mesure (61), les déformations virtuelles de cisaillement s'écrivent alors :

$$(62)\qquad \delta\overline{\gamma} = \begin{Bmatrix} \delta\overline{\gamma}_1, \\ \delta\overline{\gamma}_2 \end{Bmatrix} = \mathbf{B}_\xi \boldsymbol{\delta\Phi},$$

où la matrice $\mathbf{B}_\xi$, qui remplace $\mathbf{B}_{cu}$ et $\mathbf{B}_{cr}$ dans (50), est donnée par :

$$(63)\qquad \mathbf{B}_\xi = \frac{1}{3}\begin{bmatrix} -(2\mathbf{d}_B + \mathbf{d}_A)^T & (\mathbf{a}_1 + \frac{1}{2}\mathbf{a}_2)^T\widetilde{\mathbf{Q}}_1 & (2\mathbf{d}_B + \mathbf{d}_C)^T \\ -(2\mathbf{d}_A + \mathbf{d}_B)^T & (\mathbf{a}_2 + \frac{1}{2}\mathbf{a}_1)^T\widetilde{\mathbf{Q}}_1 & (\mathbf{d}_B + \mathbf{d}_C)^T \end{bmatrix}, \\ \begin{bmatrix} \frac{1}{2}(3 - \mathbf{a}_2)^T\widetilde{\mathbf{Q}}_2 & (\mathbf{d}_A + \mathbf{d}_C)^T & \frac{1}{2}(\mathbf{a}_1)^T\widetilde{\mathbf{Q}}_3 \\ \frac{1}{2}(\mathbf{a}_2)^T\widetilde{\mathbf{Q}}_2 & (2\mathbf{d}_A + \mathbf{d}_C)^T & (3\mathbf{a}_2 - \mathbf{a}_1)^T\widetilde{\mathbf{Q}}_3 \end{bmatrix}.$$

La contribution de la partie cisaillement au résidu total est donnée par :

$$(64)\qquad \mathbf{R}_c = \int_A \mathbf{B}_\xi^T\mathbf{T}\,dA = \frac{\sqrt{A}}{2}\mathbf{B}_\xi^T\mathbf{T}.$$

La matrice tangente géométrique en cisaillement relative au couple de nœuds (i,j) est donnée par :

$$(65)\qquad \mathbf{G}K_c^{ij} = \frac{\sqrt{A}}{6}\begin{bmatrix} \mathbf{0} & cu_{ij}\widetilde{\mathbf{Q}}_j \\ cu_{ji}\widetilde{\mathbf{Q}}_i^T & cr_{ij}\mathbf{I} \end{bmatrix},$$

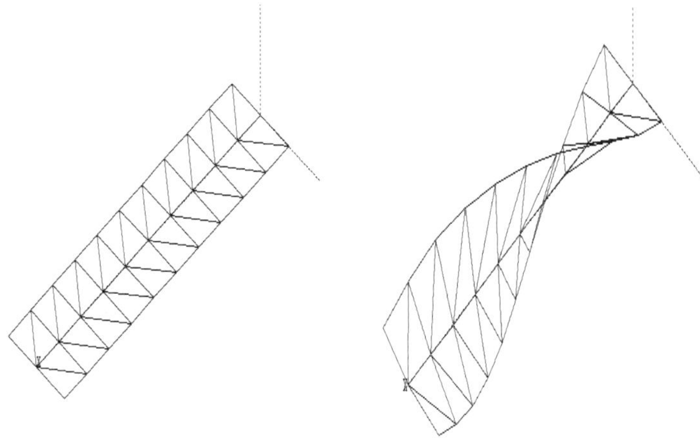

Configuration initiale Déformée réelle

FIGURE 3. Poutre sous moment de torsion.

où les termes cu_{ij} peuvent être regroupés pour former la matrice suivante :

$$[cu] = \begin{bmatrix} -\alpha-\beta & -\beta & -\alpha \\ \beta & \beta-\gamma & -\gamma \\ \alpha & \gamma & \alpha+\gamma \end{bmatrix}, \tag{66}$$

où les constantes α, β et γ sont définis par :

$$\alpha = T_2 + 0.5T_1; \quad \beta = T_1 + 0.5T_2; \quad \gamma = 0.5\,(T_2 - T_1). \tag{67}$$

Les termes cr_{ij} de l'équation (66) sont nuls pour $i \neq j$. Les trois autres termes non-nuls (cr_{ii}) peuvent former le vecteur colonne suivant :

$$\{cr\} = -\left\{ \begin{array}{c} (\beta\mathbf{a}_1 + \alpha\mathbf{a}_2)\cdot\mathbf{d}_1 \\ \big((\beta-\gamma)\mathbf{a}_1 + \gamma\mathbf{a}_2\big)\cdot\mathbf{d}_2 \\ \big(-\gamma\mathbf{a}_1 + (\alpha+\gamma)\mathbf{a}_2\big)\cdot\mathbf{d}_3 \end{array} \right\}. \tag{68}$$

6. Résultats numériques

Les performances de la formulation numérique proposée sont illustrées à travers quelques exemples numériques incluant les grandes rotations et la plasticité.

6.1. Poutre sous moment de torsion. Le but de ce test est de démontrer que la formulation proposée peut capter des grandes rotations et grands déplacements. Ce test consiste en une poutre encastrée d'un côté et soumise à un couple de torsion selon son axe à son extrémité libre l'entraînant à un angle de 180 degrés. Ce test a été traité dans Simo *et al.* [**S2**] et dans Parisch [**P1**] avec un élément quadrilatère à quatre nœuds. La poutre est de longueur $L = 1.0$, de largeur $b = 0.25$ et d'épaisseur $h = 0.1$. Les propriétés matérielles son : $E = 12 \times 10^6$ et $\nu = 0.3$. La poutre est discrétisée avec 2×10 quadrilatères soit un total de 40 triangles. Tout comme dans Simo *et al.* [**S2**] et dans Parisch [**P1**], la configuration finale (figure 3) est atteinte en trois pas de chargement.

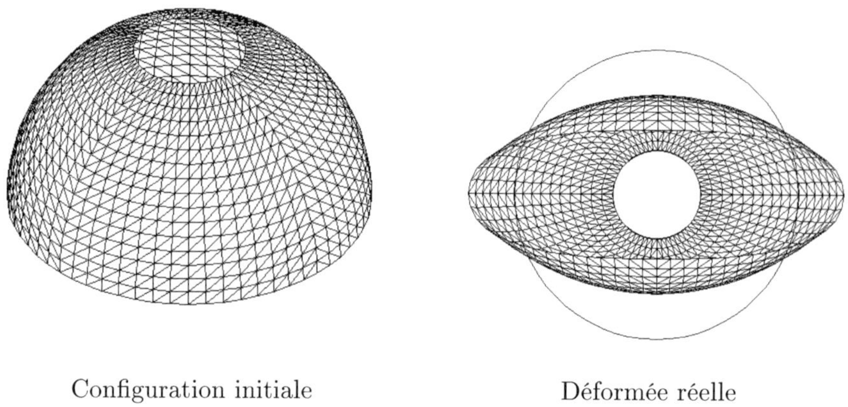

FIGURE 4. Sphère trouée.

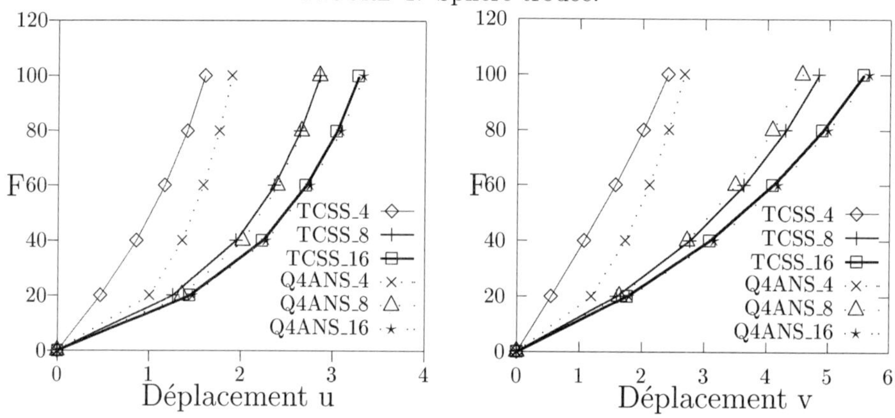

FIGURE 5. Sphère trouée. Résultats.

6.2. Sphère trouée. Ce problème de la sphère trouée est devenu récemment un test standard pour l'évaluation des éléments de coques avec non-linéarité géométrique. L'objectif de ce test, qui a été traité dans plusieurs travaux [**D1, P1, S2**], est de vérifier l'aptitude des différents éléments à représenter correctement les mouvements de corps rigide (rotations et translations) qui sont en abondance dans ce test. Les propriétés géométriques et matérielles de la sphère sont données dans les références suivantes [**D1, P1, S2**]. Trois maillages réguliers du huitième de la sphère ont été considérés : 4×4, 8×8 et 16×16. La charge totale de $f = 100$ est appliquée en cinq pas égaux. Les courbes charge-déplacement aux deux points d'application de la charge sont données à la figure 4. Nos résultats, obtenus avec le TCSS sont comparés avec ceux obtenus avec l'élément Q4ANS dont la formulation est basée sur le modèle de cisaillement de Bathe et Dvorkin [**B1**].

Selon la figure 5, nous remarquons que seuls les résultats obtenus avec le maillage 4×4 ne sont pas bons. Avec le maillage 8×8 et 16×16 les résultats obtenus avec l'élément TCSS et Q4ANS sont presque identiques. Le nombre total d'itérations est aussi identique pour les deux éléments et les deux derniers maillages

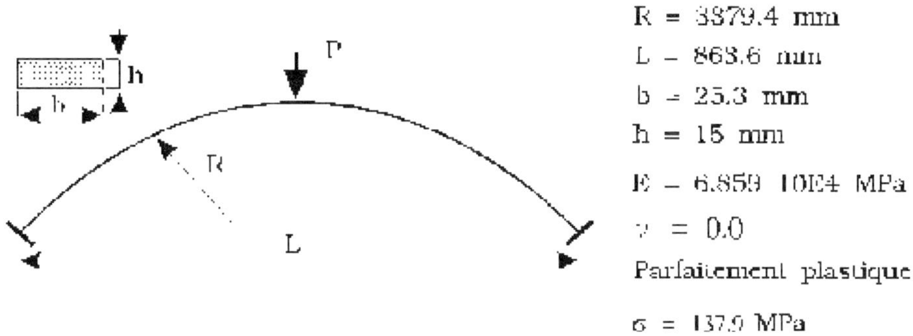

FIGURE 6. Arc mince. Description.

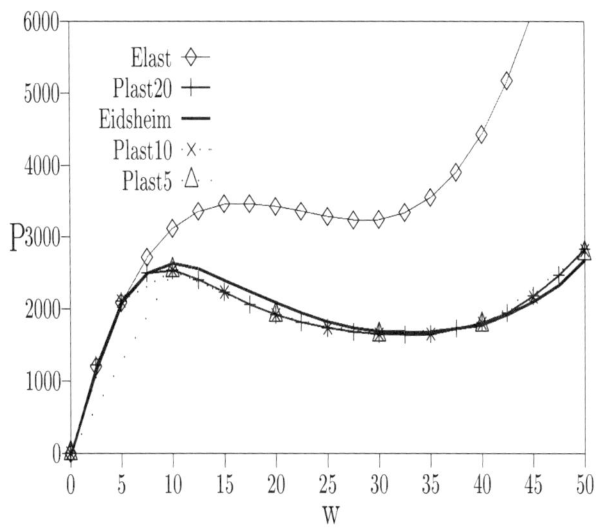

FIGURE 7. Arc mince. Description.

(44 itérations). Cependant, une nette réduction du temps de calcul élémentaire est observée pour l'élément TCSS.

6.3. Arc mince encastré soumis à une charge concentrée. Dans ce test, les deux types de non-linéarités géométriques et matérielles sont considérés. Les caractéristiques géométriques et matérielles de l'arc mince sont données à la figure 6. Étant donné la symétrie du problème, la moitié de l'arc est discrétisé avec un maillage de 20 éléments TCSS (10 quadrilatères). Une intégration de 7 points de Gauss est utilisée sur l'épaisseur. La méthode de Newton Raphson avec pilotage en déplacement est utilisée avec une erreur de 10^{-10} sur l'énergie. Un déplacement total de 50 mm est imposé au centre de l'arc respectivement en 5, 10, et 20 pas. Une comparaison de nos résultats avec ceux obtenu par Eidsheim et Larsen [**E1**] est donnée à la figure 7. Nos résultats sont très peu sensibles aux variations du nombre de pas et sont en accord avec ceux d'Eidsheim et Larsen.

7. Conclusion

Un élément triangulaire non-linéaire simple et efficace a été développé. Puisque la formulation de cet élément suppose un champ de déformation de cisaillement constant sur l'élément, un seul point d'intégration est suffisant pour l'intégration exacte de la matrice tangente et du résidu. Cet élément à un mode parasite du à la distribution constante de cisaillement. Cependant, ce mode parasite disparaît après assemblage de deux ou plusieurs éléments. Cet élément à trois nœuds peut présenter une grande importance dans le domaine de la mise en forme des métaux surtout en présence d'une procédure de maillage adaptative basée sur un estimateur d'erreur adéquat.

Références

[A1] R. Abraham and J. E. Marsden, *Foundation of mechanics*, 2nd ed., Addison-Wesley, 1978.

[B1] K. J. Bathe and E. Dorkin, *A four-node plate bending element based on Mindlin/Reissner plate theory and a mixed interpolation*, Internat. J. Numer. Methods Engrg. **21** (1985), 367–383.

[B2] J. L. Batoz, K. J. Bathe, and L. W. Ho, *A study of three-node plate bending elements*, Int. J. Numer. Methods Engrg. **15** (1980), 1771–1812.

[B3] T. Belytschko, H. Stolarski, and N. Carpenter, *A C^0 triangular plate element with one point quadrature*, Int. J. Numer. Methods Engrg. **20** (1984), 787–802.

[B4] P. Boisse, J. L. Daniel, and J. C. Gelin, *A simple isoparametric three-node shell finite element*, Comput. & Structures **44** (1992), 12633–1273.

[C1] A. Cardona and M. Geradin, *A beam finite element nonlinear theory with finite rotations*, Int. J. Numer. Methods Engrg. **26** (1988), 2403–2438.

[D1] F. Dammak, *Formulation isoparamétrique généralisée en analyse linéaire et non-linéaire des coques par éléments finis*, Ph.D. thesis, Univ. Laval, Québec, Canada , 1996.

[D2] F. Dammak, S. Chamlal, A. Gakwaya, and G. Dhatt, *Analyse non linéaire avec grandes rotations des coques composites multicouches*, CSME (Montréal, 1994), vol. I, 1994, pp. 75–86.

[E1] O. M. Eidsheim and P. K. Larsen, *Nonlinear analysis of elasto-plastic shells by hybrid stress finite elements*, Comput. Methods Appl. Mech. Engrg. **34** (1982), 989–1018.

[H1] T. J. R. Hughes, *The finite element method: linear static and dynamic finite element analysis*, Prentice Hall, 1987.

[P1] H. Parisch, *An investigation of a finite rotation four node assumed strain shell element*, Internat. J. Numer. Methods Engrg. **31** (1991), 127–150.

[S1] J. C. Simo and D. D. Fox, *On a stress resultant geometrically exact shell model.* I. *Formulation and optimal parametrisation*, Comput. Methods Appl. Mech. Engrg. **72** (1989), 267–304.

[S2] J. C. Simo, M. S. Rifai, and D. D. Fox, *On a stress resultant geometrically exact shell model.* III. *Computational aspects of the nonlinear theory*, Comput. Methods Appl. Mech. Engrg. **79** (1990), 21–70.

[S3] J. C. Simo and R. L. Taylor, *A return mapping algorithm for plane stress elastoplasticity*, Internat. J. Numer. Methods Engrg. **22** (1986), 649–670.

Département de Génie Mécanique , Université Laval, Ste Foy (Qc), G1K 7P4, Canada

E-mail address, F. Dammak: `fdammak@gmc.ulaval.ca`

E-mail address, A. Gakwaya: `Augustin.Gakwaya@gmc.ulaval.ca`

Département de Génie Mécanique, INSA de ROUEN, Place Emile Blondel B.P 8, 76131 Mont Saint Aignan Cedex, France

E-mail address: `Gouri.Dhatt@utc.fr`

Centre de Recherches Mathématiques
CRM Proceedings and Lecture Notes
Volume **21**, 1999

Intrinsic Nonlinear Models of Shells

Michel C. Delfour and Jiabin Zhao

ABSTRACT. This paper is devoted to the analysis of nonlinear shell problems using the same approximation as for the intrinsic linear model considered in [**13**], that is a $P(1,2)$ approximation in the terminology of [**14, 15**]. First, using the tangential differential calculus and the oriented boundary distance function, we derive two intrinsic nonlinear models for thin shells made up of a Saint Venant-Kirchhoff material or a homogeneous, isotropic, elastic material. Next, we prove the differentiability of the nonlinear operators associated with the nonlinear models in the associated spaces. Finally, we prove existence and uniqueness theorems for the solution of the two nonlinear models using the inverse function theorem. This analysis clearly applies to the $P(1,1)$ approximation in [**15**] and very likely extends to richer nonlinear models associated with the linear $P(2_n,1)$ or $P(2,1)$ models in [**15, 8**].

1. Introduction

Research in shell theory has currently been stimulated by questions arising from a number of applications in control and design of large space structures, flexible robots, composite materials, etc, where intrinsic and mathematically more tractable models would be much preferred. Delfour and Zolésio introduced in [**12, 13**] a new way to deal with the differential geometry, to express tangential differential operators, and to do the differential calculus on submanifolds of the n-dimensional Euclidean space. Their early natural and Love-Kirchhoff theories of static and dynamical linear thin shells in [**12, 13**] correspond to a $P(1,2)$ approximation in the terminology of [**14, 15**]. They further refined and expanded their techniques to study $P(1,1)$, $P(2_n,1)$ and $P(2,1)$ approximations in [**14, 15, 8**]. For a recent discussion of their relative merits the reader is referred to [**8**]. All those intrinsic linear models are expressed in terms of intrinsic tangential differential operators such as the tangential gradient, divergence, strain tensor. They also show in [**14**] that the classical linear models of Naghdi and Koiter can be rewritten in terms of these intrinsic tangential differential operators.

1991 *Mathematics Subject Classification.* Primary: 73K15, 35J55; Secondary: 49Q10, 73C50.
This research has been supported in part by National Sciences and Engineering Research Council of Canada research grant A-8730 and by a FCAR grant from the Ministère de l'Éducation du Québec.
This is the final form of the paper.

The main objective of this paper is to use this approach to extend the $P(1, 2)$ approximation of linear thin shells in [**12, 13**] to a large class of nonlinear shells made up of a Saint Venant-Kirchhoff material or a homogeneous, isotropic, elastic material. This yields two intrinsic second order nonlinear tangential differential systems which are significantly different from more traditional models of thin shells. This is one of the main results of this paper. We prove that the nonlinear operators related to these systems are differentiable. A direct calculation shows that the derivatives of the nonlinear operators, i.e. their linear part, coincide with the linear operator in [**13**]. Using the regularity result of the weak solutions of the linear $P(1, 2)$ model and [**9**], which implies that the derivative of the nonlinear operator is an isomorphism between the associated function spaces, the existence and uniqueness of solutions of the nonlinear models follow from the inverse function theorem.

Few authors have studied questions of existence for nonlinear models of shells, largely because of the geometrically nonlinear behavior of shells. We mention, for example, Bernadou and Oden [**4**] for Koiter's models (one may find more references in this paper) and Figueiredo [**19**] for Donnell-Mushtari-Vlasov's models. The pseudo-monotone operators are the main tools in [**4**], and the implicit function theorem is used in [**19**]. For the fully nonlinear case, We mention that Le Dret and Raoult [**24**] (one may find more references in this paper) study membrane shell model of superelastic materials, and use a variational asymptotic method.

Again the main advantage of the approach described in this paper is that it avoids the necessity of introducing local coordinates tailored to the geometry of each individual shell and permits the development of a theory that parallels, to a large extent, the much better understood theory of thin plates (cf. for instance Germain [**17**], Dautray and Lions [**7**]). Further, when the equilibrium equations are written in terms of quantities intrinsic to the mean surface, their basic structure is much more apparent, and the equations easier to analyze, than equations written in terms of local coordinate system and Christoffel symbols.

2. Linear Theory

We use the notation, definitions, some technical results about the tangential differential calculus and the oriented boundary distance functions from [**13, 14**]. We recall the $P(1, 2)$ linear model and the existence and uniqueness result and regularity result of the weak solutions in [**13**].

The inner product in $\mathbb{R}^3$ and the double inner product in $\mathcal{L}(\mathbb{R}^3; \mathbb{R}^3)$ (space of 3×3 matrices or tensors) are denoted as

$$x \cdot y = \sum_{i=1}^{3} x_i\, y_i, \quad A \cdot\cdot B = \sum_{i=1}^{3} \sum_{j=1}^{3} A_{ij}\, B_{ij}.$$

${}^*M$ denotes the transpose of of an arbitrary matrix M.

The basic constructions are as follows. The submanifold Γ of codimension one is specified as the boundary of a subset Ω of $\mathbb{R}^3$. It is assumed that the *midsurface* ω is a bounded (relatively) open subset of Γ and that Γ is of class $C^{1,1}$ in some neighborhood of ω. This is equivalent to say that the algebraic distance function $b = b_\Omega$ of Ω is $C^{1,1}$ in that neighborhood. Its gradient ∇b coincides with the unit exterior normal n to Γ on ω and its Hessian matrix $D^2 b$ to the second fundamental form. The *projection* p onto Γ is given as

$$p(x) \stackrel{\text{def}}{=} x - b(x)\,\nabla b(x).$$

The *shell of thickness* $2h$ is defined as the bounded open domain

$$S_h(\omega) \stackrel{\text{def}}{=} \left\{x \in \mathbb{R}^3 \,:\, |b(x)| < h \text{ and } p(x) \in \omega\right\}.$$

Further assume that the (relative) boundary γ of ω is non-empty and that the *lateral boundary* of $S_h(\omega)$,

$$\Sigma_h(\omega) \stackrel{\text{def}}{=} \left\{x \in \mathbb{R}^3 \,:\, |b(x)| < h \text{ and } p(x) \in \gamma\right\},$$

is Lipschitzian.

2.1. Linear $P(1,2)$ model. The first assumption for the $P(1,2)$ model is that the three-dimensional *displacement vector* is approximated by a first order development with respect to the local normal coordinate to the mean surface (cf. for instance Germain [**17**], Dautray and Lions [**7**]). Delfour and Zolésio [**13**] observed that this is equivalent to a first order development with respect to $b(x)$. For the techniques of tangential calculus and integration by parts on ω, see [**14**].

ASSUMPTION 1. At each point x of the shell $S_h = S_h(\omega)$ the *approximate displacement vector* $V(x)$ is of the form

$$V(x) = e \circ p(x) + b(x)\ell \circ p(x), \quad x \in S_h, \tag{1}$$

for vector-valued mappings e and ℓ from ω to $\mathbb{R}^3$.

REMARK 1. In [**13**], Delfour and Zolésio introduced another assumption by adding the condition $\ell(X) \cdot n(X) = 0$ on ω in Assumption 1. We may add this condition later, and get the same result by introducing appropriate spaces (cf. [**9**]) since the proof of the generalized Korn's inequality [**9, 13, 14**] does not use the fact that ℓ is tangent.

For the *strain tensor* $\varepsilon(V)$ defined by

$$2\varepsilon(V) = D(V) + {}^*D(V),$$

we have, in the (X,z) coordinates system

$$\varepsilon(V) \circ T_z = \sum_{i=0}^{\infty} \varepsilon^i(e,\ell) z^i$$

where, for 0, 1, 2 and $i > 2$

$$\begin{aligned}
2\varepsilon^0(e,\ell) &= D_\Gamma e + \ell^* n + {}^*D_\Gamma e + n^*\ell,\\
2\varepsilon^1(e,\ell) &= D_\Gamma \ell - D_\Gamma e D^2 b + {}^*D_\Gamma \ell - D^2 b^* D_\Gamma e,\\
2\varepsilon^2(e,\ell) &= (D_\Gamma \ell - D_\Gamma e D^2 b)(-D^2 b) + (-D^2 b)({}^*D_\Gamma \ell - D^2 b^* D_\Gamma e)\\
2\varepsilon^i(e,\ell) &= (D_\Gamma \ell - D_\Gamma e D^2 b)(-D^2 b)^{i-1} + (-D^2 b)^{i-1}({}^*D_\Gamma \ell - D^2 b^* D_\Gamma e).
\end{aligned}$$

The second assumption for the $P(1,2)$ model is that

ASSUMPTION 2. There exists β, $0 \le \beta < 1$, such that

$$\forall z,\ |z| \le h, \text{ and } \forall x \in \omega, \quad \left\|z D^2 b(x)\right\| \le \beta \tag{2}$$

and the *approximate strain tensor* is chosen as

$$\tilde{\varepsilon}(V) \circ T_z \stackrel{\text{def}}{=} \varepsilon^0 + \varepsilon^1 z + \varepsilon^2 z^2. \tag{3}$$

The first part of the assumption says that the dimensionless quantity $\|zD^2b\|$ is small compared to one or equivalently that the shell is either thin (h small) or shallow ($\|D^2b\|$ small) or both. Note that the notation $P(1,2)$ is in line with the terminology later introduced in [**14, 15, 8**]. The $P(1,2)$ model is not to be confused with the $P(2,1)$ model.

The associated natural function spaces are

$$\mathcal{H} = L^2(\omega)^3 \times L^2(\omega)^3, \quad \mathcal{V} = \left\{(e,\ell) \in \mathcal{H} \colon \varepsilon^i \in L^2(\omega)^{3\times 3},\ 0 \le i \le 2\right\}$$

and from a generalized Korn's inequality (cf. Theorem 7.2 in [**13**])

$$\mathcal{V} = H^1(\omega)^3 \times H^1(\omega)^3.$$

Under the following *constitutive law* for isotropic materials

$$\sigma = \lambda \operatorname{tr} \tilde{\varepsilon}(V) I + 2\mu \tilde{\varepsilon}(V),$$

where $\lambda \ge 0$ and $\mu > 0$ are the Lamé's coefficients, we have the strong form of the equation of shell

$$\begin{cases} -\overrightarrow{\operatorname{div}}_\Gamma L_1 + HL_1 n = -\alpha_0 f \quad \text{on } \omega \\ -\overrightarrow{\operatorname{div}}_\Gamma L_2 + HL_2 n + L_3 = -\alpha_0 m - \alpha_1 f \quad \text{on } \omega \end{cases} \tag{4}$$

$$e = \ell = 0 \quad \text{on } \gamma, \tag{5}$$

where

$$\alpha_m(X) = \int_{-h}^{h} j(z)(X) z^m \, dz, \quad X \in \omega,\ 0 \le m \le 4,$$

$$L_1 = A^0 - A^1 D^2 b + A^2 (D^2 b)^2, \quad L_2 = A^1 - A^2 D^2 b, \quad L_3 = A^0 n$$

$$A^i(e,\ell) = 2\mu \sum_{j=0}^{2} \alpha_{i+j} \varepsilon^j(e,\ell) + \lambda \operatorname{tr} \sum_{j=0}^{2} \alpha_{i+j} \varepsilon^j(e,\ell) I$$

2.2. Existence and smoothness of solutions. It is convenient to rewrite system (4)–(5) in the form: to find vector fields $(e,\ell)\colon \omega \to \mathbb{R}^3 \times \mathbb{R}^3$ that verify

$$\begin{cases} \mathcal{L}(e,\ell) = \vec{G} \quad \text{in } \omega \\ (e,\ell) = 0 \quad \text{on } \gamma \end{cases} \tag{6}$$

where

$$\mathcal{L}(e,\ell) \stackrel{\text{def}}{=} (-\overrightarrow{\operatorname{div}}_\Gamma L_1 + HL_1 n, -\overrightarrow{\operatorname{div}}_\Gamma L_2 + HL_2 n + L_3)$$

and

$$\vec{G} \stackrel{\text{def}}{=} (-\alpha_0 f, -\alpha_0 m - \alpha_1 f).$$

The following theorem is Theorem 4.3 in [**13**].

THEOREM 1. *There exists $\bar{h} > 0$, such that for all h, $0 < h < \bar{h}$, there exists a unique weak solution $(e,\ell) \in \mathcal{V}_0(\omega) = H_0^1(\omega)^3 \times H_0^1(\omega)^3$ to the boundary problem* (4) *and* (5).

The following theorem from [**9**] is about regularity property of the weak solutions of system (4)–(5). It is essential for the analysis of nonlinear shell problem. There are few results about regularity property of the weak solutions for linear models of shells. We mention that, for example, H^2 regularity property of the weak solutions for linear Donnell-Mushtari-Vlasov's models is given by Figueiredo [**19**] by using Guseva's results [**22**]. It will be useful to introduce the following subspaces for an arbitrary $p \geq 1$

$$\mathcal{W}^p(\omega) = \{(e,\ell) \in W^{2,p}(\omega)^3 \times W^{2,p}(\omega)^3 \colon (e,\ell) \in \mathcal{V}_0(\omega)\}$$
$$\mathcal{L}^p(\omega) = L^p(\omega)^3 \times L^p(\omega)^3.$$

THEOREM 2. *Let Ω be a domain in $\mathbb{R}^3$ with a boundary Γ which is a $C^{2,1}$ submanifold of $\mathbb{R}^3$, let ω be a $C^{1,1}$ domain in the submanifold Γ, and $f, m \in L^p(\omega)^3$, $p \geq 2$. Then the weak solution $(e,\ell) \in \mathcal{V}_0(\omega)$ of equations* (4) *belongs to the space $\mathcal{W}^p(\omega)$, and satisfies* (4) *in $\mathcal{L}^p(\omega)$.*

REMARK 2. By the last theorem, we get that the problem (6) given in term of the operator $\mathcal{L}$ has one and only one solution $(e,\ell) \in \mathcal{W}^p(\omega)$ for any $(m,f) \in \mathcal{L}^p(\omega)$.

3. Nonlinear Models for Saint Venant-Kirchhoff Material

For simplicity we assume that the shell S_h has constant thickness. As in the linear case Assumptions 1 and an extension of Assumptions 2 will be the basic assumptions for the nonlinear case.

3.1. Strain and stress tensors. Under Assumption 1, consider the *Green Saint Venant strain tensor* $\varepsilon(V)$ over S_h associated with the displacement field $V(x)$ (cf. Ciarlet [**5**], Marsden and Hughes [**25**]),

$$2\varepsilon(V) = DV + {}^*DV + {}^*DV\,DV. \tag{7}$$

In terms of the coordinates (X, z),

$$DV \circ T_z = \sum_{i=0}^{\infty} v^i(e,\ell) z^i$$

where

$$v^0(e,\ell) = D_\Gamma e + \ell^* n \text{ and for } i \geq 1, \quad v^i(e,\ell) = [D_\Gamma \ell - D_\Gamma e D^2 b](-D^2 b)^{i-1}.$$

Then, we have

$$2\varepsilon(V) \circ T_z = \varepsilon^0 + \eta^0 + (\varepsilon^1 + \eta^1)z + (\varepsilon^2 + \eta^2)z^2 + O(z^3), \tag{8}$$

where, for $i = 0, 1, 2$

$$2\varepsilon^i(e,\ell) \stackrel{\text{def}}{=} v^i(e,\ell) + {}^*v^i(e,\ell),$$

and

$$\begin{aligned}
2\eta^0(e,\ell) &= {}^*v^0 v^0 = {}^*[D_\Gamma e + \ell^* n][D_\Gamma e + \ell^* n],\\
2\eta^1(e,\ell) &= {}^*v^0 v^1 + {}^*v^1 v^0\\
&= {}^*[D_\Gamma e + \ell^* n][D_\Gamma \ell - D_\Gamma e D^2 b] + {}^*[D_\Gamma \ell - D_\Gamma e D^2 b][D_\Gamma e + \ell^* n],\\
2\eta^2(e,\ell) &= {}^*v^0 v^2 + {}^*v^1 v^1 + + {}^*v^2 v^0\\
&= -{}^*[D_\Gamma e + \ell^* n][D_\Gamma \ell - D_\Gamma e D^2 b] D^2 b\\
&\qquad + {}^*[D_\Gamma \ell - D_\Gamma e D^2 b][D_\Gamma \ell - D_\Gamma e D^2 b]\\
&\qquad\qquad - D^2 b^* [D_\Gamma \ell - D_\Gamma e D^2 b][D_\Gamma e + \ell^* n].
\end{aligned}$$

Comparing with Assumption 2, here we introduce the following approximation of the strain tensor.

ASSUMPTION 3. There exists β, $0 \le \beta < 1$, such that

$$\forall z,\ |z| \le h, \text{ and } x \in \omega, \quad \|zD^2b(x)\| \le \beta \tag{9}$$

and the *approximate Green Saint Venant strain tensor* is chosen as

$$\tilde{\varepsilon}(V) \circ T_z = \varepsilon^0 + \eta^0 + (\varepsilon^1 + \eta^1)z + (\varepsilon^2 + \eta^2)z^2. \tag{10}$$

We assume that S_h is made up of a Saint Venant-Kirchhoff material which obeys the following constitutive law (cf. Marsden and Hughes [**25**] and Ciarlet [**5**]),

$$\sigma = \lambda \operatorname{tr} \tilde{\varepsilon}(V) I + 2\mu \tilde{\varepsilon}(V), \tag{11}$$

where $\lambda \ge 0$ and $\mu > 0$ are the Lamé coefficients. Hence

$$\begin{aligned}\sigma \circ T_z \cdot\cdot\, \tilde{\varepsilon}(V) \circ T_z &= \left[\lambda \operatorname{tr} \tilde{\varepsilon}(V) \circ T_z I + 2\mu \tilde{\varepsilon}(V) \circ T_z\right] \cdot\cdot\, \tilde{\varepsilon}(V) \circ T_z \\ &= \lambda \left|\operatorname{tr}\, \tilde{\varepsilon}(V) \circ T_z\right|^2 + 2\mu \left\|\tilde{\varepsilon}(V) \circ T_z\right\|^2.\end{aligned}$$

3.2. Strain energy and nonlinear equation. We now follow the standard approach and compute the total energy made up of the strain energy plus the work of the external forces. External forces mean forces and torques applied to the mean surface of the shell. The shell equation corresponds to a stationary point of the total energy functional.

Define the *strain energy* $\mathcal{P}$ as follow:

$$\mathcal{P} = \frac{1}{2} \int_{\Omega_h} \sigma \cdot\cdot\, \tilde{\varepsilon}(V)\, dx. \tag{12}$$

Using Federer's decomposition of the measure (cf. Evans and Gariepy [**16**]) along the level curves of the oriented distance function b for which $\left|\nabla b(x)\right| = 1$ in S_h

$$\begin{aligned}\mathcal{P} &= \frac{1}{2} \int_{\Omega_h} \sigma \cdot\cdot\, \tilde{\varepsilon}(V)\, dx \\ &= \frac{1}{2} \int_\omega \int_{-h}^{h} dz\, d\Gamma \left[\lambda \left|\operatorname{tr} \tilde{\varepsilon}(V) \circ T_z\right|^2 + 2\mu \left\|\tilde{\varepsilon}(V) \circ T_z\right\|^2\right] j(z) \\ &= \frac{1}{2} \int_\omega \sum_{i=0}^{2} B^i(e, \ell) \cdot\cdot \left[\varepsilon^i(e, \ell) + \eta^i(e, \ell)\right] d\Gamma,\end{aligned}$$

where

$$B^i(e, \ell) = 2\mu \sum_{j=0}^{2} \alpha_{i+j} \left(\varepsilon^j(e, \ell) + \eta^j(e, \ell)\right) + \lambda \operatorname{tr} \sum_{j=0}^{2} \alpha_{i+j} \left(\varepsilon^j(e, \ell) + \eta^j(e, \ell)\right) I$$

are symmetrical tensors. From [**13**], we know the work $\mathcal{W}$ of the external forces and torques are in the form

$$\begin{aligned}\mathcal{W} &= \int_{-h}^{h} dz \int_\omega d\Gamma \left[F \circ T_z \cdot V \circ T_z + M \circ T_z \cdot (\ell \circ p \circ T_z)\right] j(z) \\ &= \int_\omega \left[\alpha_0 (f \cdot e + m \cdot \ell) + \alpha_1 f \cdot \ell\right] d\Gamma.\end{aligned}$$

Then the total energy is

$$\begin{aligned}\mathcal{E} &= \mathcal{P} + \mathcal{W} \\ &= \frac{1}{2} \int_\omega \sum_{i=0}^{2} B^i(e, \ell) \cdot\cdot \left[\varepsilon^i(e, \ell) + \eta^i(e, \ell)\right] d\Gamma + \int_\omega \left[\alpha_0 (f \cdot e + m \cdot \ell) + \alpha_1 f \cdot \ell\right] d\Gamma.\end{aligned}$$

The *variational principle* may now be used to obtain the equations of the displacement. A straightforward calculation of the directional derivative $d\mathcal{P}(e,\ell;\bar{e},\bar{\ell})$ at (e,ℓ) in the direction $(\bar{e},\bar{\ell})$ gives

$$d\mathcal{P}(e,\ell;\bar{e},\bar{\ell}) = \int_\omega \sum_{i=0}^{2} B^i(e,\ell)\cdot\cdot\left[\varepsilon^i(\bar{e},\bar{\ell}) + d\eta^i(e,\ell;\bar{e},\bar{\ell})\right] d\Gamma. \tag{13}$$

where

$$2d\eta^i(e,\ell;\bar{e},\bar{\ell}) = \sum_{j=0}^{i} {}^*v^j\bar{v}^{i-j} + {}^*\bar{v}^j v^{i-j}$$
$$2\varepsilon^i(\bar{e},\bar{\ell}) = \bar{v}^i + {}^*\bar{v}^i.$$

we can rearrange again to isolate $\bar{l}$, $D_\Gamma\bar{e}$, and $D_\Gamma\bar{l}$

$$d\mathcal{P}(e,\ell;\bar{e},\bar{\ell}) = \int_\omega [N_1\cdot\cdot D_\Gamma\bar{e} + N_2\cdot\cdot D_\Gamma\bar{\ell} + N_3\cdot\bar{\ell}]\, d\Gamma,$$

where

$$N_1 = \sum_{i=0}^{2}\left[B^i + \sum_{j=i}^{2} v^{j-i}B^j\right](-D^2b)^i$$
$$N_2 = \sum_{i=1}^{2}\left[B^i + \sum_{j=i}^{2} v^{j-i}B^j\right](-D^2b)^{i-1}$$
$$N_3 = \left[B^0 + \sum_{j=0}^{2} v^j B^j\right]n.$$

This yields the following variational equation

$$d\mathcal{P}(e,\ell;\bar{e},\bar{\ell}) + d\mathcal{W}(e,\ell;\bar{e},\bar{\ell}) = \int_\gamma g\cdot\bar{e} + q\cdot\bar{\ell}\, ds$$

for some appropriate vector functions $g\colon \gamma\to\mathbb{R}^3$ and $q\colon\gamma\to\mathbb{R}^3$. From this we get the following nonlinear boundary problem for the shells

$$\begin{cases} -\overrightarrow{\mathrm{div}}_\Gamma N_1 + HN_1 n = -\alpha_0 f \quad \text{on } \omega \\ -\overrightarrow{\mathrm{div}}_\Gamma N_2 + HN_2 n + N_3 = -\alpha_0 m - \alpha_1 f \quad \text{on } \omega \end{cases} \tag{14}$$

$$\begin{cases} N_1\nu = g \quad \text{on } \gamma_1 \\ N_2\nu = q \quad \text{on } \gamma_1 \\ \ell = e = 0 \quad \text{on } \gamma_2 \end{cases} \tag{15}$$

where $\gamma = \gamma_1\cup\gamma_2$ and $\mathring{\gamma}_1\cap\mathring{\gamma}_2 = \emptyset$. But here we shall only consider the clamped case, that is $\gamma_1 = \emptyset$.

3.3. Existence of solutions to the nonlinear equations. We first compare the linear and the nonlinear equations in this section. For convenience, we write the nonlinear equations as follows: to find vector field $(e,\ell) : \omega\to\mathbb{R}^3\times\mathbb{R}^3$ that satisfies

$$\begin{cases} \mathcal{N}(e,\ell) = \vec{G}, \\ (e,\ell) = 0 \quad \text{on } \gamma, \end{cases} \tag{16}$$

where

$$\mathcal{N}(e,\ell) \stackrel{\text{def}}{=} (-\overrightarrow{\operatorname{div}}_\Gamma N_1 + HN_1 n, -\overrightarrow{\operatorname{div}}_\Gamma N_2 + HN_2 n + N_3)$$

and

$$\overrightarrow{G} \stackrel{\text{def}}{=} (-\alpha_0 f, -\alpha_0 m - \alpha_1 f).$$

A direct computation gives the following result, which relates the operators $\mathcal{N}$ and $\mathcal{L}$. Denote by $\mathcal{T}$ the grouping of "terms which are at least of second order". For simplicity we use this as a generic notation in different equations.

THEOREM 3.

$$\mathcal{N}(e,\ell) = \mathcal{L}(e,\ell) + \mathcal{T}(e,\ell,D_\Gamma e, D_\Gamma \ell, \overrightarrow{\operatorname{div}}_\Gamma(D_\Gamma e), \overrightarrow{\operatorname{div}}_\Gamma(D_\Gamma \ell)). \tag{17}$$

We then prove the differentiability of the nonlinear operator $\mathcal{N}$ associated with the nonlinear model. The key observation is that the Sobolev space $W^{1,p}(\omega)$ is a Banach algebra for $p > 2$ (see Adams [**1**]). As a consequence, the nonlinear operator $\mathcal{N}$ mapping the space $\mathcal{W}^p(\omega)$ into the space $\mathcal{L}^p(\omega)$ is well-defined and infinitely differentiable between these two spaces.

THEOREM 4. *Assume that $p > 2$. Then the associated nonlinear operator*

$$(e,\ell) \mapsto \mathcal{N}(e,\ell) \colon \mathcal{W}^p(\omega) \to \mathcal{L}^p(\omega) \tag{18}$$

is well-defined and differentiable, and

$$d\mathcal{N}(0,0;e,\ell) = \mathcal{L}(e,\ell) \tag{19}$$

in $\mathcal{W}^p(\omega)$.

PROOF. From the definitions, we know that N_1, N_2 and N_3 are sums of continuous linear, bilinear, and trilinear mappings of $e, \ell, D_\Gamma e$, and $D_\Gamma \ell$. Since $p > 2$, the Sobolev space $W^{1,p}(\omega)$ is a Banach algebra. Therefore the nonlinear mappings

$$(e,\ell) \mapsto \mathcal{N}(e,\ell) \colon \mathcal{W}^p(\omega) \to \mathcal{L}^p(\omega)$$

is well-defined. Moreover the mapping

$$e \mapsto D_\Gamma e \colon W^{2,\,p}(\omega) \to W^{1,p}(\omega)^{3\times 3},$$
$$\ell \mapsto D_\Gamma \ell \colon W^{2,\,p}(\omega) \to W^{1,p}(\omega)^{3\times 3},$$

are smooth (C^∞), and the mapping

$$G \mapsto \overrightarrow{\operatorname{div}}_\Gamma G \colon W^{1,p}(\omega)^{3\times 3} \to L^p(\omega)^3$$

is also smooth (C^∞). Therefore by definition of $\mathcal{N}$, we get that $\mathcal{N}(e,\ell)$ is differentiable in $\mathcal{W}^p(\omega)$. Then, from Theorem 3, a direct computation shows that

$$d\mathcal{N}(0,0;e,\ell) = \mathcal{L}(e,\ell). \quad \square$$

We are now in a position to establish the existence of solutions to the nonlinear equation. The objective is to apply the inverse function theorem (cf. for instance Hörmander [**23**]) in neighborhoods of the origins in the spaces $\mathcal{L}^p(\omega)$ and $\mathcal{W}^p(\omega)$.

THEOREM 5. *Let $\bar{h}$ be as in Theorem 1, and h, $0 < h < \bar{h}$ be fixed. Let Ω be a domain in $\mathbb{R}^3$ with a boundary Γ which is a $C^{2,1}$ submanifold of $\mathbb{R}^3$ of co-dimension one. Let ω be a C^2 domain in the submanifold Γ. Then for each $p > 2$, there exist two neighborhoods $\mathcal{L}^p_0$ and $\mathcal{W}^p_0$ of the origins in $\mathcal{L}^p(\omega)$ and $\mathcal{W}^p(\omega)$, respectively, such that for each $(m,f) \in \mathcal{L}^p_0$, the boundary value problem*

$$\mathcal{N}(e,\ell) = \overrightarrow{G}, \tag{20}$$

for

$$\overrightarrow{G} = (-\alpha_0 f, -\alpha_0 m - \alpha_1 f) \in \mathcal{L}_0^p, \tag{21}$$

has exactly one solution (e, ℓ) *in* $\mathcal{W}_0^p \subset \mathcal{W}^p(\omega)$.

PROOF. From Theorem 5,

$$d\mathcal{N}(0, 0; e, \ell) = \mathcal{L}(e, \ell).$$

Since ω is a C^2 domain in the submanifold Γ, the regularity result of Theorem 2 implies that, for each $\overrightarrow{G} \in \mathcal{L}^p(\omega)$, $p > 2$, the problem

$$\mathcal{L}(e, \ell) = \overrightarrow{G}$$

has one and only one solution (e, ℓ) in the space $\mathcal{W}^p(\omega)$ for each $\overrightarrow{G} \in \mathcal{L}^p(\omega)$, $p > 2$. Hence the linear bounded operator

$$\mathcal{N}'(0, 0) = \mathcal{L}\colon \mathcal{W}^p(\omega) \to \mathcal{L}^p(\omega)$$

is bijective, where $\mathcal{N}'(0, 0)$ denotes the linear mapping $(e, \ell) \mapsto d\mathcal{N}(0, 0; e, \ell)$. Thus, its inverse is also continuous from the closed graph theorem. Therefore the operator $\mathcal{N}'(0, 0)$ is an isomorphism between the spaces $\mathcal{W}^p(\omega)$ and $\mathcal{L}^p(\omega)$ at the origin. Then the inverse function theorem [**23**] implies that there exists a neighborhood $\mathcal{L}_0^p$ of the origin of $\mathcal{L}^p(\omega)$ and a neighborhood $\mathcal{W}_0^p$ of $\mathcal{W}^p(\omega)$, such that the problem

$$\mathcal{N}(e, \ell) = \overrightarrow{G}$$

has exactly one solution (e, ℓ) in $\mathcal{W}_0^p$ for each $\overrightarrow{G} \in \mathcal{L}_0^p$. □

4. Intrinsic Shell Models of Nonlinear Elasticity

In this section, we derive intrinsic nonlinear models of thin shells from three-dimensional nonlinear elasticity theory. For simplicity we still assume that the shell S_h has constant thickness. As in the linear case Assumption 1 and Assumption 2 will be the basic assumptions.

4.1. Strain and stress tensors, strain energy. Under Assumption 1, consider the *linear elastic strain tensor* $\varepsilon(V)$ over S_h associated with the displacement field $V(x)$. Again, we shall use the approximate strain tensor $\tilde{\varepsilon}$ of the strain tensor $\varepsilon(V)$ defined in Assumption 2.

Assume that the shell is made up of a homogeneous, isotropic, elastic material, then there exists a response function $\sigma\colon \mathcal{M}^3 \to \mathcal{S}^3$ which is of the form, for any $A \in \mathcal{M}^3$,

$$\sigma(A) = \lambda tr\, AI + 2\mu A + o(A), \tag{22}$$

where $\lambda \geq 0$ and $\mu > 0$ are the Lamé coefficients, such that the second Piola-Kirchhoff stress tensor $\tilde{\sigma}$ is of the form

$$\tilde{\sigma} = \sigma\big(\tilde{\varepsilon}(V)\big), \tag{23}$$

(cf. Ciarlet [**5**], Marsden and Hughes [**25**]). From now on, we assume that σ is smooth enough for our computation. We will give the assumption about σ in the following discussion.

REMARK 3. For simplification, we use the linear elastic strain tensor in the following discussion. However, the following results are true for the Green-Saint Venant strain tensor.

For the *strain energy* $\mathcal{P}$

$$\mathcal{P} = \frac{1}{2}\int_{S_h} \sigma\big(\tilde{\varepsilon}(V)\big)\cdot\cdot\tilde{\varepsilon}(V)\,dx, \tag{24}$$

using Federer's decomposition of the measure (cf. Evans and Gariepy [**16**]) along the level curves of the oriented distance function b for which $|\nabla b(x)| = 1$ in S_h.

$$\begin{aligned}\mathcal{P} &= \frac{1}{2}\int_{-h}^{h} dz \int_\omega \sigma(\tilde{\varepsilon}(V)\circ T_z)\cdot\cdot\tilde{\varepsilon}(V)\circ T_z j(z)\,d\Gamma\\ &= \int_\omega \sum_{i=0}^{2}\sigma_i(\varepsilon^0,\varepsilon^1,\varepsilon^2)\cdot\cdot\varepsilon^i\,d\Gamma,\end{aligned}$$

where for $i = 0, 1, 2$

$$\sigma_i(\varepsilon^0,\varepsilon^1,\varepsilon^2) = \frac{1}{2}\int_{-h}^{h}\sigma(\varepsilon^0+\varepsilon^1 z+\varepsilon^2 z^2)j(z)z^i\,dz.$$

4.2. Fully nonlinear equations of shells. The *variational principle* may now be used to obtain the equations of the displacement. A straightforward calculation of the directional derivative $d\mathcal{P}(e,\ell;\bar{e},\bar{\ell})$ at (e,ℓ) in the direction $(\bar{e},\bar{\ell})$ gives

$$\begin{aligned}d\mathcal{P}(e,\ell;\bar{e},\bar{\ell}) &= \sum_{i=0}^{2}\int_\omega\left[\sum_{j=0}^{2}\sigma_{i,j}(\varepsilon^0,\varepsilon^1,\varepsilon^2)\bar{\varepsilon^j}\cdot\cdot\varepsilon^i + \sigma_i\cdot\cdot\bar{\varepsilon^i}\right]d\Gamma\\ &= \sum_{i=0}^{2}\int_\omega\left[\sum_{j=0}^{2}{}^*\sigma_{j,i}(\varepsilon^0,\varepsilon^1,\varepsilon^2)\varepsilon^j + \sigma_i\right]\cdot\cdot\bar{\varepsilon^i}\,d\Gamma,\end{aligned}$$

where

$$\sigma_{i,j} = \partial\sigma_i/\partial\varepsilon^j,$$

is a linear mapping from $\mathcal{M}^3$ to $\mathcal{M}^3$,

$$\bar{\varepsilon^i} = \varepsilon^i(\bar{e},\bar{\ell}),$$

for $i,j = 0,1,2$, and ${}^*\sigma_{i,j}$ is the conjugate of $\sigma_{i,j}$ with respect to the inner product $\langle\cdot,\ \cdot\rangle$ ($\langle A,\ B\rangle = A\cdot\cdot B$) in the space $\mathcal{M}^3$.

Using integration by parts, we can rearrange to isolate $\bar{\ell}$, $D_\Gamma\bar{e}$, and $D_\Gamma\bar{\ell}$

$$d\mathcal{P}(e,\ell;\bar{e},\bar{\ell}) = \int_\omega [N_1\cdot\cdot D_\Gamma\bar{e} + N_2\cdot\cdot D_\Gamma\bar{\ell} + N_3\cdot\bar{\ell}]\,d\Gamma$$

where (for simplify, we use the same notation as the case of the Saint Venant-Kirchhoff material),

$$\begin{aligned}N_1 = &\left[\sum_{j=0}^{2}{}^*\sigma_{j,0}(\varepsilon^0,\varepsilon^1,\varepsilon^2)\varepsilon^j+\sigma_0\right] - \left[\sum_{j=0}^{2}{}^*\sigma_{j,1}(\varepsilon^0,\varepsilon^1,\varepsilon^2)\varepsilon^j+\sigma_1\right]D^2b\\ &+\left[\sum_{j=0}^{2}{}^*\sigma_{j,2}(\varepsilon^0,\varepsilon^1,\varepsilon^2)\varepsilon^j+\sigma_2\right]D^2b^2\end{aligned}$$

$$N_2 = \left[\sum_{j=0}^{2}{}^*\sigma_{j,1}(\varepsilon^0,\varepsilon^1,\varepsilon^2)\varepsilon^j+\sigma_1\right] - \left[\sum_{j=0}^{2}{}^*\sigma_{j,2}(\varepsilon^0,\varepsilon^1,\varepsilon^2)\varepsilon^j+\sigma_2\right]D^2b,$$

$$N_3 = \left[\sum_{j=0}^{2} {}^*\sigma_{j,0}(\varepsilon^0, \varepsilon^1, \varepsilon^2)\varepsilon^j + \sigma_0\right] n.$$

Using the same assumption on the work of the external forces in Section 3, we get the following nonlinear boundary problem for the shells

$$\begin{cases} -\overrightarrow{\operatorname{div}}_\Gamma N_1 + HN_1 n = -\alpha_0 f & \text{on } \omega \\ -\overrightarrow{\operatorname{div}}_\Gamma N_2 + HN_2 n + N_3 = -\alpha_0 m - \alpha_1 f & \text{on } \omega \end{cases} \tag{25}$$

$$\begin{cases} N_1 \nu = g & \text{on } \gamma_1 \\ N_2 \nu = q & \text{on } \gamma_1 \\ \ell = e = 0 & \text{on } \gamma_2 \end{cases} \tag{26}$$

where $\gamma = \gamma_1 \cup \gamma_2$ and $\mathring{\gamma}_1 \cap \mathring{\gamma}_2 = \emptyset$. But, here we shall just consider the clamped case, i.e., $\gamma_1 = \emptyset$.

Rewrite the system (25), and (26) in the form: to find vector fields $(e, \ell)\colon \omega \to \mathbb{R}^3 \times \mathbb{R}^3$ that verify

$$\begin{cases} \mathcal{N}(e, \ell) = \overrightarrow{G} & \text{in } \omega \\ (e, \ell) = 0 & \text{on } \gamma \end{cases} \tag{27}$$

where

$$\mathcal{N}(e, \ell) \stackrel{\text{def}}{=} (-\overrightarrow{\operatorname{div}}_\Gamma N_1 + HN_1 n, -\overrightarrow{\operatorname{div}}_\Gamma N_2 + HN_2 n + N_3)$$

and

$$\overrightarrow{G} \stackrel{\text{def}}{=} (-\alpha_0 f, -\alpha_0 m - \alpha_1 f).$$

4.3. Differentiability of nonlinear operators. For the shell made up of a Saint Venant-Kirchhoff material, we can get differentiability of nonlinear operators $\mathcal{N}$ directly from the fact that the Sobolev space $W^{1,p}(\omega)$ is a Banach algebra for $p > 2$, since N_1, N_2 and N_3 in the nonlinear operator $\mathcal{N}$ are sums of continuous linear, bilinear, and trilinear mappings of $e, \ell, D_\Gamma e$, and $D_\Gamma \ell$. But for the shell of nonlinear elasticity, the situation is much more complicated since σ is a general nonlinear function. We first prove the following result about the differentiability of nonlinear operators on $W^{1,p}(\omega)$. The result is an extension of Valent's result [**28**] to shells. The proof essentially relies on the integration formula on shells, and the fact that the Sobolev space $W^{1,\,p}(\omega)$ is continuously embedded in the space $C^0(\bar{\omega})$ and is a Banach algebra for $p > 2$ (cf. Adams [**1**]). The result is about any nonlinear operators on $W^{1,\,p}(\omega)$. It can be used for the analysis of nonlinear shell problems.

THEOREM 6. *Assume that* $p > 2$, $\mathcal{F} \in C^2(\omega \times \mathbb{R}^3 \times \mathbb{R}^{3\times 3}; \mathbb{R}^3)$, *then, for* $\ell \in W^{1,p}(\omega)^3$, $G \in W^{1,p}(\omega)^{3\times 3}$, *the vector-valued function*

$$\mathcal{F}(\cdot, \ell, G)\colon x \in \omega \to \mathcal{F}\big(x, \ell(x), G(x)\big)$$

belongs to the space $W^{1,p}(\omega)^3$, *and*

$$\nabla_\Gamma f_k(\cdot, \ell, G)(x) = \nabla_{\Gamma,1} f_k\big(x, \ell(x), G(x)\big) + \frac{\partial f_k}{\partial \ell_i}\big(x, \ell(x), G(x)\big)\nabla_\Gamma \ell_i(x) + \frac{\partial f_k}{\partial g_{ij}}\big(x, \ell(x), G(x)\big)\nabla_\Gamma g_{ij}(x) \tag{28}$$

for almost all $x \in \omega$, *where* $\mathcal{F} = (f_k), \ell = (\ell_i), G = (g_{ij})$. *Moreover, the mapping*

$$\mathcal{F}\colon (\ell, G) \in W^{1,p}(\omega)^3 \times W^{1,p}(\omega)^{3\times 3} \to \mathcal{F}(\cdot, \ell, G) \in W^{1,p}(\omega)^3$$

defined in this way is of class C^1 *and its derivative is continuous and given by*

$$f_k'(\cdot, \ell, G)(m, F) = \frac{\partial f_k}{\partial \ell_i}(\cdot, \ell, G) m_i + \frac{\partial f_k}{\partial g_{ij}}(\cdot, \ell, G) f_{ij}, \tag{29}$$

where $m = (m_i), F = (f_{ij}), 1 \leq i, j \leq 3$.

With the above theorem, we can easily prove the following corollary. It will be used to prove differentiability of the nonlinear operator $\mathcal{N}$ associated with the fully nonlinear models.

COROLLARY 1. *Let* L, M *be two integers. Assume that* $p > 2$, $\mathcal{F} \in C^2(\omega \times \mathbb{R}^3 \cdots \times \mathbb{R}^3 \times \mathbb{R}^{3\times 3} \cdots \times \mathbb{R}^{3\times 3}; \mathbb{R}^3)$, *then, for* $\ell^\alpha \in W^{1,p}(\omega)^3, \alpha = 1, \ldots, L$, $G^\beta \in W^{1,p}(\omega)^{3\times 3}$, $\beta = 1, \ldots, M$, *the vector-valued function*

$$\mathcal{F}(\cdot, \ell^1, \ldots, \ell^L, G^1, \ldots, G^M)\colon x \in \omega \to \mathcal{F}\big(x, \ell^1(x), \ldots, \ell^L(x), G^1(x), \ldots, G^M(x)\big)$$

belongs to the space $W^{1,p}(\omega)^3$, *and*

$$\begin{aligned} &\nabla_\Gamma f_k(\cdot, \ell^1, \ldots, \ell^L, G^1, \ldots, G^M)(x) \\ &\quad = \nabla_{\Gamma,1} f_k\big(x, \ell^1(x), \ldots, \ell^L(x), G^1(x), \ldots, G^M(x)\big) \\ &\qquad + \sum_{i,\alpha} \frac{\partial f_k}{\partial \ell_i^\alpha}\big(x, \ell^1(x), \ldots, \ell^L(x), G^1(x), \ldots, G^M(x)\big) \nabla_\Gamma l_i^\alpha(x) \\ &\qquad + \sum_{i,j,\beta} \frac{\partial f_k}{\partial g_{ij}^\beta}\big(x, \ell^1(x), \ldots, \ell^L(x), G^1(x), \ldots, G^M(x)\big) \nabla_\Gamma g_{ij}^\beta(x) \end{aligned} \tag{30}$$

for almost all $x \in \omega$, *where* $\mathcal{F} = (f_k)$, $\ell^\alpha = (\ell_i^\alpha)$, $G^\beta = (g_{ij}^\beta)$, $1 \leq i, j, k \leq 3$. *Moreover, the mapping*

$$\begin{aligned} \mathcal{F}\colon (\ell^1, \ldots, \ell^L, G^1, \ldots, G^M) &\in \left[W^{1,p}(\omega)^3\right]^L \times \left[W^{1,p}(\omega)^{3\times 3}\right]^M \\ &\to \mathcal{F}(\cdot, \ell^1, \ldots, \ell^L, G^1, \ldots, G^M) \in W^{1,p}(\omega)^3 \end{aligned}$$

defined in this way is of the class C^1 *and its derivative is continuous and given by*

$$\begin{aligned} &f_k'(\cdot, \ell^1, \ldots, \ell^L, G^1, \ldots, G^M)(m^1, \ldots, m^L, F^1, \ldots, F^M) \\ &= \sum_{i,\alpha} \frac{\partial f_k}{\partial \ell_i^\alpha}(\cdot, \ell^1, \ldots, \ell^L, G^1, \ldots, G^M) m_i^\alpha + \sum_{i,j,\beta} \frac{\partial f_k}{\partial g_{ij}^\beta}(\cdot, \ell^1, \ldots, \ell^L, G^1, \ldots, G^M) f_{ij}^\beta, \end{aligned}$$

where $m^\alpha = (m_i^\alpha)$, $F^\beta = (f_{ij}^\beta)$, $1 \leq i, j, k \leq 3$.

4.4. Existence and uniqueness of solutions. First of all, a direct computation gives the following result, which relates the operator $\mathcal{N}$ and $\mathcal{L}$, and gives the linear part of the fully nonlinear operator $\mathcal{N}$.

THEOREM 7.

$$\mathcal{N}(e, \ell) = \mathcal{L}(e, \ell) + o\big(e,\ \ell,\ D_\Gamma e,\ D_\Gamma \ell, \overrightarrow{\operatorname{div}}_\Gamma(D_\Gamma e), \overrightarrow{\operatorname{div}}_\Gamma(D_\Gamma \ell)\big). \tag{31}$$

We now prove the differentiability of the nonlinear operator $\mathcal{N}$ associated with the nonlinear model of shells. The corollary in the last Chapter will be used here. As a consequence, the nonlinear operator $\mathcal{N}$ mapping the space $\mathcal{W}^p(\omega)$ into the space $\mathcal{L}^p(\omega)$ is well-defined and differentiable between these two spaces. Then we compute the derivative of $\mathcal{N}$.

THEOREM 8. *Assume that σ is C^3, and $p > 2$. Then the associated nonlinear operator*

$$(e, \ell) \mapsto \mathcal{N}(e, \ell) \colon \mathcal{W}^p(\omega) \to \mathcal{L}^p(\omega) \tag{32}$$

is well-defined and differentiable, and

$$d\mathcal{N}(0, 0; e, \ell) = \mathcal{L}(e, \ell) \tag{33}$$

in $\mathcal{W}^p(\omega)$.

PROOF. By the definitions, we know that N_1, N_2 and N_3 are C^2. Since $p > 2$, from the Corollary, the nonlinear mapping

$$(e, \ell) \mapsto \mathcal{N}(e, \ell) \colon \mathcal{W}^p(\omega) \to \mathcal{L}^p(\omega)$$

is well defined. Moreover, for any $(e,\ \ell) \in \mathcal{W}^p(\omega) \subset W^{2,p}(\omega)^3 \times W^{2,p}(\omega)^3$, the following mappings:

$$e \mapsto D_\Gamma e \colon W^{2,\ p}(\omega) \to W^{1,p}(\omega)^{3\times 3},$$
$$\ell \mapsto D_\Gamma \ell \colon W^{2,\ p}(\omega) \to W^{1,p}(\omega)^{3\times 3},$$

are smooth (C^∞), and the mapping

$$G \mapsto \overrightarrow{\operatorname{div}_\Gamma} G \colon W^{1,p}(\omega)^{3\times 3} \to L^p(\omega)^3$$

is also smooth (C^∞). Therefore, by the definition of $\mathcal{N}$, we get that $\mathcal{N}(e,\ \ell)$ is well defined and differentiable in $\mathcal{W}^p(\omega)$. Next a direct calculation concludes the proof. □

We are now in the position to establish the existence and uniqueness result for the fully nonlinear equations. The objective is to apply the regularity result of weak solutions of the linear models, and the inverse function theorem (cf. Hörmander [**23**]) in neighborhoods of the origins in the spaces $\mathcal{L}^p(\omega)$ and $\mathcal{W}^p(\omega)$.

THEOREM 9. *Let $\bar{h}$ be as in Theorem 8, and h, $0 < h < \bar{h}$ be fixed. Let Ω be a domain in $\mathbb{R}^3$ of co-dimension one with a boundary Γ which is a $C^{2,1}$ submanifold of $\mathbb{R}^3$ and let ω be a C^2 domain in the submanifold Γ. Assume σ is C^3. Then for each $p > 2$, there exist two neighborhoods $\mathcal{L}^p_0$ and $\mathcal{W}^p_0$ of the origins of $\mathcal{L}^p(\omega)$ and $\mathcal{W}^p(\omega)$ respectively, such that, for each $(m,\ f) \in \mathcal{L}^p_0$, the boundary value problem*

$$\mathcal{N}(e,\ \ell) = \overrightarrow{G}, \tag{34}$$

where

$$\overrightarrow{G} = (-\alpha_0 f, -\alpha_0 m - \alpha_1 f) \in \mathcal{L}^p_0, \tag{35}$$

has exactly one solution $(e,\ \ell)$ in $\mathcal{W}^p_0 \subset \mathcal{W}^p(\omega)$.

PROOF. From Theorem 5,

$$d\mathcal{N}(0, 0; e, \ell) = \mathcal{L}(e, \ell).$$

Since ω is a C^2 domain in the submanifold Γ, the regularity result of Theorem 2 implies that, for each $\overrightarrow{G} \in \mathcal{L}^p(\omega)$, $p > 2$, the problem

$$\mathcal{L}(e, \ell) = \overrightarrow{G}$$

has one and only one solution (e, ℓ) in the space $\mathcal{W}^p(\omega)$ for each $\overrightarrow{G} \in \mathcal{L}^p(\omega)$, $p > 2$. Hence the linear bounded operator

$$\mathcal{N}'(0,0) = \mathcal{L}\colon \mathcal{W}^p(\omega) \to \mathcal{L}^p(\omega)$$

is bijective, where $\mathcal{N}'(0,0)$ denotes the linear mapping $(e, \ell) \mapsto d\mathcal{N}(0,0; e, \ell)$. Thus, its inverse is also continuous from the closed graph theorem. Therefore the operator $\mathcal{N}'(0,0)$ is an isomorphism between the spaces $\mathcal{W}^p(\omega)$ and $\mathcal{L}^p(\omega)$ at the origin. Then the inverse function theorem [**23**] implies that there exists a neighborhood $\mathcal{L}^p_0$ of the origin of $\mathcal{L}^p(\omega)$ and a neighborhood $\mathcal{W}^p_0$ of $\mathcal{W}^p(\omega)$, such that the problem

$$\mathcal{N}(e, \ell) = \overrightarrow{G}$$

has exactly one solution (e, ℓ) in $\mathcal{W}^p_0$ for each $\overrightarrow{G} \in \mathcal{L}^p_0$. □

References

1. R. A. Adams, *Sobolev spaces*, Academic Press, New York, 1975.
2. S. Agmon, A. Douglis, and L. Nirenberg, *Estimates near the boundary for solutions of elliptic partial differential equations satisfying general boundary condition.* II, Comm. Pure Appl. Math. **17** (1962), 35–92.
3. M. Bernadou, *Méthodes d'éléments finis pour les problèmes de coques minces*, Masson, 1994.
4. M. Bernadou and J. T. Oden, *An existence theorem for a class of nonlinear shallow shell problems*, J. Math. Pures Appl. **60** (1981), 285–308.
5. P. G. Ciarlet, *Mathematical elasticity.* I. *Three-dimensional elasticity*, North-Holland Publ. Co., Amsterdam, 1988.
6. ———, *Plates and junctions in elastic multi-structures: An asymptotic analysis*, Research in Applied Mathematic, vol. 14, Masson, Paris; Springer-Verlag, Berlin, 1990.
7. R. Dautray and J. L. Lions, *Mathematical analysis and numerical methods for science and technology*, vol. I, Springer-Verlag, Berlin, 1994.
8. M. C. Delfour, *Intrinsic differential geometry methods in the asymptotic analysis of linear thin shells*, Boundaries, Interfaces and Transitions (M. Delfour, ed,) (Banff, 1995), CRM Proc. Lecture Notes, vol. 13, Amer. Math. Soc., Providence, RI, 1998, pp. 19–90.
9. M. C. Delfour and J. Zhao, *The nonlinear shell models of Saint Venant-Kirchhoff materials*, CRM–2355, Univ. de Montréal, 1996.
10. M. C. Delfour and J. P. Zolésio, *Shape analysis via oriented distance*, J. Funct. Anal. **123** (1994), 105–201.
11. ———, *On a variational equation for thin shell*, Control and Optimal Design of Distributed Parameter Systems (J. Lagnese, D. L. Russel, and L. White, eds.) (Minneapolis, 1992), Springer, New York, 1995, pp. 25–37.
12. ———, *A boundary differential equations of thin shells*, J. Differential Equations **119** (1995), no. 2, 426–449.
13. ———, *Tangential differential equations for dynamical thin/shallow shells*, J. Differential Equations **128** (1996), 125–167.
14. ———, *Differential equations for linear shells: comparison between intrinsic and classical methods*, Advances in Mathematical Sciences: CRM's 25 Years (L. Vinet, ed.) (Montréal, 1994) CRM Proc. Lecture Notes, vol. 11, Amer. Math. Soc., Providence, RI, 1997, pp. 41–124.
15. ———, *Convergence of the linear* $P(1,1)$ *and* $P(2,1)$ *thin shells to asymptotic shells*, Plates and Shells (M. Fortin, ed.) (Québec, 1996), CRM Proc. Lecture Notes, vol. 21, Amer. Math. Soc., Providence, RI, 1999, pp. 113-146.
16. L. C. Evans and R. F. Gariepy, *Measure theory and fine properties of functions*, Stud. Adv. Math., CRC Press, Boca Raton, FL, 1992.
17. P. Germain, *Mécanique,* I *and* II, Ellipses, Paris, 1986.

18. G. Fichera, *Linear elliptic differential systems and eigenvalue problems*, Lecture Notes in Math., vol. 8, Springer-Verlag, Berlin, 1965.
19. I. M. N. Figueiredo, *Local existence and regularity of the solutions of the nonlinear thin shell model of Donnell-Mushtari-Vlasov*, Appl. Anal. **36** (1990), 221–234.
20. G. Geymonat, *Sui problemi ai limiti per i sistemi lieari ellitici*, Ann. Mat. Pura Appl. **69** (1965), 207–284.
21. D. Gilbarg and N. Trudinger, *Elliptic partial differential equations of second order*, Springer-Verlag, Berlin, New York, Heidelberg, Tokyo, 1983.
22. O. V. Guseva, *On boundary problems for strongly elliptic systems*, Dokl. Akad. Nauk SSSR **102** (1955), 1069–1072.
23. L. Hörmander, *The analysis of linear partial differential operators* I, Springer-Verlag, Berlin, New York, Heidelberg, Tokyo, 1983.
24. H. Le Dret and A. Raoult, *The membrane shell model in nonlinear elasticity: A variational asymptotic derivation*, J. Nonlinear Sci. **6** (1996), 59–84.
25. J. E. Marsden and T. Hughes, *Mathematical foundations of elasticity*, Mathematical foundations of elasticity, Corrected reprint of the 1983 original, Dover Publications, Inc., New York, 1994.
26. C. B. Morrey Jr., *Multiple integrals in the calculus of variations*, Die Grundlehren der mathematischen Wissenschaften, Band 130 Springer-Verlag New York, Inc., New York 1966.
27. J. Sokolowski and J. P. Zolésio, *Introduction to shape optimization: Shape sensitivity analysis*, Springer Ser. Comput. Math., vol. 16. Springer-Verlag, Berlin, 1992.
28. T. Valent, *Teoremi di esistenza e unicitá in elastostatica finita,* Rend. Sem. Mat. Univ. Padova **60** (1978), 165–181.

Centre de recherches mathématiques et Département de mathématiques et de statistique, Université de Montréal, C.P. 6128, Succ. Centre-ville, Montréal (Qc), Canada H3C 3J7

E-mail address: delfour@CRM.UMontreal.CA

BMG, Tellabs Operation Inc., 1000 Remington Blvd, Bolingbrook, IL 60440, USA

E-mail address: jzhao@tellabs.com

Centre de Recherches Mathématiques
CRM Proceedings and Lecture Notes
Volume **21**, 1999

Convergence of the Linear $P(1,1)$ and $P(2,1)$ Thin Shells to Asymptotic Shells

Michel C. Delfour and Jean-Paul Zolésio

ABSTRACT. Given a midsurface with Lipschitzian boundary in a $C^{1,1}$ submanifold of $\mathbf{R}^N$, we show that solutions of the intrinsic linear $P(1,1)$ and $P(2_n,1)$ models of thin shells studied by Delfour and Zolésio converge to solutions of *asymptotic shell models* which consist of a coupled system of two variational equations. The first one is the asymptotic $P(1,0)$ model which yields the generally accepted classical asymptotic membrane equation and the Love-Kirchhoff condition. The second is a generalized asymptotic bending equation. In the bending dominated case, the quadratic term of the second equation of the asymptotic $P(2_n,1)$ model is the one of the classical bending dominated equation. Strong convergence is established in the natural spaces and norms under an assumption on the asymptotic behavior of the constant of continuity of the right-hand side for shells without boundary or shells with homogeneous Neumann (quotient space) or partial Dirichlet boundary conditions.

1. Introduction

We consider linear $P(k,\ell)$ models of thin shells of thickness $2h$ around a bounded open connected midsurface ω (with Lipschitzian boundary γ when ω has a nonempty boundary). The domain ω is contained in a $C^{1,1}$ submanifold Γ of $\mathbf{R}^N$. $P(k,\ell)$ means a k-th order polynomial approximation of the displacement vector with respect to the variable z normal to Γ, and an ℓ-th order approximation of the associated linear strain tensor. Such models arise from expressions obtained through infinite series expansions of the displacement vector and the linear strain tensor [**27, 28, 32**]. They are somewhat related to the mixed approximations studied in [**3**] for plates.

The *object of this paper* is to show that the solutions of the intrinsic linear $P(1,1)$ and $P(2_n,1)$ models converge to the solutions of *asymptotic shell models*

1991 *Mathematics Subject Classification.* Primary: 73K15, 73O62; Secondary: 73K10, 49Q.

The research of the first author has been supported by National Sciences and Engineering Research Council of Canada research grant A-8730 and by a FCAR grant from the Ministère de l'Éducation du Québec. He would like to thank Professors D. Arnold (Pennsylvania State University), A. Raoult (LMC, Université J. Fourier, Grenoble) and D. Chapelle (Laboratoire Central des Ponts et Chaussées, Paris) for enlightening discussions.

This is the final form of the paper.

with zero thickness which consist of two coupled variational equations. The convergence of the $P(2,1)$ model and a more detailed analysis of the asymptotic models can be found in [**23**]. The first equation is the one which characterizes the asymptotic $P(1,0)$ model. It is common to the three models. It yields the generally accepted asymptotic membrane equation and the Love-Kirchhoff condition. Strong convergence in the naturally associated quotient spaces for the *extended membrane energy* was established in the companion paper [**33**] and expended in [**23**]. To our knowledge the second equation is new for shells and extends results obtained in special cases such as the bending dominated case where the extended membrane energy is assumed to be zero or in the case of the plate. It explicits the role of the *mean* and *Gauss curvatures* in the right-hand side of the equation, but it is slightly different for the two models.

In the *bending dominated* case for the special constitutive law specified by two Lamé constants, the second equation of the asymptotic $P(2_n,1)$ and $P(2,1)$ models yields the *bending dominated equation* which is obtained from the asymptotic analysis of the three-dimensional problem [**17, 18, 15**]; for the $P(1,1)$ model the equation is similar but one of the coefficients of the quadratic terms is different.

Strong convergence is established in the natural spaces and norms associated with the asymptotic models under an assumption on the asymptotic behavior of the constant of continuity of the right-hand side of the N-dimensional equation of linear elasticity (in practice $N = 3$) as the thickness h goes to zero without a priori assumptions on the stress tensor. In fact explicit expressions of the normal and/or plane stress are obtained in terms of the loading vectors. They show that the cases of *zero normal stress* and *plane stress* arise from simple assumptions on the loading and not from assumptions on the shell itself. The convergence is true for shells without boundary or shells with a Lipschitzian boundary γ and homogeneous Neumann (quotient space with respect to the rigid displacements) or homogeneous Dirichlet boundary condition on a part γ_0 of or the whole boundary γ. Our results do not use changes of variable to normalize the thickness or Γ-convergence techniques, but obviously the final mathematical results are or should be the same and it is or should only be a matter of choice of tools.

The convergence of the solutions of the $P(1,1)$, $P(2_n,1)$ or $P(2,1)$ models to those of the asymptotic shell model is one step in the convergence of the solutions of the N-dimensional model to those of the asymptotic shell (cf. for instance [**19, 11, 12, 15, 17, 18**]). The other step involves interpolation techniques. For plates the reader is referred to the early work of [**43**] in 1959, [**35, 9**], and [**3**]. Ph. Destuynder established in [**34**] (1980) that the order of convergence of the solution of the three-dimensional model to the asymptotic model is $h^{1/2}$. This rate was also obtained for mixed formulations of the Hellinger-Reissner type by [**3**]. A slightly better rate can be found in [**5**, pp. 80–86]. A convergence rate of $h^{1/6}$ is given in [**42**] (1996) for shells with homogeneous Dirichlet boundary conditions on the whole lateral boundary $\Sigma_h(\omega)$, a midsurface $\omega \subset \Gamma$ which is uniformly elliptic with a C^5 boundary γ and a C^3 submanifold Γ of $\mathbf{R}^3$.

For a Γ convergence approach the reader is referred to the pioneering paper [**1**] of Acerbi et als for an elastic string. A rigorous justification of the Γ-convergence is given in [**38, 39**] for the asymptotic nonlinear membrane model. They successively consider as initial configuration a cylinder and an arbitrary midsurface, and they provide justifications for the first of the nonlinear membrane models which seems to have been first introduced in [**37**]. For the use of the Γ-convergence and asymptotic analysis for linear thin plates see [**4**].

2. Notation and Background Material

The inner product in $\mathbf{R}^N$ and the double inner product of $N \times N$ matrices or tensors A and B are denoted as

$$x \cdot y = \sum_{i=1}^{N} x_i y_i, \quad A{\cdot\cdot}B = \sum_{i=1}^{N} \sum_{j=1}^{N} A_{ij} B_{ij}.$$

${}^*A$ denotes the transpose of A. Given Ω in $\mathbf{R}^N$, $\Omega \neq \varnothing$ (resp. $\Gamma \stackrel{\text{def}}{=} \partial\Omega \neq \varnothing$) the *distance function* (resp. *oriented distance function*) is defined as

$$d_\Omega(x) \stackrel{\text{def}}{=} \inf_{y \in \Omega} |y - x| \quad (\text{resp. } b_\Omega(x) \stackrel{\text{def}}{=} d_\Omega(x) - d_{\mathbf{R}^N - \Omega}(x)).$$

When Ω is a domain of class $C^{1,1}$ in $\mathbf{R}^N$, $b = b_\Omega$ is $C^{1,1}$ in a neighborhood of every point of Γ and the converse is true. Its gradient ∇b coincides with the exterior unit normal n to the boundary on Γ. The *projection* p onto Γ and the *orthogonal projection* P onto the *tangent plane* $T_x\Gamma$ are given by

$$p(x) \stackrel{\text{def}}{=} x - b(x)\nabla b(x), \quad P(x) \stackrel{\text{def}}{=} I - \nabla b(x)^* \nabla b(x),$$

where ${}^*V$ denotes the transpose of a column vector V in $\mathbf{R}^N$. Given $h > 0$ and an open domain ω in Γ, a *shell* is the bounded open connected domain

$$S_h(\omega) \stackrel{\text{def}}{=} \{x \in \mathbf{R}^N : |b_\Omega(x)| < h, p(x) \in \omega\}$$

in $\mathbf{R}^N$. When $\omega = \Gamma$ the shell has *no boundary*; otherwise we denote by γ the (relative) *boundary* of ω in Γ and by

$$\Sigma_h(\gamma) \stackrel{\text{def}}{=} \{x \in \mathbf{R}^N : |b_\Omega(x)| < h, p(x) \in \gamma\}$$

its *lateral boundary*. For $b \in C^{1,1}(S_h(\omega))$, $|z| < h$, $X \in \omega$, define

$$T_z(X) \stackrel{\text{def}}{=} X + z\nabla b(X), \; j_z \stackrel{\text{def}}{=} \det DT_z(X) = \det[I + zD^2 b(X)] = \sum_{i=0}^{N-1} \kappa_i(X) z^i$$

where the κ_i's are the coefficients of the polynomial j_z of degree $(N-1)$ in z. For $n \geq 0$ also define the following functions of $X \in \omega$ and their averages

$$(2.1) \qquad \boxed{\alpha_n(h) \stackrel{\text{def}}{=} \int_{-h}^{h} j_z z^n \, dz = \sum_{j=n}^{n+N-1} \left[1 - (-1)^{j+1}\right] \frac{h^{j+1}}{j+1} \kappa_{j-n}, \quad \overline{\alpha}_n(h) \stackrel{\text{def}}{=} \frac{\alpha_n(h)}{2h}.}$$

The $\alpha_n(h)$'s are polynomials in odd powers of h. As h goes to zero

$$\frac{\overline{\alpha}_0(h) - 1}{h^2} \to \frac{1}{3}\kappa_2, \quad \frac{\overline{\alpha}_1(h)}{h^2} \to \frac{1}{3}\kappa_1, \quad \frac{\overline{\alpha}_2(h)}{h^2} \to \frac{1}{3}, \quad \frac{\overline{\alpha}_3(h)}{h^4} \to \frac{1}{5}\kappa_1.$$

For $N = 3$, $\kappa_1 = \Delta b$ is twice the *mean curvature* and κ_2 the *Gauss curvature*.

ASSUMPTION 2.1. Let C be a linear bijective transformation which maps symmetrical $N \times N$ matrices or tensors onto symmetrical $N \times N$ matrices or tensors, and assume that there exists $\alpha > 0$ such that $C^{-1}\tau{\cdot\cdot}\tau \geq \alpha\tau{\cdot\cdot}\tau$ for all τ. Further assume that C is symmetrical.

For instance for the Lamé constants $\mu > 0$ and $\lambda \geq 0$, $C^{-1}\tau = 2\mu\tau + \lambda \operatorname{tr} \tau I$ verifies Assumption 2.1 with $\alpha = 2\mu$. We separately deal with a shell without boundary or Neumann boundary conditions, and a shell with homogeneous Dirichlet boundary conditions on part of the or on the whole boundary, but the constructions and the proofs are identical.

3. Shell Without Boundary or Neumann Boundary Conditions

The N-dimensional variational equation of linear elasticity

$$\boxed{\begin{aligned}&\exists V(h) \in H^1\big(S_h(\omega)\big)^N \text{ such that } \forall V \in H^1\big(S_h(\omega)\big)^N \\ &\qquad \int_{S_h(\omega)} C^{-1}\varepsilon(V(h)) \cdot\cdot\, \varepsilon(V) - F \cdot V - G \cdot DV\nabla b \, dx = 0\end{aligned}} \tag{3.1}$$

makes sense as a variational equation in $H^1\big(S_h(\omega)\big)^N / \ker \varepsilon$ for F and G in $L^2\big(S_h(\omega)\big)^N$ verifying the condition

$$\forall V \in \ker \varepsilon, \quad \int_{S_h(\omega)} F \cdot V + G \cdot DV\nabla b \, dx = 0, \quad \varepsilon(V) = \frac{1}{2}(DV + {}^*DV). \tag{3.2}$$

From this condition there exists a constant $c(h) = c\big(S_h(\omega)\big) > 0$ such that

$$\boxed{\forall V \in H^1\big(S_h(\omega)\big)^N, \quad \left| \int_{S_h(\omega)} F \cdot V + G \cdot DV\nabla b \, dx \right| \leq c(h) \big\| \varepsilon(V) \big\|_{L^2(S_h(\omega))}.} \tag{3.3}$$

3.1. Displacement vector and linear strain tensor in local coordinates. The asymptotic behavior of the solution $V(h)$ of (3.1) will depend on the smoothness of F and G in the direction of ∇b and the uniformity of condition (3.3) with respect to h. We first give characterizations of Sobolev spaces and an approximation result in local coordinates.

THEOREM 3.1. *Assume that Γ is the boundary of a $C^{1,1}$ domain in $\mathbf{R}^N$ and let ω be a bounded open domain with a Lipschitzian boundary γ in Γ.*

(i) *The map*

$$T : \overline{\omega} \times [-h, h] \to \overline{S_h(\omega)}, \quad T(X, z) = X + z\nabla b(X)$$

and its inverse $T^{-1}(x) = \big(p(x), b(x)\big)$ are both Lipschitzian.

(ii) *The map*

$$F \mapsto f = F \circ T : H^1_n\big(S_h(\omega)\big)^N \to H^1\big(-h, h; L^2(\omega)^N\big) \tag{3.4}$$

is a continuous linear bijection for the space

$$H^1_n\big(S_h(\omega)\big)^N \stackrel{\text{def}}{=} \big\{F \in L^2\big(S_h(\omega)\big)^N : DF\nabla b \in L^2\big(S_h(\omega)\big)^N\big\}. \tag{3.5}$$

(iii) *The map*

$$F \mapsto f = F \circ T : H^1\big(S_h(\omega)\big)^N \to H^1\big(-h, h; L^2(\omega)^N\big) \cap L^2\big(-h, h; H^1(\omega)^N\big)$$

is a continuous linear bijection. Moreover

$$(DF\nabla b) \circ T = \frac{\partial f}{\partial z}, \quad (DFP) \circ T = D_\Gamma f[I + zD^2 b]^{-1} \tag{3.6}$$

$$DF \circ T = \frac{\partial f}{\partial z}{}^* n + D_\Gamma f[I + zD^2 b]^{-1} = \left[\frac{\partial f}{\partial z}{}^* n + D_\Gamma f\right][I + zD^2 b]^{-1}. \tag{3.7}$$

In fact

$$H^1\big(\omega\times]-h,h[\big)^N = H^1\big(-h,h;L^2(\omega)^N\big)\cap L^2\big(-h,h;H^1(\omega)^N\big). \tag{3.8}$$

PROOF. Cf. [**33**]. □

For small $h>0$ we now work in local coordinates where we do the approximations with respect to the variable z normal to the midsurface. Assume that the displacement vector can be developed as an infinite sum

$$v(X,z)=\sum_{i=0}^{\infty} v^i(X)z^i, \quad v^i:\omega\to\mathbf{R}^N.$$

By using the transformation T we get the explicit expression for V in $S_h(\omega)$

$$V(x)=(v\circ T^{-1})(x)=\sum_{i=0}^{\infty}(v^i\circ p)(x)b(x)^i.$$

From the tangential calculus, we have $D(v^i\circ p)$ in terms of $D_\Gamma(v^i)$ and the derivatives of b (cf. [**27, 28, 32**]). Now introduce

ASSUMPTION 3.1. Assume that $\exists \bar{h}>0$ such that $b\in C^{1,1}\big(S_{\bar{h}}(\omega)\big)$ and

$$\exists\beta, 0<\beta<1, \quad \forall X\in\omega, \quad \bar{h}\big\|D^2b(X)\big\|\le\beta.$$

The matrix $[I+zD^2b]^{-1}$ can now be expressed as the infinite sum

$$[I+zD^2b]^{-1}=\sum_{i=0}^{\infty}(-D^2b)^i z^i \tag{3.9}$$

and after substitution in the expression of $DV\circ T$

$$\boxed{DV\circ T_z=\sum_{i=0}^{\infty} z^i\Big\{(i+1)(v^{i+1*}n)+\sum_{k=0}^{i} D_\Gamma(v^k)(-D^2b)^{i-k}\Big\}}$$

and the linear strain tensor $2\varepsilon(V)=D(V)+{}^*DV$ is given by

$$\boxed{\varepsilon(V)\circ T_z=\sum_{i=1}^{\infty}\varepsilon^i(v)z^i} \tag{3.10}$$

$$\boxed{\begin{aligned}2\varepsilon^i(v) &\stackrel{\text{def}}{=} (i+1)[v^{i+1*}n+n^*v^{i+1}]\\ &+\sum_{k=0}^{i}\big[D_\Gamma(v^k)(-D^2b)^{i-k}+(-D^2b)^{i-k*}D_\Gamma(v^k)\big].\end{aligned}} \tag{3.11}$$

Notice that $\varepsilon^i(v)=\varepsilon^i(v^0,\dots,v^{i+1})$. Next use Federer's formula to compute the strain energy. With $v(h)=V(h)\circ T$ and $v=V\circ T$, the first term is

$$\int_{S_h(\omega)} C^{-1}\varepsilon\big(V(h)\big)\cdot\cdot\,\varepsilon(V)\,dx=\sum_{i=0}^{\infty}\int_\omega \alpha_i(h)\sum_{j=0}^{i} C^{-1}\varepsilon^j\big(v(h)\big)\cdot\cdot\,\varepsilon^{i-j}(v)\,d\Gamma$$

where $\alpha_n(h)$ is given by (2.1) for $n \geq 0$. Assume that the force F is of the form

$$F = f \circ T^{-1}, \quad f = \sum_{i=0}^{\infty} f^i z^i, \quad f^i \in L^2(\omega)^N, \quad \sum_{i=0}^{\infty} h^{2i} \|f^i\|^2_{L^2(\omega)} < \infty \tag{3.12}$$

$$\implies \int_{S_h(\omega)} F \cdot V \, dx = \sum_{i=0}^{\infty} \int_{\omega} \left\{ \alpha_i(h) \sum_{j=0}^{i} f^{i-j} \cdot v^j \, dz \right\} d\Gamma. \tag{3.13}$$

Assume also that G is of the form

$$G = g \circ T^{-1}, \quad g = \sum_{i=0}^{\infty} g^i z^i, \quad g^i \in L^2(\omega)^N, \quad \sum_{i=0}^{\infty} h^{2i} \|g^i\|^2_{L^2(\omega)} < \infty. \tag{3.14}$$

$$\implies \int_{S_h(\omega)} G \cdot DV \nabla b \, dx = \sum_{i=0}^{\infty} \int_{\omega} \left\{ \alpha_i(h) \sum_{j=0}^{i} (j+1) g^{i-j} \cdot v^{j+1} \, dz \right\} d\Gamma. \tag{3.15}$$

Finally

$$\begin{aligned} \int_{S_h(\omega)} & C^{-1}\varepsilon(V(h)) \cdot\cdot \varepsilon(V) - F \cdot V - G \cdot DV \nabla b \, dx \\ &= \sum_{i=0}^{\infty} \int_{\omega} \alpha_i(h) \sum_{j=0}^{i} \{ C^{-1} \varepsilon^{i-j}(v(h)) \cdot\cdot \varepsilon^j(v) \\ & \qquad - f^{i-j} \cdot v^j - (j+1) g^{i-j} \cdot v^{j+1} \} \, d\Gamma. \end{aligned} \tag{3.16}$$

Define

$$B^i_j \stackrel{\text{def}}{=} C^{-1} \varepsilon^{i-j}(v(h)) \cdot\cdot \varepsilon^j(v) - f^{i-j} \cdot v^j - (j+1) g^{i-j} \cdot v^{j+1}.$$

For instance for $N = 3$ we can write equation (3.16) explicitly as

$$\begin{aligned} \int_{\omega} & 2h B^0_0 \, d\Gamma \\ & + \sum_{m=0}^{\infty} \int_{\omega} 2 \frac{h^{2m+3}}{2m+3} \left[\kappa_2 \sum_{j=0}^{2m} B^{2m}_j + \kappa_1 \sum_{j=0}^{2m+1} B^{2m+1}_j + \sum_{j=0}^{2m+2} B^{2m+2}_j \right] d\Gamma. \end{aligned} \tag{3.17}$$

This identity must be zero for all v.

3.2. Polynomial models and choice of the space of solutions. We proceed by polynomial approximation $P(k, \ell)$ of order k for the *displacement vector* $v = V \circ T_z$ and of order ℓ for the *linear strain tensor* $\varepsilon(V) \circ T_z$. An intermediary case is the $P(k_n, \ell)$ model where the displacement vector $v = V \circ T_z$ is approximated by a polynomial of order $k - 1$ for the tangential displacement v_Γ and a polynomial of order k for the normal displacement v_n. In this paper we only consider the first two of the following three models

$$V \circ T_z \simeq v^0 + z v^1 \text{ and } \varepsilon(V) \circ T_z \simeq \varepsilon^0(v^0, v^1) + z \varepsilon^1(v^0, v^1, 0) \text{ for } P(1,1)$$

$$\left. \begin{aligned} V \circ T_z \simeq v^0 + z v^1 + z^2 v^2_n n \\ \text{and } \varepsilon(V) \circ T_z \simeq \varepsilon^0(v^0, v^1) + z \varepsilon^1(v^0, v^1, v^2_n n) \end{aligned} \right\} \text{ for } P(2_n, 1)$$

$$V \circ T_z \simeq v^0 + z v^1 + z^2 v^2 \text{ and } \varepsilon(V) \circ T_z \simeq \varepsilon^0(v^0, v^1) + z \varepsilon^1(v^0, v^1, v^2) \text{ for } P(2,1)$$

where from expressions (3.11) for the $\varepsilon^i(v)$'s

$$\varepsilon^0(v^0, v^1) \stackrel{\text{def}}{=} \varepsilon^0(v) = \frac{1}{2}(v^{1*} n + n^* v^1) + \varepsilon_\Gamma(v^0) \tag{3.18}$$

$$\varepsilon^1(v^0, v^1, v^2) \stackrel{\text{def}}{=} \varepsilon^1(v) = [v^{2*} n + n^* v^2] + \varepsilon_\Gamma(v^1) - \frac{1}{2}[D_\Gamma(v^0) D^2 b + D^2 b^* D_\Gamma(v^0)]$$

With the common notation $v = (v^0, v^1, 0)$ for the $P(1,1)$-model, $v = (v^0, v^1, v^2_n n)$ for the $P(2_n, 1)$-model, and $v = (v^0, v^1, v^2)$ for the $P(2,1)$-model, we simultaneously deal with the three models since the corresponding variational equations have the same structure in term of $\varepsilon^0(v)$ and $\varepsilon^1(v)$. In the three cases $\varepsilon^0(v) = \varepsilon^0(v^0, v^1)$ and $\varepsilon^1(v)$ respectively denotes $\varepsilon^1(v^0, v^1, 0)$ for the $P(1,1)$-model, $\varepsilon^1(v^0, v^1, v^2_n n)$ for the $P(2_n, 1)$-model, and $\varepsilon^1(v^0, v^1, v^2)$ for the $P(2,1)$-model.

Recall from **[28]** and **[32]** that for the $P(1,1)$ model the set of *rigid displacements* is the finite dimensional subspace

$$\boxed{\begin{aligned} K \stackrel{\text{def}}{=} \big\{(v^0, v^1) \colon & (v^0, v^1) \in H^1(\omega)^N \times H^1(\omega)^N \\ & \varepsilon^0(v^0, v^1) = 0 \text{ and } \varepsilon^1(v^0, v^1, 0) = 0\big\} \\ = \big\{(v^0, v^1) \colon & v^0(X) = a + AX, \quad v^1(X) = An(X) \\ & \forall a \in \mathbf{R}^{\mathrm{N}} \text{ and all tensors } A \text{ such that } A + {}^*A = 0\big\}. \end{aligned}}$$

For the $P(2_n, 1)$ model

$$\begin{aligned} \ker \varepsilon^0 \cap \ker \varepsilon^1 &= \big\{(v^0, v^1, v^2_n) \colon (v^0, v^1, v^2_n) \in H^1(\omega)^N \times H^1(\omega)^N \times L^2(\omega) \\ &\qquad \varepsilon^0(v^0, v^1) = 0 \text{ and } \varepsilon^1(v^0, v^1, v^2_n n) = 0\big\} \\ &= \big\{(v^0, v^1, 0) \colon (v^0, v^1) \in K\big\}. \end{aligned}$$

The variational equations associated with the two models will only involve the operators ε^0 and ε^1 and the solutions will naturally belong to a quotient space whose topology will be specified by the $L^2(\omega)$-norms of $\|\varepsilon^0(v)\|$ and $\|\varepsilon^1(v)\|$.

THEOREM 3.2. (i) ($P(1,1)$ model)

$$\Big\{\big\|\varepsilon^0(v^0, v^1)\big\|^2_{L^2(\omega)} + \big\|\varepsilon^1(v^0, v^1, 0)\big\|^2_{L^2(\omega)} + \|v^0\|^2_{L^2(\omega)} + \|v^1_\Gamma\|^2_{L^2(\omega)}\Big\}^{1/2}$$

is a norm equivalent to the standard norm on $H^1(\omega)^N \times H^1(\omega)^N$ *and*

$$\Big\{\big\|\varepsilon^0(v^0, v^1)\big\|^2_{L^2(\omega)} + \big\|\varepsilon^1(v^0, v^1, 0)\big\|^2_{L^2(\omega)}\Big\}^{1/2} \tag{3.19}$$

is a norm equivalent to the canonical quotient norm on

$$\frac{H^1(\omega)^N \times H^1(\omega)^N}{K}.$$

(ii) ($P(2_n, 1)$ model)

$$\Big\{\big\|\varepsilon^0(v^0, v^1)\big\|^2_{L^2(\omega)} + \big\|\varepsilon^1(v^0, v^1, v^2_n n)\big\|^2_{L^2(\omega)} + \|v^0\|^2_{L^2(\omega)} + \|v^1_\Gamma\|^2_{L^2(\omega)}\Big\}^{1/2}$$

is a norm equivalent to the standard norm on $H^1(\omega)^N \times H^1(\omega)^N \times L^2(\omega)$ *and, as for the* $P(1,1)$ *model,*

$$\Big\{\big\|\varepsilon^0(v^0, v^1)\big\|^2_{L^2(\omega)} + \big\|\varepsilon^1(v^0, v^1, v^2_n n)\big\|^2_{L^2(\omega)}\Big\}^{1/2} \tag{3.20}$$

is a norm which is equivalent to the canonical quotient norm on

$$\frac{H^1(\omega)^N \times H^1(\omega)^N \times L^2(\omega)}{\ker \varepsilon^0 \cap \ker \varepsilon^1} = \frac{H^1(\omega)^N \times H^1(\omega)^N}{K} \times L^2(\omega).$$

Keeping in mind the differences, we shall use the common notation E^{01} for the two quotient spaces. The index ^{01} indicates that this Hilbert space is endowed with the topology generated by the inner product

$$\int_\omega \varepsilon^0(v)\cdot\cdot\varepsilon^0(w) + \varepsilon^1(v)\cdot\cdot\varepsilon^1(w)\, d\Gamma. \tag{3.21}$$

Recall the decomposition of a tensor into its tangential and normal parts along the level sets of b [**28, 32**]

$$\varepsilon^i = \varepsilon^{iP} + (P\varepsilon^i n)^* n + n^*(P\varepsilon^i n) + (\varepsilon^i n\cdot n)n^*n \tag{3.22}$$

and the identities

$$A^P \stackrel{\text{def}}{=} PAP, \quad D_\Gamma^P(v) = PD_\Gamma(v), \quad \varepsilon_\Gamma^P(v) = P\varepsilon_\Gamma(v)P, \quad \varepsilon^{iP} = P\varepsilon^i P \tag{3.23}$$

By direct computation

$$\varepsilon^{0P}(v^0, v^1) = \varepsilon_\Gamma^P(v^0) \tag{3.24}$$

$$P\varepsilon^0(v^0,v^1)n = \frac{1}{2}\big[v_\Gamma^1 + {}^*D_\Gamma(v^0)n\big], \quad \varepsilon^0(v^0,v^1)n\cdot n = v_n^1 \tag{3.25}$$

$$\varepsilon^{1P}(v^0,v^1,v^2) = \varepsilon^{1P}(v^0,v^1,0) \tag{3.26}$$

$$\begin{aligned} P\varepsilon^1(v^0,v^1,v^2)n &= v_\Gamma^2 + \frac{1}{2}\big[{}^*D_\Gamma(v^1)n - D^2b^*D_\Gamma(v^0)n\big] \\ &= v_\Gamma^2 + \frac{1}{2}\big[\nabla_\Gamma v_n^1 - D^2b(v_\Gamma^1 + {}^*D_\Gamma(v^0)n)\big] \\ &= v_\Gamma^2 + \frac{1}{2}\nabla_\Gamma v_n^1 - D^2b\big(\varepsilon^0(v^0,v^1)n\big) \\ &= v_\Gamma^2 + \frac{1}{2}\nabla_\Gamma\big(\varepsilon^0(v^0,v^1)n\cdot n\big) - D^2b\big(\varepsilon^0(v^0,v^1)n\big) \end{aligned} \tag{3.27}$$

$$\varepsilon^1(v^0,v^1,v^2)n\cdot n = 2v_n^2 = \varepsilon^1(0,0,v_n^2 n)n\cdot n. \tag{3.28}$$

For $(v^0,v^1)\in H^1(\omega)^N \times L^2(\omega)^N$

$$\big\|\varepsilon^0(v^0,v^1)\big\|^2 = \big\|\varepsilon_\Gamma^P(v^0)\big\|^2 + \frac{1}{2}\|v_\Gamma^1 + \nabla_\Gamma v_n^0 - D^2bv_\Gamma^0\|^2 + \|v_n^1\|^2. \tag{3.29}$$

The first term is associated with the *membrane energy*, the second with the *shearing energy* and the third with the *pinching energy*. The other part is the *extended bending energy* $\big\|\varepsilon^1(v^0,v^1,v^2)\big\|^2$. For $(v^0,v^1,v_n^2)\in H^1(\omega)^N\times H^1(\omega)^N\times L^2(\omega)$

$$\begin{aligned} \big\|\varepsilon^1(v^0,v^1,v_n^2 n)\big\|^2 &= \left\|\varepsilon_\Gamma^P(v^1) - \frac{1}{2}\big[D_\Gamma^P(v^0)D^2b + D^2b^*D_\Gamma^P(v^0)\big]\right\|^2 \\ &\quad + \left\|\frac{1}{2}\big[\nabla_\Gamma v_n^1 - D^2b(v_\Gamma^1 + \nabla_\Gamma v_n^0 - D^2bv_\Gamma^0)\big]\right\|^2 + \|2v_n^2\|^2. \end{aligned}$$

The first term is a *bending energy norm* while the second is again a *shearing energy norm* where the shearing energy term of $\varepsilon^0(v^0,v^1)$ is multiplied by D^2b.

The two models are characterized by the following variational equation: to find v_h in E^{01} such that for all v in E^{01}

$$\boxed{\begin{aligned} \int_\omega &\overline{\alpha}_0(h)C^{-1}\varepsilon_h^0\cdot\cdot\varepsilon^0 + \overline{\alpha}_1(h)[C^{-1}\varepsilon_h^1\cdot\cdot\varepsilon^0 + C^{-1}\varepsilon_h^0\cdot\cdot\varepsilon^1] + \overline{\alpha}_2(h)C^{-1}\varepsilon_h^1\cdot\cdot\varepsilon^1 \\ &- F_h^0\cdot v^0 - F_h^1\cdot v^1 - F_h^2\cdot nv_n^2 - G_h^0\cdot v^1 - G_h^1\cdot nv_n^2, \quad d\Gamma = 0. \end{aligned}} \tag{3.30}$$

where $\varepsilon_h^i \overset{\text{def}}{=} \varepsilon^i(v_h)$, $\varepsilon^i \overset{\text{def}}{=} \varepsilon^i(v)$, and

$$\int_\omega F_h^0 \cdot v^0 + F_h^1 \cdot v^1 + F_h^2 \cdot v^2 \, d\Gamma \overset{\text{def}}{=} \frac{1}{2h} \int_{S_h(\omega)} F \cdot (v^0 \circ p + bv^1 \circ p + b^2 v^2 \circ p) \, dx \tag{3.31}$$

$$\int_\omega G_h^0 \cdot v^1 + G_h^1 \cdot v^2 \, d\Gamma \overset{\text{def}}{=} \frac{1}{2h} \int_{S_h(\omega)} G \cdot D(v^0 \circ p + bv^1 \circ p + b^2 v^2 \circ p) \nabla b \, dx \tag{3.32}$$

since $D(v^0 \circ p + bv^1 \circ p + b^2 v^2 \circ p)\nabla b = v^1 \circ p + 2bv^2 \circ p$. For the $P(1,1)$ model (v_h^0, v_h^1) and (v^0, v^1) belong to $\left(H^1(\omega)^N \times H^1(\omega)^N\right)/K$, and for the $P(2_n, 1)$ model (v_h^0, v_h^1, v_{hn}^2) and (v^0, v^1, v_n^2) belong to $\left(H^1(\omega)^N \times H^1(\omega)^N\right)/K \times L^2(\omega)$.

In the two cases let Assumptions 2.1 and 3.1 be verified. For the $P(1,1)$ model, if F and G in $L^2\left(S_h(\omega)\right)^N$ verify the condition

$$\forall v = (v^0, v^1) \in K, \quad \int_\omega F_h^0 \cdot v^0 + (F_h^1 + G_h^0) \cdot v^1 \, d\Gamma = 0 \tag{3.33}$$

equation (3.30) has a solution $v_h = (v_h^0, v_h^1)$ in $H^1(\omega)^N \times H^1(\omega)^N$ unique up to rigid displacements in K. For the $P(2_n, 1)$ model, if F and G in $L^2\left(S_h(\omega)\right)^N$ verify

$$\forall v = (v^0, v^1) \in K, \quad \int_\omega F_h^0 \cdot v^0 + (F_h^1 + G_h^0) \cdot v^1 + (F_h^2 + G_h^1) \cdot n v_n^2 \, d\Gamma = 0 \tag{3.34}$$

the variational equation (3.30) has a solution $v_h = (v_h^0, v_h^1, v_{hn}^2)$ in $H^1(\omega)^N \times H^1(\omega)^N \times L^2(\omega)$ which is unique up to additive rigid displacements in K.

4. First Part of the Asymptotic Shell Model: Asymptotic $P(1,0)$ model

The variational equation obtained from the term B_0^0 in expression (3.17)

$$\boxed{\int_\omega C^{-1}\varepsilon^0(\hat{v}) \cdot\cdot\, \varepsilon^0(v) - f^0 \cdot v^0 - g^0 \cdot v^1 \, d\Gamma = 0, \quad \forall (v^0, v^1)} \tag{4.1}$$

coincides with the asymptotic $P(1,0)$ model. It yields the following two variational equations for the respective vector test functions v^0 and v^1

$$\int_\omega \left\{ C^{-1}\varepsilon^0(\hat{v}) \cdot\cdot\, \frac{1}{2}[v^{1*}n + n^* v^1] - g^0 \cdot v^1 \right\} d\Gamma = 0, \quad \forall v^1,$$

$$\int_\omega \left\{ C^{-1}\varepsilon^0(\hat{v}) \cdot\cdot\, \varepsilon_\Gamma(v^0) - f^0 \cdot v^0 \right\} d\Gamma = 0, \quad \forall v^0.$$

Again using the decomposition of a tensor into its tangential and normal parts along the level sets of b in the second equation, the above system is equivalent to the following new system of equation and variational equation.

$$\boxed{\begin{gathered} \left[C^{-1}\varepsilon^0(\hat{v})\right]n = g^0 \\ \forall v^0, \quad \int_\omega \left\{ \left[C^{-1}\varepsilon^0(\hat{v})\right]^P \cdot\cdot\, \varepsilon_\Gamma^P(v^0) - f^0 \cdot v^0 + 2g_\Gamma^0 \cdot \varepsilon_\Gamma(v^0)n \right\} d\Gamma = 0. \end{gathered}}$$

If we introduce the *stress tensor* $\sigma^0 = C^{-1}\varepsilon^0$, then

$$\boxed{\sigma^0(\hat{v})n = g^0 \text{ and } \sigma^0(\hat{v})n \cdot n = g_n^0.} \tag{4.2}$$

Therefore the case $g^0 = 0$ corresponds to *zero normal stress*, and $g_n^0 = 0$ to *plane stress*. So these two familiar cases are related to an assumption on the loading g^0 and not on the shell itself. In the literature for the usual constitutive law and $g^0 = 0$ we have the plane stress assumption and the Love-Kirchhoff condition.

For the special *constitutive law* $C^{-1}\varepsilon = 2\mu\varepsilon + \lambda \operatorname{tr}\varepsilon I$, assuming that $\hat{v}^0$, $\hat{v}^1$ and Γ are *sufficiently smooth*,

$$\left[C^{-1}\varepsilon^0(\hat{v})\right]n = 2\mu\left[\varepsilon_\Gamma(\hat{v}_0)n + \frac{1}{2}v_\Gamma^1\right] + \left[\lambda \operatorname{tr}\varepsilon_\Gamma(\hat{v}_0) + (2\mu+\lambda)\hat{v}_n^1\right]n$$

and the condition $\left[C^{-1}\varepsilon^0(\hat{v})\right]n = g^0$ yields the following explicit expressions of the normal and tangential components of $\hat{v}^1$ in term of $\hat{v}^0$

$$\tag{4.3} \boxed{\begin{aligned} \hat{v}_n^1 &= -\frac{\lambda}{2\mu+\lambda}\operatorname{tr}\varepsilon_\Gamma^P(\hat{v}^0) + \frac{1}{2\mu+\lambda}g_n^0 \\ &= -\frac{\lambda}{2\mu+\lambda}\left[\kappa_1\hat{v}_n^0 + \operatorname{div}_\Gamma(\hat{v}_\Gamma^0)\right] + \frac{1}{2\mu+\lambda}g_n^0 \\ \hat{v}_\Gamma^1 &= -2\varepsilon_\Gamma(\hat{v}^0)n + \frac{1}{\mu}g_\Gamma^0 \\ &= -\nabla_\Gamma(\hat{v}_n^0) + D^2b\hat{v}_\Gamma^0 + \frac{1}{\mu}g_\Gamma^0 \quad \text{(Love-Kirchhoff term).} \end{aligned}}$$

The substitution of the value of $\hat{v}^1$ into the term

$$\begin{aligned} \left[C^{-1}\varepsilon^0(\hat{v})\right]^P \cdot\cdot\, \varepsilon_\Gamma^P(v^0) = 2\mu\varepsilon_\Gamma^P(\hat{v}^0)\cdot\cdot\,\varepsilon_\Gamma^P(v^0) + \frac{2\mu\lambda}{2\mu+\lambda}\operatorname{tr}\varepsilon_\Gamma^P(\hat{v}^0)\operatorname{tr}\varepsilon_\Gamma^P(v^0) \\ + \frac{\lambda}{2\mu+\lambda}g_n^0\operatorname{tr}\varepsilon_\Gamma^P(v^0) \end{aligned}$$

yields the second equation for $\hat{v}^0$: for all v^0

$$\tag{4.4} \boxed{\begin{aligned} \int_\omega 2\mu\varepsilon_\Gamma^P(\hat{v}^0)\cdot\cdot\,\varepsilon_\Gamma^P(v^0) + \frac{2\mu\lambda}{2\mu+\lambda}\operatorname{tr}\varepsilon_\Gamma^P(\hat{v}^0)\operatorname{tr}\varepsilon_\Gamma^P(v^0) \\ + \frac{\lambda}{2\mu+\lambda}g_n^0\operatorname{tr}\varepsilon_\Gamma^P(v^0) + 2g_\Gamma^0\cdot\varepsilon_\Gamma(v^0)n - f^0\cdot v^0\, d\Gamma = 0. \end{aligned}}$$

This is the *asymptotic membrane shell equation* with the right coefficients and the tensor $\varepsilon_\Gamma^P(\hat{v}^0)$ is the *linearized change of metric tensor* $\gamma_{\alpha\beta}$, of Γ (cf. [**28, 32**]). The asymptotic $P(1,0)$ model simultaneously yields $\hat{v}^0$ and $\hat{v}^1$.

The solution $(\hat{v}^0, \hat{v}^1)$ of equation (4.1) is such that the *extended membrane energy* $\left\|\varepsilon^0(\hat{v}^0, \hat{v}^1)\right\|$ is finite. We have seen that for the special constitutive law $C^{-1}\varepsilon = 2\mu\varepsilon + \lambda\operatorname{tr}\varepsilon I$, equation (4.1) formally yields the explicit expressions (4.3) for $\hat{v}^1$ and the membrane shell equation (4.4) for $\hat{v}^0$. If we introduce the *effective constitutive law* $C_e^{-1}\varepsilon = 2\mu\varepsilon + 2\mu\lambda/(2\mu+\lambda)\operatorname{tr}\varepsilon I$, the bilinear form associated with (4.4) takes the more compact form

$$\tag{4.5} \int_\omega \left[C_e^{-1}\varepsilon_\Gamma^P(u^0)\right]\cdot\cdot\,\varepsilon_\Gamma^P(v^0)\, d\Gamma$$

and it is easy to verify that this bilinear form is V-H-coercive for the spaces

$$\tag{4.6} V \stackrel{\text{def}}{=} \left\{v \in L^2(\omega)^N : v_\Gamma \in H^1(\omega)^N\right\} \subset H \stackrel{\text{def}}{=} L^2(\omega)^N.$$

As in the case of $\varepsilon^0(v^0, v^1)$, the variational equation associated with the bilinear form (4.5) has a unique solution in the *completion* E_Γ^P of the quotient space $V/\ker\varepsilon_\Gamma^P$

with respect to the norm $\|\varepsilon_\Gamma^P(v)\|$ for a right-hand side continuous with respect to that norm. This raises general issues about the nature of this completion and the classification of the different cases according to the geometry of the underlying submanifold and the choice of boundary conditions (cf. [**19, 20, 50**]).

4.1. Existence of solutions to the asymptotic $P(1,0)$ model. We first give an existence result for the variational equation (4.1) by introducing an appropriate quotient space for the solution. Recall from [**28**] and [**32**] that

$$\left\{\left\|\varepsilon^0(v^0,v^1)\right\|^2_{L^2(\omega)} + \|v^0\|^2_{L^2(\omega)} + \|v^1_\Gamma\|^2_{L^2(\omega)}\right\}^{1/2}$$

is a norm equivalent to the standard norm on $H^1(\omega)^N \times L^2(\omega)^N$ and $\left\|\varepsilon^0(v^0,v^1)\right\|$ is a norm on the quotient space

$$\left(H^1(\omega)^N \times L^2(\omega)^N\right)/\ker\varepsilon^0$$

which generally generates a topology *weaker* than the one generated by the canonical norm on the quotient space. Denote by E^0 the *completion* of this quotient space with respect to the norm generated by the inner product

$$\int_\omega \varepsilon^0(v^0,v^1)\cdot\cdot\,\varepsilon^0(w^0,w^1)\,d\Gamma. \tag{4.7}$$

For the homogeneous Dirichlet problem on a part γ_0 of γ, denote by $E^0_{\gamma_0}$ the completion of the quotient space

$$\left(H^1_{\gamma_0}(\omega)^N \times L^2(\omega)^N\right)/\ker\varepsilon^0, \quad H^1_{\gamma_0}(\omega) \stackrel{\text{def}}{=} \left\{v\in H^1(\omega)\colon v=0 \text{ on } \gamma_0\right\}$$

with respect to the norm generated by the scalar product (4.7).

THEOREM 4.1. *Under Assumption* 2.1 *on* C, *let* ω *be a Lipschitzian domain in* Γ *which is itself the boundary of a set of class* $C^{1,1}$ *in* $\mathbf{R}^N$.

(i) *Given* $\ell^0 \in (E^0)'$, *the variational equation*

$$\forall (v^0,v^1)\in E^0, \quad \int_\omega \left[C^{-1}\varepsilon^0(\hat v^0,\hat v^1)\right]\cdot\cdot\,\varepsilon^0(v^0,v^1)\,d\Gamma = \ell^0(v^0,v^1), \tag{4.8}$$

has a unique solution $(\hat v^0,\hat v^1)$ *in* E^0.

(ii) *Assume that* ω *is connected and that* γ_0 *is a subset of* γ *with strictly positive* $(N-1)$ *capacity. Given* $\ell^0\in(E^0_{\gamma_0})'$, *the variational equation*

$$\forall (v^0,v^1)\in E^0_{\gamma_0}, \quad \int_\omega \left[C^{-1}\varepsilon^0(\hat v^0,\hat v^1)\right]\cdot\cdot\,\varepsilon^0(v^0,v^1)\,d\Gamma = \ell^0(v^0,v^1), \tag{4.9}$$

has a unique solution $(\hat v^0,\hat v^1)$ *in* $E^0_{\gamma_0}$.

An important challenging issue is to identify the elements of E^0 and $E^0_{\gamma_0}$ and characterize their regularity for different geometries and boundary conditions.

4.2. Convergence to the asymptotic $P(1,0)$ model. For simplicity we specialize to vector functions F and G of the form

$$f = F\circ T = f^0 + zf^1 + z^2 f^2 \text{ and } g = G\circ T = g^0 + zg^1 + z^2 g^2 \tag{4.10}$$

for vector functions f^0, f^1, f^2, g^0, g^1 and g^2 in $L^2(\omega)^N$. The results can be readily extended to sufficiently differentiable f and g with respect to z with $f^0(X) = f(X,0)$, $f^1(X) = \partial f/\partial z(X,0)$, $f^2(X) = (\partial^2 f/\partial^2 z)(X,0)/2$, etc ...

ASSUMPTION 4.1. There exists $\bar{h} > 0$ and a constant $c > 0$ such that for all h, $0 < h \le \bar{h}$, and all $(v^0, v^1, v_n^2) \in H^1(\omega)^N \times H^1(\omega)^N \times L^2(\omega)$, the following condition is verified

$$\left|\frac{1}{2h}\int_{S_h(\omega)} F\cdot\big(v^0\circ p + bv^1\circ p + b^2(v_n^2 n)\circ p\big) + G\cdot D\big(v^0\circ p + bv^1\circ p + b^2(v_n^2 n)\circ p\big)\nabla b\, dx\right| \tag{4.11}$$

$$\le c\left\{\int_\omega \Big[\big\|\varepsilon^0(v^0, v^1)\big\|^2 + h^2\|\varepsilon^1(v^0, v^1, v_n^2 n)\|^2\Big]\, d\Gamma\right\}^{1/2}. \tag{4.12}$$

Same assumption for the $P(1,1)$ model with $v^2 = 0$.

This is an assumption on the continuity constant of the average of the right hand-side of the variational equation of the three-dimensional problem. Under this assumption, the variational equation (3.30) makes sense and has a unique solution in E^{01} for the $P(2_n, 1)$ model and the $P(1,1)$ model ($v^2 = 0$).

It will be convenient to introduce the following notation

$$\begin{aligned}
I(h)v \overset{\text{def}}{=} &\frac{1}{2h}\int_{S_h(\omega)} F\cdot(v^0\circ p + bv^1\circ p + b^2v^2\circ p)\\
&+ G\cdot D(v^0\circ p + bv^1\circ p + b^2v^2\circ p)\nabla b\, dx\\
= &\int_\omega \overline{\alpha}_0(h)(f^0\cdot v^0 + g^0\cdot v^1)\\
&+ \overline{\alpha}_1(h)\big(f^1\cdot v^0 + (f^0+g^1)\cdot v^1 + 2g^0\cdot v^2\big)\\
&+ \overline{\alpha}_2(h)\big(f^2\cdot v^0 + (f^1+g^2)\cdot v^1 + (f^0+2g^1)\cdot v^2\big)\\
&+ \overline{\alpha}_3(h)\big(f^2\cdot v^1 + (f^1+2g^2)\cdot v^2\big) + \overline{\alpha}_4(h) f^2\cdot v^2\, d\Gamma
\end{aligned} \tag{4.13}$$

$$\begin{aligned}
J(h)v \overset{\text{def}}{=} &\int_\omega \frac{\overline{\alpha}_0(h)-1}{h^2}(f^0\cdot v^0 + g^0\cdot v^1)\\
&+ \frac{\overline{\alpha}_1(h)}{h^2}\big(f^1\cdot v^0 + (f^0+g^1)\cdot v^1 + 2g^0\cdot v^2\big)\\
&+ \frac{\overline{\alpha}_2(h)}{h^2}\big(f^2\cdot v^0 + (f^1+g^2)\cdot v^1 + (f^0+2g^1)\cdot v^2\big)\\
&+ \frac{\overline{\alpha}_3(h)}{h^2}\big(f^2\cdot v^1 + (f^1+2g^2)\cdot v^2\big) + \frac{\overline{\alpha}_4(h)}{h^2} f^2\cdot v^2\, d\Gamma
\end{aligned} \tag{4.14}$$

$$\begin{aligned}
K(h)v \overset{\text{def}}{=} &\int_\omega \frac{1}{h^2}\left[\frac{\overline{\alpha}_0(h)-1}{h^2} - \frac{\kappa_2}{3}\right](f^0\cdot v^0 + g^0\cdot v^1)\\
&+ \frac{1}{h^2}\left[\frac{\overline{\alpha}_1(h)}{h^2} - \frac{\kappa_1}{3}\right]\big(f^1\cdot v^0 + (f^0+g^1)\cdot v^1 + 2g^0\cdot v^2\big)\\
&+ \frac{1}{h^2}\left[\frac{\overline{\alpha}_2(h)}{h^2} - \frac{1}{3}\right]\big(f^2\cdot v^0 + (f^1+g^2)\cdot v^1 + (f^0+2g^1)\cdot v^2\big)\\
&+ \frac{1}{h^2}\left[\frac{\overline{\alpha}_3(h)}{h^2}\right]\big(f^2\cdot v^1 + (f^1+2g^2)\cdot v^2\big)\\
&+ \frac{1}{h^2}\left[\frac{\overline{\alpha}_4(h)}{h^2}\right] f^2\cdot v^2\, d\Gamma.
\end{aligned} \tag{4.15}$$

(4.16)

$$L(h)v \stackrel{\text{def}}{=} \int_\omega \frac{1}{h^2}\left\{\frac{1}{h^2}\left[\frac{\overline{\alpha}_0(h)-1}{h^2}-\frac{\kappa_2}{3}\right]-\frac{\kappa_4}{5}\right\}(f^0\cdot v^0+g^0\cdot v^1)$$
$$+\frac{1}{h^2}\left\{\frac{1}{h^2}\left[\frac{\overline{\alpha}_1(h)}{h^2}-\frac{\kappa_1}{3}\right]-\frac{\kappa_3}{5}\right\}\big(f^1\cdot v^0+(f^0+g^1)\times v^1+2g^0\cdot v^2\big)$$
$$+\frac{1}{h^2}\left\{\frac{1}{h^2}\left[\frac{\overline{\alpha}_2(h)}{h^2}-\frac{1}{3}\right]-\frac{\kappa_2}{5}\right\}\big(f^2\cdot v^0+(f^1+g^2)\times v^1+(f^0+2g^1)\cdot v^2\big)$$
$$+\frac{1}{h^2}\left\{\frac{\overline{\alpha}_3(h)}{h^4}-\frac{\kappa_1}{5}\right\}\big(f^2\cdot v^1+(f^1+2g^2)\cdot v^2\big)$$
$$+\frac{1}{h^2}\left\{\frac{\overline{\alpha}_4(h)}{h^4}-\frac{1}{5}\right\}f^2\cdot v^2\,d\Gamma.$$

By definition Assumption 4.1 is equivalent to

$$(4.17)\qquad \big|I(h)(v^0,v^1,v_n^2n)\big| \le c\left\{\int_\omega \big\|\varepsilon^0(v^0,v^1)\big\|^2+h^2\big\|\varepsilon^1(v^0,v^1,v_n^2n)\big\|^2\,d\Gamma\right\}^{1/2}.$$

Note that $I(h)-I(0)=h^2J(h)$, $J(h)-J(0)=h^2K(h)$, $K(h)-K(0)=h^2L(h)$

$$(4.18)\qquad \begin{aligned} I(h) &= I(0)+h^2J(0)+h^4K(h)\\ &= I(0)+h^2J(0)+h^4K(0)+h^4\big(K(h)-K(0)\big)\\ &= I(0)+h^2J(0)+h^4K(0)+h^6L(0)+h^6\big(L(h)-L(0)\big).\end{aligned}$$

In fact since j_z is a polynomial of degree $N-1$ this series expansion stops in a finite number of steps with the coefficients $I(0)$, $J(0)$, $K(0)$, $L(0)$, etc ... For $N=3$

$$I(h)v = I(0)+h^2J(0)+h^4K(0)+h^6L(0)$$

where

$$I(0)v=\int_\omega (f^0\cdot v^0+g^0\cdot v^1)\,d\Gamma,\quad L(0)v=\frac{1}{7}\int_\omega \kappa_2 f^2\cdot v^2\,d\Gamma$$
$$J(0)v=\big\{\kappa_2(f^0\cdot v^0+g^0\cdot v^1)+\kappa_1\big(f^1\cdot v^0+(f^0+g^1)\cdot v^1+2g^0\cdot v^2\big)$$
$$+\big(f^2\cdot v^0+(f^1+g^2)\cdot v^1+(f^0+2g^1)\cdot v^2\big)\big\}\,d\Gamma$$
$$K(0)v=\frac{1}{5}\int_\omega \big\{\kappa_2(f^2\cdot v^0+(f^1+g^2)\cdot v^1+\big(f^0+2g^1)\cdot v^2\big)$$
$$+\kappa_1\big(f^2\cdot v^1+(f^1+2g^2)\cdot v^2\big)+f^2\cdot v^2\big\}\,d\Gamma.$$

We summarize the key results (with $v^2=0$ for the $P(1,1)$ model).

LEMMA 4.1. (i) *Assume that there exists* $\bar{h}>0$ *and constants* $c>0$ *such that for all* h, $0<h\le\bar{h}$, *the following three conditions are verified*:

condition on I(0): *for all* (v^0,v^1) *in* $H^1(\omega)^N\times L^2(\omega)^N$

$$(4.19)\qquad \big|I(0)(v^0,v^1,0)\big|\le c\big\|\varepsilon^0(v^0,v^1)\big\|_{L^2(\omega)}$$

condition on J(0): *for all* $v=(v^0,v^1,v_n^2)\in H^1(\omega)^N\times H^1(\omega)^N\times L^2(\omega)$

$$(4.20)\qquad \big|J(0)v\big|\le c\left\{\int_\omega \big\|\varepsilon^0(v^0,v^1)\big\|^2+\big\|\varepsilon^1(v^0,v^1,v_n^2n)\big\|^2\,d\Gamma\right\}^{1/2}$$

condition on K(h): *for all* $v = (v^0, v^1, v^2_n) \in H^1(\omega)^N \times H^1(\omega)^N \times L^2(\omega)$

$$|K(h)v| \le c\left\{\int_\omega \|\varepsilon^0(v^0,v^1)\|^2 + \|\varepsilon^1(v^0,v^1,v^2_n n)\|^2\, d\Gamma\right\}^{1/2}, \tag{4.21}$$

then Assumption 4.1 is verified.

(ii) *As $h > 0$ goes to zero, there exists $c > 0$ such that for all (v^0, v^1, v^2_n) in E^{01}*

$$|I(h)v - I(0)v| \le ch^2\left\{\int_\omega \|\varepsilon^0(v^0,v^1)\|^2 + \|\varepsilon^1(v^0,v^1,v^2_n n)\|^2\, d\Gamma\right\}^{1/2} \tag{4.22}$$

$$|J(h)v - J(0)v| \le ch^2\left\{\|\varepsilon^0(v^0,v^1)\|^2 + \|\varepsilon^1(v^0,v^1,v^2_n n)\|^2\, d\Gamma\right\}^{1/2}. \tag{4.23}$$

(iii) *For the $P(1,1)$ model ($v^2 = 0$) and the $P(2_n,1)$ model ($v^2_\Gamma = 0$) conditions (4.20) and (4.21) are equivalent to $J(0) = 0$ and $K(h) = 0$, $0 < h \le \bar{h}$, on K. Moreover condition (4.17) is equivalent to the three conditions (4.19), (4.20), and (4.21).*

In particular for $N = 3$ Assumption 4.1 is equivalent to the condition (4.19) and the two conditions: for all $(v^0, v^1) \in K$

$$\int_\omega \left(f^2\cdot v^0 + (f^1+g^2)\cdot v^1\right) + \kappa_1\left(f^1\cdot v^0 + (f^0+g^1)\cdot v^1\right) + \kappa_2(f^0\cdot v^0 + g^0\cdot v^1)\, d\Gamma = 0 \tag{4.24}$$

$$\int_\omega \kappa_2\left(f^2\cdot v^0 + (f^1+g^2)\cdot v^1\right) + \kappa_1 f^2\cdot v^1\, d\Gamma = 0. \tag{4.25}$$

PROOF. (i) This follows from conditions (4.19), (4.20), (4.21) and the identity

$$I(h) = I(0) + h^2 J(0) + h^4 K(h)$$
$$\Longrightarrow |I(h)| \le c\|\varepsilon^0\| + ch^2\left[\|\varepsilon^0\| + \|\varepsilon^1\|\right] + ch^4\left[\|\varepsilon^0\| + \|\varepsilon^1\|\right] \le c\left[\|\varepsilon^0\| + h^2\|\varepsilon^1\|\right].$$

(ii) Condition (4.22) follows from (4.20) and (4.21) for all $v \in H^1(\omega)^N \times H^1(\omega)^N \times L^2(\omega)$

$$I(h)v - I(0)v = h^2 J(h)v = h^2\left[J(0) + h^2 K(h)\right]v$$
$$\Longrightarrow |I(h)v - I(0)v| \le ch^2\left[\|\varepsilon^0(v)\| + \|\varepsilon^1(v)\|\right].$$

Similarly from (4.21)

$$J(h) - J(0) = h^2 K(h) \Longrightarrow |J(h) - J(0)| \le ch^2\left[\|\varepsilon^0(v)\| + \|\varepsilon^1(v)\|\right]$$

and this gives (4.23).

(iii) We only do the $P(1,1)$ case. The $P(2_n,1)$ case is similar. Recall that for the $P(1,1)$ case, the canonical quotient norm on $\left(H^1(\omega)^N \times H^1(\omega)^N\right)/K$ is equivalent to the norm $\|\varepsilon^0(v^0,v^1)\| + \|\varepsilon^1(v^0,v^1,0)\|$. As a result a continuous linear functional on $L^2(\omega)^N \times L^2(\omega)^N$ is continuous on the quotient space if and only if it is zero on K. Moreover by definition of the $\overline{\alpha}_i(h)$'s the coefficients in the expressions of

$J(h)v$ and $K(h)v$ are bounded in $\overline{\omega}$

$$\frac{\overline{\alpha}_0(h)-1}{h^2} \to \frac{\kappa_2}{3}, \quad \frac{\overline{\alpha}_1(h)}{h^2} \to \frac{\kappa_1}{3}, \quad \frac{\overline{\alpha}_2(h)}{h^2} \to \frac{1}{3}, \quad \frac{\overline{\alpha}_3(h)}{h^2} \to 0, \quad \frac{\overline{\alpha}_4(h)}{h^2} \to 0,$$
$$\frac{1}{h^2}\left[\frac{\overline{\alpha}_0(h)-1}{h^2} - \frac{\kappa_2}{3}\right], \quad \frac{1}{h^2}\left[\frac{\overline{\alpha}_1(h)}{h^2} - \frac{\kappa_1}{3}\right], \quad \frac{1}{h^2}\left[\frac{\overline{\alpha}_2(h)}{h^2} - \frac{1}{3}\right], \quad \frac{\overline{\alpha}_3(h)}{h^4}, \quad \frac{\overline{\alpha}_4(h)}{h^4}.$$

They also strongly converge in the L^∞ norm. Therefore

$$\exists c > 0, \forall h,\ 0 < h \le \bar{h}, \quad |J(h)v| \le c\|v\|_{L^2}, |K(h)v| \le c\|v\|_{L^2}.$$

If, in addition, for all h, $0 < h \le \bar{h}$, $J(h)v = 0 = K(h)v$ for all v in K, then for all w in K

$$|J(h)v| = |J(h)(v+w)| \le c\|v+w\|_{L^2} \le c\|v+w\|_{H^1}$$
$$\implies |J(h)v| \le \inf_{w\in K} \|v+w\|_{H^1} \le c'[\|\varepsilon^0(v)\|^2 + \|\varepsilon^1(v)\|^2]^{1/2}$$

by equivalence of norms. The same is true for $K(h)$ and

$$\exists c > 0,\ \forall h,\ 0 < h \le \bar{h}, \quad |J(h)v| \le c\Big[\|\varepsilon^0(v)\|^2 + \|\varepsilon^1(v)\|^2\Big]^{1/2},$$
$$|K(h)v| \le c\Big[\|\varepsilon^0(v)\|^2 + \|\varepsilon^1(v)\|^2\Big]^{1/2}.$$

Therefore conditions (4.20) and (4.21) are verified if and only if

$$J(0)v = 0 \quad \text{and} \quad K(h)v = 0 \text{ on } K.$$

We have already shown in part (i) that conditions (4.19), (4.20), (4.21) imply condition (4.17). Conversely the condition

$$\exists c > 0,\ \forall h,\ 0 < h \le \bar{h}, \quad |I(h)v| \le c\Big[\|\varepsilon^0(v)\|^2 + h^2\|\varepsilon^1(v)\|^2\Big]^{1/2},$$

implies (4.19) by continuity of $I(h)v$ at $h = 0$. Then

$$\forall h,\ 0 < h \le \bar{h}, \quad J(h) = \frac{1}{h^2}\big[I(h) - I(0)\big] \implies J(h)v = 0 \text{ on } K.$$

Again by strong continuity of the coefficient in L^∞

$$\forall v \in H^1(\omega)^N \times H^1(\omega)^N, \quad J(h)v \to J(0)v \implies J(0) = 0 \text{ on } K.$$

From the first part of the proof $J(0)$ verifies condition (4.20). Next

$$\forall h,\ 0 < h \le \bar{h}, \quad K(h) = \frac{1}{h^2}\big[J(h) - J(0)\big] \implies K(h) = 0 \text{ on } K$$

and by the same argument condition (4.21) is verified for $0 < h \le \bar{h}$. For $N = 3$, $j_z = 1 + \kappa_1 z + \kappa_2 z^2$, and

$$\alpha_0(h) = 2h + \kappa_2\frac{2h^3}{3}, \quad \alpha_1(h) = \kappa_1\frac{2h^3}{3}, \quad \alpha_2(h) = \frac{2h^3}{3} + \kappa_2\frac{2h^5}{5}$$
$$\alpha_3(h) = \kappa_1\frac{2h^5}{5}, \quad \alpha_4(h) = \frac{2h^5}{5} + \kappa_2\frac{2h^7}{7}.$$

So for the first two terms in (4.21) are zero and

$$\frac{1}{h^2}\left[\frac{\overline{\alpha}_2(h)}{h^2} - \frac{1}{3}\right] = \frac{\kappa_2}{5}, \quad \frac{\overline{\alpha}_3(h)}{h^4} = \frac{\kappa_1}{5}, \quad \frac{\overline{\alpha}_4(h)}{h^4} = \frac{1}{5} + \kappa_2\frac{h^2}{7}.$$

Recalling that $v^2 = 0$ then we get (4.25) and $L(0)v = 0$ on K. □

We have the strong convergence in E^0.

THEOREM 4.2. *Let Assumptions* 2.1, 3.1 *and* 4.1 *be verified, and assume that* ω *is a bounded open Lipschitzian domain in* Γ. *As* h *goes to zero, the solution*

$$(v_h^0, v_h^1, v_{hn}^2) \quad (\text{resp. } (v_h^0, v_h^1)) \in E^{01}$$

of the $P(2_n, 1)$ (*resp.* $P(1,1)$) *model* (3.30) *strongly converges to the solution* $(\hat{v}^0, \hat{v}^1) \in E^0$ *of the asymptotic* $P(1,0)$ *model* (4.1). *Moreover, as* h *goes to zero,*

$$\int_\omega \left\|\varepsilon^0(v_h^0 - \hat{v}^0, v_h^1 - \hat{v}^1)\right\|^2 + h^2 \left\|\varepsilon^1(v_h^0, v_h^1, v_{hn}^2 n)\right\|^2 d\Gamma \to 0. \tag{4.26}$$

The same results hold for the $P(1,1)$ *model with* $v_h^2 = 0$.

PROOF. (i) **Weak convergence:** By Assumption 2.1 first notice that

$$\int_\omega C^{-1}\varepsilon^0(u^0, u^1)\cdot\cdot\varepsilon^0(w^0, w^1)\, d\Gamma \quad \text{and} \quad \left\{\int_\omega C^{-1}\varepsilon^0(u^0, u^1)\cdot\cdot\varepsilon^0(u^0, u^1)\, d\Gamma\right\}^{1/2}$$

are equivalent inner product and norm on E^0. By Assumption 4.1 and coercivity of C^{-1} there exist $\alpha > 0$ and $c > 0$ such that

$$\sum_{i,j=0}^{1} \int_\omega \overline{\alpha}_{i+j}(h) C^{-1}\varepsilon_h^i\cdot\cdot\varepsilon_h^j\, d\Gamma = \int_\omega \sum_{i=0}^{2} F_h^i \cdot v_h^i + \sum_{i=0}^{1} G_h^i \cdot v_h^{i+1}\, d\Gamma$$

$$\Longrightarrow \alpha\left\{\|\varepsilon_h^0\|_{L^2}^2 + h^2\|\varepsilon_h^1\|_{L^2}^2\right\} \le c\left\{\|\varepsilon_h^0\|_{L^2}^2 + h^2\|\varepsilon_h^1\|_{L^2}^2\right\}^{1/2}.$$

Then as h goes to zero

$$\left\|\varepsilon^0(v_h^0, v_h^1)\right\|_{L^2}^2 + h^2\left\|\varepsilon^1(v_h^0, v_h^1, v_{hn}^2 n)\right\|_{L^2}^2 = \|\varepsilon_h^0\|_{L^2}^2 + h^2\|\varepsilon_h^1\|_{L^2}^2 \le c^2/\alpha^2.$$

Therefore $\varepsilon^0(v_h^0, v_h^1)$ and $h\varepsilon^1(v_h^0, v_h^1, v_{hn}^2 n)$ are bounded in L^2. But (v_h^0, v_h^1) is a well-defined element of E^0 and we can conclude that (v_h^0, v_h^1) is bounded in E^0 since $\left\|\varepsilon^0(v^0, v^1)\right\|$ is a norm in E^0. Therefore there exists $w \in E^0$ and a subsequence such that v_h weakly converges to w in E^0. By going to the limit in (3.30)

$$\int_\omega (\overline{\alpha}_0(h) - 1)C^{-1}\varepsilon_h^0\cdot\cdot\varepsilon^0 + \overline{\alpha}_1(h)[C^{-1}\varepsilon_h^1\cdot\cdot\varepsilon^0 + C^{-1}\varepsilon_h^0\cdot\cdot\varepsilon^1]$$
$$+ \overline{\alpha}_2(h)C^{-1}\varepsilon_h^1\cdot\cdot\varepsilon^1 + C^{-1}\varepsilon_h^0\cdot\cdot\varepsilon^0\, d\Gamma - I(h)v = 0$$

we get that w is solution of (4.1). Indeed the integral of $C^{-1}\varepsilon_h^0\cdot\cdot\varepsilon^0$ converges to the integral of $C^{-1}\varepsilon^0(w)\cdot\cdot\varepsilon^0(v)$. For all v in $H^1(\omega)^N \times H^1(\omega)^N \times L^2(\omega)$, $I(h)v$ goes to $I(0)v$. All the other terms go to zero by boundedness of $\|\varepsilon_h^0\|$ and $h\|\varepsilon_h^1\|$ and the fact that since $\left\|D^2 b(x)\right\|$ is uniformly bounded, the coefficients

$$\frac{\overline{\alpha}_0(h) - 1}{h^2} \to \frac{\kappa_2}{3}, \quad \frac{\overline{\alpha}_1(h)}{h^2} \to \frac{\kappa_1}{3}, \quad \frac{\overline{\alpha}_2(h)}{h^2} \to \frac{1}{3}$$

strongly converge as h goes to zero. By uniqueness in E^0, $w = \hat{v}$ in E^0. Therefore the whole sequence $\{v_h\}$ weakly converges to $\hat{v}$ in E^0.

(ii) **Strong convergence:** We now prove that, as h goes to zero,

$$\delta_h \stackrel{\text{def}}{=} \int_\omega C^{-1}(\varepsilon_h^0 - \hat{\varepsilon}^0)\cdot\cdot(\varepsilon_h^0 - \hat{\varepsilon}^0) + \overline{\alpha}_2(h)C^{-1}\varepsilon_h^1\cdot\cdot\varepsilon_h^1\, d\Gamma$$

goes to zero. This implies that $v_h \to \hat{v}$ in E^0-strong and that $h\|\varepsilon_h^1\|_{L^2(\omega)}$ goes to zero. By definition

$$\delta_h = -\int_\omega C^{-1}(\varepsilon_h^0 - \hat{\varepsilon}^0)\cdot\cdot\hat{\varepsilon}^0\,d\Gamma + \int_\omega C^{-1}(\varepsilon_h^0 - \hat{\varepsilon}^0)\cdot\cdot\varepsilon_h^0 + \overline{\alpha}_2(h)C^{-1}\varepsilon_h^1\cdot\cdot\varepsilon_h^1\,d\Gamma.$$

The first integral goes to zero by weak convergence in E^0. The second one can be computed from equations (3.30) and (4.1) with $v = v_h$:

$$\begin{gathered}\int_\omega C^{-1}\varepsilon_h^0\cdot\cdot\varepsilon_h^0 + (\overline{\alpha}_0(h)-1)C^{-1}\varepsilon_h^0\cdot\cdot\varepsilon_h^0 + 2\overline{\alpha}_1(h)C^{-1}\varepsilon_h^1\cdot\cdot\varepsilon_h^0 + \overline{\alpha}_2(h)C^{-1}\varepsilon_h^1\cdot\cdot\varepsilon_h^1 \\ -(F_h^0\cdot v_h^0 + F_h^1\cdot v_h^1 + F_h^2\cdot n v_{hn}^2) - (G_h^0\cdot v_h^1 + G_h^1\cdot n v_{hn}^2)\,d\Gamma = 0 \\ \int_\omega C^{-1}\hat{\varepsilon}^0\cdot\cdot\varepsilon_h^0 - f^0\cdot v_h^0 - g^0\cdot v_h^1\,d\Gamma = 0.\end{gathered}$$

By subtracting the two equations

$$\delta_h = -\int_\omega (\overline{\alpha}_0(h)-1)C^{-1}\varepsilon_h^0\cdot\cdot\varepsilon_h^0 + 2\overline{\alpha}_1(h)C^{-1}\varepsilon_h^0\cdot\cdot\varepsilon_h^1\,d\Gamma + I(h)v_h - I(0)v_h.$$

Since $\|\varepsilon_h^0\|$ and $\|h\varepsilon_h^1\|$ are both bounded and the coefficients strongly converge in h^2, the first integral is bounded by $ch^2 + ch$ and goes to zero. By Lemma 4.1 (ii) the second integral is also bounded by ch. The final result follows by equivalence of the seminorms and norms indicated at the beginning of part (i). □

5. Second Part of the Asymptotic Shell Model

The asymptotic $P(1,0)$ model has its solution in the completion E^0 of the quotient space $\left(H^1(\omega)^N \times L^2(\omega)^N\right)/\ker\varepsilon^0$. In general $\ker\varepsilon^0$ can be quite big and it is natural to wonder if some critical information has been forgotten in the limiting process and if $\ker\varepsilon^0$ can be reduced to the smaller set $\ker\varepsilon^0 \cap \ker\varepsilon^1$ or K of rigid displacements. In other words can the asymptotic model be the one of a shell with zero thickness? It turns out that under Assumptions 2.1, 3.1, and 4.1, the variational equation (4.1) and the condition (4.19) can be *complemented* with a second variational equation involving the operator ε^1 on $\ker\varepsilon^0$. The new system of equations characterizes an *asymptotic shell* with *finite total energy* measured in an appropriate norm and its solution is unique up to the "*rigid displacements*" for a shell without boundary or with homogeneous Neumann boundary conditions.

5.1. Second asymptotic (bending) equation. In order to better understand how the second asymptotic equation arises, we first proceed formally with the $P(2_n,1)$ model. The $P(1,1)$ model yields the same results with $v_h^2 = v^2 = \hat{v}^2 = 0$. Go back to the variational equation (3.30) which we rewrite as follows

$$\begin{aligned}\int_\omega C^{-1}\varepsilon_h^0\cdot\cdot\varepsilon^0 - f^0\cdot v^0 - g^0\cdot v^1\,d\Gamma + \int_\omega &(\overline{\alpha}_0(h)-1)\left(C^{-1}\varepsilon_h^0\cdot\cdot\varepsilon^0 - f^0\cdot v^0 - g^0\cdot v^1\right) \\ &+ \overline{\alpha}_1(h)\left(C^{-1}\varepsilon_h^0\cdot\cdot\varepsilon^1 + C^{-1}\varepsilon_h^1\cdot\cdot\varepsilon^0 - f^1\cdot v^0 - (f^0+g^1)\cdot v^1 - 2g^0\cdot v^2\right) \\ &+ \overline{\alpha}_2(h)\left(C^{-1}\varepsilon_h^1\cdot\cdot\varepsilon^1 - f^2\cdot v^0 - (f^1+g^2)\cdot v^1 - (f^0+2g^1)\cdot v^2\right) \\ &- \overline{\alpha}_3(h)\left(f^2\cdot v^1 + (f^1+2g^2)\cdot v^2\right) - \overline{\alpha}_4(h)f^2\cdot v^2\,d\Gamma = 0\end{aligned}$$

for all $v = (v^0, v^1, v_n^2) \in H^1(\omega)^N \times H^1(\omega)^N \times L^2(\omega)$, $\varepsilon^i = \varepsilon^i(v)$, where $v_h = (v_h^0, v_h^1, v_{hn}^2)$ is the solution, $\varepsilon_h^i = \varepsilon^i(v_h)$. Under condition (4.19) for v such that

$\varepsilon^0(v) = 0$ the first integral is zero. Divide the remaining terms by $h^2/3$

$$\begin{aligned}\int_\omega 3&\frac{\overline{\alpha}_0(h)-1}{h^2}(-f^0\cdot v^0 - g^0\cdot v^1)\\ &+3\frac{\overline{\alpha}_1(h)}{h^2}(C^{-1}\varepsilon_h^0\cdot\cdot\varepsilon^1 - f^1\cdot v^0 - (f^0+g^1)\cdot v^1 - 2g^0\cdot v^2)\\ &+3\frac{\overline{\alpha}_2(h)}{h^2}(C^{-1}\varepsilon_h^1\cdot\cdot\varepsilon^1 - f^2\cdot v^0 - (f^1+g^2)\cdot v^1 - (f^0+2g^1)\cdot v^2)\\ &-3\frac{\overline{\alpha}_3(h)}{h^2}(f^2\cdot v^1 + (f^1+2g^2)\cdot v^2) - 3\frac{\overline{\alpha}_4(h)}{h^2}f^2\cdot v^2\,d\Gamma = 0.\end{aligned}$$

By going to the limit as h goes to zero

$$\frac{\overline{\alpha}_0(h)-1}{h^2}\to\frac{1}{3}\kappa_2,\quad \frac{\overline{\alpha}_1(h)}{h^2}\to\frac{1}{3}\kappa_1,\quad \frac{\overline{\alpha}_2(h)}{h^2}\to\frac{1}{3},\quad \frac{\overline{\alpha}_3(h)}{h^2}\to 0,\quad \frac{\overline{\alpha}_4(h)}{h^2}\to 0$$

in $L^\infty(\omega)$-strong. We have seen that the term $C^{-1}\varepsilon_h^0$ weakly converges to $C^{-1}\hat{\varepsilon}^0$, but we shall see later in section 5.5 that the term $C^{-1}\varepsilon_h^1$ does not generally weakly converge to $C^{-1}\hat{\varepsilon}^1$. For this reason we have to introduce the *projection operator* π onto the subspace $\ker\varepsilon^0 = \{(v^0,v^1,v^2)\in E^{01}\colon \varepsilon^0(v^0,v^1)=0\}$ of E^{01} for the equivalent norm associated with the scalar product

$$\int_\omega C^{-1}\varepsilon^0(v)\cdot\cdot\varepsilon^0(w) + C^{-1}\varepsilon^1(v)\cdot\cdot\varepsilon^1(w)\,d\Gamma.$$

The *projection* πv of an element $v\in E^{01}$ is completely characterized by the variational equation

$$\int_\omega C^{-1}\varepsilon^0(v-\pi v)\cdot\cdot\varepsilon^0(w) + C^{-1}\varepsilon^1(v-\pi v)\cdot\cdot\varepsilon^1(w)\,d\Gamma = 0,\quad \forall w\in\ker\varepsilon^0$$

and since $\varepsilon^0(w)=0$

$$\int_\omega C^{-1}\varepsilon^1(v-\pi v)\cdot\cdot\varepsilon^1(w)\,d\Gamma = 0,\quad \forall w\in\ker\varepsilon^0$$

which is itself equivalent to

$$\int_\omega C^{-1}\varepsilon^1(v-\pi v)\cdot\cdot\varepsilon^1(\pi u)\,d\Gamma = 0,\quad \forall u\in E^{01}.$$

Assuming that $C^{-1}\varepsilon^1(\pi v_h)$ weakly converges to $C^{-1}\varepsilon^1(\pi\hat{v})$, we get

$$\begin{aligned}\int_\omega C^{-1}&\hat{\varepsilon}^1\pi\cdot\cdot\varepsilon^1 - f^2\cdot v^0 - (f^1+g^2)\cdot v^1 - (f^0+2g^1)\cdot v^2\\ &+\kappa_1(C^{-1}\hat{\varepsilon}^0\cdot\cdot\varepsilon^1 - f^1\cdot v^0 - (f^0+g^1)\cdot v^1 - 2g^0\cdot v^2)\\ &+\kappa_2(-f^0\cdot v^0 - g^0\cdot v^1)\,d\Gamma = 0,\quad \forall v\in\ker\varepsilon^0,\end{aligned}\tag{5.1}$$

where the notation $\varepsilon_h^1\pi$ and $\hat{\varepsilon}^1\pi$ stands for $\varepsilon^1(\pi v_h)$ and $\varepsilon^1(\pi\hat{v})$.

In a more compact form, $\hat{v}$ would be a solution in a space to be specified of the system

$$\begin{aligned}&\int_\omega C^{-1}\varepsilon^0(\hat{v})\cdot\cdot\varepsilon^0(v)\,d\Gamma = I(0)v,\quad \forall v\in E^0\\ &\int_\omega C^{-1}\varepsilon^1(\pi\hat{v})\cdot\cdot\varepsilon^1(v) + \kappa_1 C^{-1}\varepsilon^0(\hat{v})\cdot\cdot\varepsilon^1(v)\,d\Gamma = 3J(0)v,\quad \forall v\in\ker\varepsilon^0.\end{aligned}\tag{5.2}$$

5.2. Choice of space of solution. The second equation (5.2) has test vector functions in $\ker \varepsilon^0$ rather than in the whole space E^{01}. It can be rewritten on the whole space E^{01} as

$$\int_\omega C^{-1}\varepsilon^1(\pi\hat{v})\cdot\cdot\varepsilon^1(\pi v) + \kappa_1 C^{-1}\varepsilon^0(\hat{v})\cdot\cdot\varepsilon^1(\pi v)\, d\Gamma = 3J(0)\pi v, \quad \forall v \in E^{01}. \tag{5.3}$$

It is an equation in the subspace $\ker \varepsilon^0$ of E^{01}. The more compact notation $\varepsilon^1\pi$ and $\hat{\varepsilon}^1\pi$ will often be used for $\varepsilon^1(\pi v)$ and $\varepsilon^1(\pi\hat{v})$ in the computations.

Observe that if $E^{01}/\ker \varepsilon^0$ is equivalent to E^0, then the first equation can be interpreted as an equation in the orthogonal complement $[\ker \varepsilon^0]^\perp$ of $\ker \varepsilon^0$, that is formally

$$\int_\omega C^{-1}\varepsilon^0(\hat{v} - \pi\hat{v})\cdot\cdot\varepsilon^0(v - \pi v)\, d\Gamma = 0, \quad \forall v \in E^{01}.$$

Unfortunately the canonical quotient norm

$$\inf_{w\in\ker \varepsilon^0} \left\{ \left\|\varepsilon^0(v+w)\right\|^2_{L^2(\omega)} + \left\|\varepsilon^1(v+w)\right\|^2_{L^2(\omega)} \right\}^{1/2}$$
$$= \left\{ \left\|\varepsilon^0(v)\right\|^2_{L^2(\omega)} + \inf_{w\in\ker \varepsilon^0} \left\|\varepsilon^1(v+w)\right\|^2_{L^2(\omega)} \right\}^{1/2}$$

is generally stronger than the norm $\left\|\varepsilon^0(v)\right\|_{L^2(\omega)}$ on the quotient space $E^{01}/\ker \varepsilon^0$. To get around this we introduce the new weaker norm

$$\left\{ \left\|\varepsilon^0(v)\right\|^2_{L^2(\omega)} + \left\|\varepsilon^1(\pi v)\right\|^2_{L^2(\omega)} \right\}^{1/2}$$

on E^{01} and denote by E^{01}_π the completion of E^{01} with respect to the norm generated by the corresponding new inner product

$$\int_\omega C^{-1}\varepsilon^0(v)\cdot\cdot\varepsilon^0(w) + C^{-1}\varepsilon^1(\pi v)\cdot\cdot\varepsilon^1(\pi w)\, d\Gamma. \tag{5.4}$$

The canonical quotient norm is now equal to the norm $\|\varepsilon^0(v)\|_{L^2(\omega)}$

$$\inf_{w\in\ker \varepsilon^0} \left\{ \left\|\varepsilon^0(v+w)\right\|^2_{L^2(\omega)} + \left\|\varepsilon^1(\pi(v+w))\right\|^2_{L^2(\omega)} \right\}^{1/2}$$
$$= \left\{ \left\|\varepsilon^0(v)\right\|^2_{L^2(\omega)} + \inf_{w\in\ker \varepsilon^0} \left\|\varepsilon^1(\pi v + w)\right\|^2_{L^2(\omega)} \right\}^{1/2} = \|\varepsilon^0(v)\|_{L^2(\omega)}.$$

Moreover E^0 coincides with $E^{01}_\pi/\ker \varepsilon^0$. The linear mapping $v \mapsto iv \colon E^{01}_\pi \to E^0$ defined through the equation

$$\int_\omega C^{-1}\varepsilon^0(iv)\cdot\cdot\varepsilon^0(w)\, d\Gamma = \int_\omega C^{-1}\varepsilon^0(v)\cdot\cdot\varepsilon^0(w)\, d\Gamma, \quad \forall w \in E^0$$

is continuous. It is also surjective: for all $\hat{v} \in E^0$, there exists a unique $u \in E^{01}_\pi$ such that

$$\int_\omega C^{-1}\varepsilon^0(u)\cdot\cdot\varepsilon^0(w) + C^{-1}\varepsilon^1(\pi u)\cdot\cdot\varepsilon^1(\pi w)\, d\Gamma = \int_\omega C^{-1}\varepsilon^0(\hat{v})\cdot\cdot\varepsilon^0(w)\, d\Gamma, \quad \forall w \in E^{01}_\pi.$$

Choose tests $w = v - \pi v$ for $v \in H^1(\omega)^N \times H^1(\omega)^N \times L^2(\omega)$. Then $\pi w = \pi(v - \pi v) = 0$, $\varepsilon^1(\pi(v - \pi v)) = 0$, $\varepsilon^0(v - \pi v) = \varepsilon^0(v)$ and

$$\int_\omega C^{-1}\varepsilon^0(u)\cdot\cdot\varepsilon^0(v)\, d\Gamma = \int_\omega C^{-1}\varepsilon^0(\hat{v})\cdot\cdot\varepsilon^0(v)\, d\Gamma$$

for all $v \in H^1(\omega)^N \times H^1(\omega)^N \times L^2(\omega)$. By definition of E^0 as the closure of $(H^1(\omega)^N \times H^1(\omega)^N)/\ker \varepsilon^0$ with respect to the norm $\|\varepsilon^0(v)\|$,

$$\int_\omega C^{-1}\varepsilon^0(iu)\cdot\cdot\varepsilon^0(v)\,d\Gamma = \int_\omega C^{-1}\varepsilon^0(u)\cdot\cdot\varepsilon^0(v)\,d\Gamma = \int_\omega C^{-1}\varepsilon^0(\hat{v})\cdot\cdot\varepsilon^0(v)\,d\Gamma, \forall v \in E^0$$

and $\hat{v} = iu$ in E^0. Since i is linear, continuous and bijective, $E^{01}_\pi/\ker\varepsilon^0 \equiv E^0$ algebraically and topologically.

The asymptotic system (5.2) is now equivalent to

$$\text{(5.5)}\quad \boxed{\begin{aligned} &\exists \hat{v} \in E^{01}_\pi \text{ such that } \int_\omega C^{-1}\varepsilon^0(\hat{v})\cdot\cdot\varepsilon^0(v)\,d\Gamma = I(0)v, \quad \forall v \in E^{01}_\pi \\ &\int_\omega C^{-1}\varepsilon^1(\pi\hat{v})\cdot\cdot\varepsilon^1(\pi v) + \kappa_1 C^{-1}\varepsilon^0(\hat{v})\cdot\cdot\varepsilon^1(\pi v)\,d\Gamma = 3J(0)\pi v, \quad \forall v \in E^{01}_\pi \end{aligned}}$$

under the conditions

$$\text{(5.6)}\quad \boxed{|I(0)v| \le c\|\varepsilon^0(v)\|_{L^2(\omega)}, \quad |J(0)\pi v| \le c\|\varepsilon^1(\pi v)\|_{L^2(\omega)}.}$$

They can also be written as

$$\boxed{\begin{aligned} &\int_\omega C^{-1}\varepsilon^0(\hat{v}-\pi\hat{v})\cdot\cdot\varepsilon^0(v-\pi v)\,d\Gamma = I(0)(v-\pi v), \quad \forall v \in E^{01}_\pi \\ &\int_\omega C^{-1}\varepsilon^1(\pi\hat{v})\cdot\cdot\varepsilon^1(\pi v) + \kappa_1 C^{-1}\varepsilon^0(\hat{v})\cdot\cdot\varepsilon^1(\pi v)\,d\Gamma = 3J(0)\pi v, \quad \forall v \in E^{01}_\pi. \end{aligned}}$$

The second equation is in $\ker\varepsilon^0$ and the first in the orthogonal complement $[\ker\varepsilon^0]^\perp$ of $\ker\varepsilon^0$ in the space E^{01}_π. The same constructions apply to the $P(1,1)$ model and we shall use the same notation E^{01}_π for the completion of the quotient space $(H^1(\omega)^N \times H^1(\omega)^N)/K$ with respect to the scalar product (5.4).

Since (4.1) has a unique solution in $(\hat{v}^0, \hat{v}^1) \in E^0$ which completely specifies $\varepsilon^0(\hat{v}^0, \hat{v}^1)$, it is natural to introduce the affine subspace

$$\text{(5.7)}\qquad \mathcal{V}(\varepsilon^0) \stackrel{\text{def}}{=} \left\{ v \in E^{01}_\pi : \int_\omega C^{-1}\varepsilon^0(v)\cdot\cdot\varepsilon^0(w)\,d\Gamma = I(0)w, \quad \forall w \in E^{01}_\pi \right\}.$$

Use the second equation (5.5) to construct a minimization problem over the space $\mathcal{V}(\varepsilon^0)$. The space $\mathcal{V}(\varepsilon^0)$ is not empty since the equation

$$\int_\omega C^{-1}\varepsilon^0(u)\cdot\cdot\varepsilon^0(v) + C^{-1}\varepsilon^1(\pi u)\cdot\cdot\varepsilon^1(\pi v)\,d\Gamma = I(0)v, \quad \forall v \in E^{01}_\pi$$

has a unique solution $u \in E^{01}_\pi$. By picking tests of the form $v = w - \pi w$, $w \in E^{01}_\pi$, we get that u is the solution in $E^{01}_\pi/\ker\varepsilon^0$ of the equation

$$\int_\omega C^{-1}\varepsilon^0(u)\cdot\cdot\varepsilon^0(w)\,d\Gamma = I(0)w, \quad \forall w \in E^{01}_\pi$$

which has a unique solution in $E^{01}_\pi/\ker\varepsilon^0$. Therefore $\mathcal{V}(\varepsilon^0)$ is not empty and we have the following existence theorem.

THEOREM 5.1. (Asymptotic shell) *Let Assumptions* 2.1 *and* 3.1 *be verified and let* f^0, f^1, f^2, g^0, g^1, *and* g^2 *in* $L^2(\omega)^N$ *verify conditions* (4.19) *and* (4.20). *Then the system of equations* (5.5) *has a unique solution* $\hat{v}$ *in* E^{01}_π. *The result extends to arbitrary linear functionals* $I(0)\colon E^{01}_\pi/\ker\varepsilon^0 \to \mathbf{R}$ *and* $J(0) : E^{01} \to \mathbf{R}$ *verifying* (5.6).

PROOF. We already know that under condition (4.19), equation (4.1) has a unique solution in $E^{01}_\pi / \ker \varepsilon^0$. By Assumption 3.1 $\|D^2 b\|$ is bounded and a fortiori the κ_i's are bounded in ω. Denote by c the upper bound of all those constants. Pick $\beta \geq 1/2 + 8c^2$ and consider the functional

$$(5.8)\quad W_\beta(v) = \beta\Big\{\int_\omega \frac{1}{2} C^{-1}\varepsilon^0(v){\cdot\cdot}\varepsilon^0(v)\,d\Gamma - I(0)v\Big\} - 3J(0)\pi v + \int_\omega \frac{1}{2} C^{-1}\varepsilon^1(\pi v){\cdot\cdot}\varepsilon^1(\pi v) + \kappa_1 C^{-1}\varepsilon^0(v){\cdot\cdot}\varepsilon^1(\pi v)\,d\Gamma.$$

Under conditions (4.19) and (4.20) this functional is well-defined and continuous on E^{01}_π. It is also coercive by choice of β. Then the functional $W_\beta(v)$ has a unique minimizing element $\hat v$ in $\mathcal{V}(\varepsilon^0)$. Moreover $\hat v$ is completely characterized by the following variational equation: $\hat v \in \mathcal{V}(\varepsilon^0)$ and for all $v \in \mathcal{V}(\varepsilon^0)$

$$(5.9)\quad \beta\Big\{\int_\omega C^{-1}\varepsilon^0(\hat v){\cdot\cdot}\varepsilon^0(v-\hat v)\,d\Gamma - I(0)\pi(v-\hat v)\Big\} - 3J(0)\pi(v-\hat v) + \int_\omega \kappa_1\big[C^{-1}\varepsilon^0(\hat v){\cdot\cdot}\varepsilon^1\big(\pi(v-\hat v)\big) + C^{-1}\varepsilon^1(\pi\hat v){\cdot\cdot}\varepsilon^0(v-\hat v)\big] + C^{-1}\varepsilon^1(\pi\hat v){\cdot\cdot}\varepsilon^1\big(\pi(v-\hat v)\big)\,d\Gamma = 0.$$

But since $\mathcal{V}(\varepsilon^0)$ is an affine space, we can use test functions in $\ker\varepsilon^0$. Thence the variational equation (5.9) is completely equivalent to find $\hat v \in \mathcal{V}(\varepsilon^0)$ such that for all $v \in \ker\varepsilon^0$

$$(5.10)\quad \beta\Big\{\int_\omega C^{-1}\varepsilon^0(\hat v){\cdot\cdot}\varepsilon^0(v)\,d\Gamma - I(0)v\Big\} - 3J(0)\pi v + \int_\omega C^{-1}\varepsilon^1(\pi\hat v){\cdot\cdot}\varepsilon^1(\pi v) + \kappa_1 C^{-1}\varepsilon^0(\hat v){\cdot\cdot}\varepsilon^1(\pi v)\,d\Gamma = 0.$$

But the expression multiplied by β is zero since $\varepsilon^0(v) = 0$ and $I(0) = 0$ on $\ker\varepsilon^0$ by condition (4.19). Therefore we get exactly the variational equation (5.1), that is the second equation of (5.2) or (5.5) together with the first equation which is specified by the space $\mathcal{V}(\varepsilon^0)$. This completes the proof. □

5.3. Decomposition of the second equation. In order to distinguish between the tensor $\varepsilon^1(v^0, v^1, 0)$ in the $P(1,1)$ model and $\varepsilon^1(v^0, v^1, v^2_n n)$ in the $P(2_n, 1)$ model, introduce the notation

$$\boxed{e^1(v^0, v^1) \overset{\text{def}}{=} \varepsilon^1(v^0, v^1, 0)}$$

and $e^1(v) = e^1(v^0, v^1)$ for $v = (v^0, v^1)$. Recall the decomposition of $\varepsilon^1(v^0, v^1, v^2_n n)$

$$\varepsilon^1(v^0, v^1, v^2_n n) = e^1(v^0, v^1) + 2v^2_n n^* n$$

$$\Longrightarrow \varepsilon^{1P}(v^0, v^1, v^2_n n) = e^{1P}(v^0, v^1), \quad P\varepsilon^1(v^0, v^1, v^2_n n)n = e^1(v^0, v^1)n,$$

$$\varepsilon^1(v^0, v^1, v^2_n n)n \cdot n = 2v^2_n.$$

Note that for $(v^0, v^1) \in \ker\varepsilon^0$, we have

$$\varepsilon^P_\Gamma(v^0) = 0, v^1_\Gamma + \nabla_\Gamma v^0_n - D^2 b v^0_\Gamma = v^1_\Gamma + 2\varepsilon_\Gamma(v^0)n = 0 \text{ and } v^1_n = 0$$

$$(5.11)\quad \Longrightarrow \boxed{\forall (v^0, v^1) \in \ker\varepsilon^0, \quad e^1(v^0, v^1)n = 0, \quad e^1(v^0, v^1) = e^{1P}(v^0, v^1).}$$

We now decompose the second variational equation in the general case by eliminating the variable $\hat{v}^2$ as in the case of the membrane shell equation obtained from the asymptotic $P(1,0)$ model. We then specialize the results to the bending dominated case. Set $q = f^0 + 2g^1 + \kappa_1 2g^0$ and start from the equation

$$\int_\omega \left[C^{-1}\varepsilon^1(\pi\hat{v}) + \kappa_1 C^{-1}\varepsilon^0(\hat{v})\right]\cdot\cdot\,\varepsilon^1(\pi v)\, d\Gamma = 3J(0)\pi v \tag{5.12}$$

for all $(v^0, v^1) \in \ker \varepsilon^0$ and $v_n^2 \in L^2(\omega)$. For test functions $(0,0,v_n^2 n)$, $v_n^2 \in L^2(\omega)$,

$$\int_\omega 2\big(\big[C^{-1}\varepsilon^1(\pi\hat{v})\big]n + \kappa_1\big[C^{-1}\varepsilon^0(\hat{v})\big]n\big)\cdot n v_n^2\, d\Gamma = 3J(0)(0,0,v_n^2 n) = \int_\omega q_n v_n^2\, d\Gamma$$

$$\Longrightarrow \boxed{\big[C^{-1}\varepsilon^1(\pi\hat{v})\big]n\cdot n + \kappa_1\big[C^{-1}\varepsilon^0(\hat{v})\big]n\cdot n = q_n/2.} \tag{5.13}$$

For test functions $(v^0, v^1, 0)$, $(v^0, v^1) \in \ker\varepsilon^0$, $\varepsilon^1(\pi v) = e^1(v) = e^{1P}(v)$ and

$$\boxed{\int_\omega \big[C^{-1}\varepsilon^1(\pi\hat{v})\big]^P\cdot\cdot\, e^{1P}(v) + \kappa_1\big[C^{-1}\varepsilon^0(\hat{v})\big]^P\cdot\cdot\, e^{1P}(v)\, d\Gamma = 3J(0)(v^0, v^1, 0).} \tag{5.14}$$

If we introduce the *stress tensor* $\sigma^1 = C^{-1}\varepsilon^1$, then

$$\sigma^1(\pi\hat{v})n\cdot n + \kappa_1\sigma^0(\hat{v})n\cdot n = \frac{1}{2}f_n^0 + g_n^1 + \kappa_1 g_n^0.$$

Now recalling identity (4.2) for $\sigma^0(\hat{v})$, we get

$$\boxed{\sigma^0(\hat{v})n = g^0 \quad\text{and}\quad \sigma^1(\pi\hat{v})n\cdot n = \frac{1}{2}f_n^0 + g_n^1.} \tag{5.15}$$

The case of *in-plane stress* $\big(\sigma^0(\hat{v})n\cdot n = 0\big)$ and the Love-Kirchhoff condition $\big(P\sigma^0(\hat{v})n = 0\big)$ corresponds to $g^0 = 0$ to the zero-th order in z and $f_n^0 + 2g_n^1 = 0$ to the first order in z $\big(\sigma^1(\pi\hat{v})n\cdot n = 0\big)$. In the literature it is usually assumed that $g^0 = 0 = g^1$. For the plate the compatibility condition (4.19) and $g_\Gamma^0 = 0$ imply that $f_n^0 = 0$. It is important to observe that these special cases arise from the choice of the loading and not from an assumption on the shell itself.

For the constitutive law $C^{-1}\varepsilon = 2\mu\varepsilon + \lambda\,\mathrm{tr}\,\varepsilon I$ assuming that the vectors $(\hat{v}^0, \hat{v}^1, \hat{v}_n^2)$ are sufficiently smooth, we have

$$\begin{aligned} &2\mu\left[2(\pi\hat{v})_n^2 n^* n + e^{1P}(\pi\hat{v}) + \kappa_1\left(\varepsilon_\Gamma(\hat{v}^0) + \frac{1}{2}(\hat{v}^{1*}n + n^*v^1)\right)\right]n\cdot n \\ &\qquad + \lambda\big[2(\pi\hat{v})_n^2 + \mathrm{tr}\, e^{1P}(\pi\hat{v}) + \kappa_1(\mathrm{tr}\,\varepsilon_\Gamma(\hat{v}^0) + \hat{v}_n^1)\big] = q_n/2 \end{aligned}$$

$$\Longrightarrow \boxed{(2\mu+\lambda)\big(2(\pi\hat{v})_n^2 + \kappa_1\hat{v}_n^1\big) + \lambda\big(\mathrm{tr}\, e^{1P}(\pi\hat{v}) + \kappa_1\,\mathrm{tr}\,\varepsilon_\Gamma(\hat{v}^0)\big) = q_n/2.} \tag{5.16}$$

Notice that since $e^1(\pi\hat{v}) = e^{1P}(\pi\hat{v})$ and $\varepsilon_\Gamma(\hat{v}^0)n\cdot n = 0$

$$\mathrm{tr}\, e^1(\pi\hat{v}) = \mathrm{tr}\, e^{1P}(\pi\hat{v}) \text{ and } \mathrm{tr}\,\varepsilon_\Gamma(\hat{v}^0) = \mathrm{tr}\,\varepsilon_\Gamma^P(\hat{v}^0).$$

Now that $(\pi\hat{v})_n^2$ has been expressed in terms of $(\hat{v}^0, \hat{v}^1)$, come back to equation (5.14) with test functions of the form $v = (v^0, v^1, 0)$, $(v^0, v^1) \in \ker\varepsilon^0$. Then

$$\begin{aligned} \int_\omega &2\mu\big[\varepsilon^{1P}(\pi\hat{v}) + \kappa_1\varepsilon^{0P}(\hat{v})\big]\cdot\cdot\, e^{1P}(v) \\ &\qquad + \lambda\big[\mathrm{tr}\,\varepsilon^{1P}(\pi\hat{v}) + \kappa_1\,\mathrm{tr}\,\varepsilon^0(\hat{v})\big]\,\mathrm{tr}\, e^{1P}(v)\, d\Gamma = 3J(0)(v^0, v^1, 0). \end{aligned}$$

By direct computation

$$\varepsilon^1(\pi\hat{v}) + \kappa_1\varepsilon^{0P}(\hat{v}) = e^{1P}(\pi\hat{v}) + \kappa_1\varepsilon_\Gamma^P(\hat{v}^0)$$

$$\begin{aligned}\operatorname{tr}\varepsilon^1(\pi\hat{v}) + \kappa_1\operatorname{tr}\varepsilon^0(\hat{v}) &= \operatorname{tr} e^{1P}(\pi\hat{v}) + \kappa_1\operatorname{tr}\varepsilon_\Gamma(\hat{v}^0) + 2(\pi\hat{v})_n^2 + \kappa_1\hat{v}_n^1 \\ &= \left[\operatorname{tr} e^{1P}(\pi\hat{v}) + \kappa_1\operatorname{tr}\varepsilon_\Gamma^P(\hat{v}^0)\right] \\ &\qquad + \frac{1}{2\mu+\lambda}\left[\frac{q_n}{2} - \lambda\left(\operatorname{tr} e^{1P}(\pi\hat{v}) + \kappa_1\operatorname{tr}\varepsilon_\Gamma(\hat{v}^0)\right)\right] \\ &= \frac{2\mu}{2\mu+\lambda}\left[\operatorname{tr} e^{1P}(\pi\hat{v}) + \kappa_1\operatorname{tr}\varepsilon_\Gamma^P(\hat{v}^0)\right] + \frac{1}{2\mu+\lambda}\frac{q_n}{2}.\end{aligned}$$

So $\hat{v}_n^2$ is given by (5.16) in term of $(\hat{v}^0, \hat{v}^1)$ and finally $\hat{v} = (\hat{v}^0, \hat{v}^1)$ is the solution of the coupled system (4.3), (4.4) and for all $(v^0, v^1) \in \ker\varepsilon^0$

$$\tag{5.17} \boxed{\begin{aligned}&\int_\omega 2\mu[e^{1P}(\pi\hat{v}) + \kappa_1\varepsilon_\Gamma^P(\hat{v}^0)]\cdot\cdot e^{1P}(v) \\ &\qquad + \frac{2\mu\lambda}{2\mu+\lambda}\left[\operatorname{tr} e^{1P}(\pi\hat{v}) + \kappa_1\operatorname{tr}\varepsilon_\Gamma^P(\hat{v}^0)\right]\operatorname{tr} e^{1P}(v)\,d\Gamma \\ &= 3J(0)(v^0, v^1, 0) - \frac{\lambda}{2\mu+\lambda}\int_\omega \frac{1}{2}[f^0 + 2g^1 + \kappa_1 2g^0]_n \operatorname{tr} e^{1P}(v)\,d\Gamma.\end{aligned}}$$

Of course one could substitute for $\hat{v}^1$ and get a coupled system in $\hat{v}^0$ only.

For the $P(1,1)$ model we can repeat the previous computations with $\hat{v}^2 = 0$ and $v^2 = 0$. Then equation (5.14) reduces to: for all $(v^0, v^1) \in \ker\varepsilon^0$,

$$\tag{5.18} \boxed{\begin{aligned}\int_\omega \left[C^{-1}e^{1P}(\pi\hat{v})\right]^P\cdot\cdot e^{1P}(v) + \kappa_1\left[C^{-1}\varepsilon^0(\hat{v})\right]^P\cdot\cdot e^{1P}(v)\,d\Gamma \\ = 3J(0)(v^0, v^1, 0).\end{aligned}}$$

Here $\sigma^0(\hat{v})n = g^0$ and there is no explicit expression for $\sigma^1(\pi\hat{v})n \cdot n$ since the test variable v_n^2 is absent. For the special constitutive law this equation becomes

$$\boxed{\begin{aligned}&\int_\omega 2\mu e^{1P}(\pi\hat{v})\cdot\cdot e^{1P}(v) + \lambda\operatorname{tr} e^{1P}(\pi\hat{v})\operatorname{tr} e^{1P}(v) \\ &\qquad \int_\omega +\kappa_1\left[2\mu\varepsilon_\Gamma^P(\hat{v}^0)\cdot\cdot e^{1P}(v) + \frac{2\mu\lambda}{2\mu+\lambda}\operatorname{tr}\varepsilon_\Gamma^P(\hat{v}^0)\operatorname{tr} e^{1P}(v)\right]d\Gamma \\ &\quad = 3J(0)(v^0, v^1, 0) - \frac{\lambda}{2\mu+\lambda}\int_\omega \kappa_1 g_n^0 \operatorname{tr} e^{1P}(v)\,d\Gamma.\end{aligned}}$$

Observe the coefficient λ of the term $\operatorname{tr} e^{1P}(\pi\hat{v})\operatorname{tr} e^{1P}(v)$ instead of $2\lambda\mu/(2\mu+\lambda)$.

5.4. Bending dominated case. The *bending dominated* asymptotic shell is a special case of the asymptotic $P(2_n, 1)$ model which is characterized by the fact that the extended membrane energy is zero, that is $\varepsilon^0(\hat{v}^0, \hat{v}^1) = 0$ (for instance $f^0 = g^0 = 0$). Under that condition the space of solutions and the space of test functions coincide for the second variational equation of system (5.5). In particular $\varepsilon^1(\pi\hat{v}) = \varepsilon^1(\hat{v})$. This case is also of independent interest since the projection $\pi\hat{v}$ is characterized by

$$\varepsilon^0(\pi\hat{v}) = 0$$

$$\int_\omega C^{-1}\varepsilon^1(\pi\hat{v})\cdot\cdot\varepsilon^1(\pi v)\,d\Gamma = 3J(0)\pi v - \int_\omega \kappa_1 C^{-1}\varepsilon^0(\hat{v})\cdot\cdot\varepsilon^1(\pi v)\,d\Gamma, \quad \forall v \in E_\pi^{01}$$

where $\hat{v} - \pi\hat{v}$ and the coupling term $\varepsilon^0(\hat{v})$ are completely characterized by the first equation of system (5.5)

$$\int_\omega C^{-1}\varepsilon^0(\hat{v})\cdot\cdot\varepsilon^0(v)\, d\Gamma = I(0)v, \quad \forall v \in E_\pi^{01}.$$

The solution $\hat{v}$ is the sum of the solutions $\hat{v} - \pi\hat{v}$ and $\pi\hat{v}$ of the two equations.

Recalling from (5.11) that for $(v^0, v^1) \in \ker \varepsilon^0$

$$e^1(v^0, v^1) = e^{1P}(v^0, v^1) = e^{1P}(v^0, -2\varepsilon_\Gamma(v^0)n) = e^{1P}(v^0, D^2bv_\Gamma^0 - \nabla_\Gamma v_n^0).$$

it is natural to introduce the tensor

$$\rho(v^0) \stackrel{\text{def}}{=} e^{1P}(v^0, -2\varepsilon_\Gamma(v^0)n). \tag{5.19}$$

It turns out that the tensor $\rho(v^0)$ is precisely the tensor

$$\rho_{\alpha\beta}(v^0) = \left\{D_\Gamma^P(D^2bv_\Gamma^0 - \nabla_\Gamma v_n^0) + {}^*D_\Gamma^P(v_\Gamma^0)D^2b + v_n^0(D^2b)^2\right\}_{\alpha\beta}$$

(with $a_3 = a^3 = -n$) appearing in the variational equation of the bending dominated case in [**17, 18**]. To show this, we introduce the temporary notation

$$\boxed{\tilde{\rho}(v^0) \stackrel{\text{def}}{=} D_\Gamma^P(D^2bv_\Gamma^0 - \nabla_\Gamma v_n^0) + {}^*D_\Gamma^P(v_\Gamma^0)D^2b + v_n^0(D^2b)^2}$$

in order to distinguish the two tensors. Use the identity

$$0 = \varepsilon_\Gamma^P(v^0) = \frac{1}{2}\left[D_\Gamma^P(v^0) + {}^*D_\Gamma^P(v^0)\right]$$

to rewrite its expression as

$$\begin{aligned}
\tilde{\rho}(v^0) &= D_\Gamma^P(D^2bv_\Gamma^0 - \nabla_\Gamma v_n^0) + {}^*D_\Gamma^P(v^0)D^2b \\
&= D_\Gamma^P(D^2bv_\Gamma^0 - \nabla_\Gamma v_n^0) - D_\Gamma^P(v^0)D^2b \\
&= D_\Gamma^P(D^2bv_\Gamma^0 - \nabla_\Gamma v_n^0) - D_\Gamma^P(v_\Gamma^0)D^2b - v_n^0(D^2b)^2.
\end{aligned}$$

For $(v^0, v^1) \in \ker \varepsilon^0$, $v_\Gamma^1 + {}^*D_\Gamma(v^0)n = 0$, $v_n^1 = 0$, and

$$\begin{aligned}
0 &= D_\Gamma(v_\Gamma^1 + {}^*D_\Gamma(v^0)n) = D_\Gamma(v_\Gamma^1) + {}^*D_\Gamma(v^0)D^2b + \sum_k n_k D_\Gamma^2(v_k^0) \\
0 &= D_\Gamma^P(v_\Gamma^1 + {}^*D_\Gamma(v^0)n) = D_\Gamma^P(v_\Gamma^1) + {}^*D_\Gamma^P(v^0)D^2b + \sum_k n_k PD_\Gamma^2(v_k^0) \\
0 &= \varepsilon_\Gamma^P(v_\Gamma^1) + \frac{1}{2}\left[{}^*D_\Gamma^P(v^0)D^2b + D^2bD_\Gamma^P(v^0)\right] + \sum_k n_k PD_\Gamma^2(v_k^0),
\end{aligned}$$

since we know from [**28**] that $PD_\Gamma^2(v_k^0) = \varepsilon_\Gamma^P(\nabla_\Gamma(v_k^0))$ is symmetrical. Therefore

$$\varepsilon_\Gamma^P(v_\Gamma^1) + \frac{1}{2}\left[{}^*D_\Gamma^P(v^0)D^2b + D^2bD_\Gamma^P(v^0)\right] = D_\Gamma^P(v_\Gamma^1) + {}^*D_\Gamma^P(v^0)D^2b = \tilde{\rho}(v^0)$$

$$\begin{aligned}
&\varepsilon_\Gamma^P(v_\Gamma^1) + \frac{1}{2}\left[{}^*D_\Gamma^P(v^0)D^2b + D^2bD_\Gamma^P(v^0)\right] \\
&\quad = \varepsilon_\Gamma^P(v_\Gamma^1) - \frac{1}{2}\left[D_\Gamma^P(v^0)D^2b + D^2b^*D_\Gamma^P(v^0)\right] + \varepsilon_\Gamma^P(v^0)D^2b + D^2b\varepsilon_\Gamma^P(v^0) \\
&\quad = \varepsilon_\Gamma^P(v^1) - \frac{1}{2}\left[D_\Gamma^P(v^0)D^2b + D^2b^*D_\Gamma^P(v^0)\right] = e^{1P}(v^0, v^1) = e^1(v^0, v^1)
\end{aligned}$$

since $\varepsilon_\Gamma^P(v^0) = 0$ and $v_n^1 = 0$. Therefore for $(v^0, v^1) \in \ker \varepsilon^0$

$$\tilde\rho(v^0) = e^{1P}(v^0, v^1) = -\sum_k n_k P D_\Gamma^2(v_k^0) = -\sum_k n_k \varepsilon_\Gamma^P(\nabla_\Gamma(v_k^0))$$

$$\Longrightarrow \boxed{\forall v^0 \in \ker \varepsilon_\Gamma^P, \quad \rho(v^0) = \tilde\rho(v^0) = -\sum_k n_k P D_\Gamma^2(v_k^0) = -\sum_k n_k \varepsilon_\Gamma^P(\nabla_\Gamma(v_k^0)).}$$

Therefore in the bending dominated case ($g^0 = f^0 = 0$) we have

$$\varepsilon^0(\hat v) = 0 \Longrightarrow \varepsilon^{1P}(\hat v) = \rho(\hat v^0) \quad \text{and} \quad \varepsilon_\Gamma^P(\hat v^0) = 0.$$

For the special constitutive law, the asymptotic $P(2_n, 1)$ model is characterized by the variational equation: to find $\hat v^0 \in \ker \varepsilon_\Gamma^P$ such that for all $v^0 \in \ker \varepsilon_\Gamma^P$

$$\boxed{\begin{aligned}\int_\omega 2\mu\rho(\hat v^0)\cdot\cdot\rho(v^0) + \frac{2\mu\lambda}{2\mu+\lambda} \operatorname{tr}\rho(\hat v^0) \operatorname{tr}\rho(v^0)\, d\Gamma \\ = 3J(0)(v^0, D^2 b v_\Gamma^0 - \nabla_\Gamma v_n^0, 0) - \frac{\lambda}{2\mu+\lambda}\int_\omega g_n^1 \operatorname{tr}\rho(v^0)\, d\Gamma\end{aligned}} \tag{5.20}$$

$$\boxed{2(\pi\hat v)_n^2 = -\frac{\lambda}{2\mu+\lambda}\operatorname{tr}\rho(\hat v^0) + \frac{1}{2\mu+\lambda} g_n^1} \tag{5.21}$$

$$\boxed{\begin{aligned}\hat v_n^1 &= -\frac{\lambda}{2\mu+\lambda}\operatorname{tr}\varepsilon_\Gamma^P(\hat v^0) = -\frac{\lambda}{2\mu+\lambda}\left[\kappa_1 \hat v_n^0 + \operatorname{div}_\Gamma(\hat v_\Gamma^0)\right] \\ \hat v_\Gamma^1 &= -2\varepsilon_\Gamma(\hat v^0) n = -\nabla_\Gamma(\hat v_n^0) + D^2 b \hat v_\Gamma^0 \quad \text{(Love-Kirchhoff condition).}\end{aligned}} \tag{5.22}$$

Moreover $\sigma^0(\hat v) n = 0$ and $\sigma^1(\hat v) n \cdot n = g_n^1$. The bilinear part of the equation (5.20) exactly coincides with the one in [**17**] and [**18**].

For the $P(1,1)$ model ($\hat v^2 = 0$ and $v^2 = 0$) equation (5.21) disappears and equation (5.20) becomes

$$\boxed{\int_\omega 2\mu\rho(\hat v^0)\cdot\cdot\rho(v^0) + \lambda \operatorname{tr}\rho(\hat v^0)\operatorname{tr}\rho(v^0)\, d\Gamma = 3J(0)(v^0, D^2 b v_\Gamma^0 - \nabla_\Gamma v_n^0, 0)}$$

with the previously mentioned difference in the coefficient of the trace term and

$$\boxed{\begin{aligned}\hat v_n^1 &= -\frac{\lambda}{2\mu+\lambda}\operatorname{tr}\varepsilon_\Gamma^P(\hat v^0) = -\frac{\lambda}{2\mu+\lambda}\left[\kappa_1 \hat v_n^0 + \operatorname{div}_\Gamma(\hat v_\Gamma^0)\right] \\ \hat v_\Gamma^1 &= -2\varepsilon_\Gamma(\hat v^0) n = -\nabla_\Gamma(\hat v_n^0) + D^2 b \hat v_\Gamma^0 \quad \text{(Love-Kirchhoff condition).}\end{aligned}}$$

Moreover $\sigma^0(\hat v) n = 0$.

5.5. Convergence of the $P(2,1)$ model to the asymptotic shell model. Now that we have guessed the second part of the asymptotic model for the special choice of vector functions F and G, it remains to show in what sense we have convergence of the solutions.

THEOREM 5.2. *Let Assumptions* 2.1, 3.1, *and* 4.1 *be verified. Let* $v_h \in E^{01}$ *and* $\hat v \in E_\pi^{01}$ *be the respective solutions of* (3.30) *and* (5.5). *As* $h > 0$ *goes to zero,*

$$\int_\omega \left\|\varepsilon^0(v_h - \hat v)\right\|^2 + \left\|\varepsilon^1\left(\pi(v_h - \hat v)\right)\right\|^2 + h^2\left\|\varepsilon^1(v_h)\right\|^2 d\Gamma \to 0, \tag{5.23}$$

where $\pi(v)$ is the projection of v onto $\ker\varepsilon^0 \overset{\text{def}}{=} \{v \in E^{01}_\pi : \varepsilon^0(v^0, v^1) = 0\}$. *The same results hold for the $P(1,1)$ model with $v^2_h = 0 = \hat{v}^2$.*

PROOF. (i) **Weak convergence:** We first show that the sequence $\{v_h\} \subset E^{01}_\pi$ is bounded in E^{01}_π as $h \to 0$. Then for any weakly convergent subsequence to some $w \in E^{01}_\pi$, we show that w is solution of (5.5). By uniqueness w coincides with the unique solution $\hat{v}$ and the whole sequence $\{v_h\}$ weakly converges to $\hat{v}$ in E^{01}_π.

By Assumption 4.1 and coercivity of C^{-1}, as h goes to zero there exist $\alpha > 0$ and $c > 0$ such that

$$\int_\omega \overline{\alpha}_0(h) C^{-1}\varepsilon^0_h \cdot\cdot \varepsilon^0_h + 2\overline{\alpha}_1(h) C^{-1}\varepsilon^0_h \cdot\cdot \varepsilon^1_h + \overline{\alpha}_2(h) C^{-1}\varepsilon^1_h \cdot\cdot \varepsilon^1_h \, d\Gamma = I(h) v_h$$

$$\Longrightarrow \alpha\big\{\|\varepsilon^0_h\|^2_{L^2} + h^2\|\varepsilon^1_h\|^2_{L^2}\big\} \le c\big\{\|\varepsilon^0_h\|^2_{L^2} + h^2\|\varepsilon^1_h\|^2_{L^2}\big\}^{1/2}.$$

Then as h goes to zero

$$\big\|\varepsilon^0(v^0_h, v^1_h)\big\|^2_{L^2} + h^2\big\|\varepsilon^1(v^0_h, v^1_h, v^2_{hn} n)\big\|^2_{L^2} = \|\varepsilon^0_h\|^2_{L^2} + h^2\|\varepsilon^1_h\|^2_{L^2} \le c^2/\alpha^2.$$

Therefore $\varepsilon^0(v_h)$ and $h\varepsilon^1(v_h)$ are bounded in L^2. Next use test functions of the form πv_h in (3.30) and divide by h^2. Then $\varepsilon^0(\pi v_h) = 0$, $\varepsilon^1\big(\pi\pi(v_h)\big) = \varepsilon^1(\pi v_h)$ and

$$\int_\omega \frac{\overline{\alpha}_2(h)}{h^2} C^{-1}\varepsilon^1(v_h)\cdot\cdot\varepsilon^1(\pi v_h) + \frac{\overline{\alpha}_1(h)}{h^2} C^{-1}\varepsilon^0(v_h)\cdot\cdot\varepsilon^1(\pi v_h)\, d\Gamma = J(h)\pi v_h.$$

By condition (4.20) and (4.21) and Lemma 4.1 (ii)

$$\big|J(h)\pi v_h\big| \le c\Big(\big\|\varepsilon^0(\pi v_h)\big\|_{L^2(\omega)} + \big\|\varepsilon^1(\pi v_h)\big\|_{L^2(\omega)}\Big) = c\big\|\varepsilon^1(\pi v_h)\big\|_{L^2(\omega)}$$

and

$$\left|\int_\omega \frac{\overline{\alpha}_1(h)}{h^2} C^{-1}\varepsilon^0(v_h)\cdot\cdot\varepsilon^1(\pi v_h)\, d\Gamma\right| \le c\big\|\varepsilon^0(v_h)\big\|_{L^2(\omega)}\big\|\varepsilon^1(\pi v_h)\big\|_{L^2(\omega)}$$
$$\le c'\big\|\varepsilon^1(\pi v_h)\big\|_{L^2(\omega)}$$

by boundedness of $\big\|\varepsilon^0(v_h)\big\|_{L^2(\omega)}$. As for the first term by definition of π

$$\begin{aligned}
T_h &\overset{\text{def}}{=} \int_\omega \frac{\overline{\alpha}_2(h)}{h^2} C^{-1}\varepsilon^1(v_h)\cdot\cdot\varepsilon^1(\pi v_h)\, d\Gamma \\
&= \int_\omega \frac{1}{3} C^{-1}\varepsilon^1(v_h)\cdot\cdot\varepsilon^1(\pi v_h)\, d\Gamma + h^2 \int_\omega \frac{1}{h^2}\left[\frac{\overline{\alpha}_2(h)}{h^2} - \frac{1}{3}\right] C^{-1}\varepsilon^1(v_h)\cdot\cdot\varepsilon^1(\pi v_h)\, d\Gamma \\
&= \int_\omega \frac{1}{3} C^{-1}\varepsilon^1(\pi v_h)\cdot\cdot\varepsilon^1(\pi v_h)\, d\Gamma + h^2 \int_\omega \frac{1}{h^2}\left[\frac{\overline{\alpha}_2(h)}{h^2} - \frac{1}{3}\right] C^{-1}\varepsilon^1(v_h)\cdot\cdot\varepsilon^1(\pi v_h)\, d\Gamma.
\end{aligned}$$

By coercivity of C^{-1} and boundedness of the coefficient in the integral, as h goes to zero there exist $\nu > 0$ and $c > 0$ such that

$$\begin{aligned}
|T_h| &\ge \nu\big\|\varepsilon^1(\pi v_h)\big\|^2_{L^2(\omega)} - ch^2\big\|\varepsilon^1(\pi v_h)\big\|_{L^2(\omega)}\big\|\varepsilon^1(v_h)\big\|_{L^2(\omega)} \\
&\ge \nu\big\|\varepsilon^1(\pi v_h)\big\|^2_{L^2(\omega)} - c'h\big\|\varepsilon^1(\pi v_h)\big\|_{L^2(\omega)}
\end{aligned}$$

since we already know that $h\big\|\varepsilon^1(v_h)\big\|_{L^2(\omega)}$ is bounded. Therefore by using the three estimates $\big\|\varepsilon^1(\pi v_h)\big\|_{L^2(\omega)}$ is bounded as h goes to zero, and from the estimate at the beginning of the proof

$$\big\|\varepsilon^0(v_h)\big\|_{L^2(\omega)} + \big\|\varepsilon^1(\pi v_h)\big\|_{L^2(\omega)} \le c$$

and $\{v_h\}$ is bounded in E_π^{01} as h goes to zero. Therefore we can extract a weakly converging subsequence to some $w \in E_\pi^{01}$.

In view of the identity $I(h) = I(0) + h^2 J(0) + h^4 K(h)$, the variational equation (3.30) can be rewritten as follows

$$(5.24)\quad \int_\omega C^{-1}\varepsilon_h^0\cdot\cdot\varepsilon^0\, d\Gamma - I(0)v \\ + \frac{h^2}{3}\Big\{\int_\omega \kappa_2 C^{-1}\varepsilon_h^0\cdot\cdot\varepsilon^0 + \kappa_1(C^{-1}\varepsilon_h^0\cdot\cdot\varepsilon^1 + C^{-1}\varepsilon_h^1\cdot\cdot\varepsilon^0) \\ + C^{-1}\varepsilon_h^1\cdot\cdot\varepsilon^1\, d\Gamma - 3J(0)v\Big\} \\ + h^4\Big\{\int_\omega \frac{1}{h^2}\Big[\frac{\overline{\alpha}_0(h)-1}{h^2} - \frac{\kappa_2}{3}\Big] C^{-1}\varepsilon_h^0\cdot\cdot\varepsilon^0 \\ + \frac{1}{h^2}\Big[\frac{\overline{\alpha}_1(h)}{h^2} - \frac{\kappa_1}{3}\Big](C^{-1}\varepsilon_h^0\cdot\cdot\varepsilon^1 + C^{-1}\varepsilon_h^1\cdot\cdot\varepsilon^0) \\ + \frac{1}{h^2}\Big[\frac{\overline{\alpha}_2(h)}{h^2} - \frac{1}{3}\Big] C^{-1}\varepsilon_h^1\cdot\cdot\varepsilon^1\, d\Gamma - K(h)v\Big\} = 0$$

for all v in E^{01}. There are three terms of respective orders 1, h^2, and h^4. The term in h^2 is bounded by

$$ch^2\big(\|\varepsilon_h^0\| + \|\varepsilon_h^1\| + c\big)\big(\|\varepsilon^0\| + \|\varepsilon^1\|\big) \le c'h\big(\|\varepsilon^0\| + \|\varepsilon^1\|\big)$$

by boundedness of $\|\varepsilon_h^0\|$ and $h\|\varepsilon_h^1\|$ as h goes to zero. Similarly the term in h^4 is bounded by

$$ch^4\big(\|\varepsilon_h^0\| + \|\varepsilon_h^1\| + c\big)\big(\|\varepsilon^0\| + \|\varepsilon^1\|\big) \le c'h^3\big(\|\varepsilon^0\| + \|\varepsilon^1\|\big)$$

and goes to zero as h goes to zero. As for the first term, by weak convergence of $\{v_h\}$ to w, it converges to

$$(5.25)\qquad \int_\omega C^{-1}\varepsilon^0(w)\cdot\cdot\varepsilon^0(v)\, d\Gamma - I(0)v = 0.$$

Now use test functions of the form πv in equation (5.24) and divide by h^2. The element πv belongs to E^{01} and since $\varepsilon^0(\pi v) = 0$ and $I(0) = 0$ on $\ker\varepsilon^0$, we get

$$\frac{1}{3}\Big\{\int_\omega \kappa_1 C^{-1}\varepsilon^0(v_h)\cdot\cdot\varepsilon^1(\pi v) + C^{-1}\varepsilon^1(v_h)\cdot\cdot\varepsilon^1(\pi v)\, d\Gamma - 3J(0)\pi v\Big\} \\ + h^2\Big\{\int_\omega \frac{1}{h^2}\Big[\frac{\overline{\alpha}_1(h)}{h^2} - \frac{\kappa_1}{3}\Big] C^{-1}\varepsilon^0(v_h)\cdot\cdot\varepsilon^1(\pi v)\, d\Gamma \\ + \int_\omega \frac{1}{h^2}\Big[\frac{\overline{\alpha}_2(h)}{h^2} - \frac{1}{3}\Big] C^{-1}\varepsilon^1(v_h)\cdot\cdot\varepsilon^1(\pi v)\, d\Gamma - K(h)\pi v\Big\} = 0.$$

By definition of π

$$(5.26)\quad \frac{1}{3}\Big\{\int_\omega \kappa_1 C^{-1}\varepsilon^0(v_h)\cdot\cdot\varepsilon^1(\pi v) + C^{-1}\varepsilon^1(\pi v_h)\cdot\cdot\varepsilon^1(\pi v)\, d\Gamma - 3J(0)\pi v\Big\} \\ + h^2\Big\{\int_\omega \frac{1}{h^2}\Big[\frac{\overline{\alpha}_1(h)}{h^2} - \frac{\kappa_1}{3}\Big] C^{-1}\varepsilon^0(v_h)\cdot\cdot\varepsilon^1(\pi v)\, d\Gamma \\ + \int_\omega \frac{1}{h^2}\Big[\frac{\overline{\alpha}_2(h)}{h^2} - \frac{1}{3}\Big] C^{-1}\varepsilon^1(v_h)\cdot\cdot\varepsilon^1(\pi v)\, d\Gamma - K(h)\pi v\Big\} = 0.$$

By weak convergence the first term goes to

$$\frac{1}{3}\left\{\int_\omega \kappa_1 C^{-1}\varepsilon^0(w)\cdot\cdot\varepsilon^1(\pi v)+C^{-1}\varepsilon^1(\pi w)\cdot\cdot\varepsilon^1(\pi v)\,d\Gamma-3J(0)\pi v\right\}.$$

The terms in h^2 are bounded by

$$ch^2\Big(\big\|\varepsilon^0(v_h)\big\|+\big\|\varepsilon^1(v_h)\big\|+c\big)\big\|\varepsilon^1(\pi v)\big\|\le c'h\big\|\varepsilon^1(\pi v)\big\|$$

by condition (4.21) and by boundedness of $\|\varepsilon^0(v_h)\|$ and $h\|\varepsilon^1(v_h)\|$ as h goes to zero. Therefore w is solution of the equation

$$\int_\omega \kappa_1 C^{-1}\varepsilon^0(w)\cdot\cdot\varepsilon^1(\pi v)+C^{-1}\varepsilon^1(\pi w)\cdot\cdot\varepsilon^1(\pi v)\,d\Gamma-3J(0)\pi v=0. \tag{5.27}$$

But the system of equations (5.27) and (5.25) is system (5.5) and by uniqueness of solution $w=\hat{v}$.

(ii) **Strong convergence:** First evaluate the term

$$\delta_h \stackrel{\text{def}}{=} \int_\omega C^{-1}(\varepsilon_h^0-\hat{\varepsilon}^0)\cdot\cdot(\varepsilon_h^0-\hat{\varepsilon}^0)\,d\Gamma.$$

By using the first equation of system (5.5)

$$\begin{aligned}\delta_h &= \int_\omega C^{-1}\varepsilon_h^0\cdot\cdot\varepsilon_h^0+C^{-1}\hat{\varepsilon}^0\cdot\cdot(\hat{\varepsilon}^0-2\varepsilon_h^0)\,d\Gamma\\ &= \int_\omega C^{-1}\varepsilon_h^0\cdot\cdot\varepsilon_h^0-I(0)v_h+I(0)(\hat{v}-v_h)\,d\Gamma.\end{aligned}$$

From (5.24) with $v=v_h$ the negative of the first term is

$$\begin{aligned}&-\left\{\int_\omega C^{-1}\varepsilon_h^0\cdot\cdot\varepsilon_h^0-I(0)v_h\right\}\\ &=\frac{h^2}{3}\left\{\int_\omega \kappa_2 C^{-1}\varepsilon_h^0\cdot\cdot\varepsilon_h^0+\kappa_1(C^{-1}\varepsilon_h^0\cdot\cdot\varepsilon_h^1+C^{-1}\varepsilon_h^1\cdot\cdot\varepsilon_h^0)+C^{-1}\varepsilon_h^1\cdot\cdot\varepsilon_h^1\,d\Gamma-3J(0)v_h\right\}\\ &\qquad+h^4\left\{\int_\omega \frac{1}{h^2}\left[\frac{\overline{\alpha}_0(h)-1}{h^2}-\frac{\kappa_2}{3}\right]C^{-1}\varepsilon_h^0\cdot\cdot\varepsilon_h^0\right.\\ &\qquad\qquad+\frac{1}{h^2}\left[\frac{\overline{\alpha}_1(h)}{h^2}-\frac{\kappa_1}{3}\right](C^{-1}\varepsilon_h^0\cdot\cdot\varepsilon_h^1+C^{-1}\varepsilon_h^1\cdot\cdot\varepsilon_h^0)\\ &\qquad\qquad\qquad+h^4\left\{\int_\omega+\frac{1}{h^2}\left[\frac{\overline{\alpha}_2(h)}{h^2}-\frac{1}{3}\right]C^{-1}\varepsilon_h^1\cdot\cdot\varepsilon_h^1\,d\Gamma-K(h)v_h\right\}.\end{aligned}$$

But we have seen in the proof of part (i) that the terms in h^2 and in h^4 are respectively bounded by ch and ch^3 as h goes to zero. Moreover the term $I(0)(\hat{v}-v_h)$ goes to zero by weak convergence in E_π^{01} and condition (4.19). Therefore $\delta_h\to 0$ as $h\to 0$.

We now evaluate the second term

$$\int_\omega C^{-1}\varepsilon^1\big(\pi(v_h-\hat{v})\big)\cdot\cdot\varepsilon^1\big(\pi(v_h-\hat{v})\big)\,d\Gamma.$$

Set the test function $v = \pi(v_h - \hat{v})$ (which belongs to E^{01}) and subtract the second equation of system (5.5) divided by 3 from (5.26)

$$\frac{1}{3}\int_\omega \kappa_1 C^{-1}\varepsilon^0(v_h - \hat{v})\cdot\cdot\varepsilon^1\big(\pi(v_h - \hat{v})\big) + C^{-1}\varepsilon^1\big(\pi(v_h - \hat{v})\big)\cdot\cdot\varepsilon^1\big(\pi(v_h - \hat{v})\big)\, d\Gamma$$
$$+ h^2\bigg\{\int_\omega \frac{1}{h^2}\left[\frac{\overline{\alpha}_1(h)}{h^2} - \frac{\kappa_1}{3}\right] C^{-1}\varepsilon^0(v_h)\cdot\cdot\varepsilon^1\big(\pi(v_h - \hat{v})\big)\, d\Gamma$$
$$+ \int_\omega \frac{1}{h^2}\left[\frac{\overline{\alpha}_2(h)}{h^2} - \frac{1}{3}\right] C^{-1}\varepsilon^1(v_h)\cdot\cdot\varepsilon^1\big(\pi(v_h - \hat{v})\big)\, d\Gamma - K(h)\pi(v_h - \hat{v})\bigg\} = 0.$$

The term in h^2 is bounded by ch since $\|\varepsilon^1(\pi(v_h - \hat{v}))\|$, $\|\varepsilon^0(v_h)\|$, and $h\|\varepsilon^1(v_h)\|$ and the coefficients are bounded. Finally the term in κ_1 goes to zero since $\|\varepsilon^1(\pi(v_h - \hat{v}))\|$ is bounded and $\|\varepsilon^0(v_h - \hat{v})\|$ goes to zero as was established in the first part of the proof of the strong convergence. Therefore the remaining term goes to zero. This completes the proof. □

6. Dirichlet Boundary Conditions on γ_0

Assume that ω is connected and that γ_0 is a subset of the boundary γ with nonzero $(N-1)$-capacity. For the N-dimensional Dirichlet boundary conditions the underlying space is

$$H^1_{\gamma_0}\big(S_h(\omega)\big)^N \overset{\text{def}}{=} \big\{V \in H^1\big(S_h(\omega)\big)^N : V|_{\Sigma_h(\gamma_0)} = 0\big\},$$
$$\text{and } \Sigma_h(\gamma_0) = \big\{x \in \mathbf{R}^N : |b_\Omega(x)| < h, \text{ and } p_{\partial\Omega}(x) \in \gamma_0\big\}$$

is the piece of the lateral boundary whose projection onto Γ is equal to γ_0. At the N-dimensional level the variational equation

$$\text{(6.1)}\qquad \boxed{\begin{aligned}&\exists V(h) \in H^1_{\gamma_0}\big(S_h(\omega)\big)^N \text{ such that } \forall V \in H^1_{\gamma_0}\big(S_h(\omega)\big)^N\\ &\int_{S_h(\omega)} C^{-1}\varepsilon\big(V(h)\big)\cdot\cdot\varepsilon(V) - F\cdot V - G\cdot DV\nabla b\, dx = 0\end{aligned}}$$

has a unique solution in the space $H^1_{\gamma_0}\big(S_h(\omega)\big)^N$ for any F and G in $L^2\big(S_h(\omega)\big)^N$. In view of the equivalence of norms this is equivalent to the condition: $\exists c(h) > 0$,

$$\forall V \in H^1_{\gamma_0}\big(S_h(\omega)\big)^N, \qquad \left|\int_{S_h(\omega)} F\cdot V + G\cdot DV\nabla b\, dx\right| \le c(h)\|\varepsilon(V)\|_{L^2(S_h(\omega))}.$$

Denote by $H^1_{\gamma_0}(\omega)$ the space $\{v \in H^1(\omega) : v|_{\gamma_0} = 0\}$ and recall that the space

$$E^{01}_{\gamma_0} \overset{\text{def}}{=} \{(v^0, v^1, v^2_n) \in H^1_{\gamma_0}(\omega)^N \times H^1(\omega)^N \times L^2(\omega) : v^1_\Gamma|_{\gamma_0} = 0\}$$

is complete for the norm **[28]**

$$\Big\{\|\varepsilon^0(v^0, v^1)\|^2_{L^2(\omega)} + \|\varepsilon^1(v^0, v^1, v^2_n n)\|^2_{L^2(\omega)} + \|v^0\|^2_{L^2(\omega)} + \|v^1_\Gamma\|^2_{L^2(\omega)}\Big\}^{1/2}.$$

It is also complete with respect to the norm

$$\|\varepsilon^0(v^0, v^1)\|_{L^2(\omega)} + \|\varepsilon^1(v^0, v^1, v^2_n n)\|_{L^2(\omega)}.$$

As before introduce the projection π onto the subspace $\ker \varepsilon^0 = \{v \in E^{01}_{\gamma_0} : \varepsilon^0(v^0, v^1) = 0\}$ and denote by $E^{01}_{\gamma_0,\pi}$ the completion of $E^{01}_{\gamma_0}$ with respect to the norm associated with the scalar product

$$\int_\omega \varepsilon^0(v)\cdot\cdot\varepsilon^0(w) + \varepsilon^1(\pi v)\cdot\cdot\varepsilon^1(\pi w)\, d\Gamma.$$

All the above considerations are also true for the subspace $H^1_{\gamma_0}(\omega)^N \times H^1_{\gamma_0}(\omega)^N \times L^2(\omega)$, but we shall only discuss the case of $E^{01}_{\gamma_0}$. The same definitions apply to the $P(1,1)$ model with the component $v^2 = 0$ and we shall use the same notation to simultaneously deal with both cases. The norm

$$\left\|\varepsilon^0(v^0, v^1)\right\|_{L^2(\omega)} + \left\|\varepsilon^1((v^0, v^1, 0))\right\|_{L^2(\omega)}.$$

is equivalent to the standard norm on the space

$$E^{01}_{\gamma_0} \stackrel{\text{def}}{=} \{(v^0, v^1) \in H^1_{\gamma_0}(\omega)^N \times H^1(\omega)^N : v^1_\Gamma|_{\gamma_0} = 0\}.$$

The same comments apply to the closed linear subspace $H^1_{\gamma_0}(\omega)^N \times H^1_{\gamma_0}(\omega)^N$.

The variational equation of the $P(2_n, 1)$ model is: to find $v_h \in E^{01}_{\gamma_0}$ such that

$$\boxed{\begin{aligned}\int_\omega \overline{\alpha}_0(h)C^{-1}\varepsilon^0_h\cdot\cdot\varepsilon^0 + \overline{\alpha}_1(h)(C^{-1}\varepsilon^1_h\cdot\cdot\varepsilon^0 + C^{-1}\varepsilon^0_h\cdot\cdot\varepsilon^1) + \overline{\alpha}_2(h)C^{-1}\varepsilon^1_h\cdot\cdot\varepsilon^1 \\ -F^0_h \cdot v^0 - F^1_h \cdot v^1 - F^2_h \cdot nv^2_n - G^0_h \cdot v^1 - G^1_h \cdot nv^2_n \quad d\Gamma = 0\end{aligned}} \tag{6.2}$$

for all $v \in E^{01}_{\gamma_0}$. The equation is the same for the $P(1,1)$ model with $v^2 = v^2_h = 0$. The asymptotic $P(1,0)$ model is: to find $(\hat{v}^0, \hat{v}^1) \in E^0_{\gamma_0} = E^{01}_{\gamma_0,\pi}/\ker\varepsilon^0$ such that

$$\boxed{\forall (v^0, v^1) \in E^{01}_{\gamma_0}, \quad \int_\omega C^{-1}\varepsilon^0(\hat{v})\cdot\cdot\varepsilon^0(v)\, d\Gamma = I(0)v,} \tag{6.3}$$

under the condition: there exists $c > 0$ such that for all (v^0, v^1) in $H^1_{\gamma_0}(\omega)^N \times L^2(\omega)^N$

$$\boxed{|I(0)v| = \left|\int_\omega f^0 \cdot v^0 + g^0 \cdot v^1\, d\Gamma\right| \le c\left\|\varepsilon^0(v^0, v^1)\right\|_{L^2(\omega)}.} \tag{6.4}$$

The second asymptotic (bending) equation is: to find $\hat{v} \in E^{01}_{\gamma_0,\pi}$ such that equation (6.3) is verified and

$$\boxed{\begin{aligned}\int_\omega C^{-1}\hat{\varepsilon}^1\pi\cdot\cdot\varepsilon^1 - f^2 \cdot v^0 - (f^1 + g^2)\cdot v^1 - (f^0 + 2g^1)\cdot nv^2_n \\ + \kappa_1\big(C^{-1}\hat{\varepsilon}^0\cdot\cdot\varepsilon^1 - f^1 \cdot v^0 - (f^0 + g^1)\cdot v^1 - 2g^0 \cdot nv^2_n\big) \\ + \kappa_2(-f^0 \cdot v^0 - g^0 \cdot v^1)\, d\Gamma = 0, \quad \forall v \in \ker\varepsilon^0.\end{aligned}} \tag{6.5}$$

Use the same definitions and assumptions for $I(h)$, $J(h)$, and $K(h)$. Then (6.3) and (6.5) are equivalent to find $\hat{v} \in E^{01}_{\gamma_0,\pi}$ such that

$$\boxed{\begin{aligned}\int_\omega C^{-1}\varepsilon^0(\hat{v})\cdot\cdot\varepsilon^0(v)\, d\Gamma = I(0)v, \quad \forall v \in E^{01}_{\gamma_0,\pi} \\ \int_\omega C^{-1}\varepsilon^1(\pi\hat{v})\cdot\cdot\varepsilon^1(\pi v) + \kappa_1 C^{-1}\hat{\varepsilon}^0(\hat{v})\cdot\cdot\varepsilon^1(\pi v)\, d\Gamma = 3J(0)\pi v, \quad \forall v \in E^{01}_{\gamma_0,\pi}.\end{aligned}} \tag{6.6}$$

ASSUMPTION 6.1. Let ω be a bounded open connected Lipschitzian domain in Γ, and let γ_0 be a subset of the boundary γ with nonzero $(N-1)$-capacity. There exists $\bar{h} > 0$ and a constant $c > 0$ such that for all h, $0 < h \leq \bar{h}$, and all $(v^0, v^1, v_n^2) \in E_{\gamma_0}^{01}$, the following condition is verified ($v^2 = 0$ for the $P(1,1)$ model)

$$\left| \frac{1}{2h} \int_{S_h(\omega)} F \cdot \left(v^0 \circ p + bv^1 \circ p + b^2(v_n^2 n) \circ p\right) \right. \tag{6.7}$$

$$\left. + G \cdot D\left(v^0 \circ p + bv^1 \circ p + b^2(v_n^2 n) \circ p\right) \nabla b \, dx \right|$$

$$\leq c \left\{ \int_\omega \left[\left\| \varepsilon^0(v^0, v^1) \right\|^2 + h^2 \left\| \varepsilon^1(v^0, v^1, v_n^2 n) \right\|^2 \right] d\Gamma \right\}^{1/2}. \tag{6.8}$$

By definition Assumption 6.1 is equivalent to

$$\left| I(h)(v^0, v^1, v_n^2 n) \right| \leq c \left\{ \int_\omega \left\| \varepsilon^0(v^0, v^1) \right\|^2 + h^2 \left\| \varepsilon^1(v^0, v^1, v_n^2 n) \right\|^2 d\Gamma \right\}^{1/2} \tag{6.9}$$

and our previous results simplify for the Dirichlet case.

LEMMA 6.1. (i) *Assumption* 6.1 *is equivalent to the condition: there exists* $c > 0$ *such that* ($v^2 = 0$ *for the* $P(1,1)$ *model*)

$$\forall (v^0, v^1) \in H_{\gamma_0}^1(\omega)^N \times L^2(\omega)^N, \quad \left| I(0)(v^0, v^1, 0) \right| \leq c \left\| \varepsilon^0(v^0, v^1) \right\|_{L^2(\omega)}. \tag{6.10}$$

When Assumption 6.1 *is verified, the variational equation* (6.2) *has a unique solution in* $E_{\gamma_0}^{01}$.

(ii) *As* $h > 0$ *goes to zero, there exists* $c > 0$ *such that for all* (v^0, v^1, v_n^2) *in* $E_{\gamma_0}^{01}$

$$\left| I(h)v - I(0)v \right| \leq ch^2 \left\{ \int_\omega \left\| \varepsilon^0(v^0, v^1) \right\|^2 + \left\| \varepsilon^1(v^0, v^1, v_n^2 n) \right\|^2 d\Gamma \right\}^{1/2} \tag{6.11}$$

$$\left| J(h)v - J(0)v \right| \leq ch^2 \left\{ \int_\omega \left\| \varepsilon^0(v^0, v^1) \right\|^2 + \left\| \varepsilon^1(v^0, v^1, v_n^2 n) \right\|^2 d\Gamma \right\}^{1/2}. \tag{6.12}$$

THEOREM 6.1 (Asymptotic shell). *Let Assumptions* 2.1 *and* 3.1 *be verified, and let* f^0, f^1, f^2, g^0, g^1 *and* g^2 *in* $L^2(\omega)^N$ *verify condition* (6.4). *Assume that* ω *is a bounded open connected Lipschitzian domain in* Γ, *and that* γ_0 *is a subset of the boundary* γ *with nonzero* $(N-1)$*-capacity. Then the system of equations* (6.6) *has a unique solution* $\hat{v}$ *in* $E_{\gamma_0,\pi}^{01}$. *The result extends to continuous linear functionals* $I(0)\colon E_{\gamma_0,\pi}^{01} / \ker \varepsilon^0 \to \mathbf{R}$ *and* $J(0)\colon E_{\gamma_0}^{01} \to \mathbf{R}$.

THEOREM 6.2. *Let Assumptions* 2.1, 3.1, *and* 6.1 *be verified. Assume that* ω *is a bounded open Lipschitzian domain in* Γ. *Further assume that* ω *is connected and that* γ_0 *is a subset of the boundary* γ *with nonzero* $(N-1)$*-capacity. Let* $v_h \in E_{\gamma_0}^{01}$ *and* $\hat{v} \in E_{\gamma_0,\pi}^{01}$ *be the respective solutions of* (6.2) *and* (6.6). *As* h *goes to zero,*

$$\int_\omega \left\| \varepsilon^0(v_h - \hat{v}) \right\|^2 + \left\| \varepsilon^1\left(\pi(v_h - \hat{v})\right) \right\|^2 + h^2 \left\| \varepsilon^1(v_h) \right\|^2 d\Gamma \to 0, \tag{6.13}$$

where $\pi(v)$ *is the projection of* v *onto* $\ker \varepsilon^0 \stackrel{\text{def}}{=} \left\{ v \in E_{\gamma_0,\pi}^{01} \colon \varepsilon^0(v^0, v^1) = 0 \right\}$. *The same results hold for the* $P(1,1)$ *model with* $v_h^2 = 0 = \hat{v}^2$.

References

1. E. Acerbi, G. Buttazzo, and D. Percivale, *A variational definition for the strain energy of an elastic string*, J. Elasticity **25** (1991) 137–148.
2. S. M. Alessandrini, *Some two-dimensional plate models: Derivation, asymptotic properties, and numerical approximation*, Ph.D. thesis, Rutgers Univ., New Brunswick, 1991.
3. S. M. Alessandrini, D. N. Arnold, R. S. Falk, and A. Madureira, *Derivation and justification of plate models by variational methods*, Plates and Shells (M. Fortin, ed.) (Québec, 1996), CRM Proc. Lecture Notes, vol. 21, Amer. Math. Soc., Providence, RI, pp. 1–20.
4. F. Bourquin, Ph. G. Ciarlet, G. Geymonat, and A. Raoult, *Γ-convergence et analyse asymptotique des plaques minces*, C. R. Acad. Sci. Paris Sér. I **315** (1992), 1017–1024.
5. F. Brezzi and M. Fortin, *Mixed and hybrid finite element methods*, Springer-Verlag, New York, Berlin, Heidelberg 1991.
6. D. Chapelle and K.-J. Bathe, *Fundamental considerations for the finite element analysis of shell structures*, Comput. & Structures (1997) (submitted).
7. C. Chen, *Asymptotic convergence rates for the Kirchhoff plate model*, Ph.D. thesis, Penn. State, University Park, Pennsylvania, 1995.
8. D. Chenais and J.-C. Paumier, *On the locking phenomenon for a class of elliptic problems*, Numer. Math. **67** (1994), 427–440.
9. Ph.G. Ciarlet, *Plates and junctions in elastic multi-structures: An asymptotic analysis*, Masson, Paris, Milano, Barcelona, Mexico, and Springer-Verlag, Berlin, New York (1990).
10. Ph. G. Ciarlet and V. Lods, *Ellipticité des équations membranaires d'une coque uniformément elliptique*, C. R. Acad. Sci. Paris Sér. I **318** (1994), 195–200.
11. ——— *Analyse asymptotique des coques linéairement élastique. I. Coques membranaires*, C. R. Acad. Sci. Paris, Sér. I **318** (1994), 863–868.
12. ——— *Analyse asymptotique des coques linéairement élastique. III. Une justification du modèle de Koiter*, C. R. Acad. Sci. Paris, Sér. I **319** (1994), 299–304.
13. ——— *On the ellipticity of linear membrane shell equations*, J. Math. Pures Appl. **75** (1996), 107–124.
14. ——— *Asymptotic analysis of linearly elastic shells. III. Justification of Koiter's shell equations*, Arch. Rational Mech. Anal. **136** (1996), 191–200.
15. ——— *Asymptotic analysis of linearly elastic shells. Generalized membrane shells*, J. Elasticity **43** (1996), 147–188.
16. ——— *Asymptotic analysis of linearly elastic shells. I. Justification of membrane shell equations*, Arch. Rational Mech. Anal. **136** (1996), 116–161.
17. Ph. G. Ciarlet, V. Lods, and B. Miara, *Analyse asymptotique des coques linéairement élastique. II. Coques en flexion*, C. R. Acad. Sci. Paris Sér. I Math. **319** (1994), 95–100.
18. ——— *Asymptotic analysis of linearly elastic shells. II. Justification of flexural shell equations*, Arch. Rational Mech. Anal. **136** (1996), 163–190.
19. Ph. G. Ciarlet and É. Sanchez-Palencia, *Un théorème d'existence et d'unicité pour les équations des coques membranaires*, C. R. Acad. Sci. Paris Sér. I **317** (1993), 801–805
20. ——— *An existence and uniqueness theorem for the two-dimensional linear membrane shell equations*, J. Math. Pures Appl. **75** (1996), 51–67.
21. M. Dauge, *Complete asymptotics in thin elastic plates and optimal estimates for Kirchhoff-Love model*, Prépublication 95-06, Institut de Recherche Mathématique de Rennes, Univ. de Rennes, février 1995.
22. M. Dauge and I. Gruais, *Développement asymptotique d'ordre arbitraire pour une plaque élastique mince encastrée*, C. R. Acad. Sc. Paris Sér. I **321** (1995), 375–380.
23. M. C. Delfour, *Intrinsic differential geometry methods in the asymptotic analysis of linear thin shells*, Boundaries, interfaces and transitions (Banff, 1995) (M. C. Delfour, ed.) CRM Proc. Lecture Notes, vol. 13, Amer. Math. Soc. Providence, RI, 1998, pp. 19–90
24. M. C. Delfour and J.-P. Zolésio, *Shape analysis via distance functions*, J. Funct. Anal. **123** (1994), 129–201.
25. ——— *On a variational equation for thin shells*, Control and Optimal Design of Distributed Parameter Systems Minneapolis, MN, 1992) (J. Lagnese, D. L. Russell, and L. White, eds.), IMA Vol. Math. Appl., vol. 70, Springer, New York, 1995, pp. 25–37.
26. ——— *A boundary differential equation for thin shells*, J. Differential Equations **119** (1995), 426–449.

27. ______ *Tangential differential equations for dynamical thin/shallow shells*, J. Differential Equations **128** (1996), 125–167.
28. ______ *Differential equations for linear shells: Comparison between intrinsic and classical models*, Advances in the Mathematical Sciences—CRM's 25 years, (Luc Vinet, ed.), CRM Proc. Lecture Notes, vol. 11, Amer. Math. Soc., Providence, RI, 1997, pp. 42–124.
29. ______ *Shape analysis via distance functions: Local theory*, Boundaries, interfaces and transitions (M. C. Delfour, ed.) (Banff, 1995) CRM Proc. Lecture Notes, vol. 13, Amer. Math. Soc. Providence, RI, 1998, pp. 91–123.
30. ______ *On the design and control of systems governed by differential equations on submanifolds*, Control Cybernet. **25** (1996), 497–514.
31. ______ *Hidden boundary smoothness for some classes of differential equations on submanifolds*, Proc. 1996 Joint Summer Research Conference on Optimization Methods in PDE's (S. Cox and I. Lasiecka, eds.) Contemporary Mathematics, Amer. Math. Soc., Providence, RI (to appear).
32. ______ *Intrinsic differential geometry and theory of thin shells*, Quaderni, Scuola Normale Superiore, Pisa (Italy) (to appear); also as lecture notes, version 1.0, August 1996.
33. ______ *Convergence to the asymptotic model for linear thin shells*, Optimization Methods in Partial Differential Equations (South Hadley, MA, 1996) (S. Cox and I. Lasiecka, eds.) Contemp. Math., vol. 209, Amer. Math. Soc., Providence, RI, 1997, pp. 75–93.
34. Ph. Destuynder, *Sur la justification des modèles de plaques et de coques par les méthodes asymptotiques*, Doctoral dissertation, Univ. Pierre et Marie Curie 1980.
35. ______ *Modélisation des coques minces élastiques*, Collection Physique Fondamentale et Applique, Masson, Paris, 1990.
36. ______ *Une théorie asymptotique des plaques minces en elasticité linéaire*, Masson, Paris, 1986.
37. D. D. Fox, A. Raoult, and J. C. Simo, *A justification of nonlinear properly invariant plate theories*, Arch. Rational Mech. Anal. **124** (1993), 157–199.
38. H. Le Dret and A. Raoult, *The nonlinear membrane model as variational limit of nonlinear three-dimensional elasticity*, J. Math. Pures Appl **74** (1995), 549–578.
39. ______ *The membrane shell model in nonlinear elasticity: A variational asymptotic derivation*, J. Nonlinear Sci. **6** (1996), 59–84.
40. K. H. Lo, R. M. Christensen, and E. M. Wu, *A high-order theory of plate deformations*, J. Appl. Mech. **46** (1977), 663–676.
41. C. Mardare, *Modèles bi-dimensionnels de coques linéairement élastiques: Estimations de l'écart entre leurs solutions*, C. R. Acad. Sc. Paris Sér. I **322** (1996), 793–796.
42. ______ *Estimation d'erreur dans l'analyse asymptotique des coques linéairement élastiques*, C. R. Acad. Sc. Paris Sér. I **322** (1996), 895–898.
43. D. Morgenstern, *Herleitung der Plattentheorie aus der dreidimensionalen Elastizitätstheorie*, Arch. Rational Mech. Anal. **4** (1959), 145–152.
44. J.-C. Paumier and A. Raoult, *Asymptotic consistency of the polynomial approximation in the linearized plate theory. Application to the Reissner-Mindlin model*, Elasticity, Viscoelasticity and Optimal Control: Theoretical and Numerical Aspects (Lyon, 1995), ESAIM Proc. vol. 2, Soc. Math. Appl. Indust., Paris, 1997, pp. 203–213 (http//www.emath.fr/Maths/Proc/Vol.2/index.htm).
45. J. Piila, *Characterization of the membrane theory of a clamped shell. The hyperbolic case*, Math. Models Methods Appl. Sci. **6** (1996), 169–194.
46. J. Pitkaranta, *The problem of membrane locking in finite element analysis of cylindrical shells*, Numer. Math. **61** (1992), 523–542.
47. É. Sanchez-Palencia, *Statique et dynamique des coques minces.* I. *Cas de flexion pure non inhibée*, C. R. Acad. Sc. Paris Sér. I **309** (1989), 411–417.
48. ______ *Statique et dynamique des coques minces.* II. *Cas de flexion pure inhibée—Approximation membranaire*, C. R. Acad. Sc. Paris Sér. I **309** (1989), 531–537.
49. ______ *Surfaces et coques élastiques minces : problèmes et défis*, La Vie des Sciences **12** (1995), no. 3 , 239–258.
50. J. Sanchez-Hubert and É. Sanchez-Palencia, *Coques élastiques minces, propriétés asymptotiques*, Masson, Paris 1997.

Centre de recherches mathématiques et Département de Math. et de statistique, Université de Montréal, C. P.6128, Succ. Centre-ville, Montréal (Qc), H3C 3J7, Canada

E-mail address: delfour@crm.UMontreal.ca

CNRS Institut Non Linéaire de Nice, 1361 route des lucioles, 06904 Sophia Antipolis Cedex, France, and Centre de Mathématiques Appl., Ecole des Mines (CMA/MEIJE, INRIA), 2004 route des Lucioles, BP 93, 06902 Sophia Antipolis Cedex, France

E-mail address: Jean-Paul.Zolesio@sophia.inria.fr

Centre de Recherches Mathématiques
CRM Proceedings and Lecture Notes
Volume **21**, 1999

Two-Level Overlapping Schwarz Methods for Plate Elements on Unstructured Meshes Using Non-Matching Coarse Grids

Qingping Deng and Xiaobing Feng

ABSTRACT. Two-level overlapping Schwarz methods are developed for solving the finite element approximations of plate bending problems on general unstructured meshes. The subdomains which make up the whole domain are allowed to have arbitrary shapes and different sizes, and the coarse grid need not to align with the fine grid. In addition, the domains defined by the coarse grid and the fine grid are not required to be identical as long as the former covers a significant part of the latter. In this general setting, it is shown that the Schwarz methods have the same optimal convergence rate as usual two-level overlapping Schwarz methods on structured meshes with matching coarse grids. The condition number of the preconditioned finite element system is shown to depend only on the overlap of the subdomains and the mesh size of the coarse grid, but to be independent of the sizes and the number of the subdomains. To handle the difficulty caused by the lack of continuity for the nonconforming plate elements, the main point is to recognize the fact that the basis functions of each plate element corresponding to different nodal parameters have different effects on the condition number of the finite element system, so it is feasible to use a reduced conforming finite element space on the coarse grid, which has less degree of freedoms, to precondition the finite element system on the fine grid. The reduced coarse grid finite element space is so chosen such that it will dampen the large eigenmodes of the fine grid finite element system.

1. Introduction

During the past ten years Schwarz methods or domain decomposition methods have emerged as a class of effective numerical methods for solving the linear systems resulted from finite element and finite difference discretizations of partial differential equations. In the case of using structured meshes, the theory and methodology of overlapping Schwarz methods are quite well developed. A considerable number of papers have been devoted to the subject. Among the vast literature, we refer to [**J,S,D,F,X,Z**] and the references therein.

On the other hand, for the problems with complicated geometries, the above well-developed Schwarz methods become inefficient. For instance, to solve a second

1991 *Mathematics Subject Classification.* Primary: 65F10, 65N55; Secondary: 65N30.
This is the final form of the paper.

or fourth order elliptic problem on a polygonal domain having one piece very thin flat boundary, if one insists to use a structured mesh, then the mesh must be very fine. Moreover, if one wants to use a matched or aligned coarse grid, this will also force to use very fine course grid, which therefore implies that the method is very expensive to use. To overcome such kind of difficulties, one idea is to use unstructured meshes and nonmatched coarse grids since they easily allow for complicated geometries and the resolution of fine scale structure in the solution. Recently, some good results have been obtained in this direction by a number of authors for second order elliptic problems, We refer to [**T**,**R**] and references therein for a detailed exposition.

The objective of this paper is to develop and analyze two-level Schwarz methods for solving the finite element approximations of the plate bending problems on unstructured meshes using nonmatched coarse grids. These methods easily allows one to handle the plate bending problems with complicated geometries and to solve the problems in parallel. This extra flexibility are due to the fact that in our methods the coarse grids are not required to align with the fine grid and the domains defined by the coarse grid and the fine grid do not need to be identical as long as the former covers a significant part of the latter. In this general setting, we show that the Schwarz methods have the same optimal convergence rate as usual two-level overlapping Schwarz methods on structured meshes with matching coarse grids, that is, the condition numbers of the preconditioned finite element systems are shown to depend only on the overlap of the subdomains and the mesh size of the coarse grid, but to be independent of the sizes and the number of the subdomains. To handle the difficulty caused by the lack of continuity for the nonconforming plate elements, our main idea is to recognize the fact that the basis functions of each plate element corresponding to different nodal parameters have different effects on the condition number of the finite element system, so it is feasible to use a reduced conforming finite element space on the coarse grid, which has less degree of freedoms, to precondition the finite element system on the fine grid. The reduced coarse grid finite element space is so chosen such that it will dampen the large eigenmodes of the fine grid finite element system.

The layout of the rest of this paper is as follows. In Section 2 we first recall the plate bending problems and their finite element approximations, and then give some preliminaries about the Schwarz method. In Section 3, two-level Schwarz algorithms are presented and analyzed abstractly. In Section 4, the application of the abstract theory to the Morley nonconforming element are presented in detail as an example.

2. Preliminaries

We consider the following clamped Kirchhoff plate bending problem:

$$\Delta^2 u = f, \qquad \text{in } \Omega, \tag{2.1.i}$$

$$u = \frac{\partial u}{\partial n} = 0, \qquad \text{on } \partial\Omega, \tag{2.1.ii}$$

where Ω is a bounded polygonal domain in $\mathbb{R}^2$, f denotes the distributed load on Ω. For a given functional f on $H_0^2(\Omega)$, the weak formulation of (2.1) is to find $u \in H_0^2(\Omega)$ such that

$$a(u, v)_\Omega = (f, v)_\Omega, \quad \forall v \in H_0^2(\Omega), \tag{2.2}$$

where

$$(2.3.i)\quad a(u,v)_\Omega = \int_\Omega \Big[\Delta u \Delta v + (1-\sigma)\Big(2\frac{\partial^2 u}{\partial x_1 \partial x_2}\frac{\partial^2 v}{\partial x_1 \partial x_2} - \frac{\partial^2 u}{\partial x_1^2}\frac{\partial^2 v}{\partial x_2^2} - \frac{\partial^2 u}{\partial x_2^2}\frac{\partial^2 v}{\partial x_1^2}\Big)\Big]\, dx,$$

$$(2.3.ii)\qquad (f,v)_\Omega = \int_\Omega f v \, dx,$$

and $0 < \sigma < \frac{1}{2}$ is the Poisson ratio. It is well known that (1.1) possesses a unique solution $u \in H_0^2(\Omega)$ (cf. [**C**,**G**]).

Let $\{\mathcal{T}_h\}$ be a family of triangular or rectangular partitions of Ω. For a easy presentation, we assume that $\overline{\Omega} = \cup_{K \in \mathcal{T}_h} \overline{K}$. Let h_K and ρ_K denote the diameter of K and the diameter of the circle inscribed in K. Introduce the usual notations $h = \bar{h} = \max_{K \in \mathcal{T}_h} h_K$ and $\underline{h} = \min_{K \in \mathcal{T}_h} h_K$. Recall the following well-known definition (cf. [**C**]).

DEFINITION 2.1. The partition $\{\mathcal{T}_h\}$ is said to be *shape regular* if there is an h-independent positive constant σ_0 such that

$$(2.4)\qquad \max_{K \in \mathcal{T}_h} \frac{h_K}{\rho_K} \le \sigma_0,$$

and to be *quasi-uniform* if it is shape regular and there is another h-independent positive constant γ such that

$$(2.5)\qquad h \le \gamma \underline{h}.$$

Throughout this paper, unless stated otherwise, we assume that $\{\mathcal{T}_h\}$ is shape regular but not necessary quasi-uniform, which means that the ratio $\bar{h}/\underline{h}$ could be very large. Suppose $\{\Omega_j\}_{j=1}^J$ are J open subdomains of Ω such that each $\partial\Omega_j$ is aligned with $\mathcal{T}_h$ and $\Omega = \cup_{j=1}^J \Omega_j$. Moreover, we assume that there exist nonnegative C^∞ functions $\varphi_1, \varphi_2, \ldots, \varphi_J$ such that

$$(2.6.i)\qquad \sum_{j=1}^J \varphi_j = 1, \text{ on } \overline{\Omega}, \quad \varphi_j = 0 \text{ on } \Omega \backslash \Omega_j,$$

$$(2.6.ii)\qquad \|\nabla \varphi_j\|_{L^\infty} \le \frac{1}{\delta_j}, \quad \|\nabla^2 \varphi_j\|_{L^\infty} \le \left(\frac{1}{\delta_j}\right)^2,$$

where $\nabla^2 \varphi$ is the Hessian of ϕ and $\{\delta_j\}_{j=1}^J$ are small positive constants which measure the amount of overlap among the subdomains Ω_j. We refer to [**L**] and [**D**] for a detailed exposition on the constructions of Ω_j and φ_j. Let $M(x)$ be the number of subdomains which contains x. We assume that $M = \max_{x\in\Omega} M(x)$ is less than a fixed constant.

Let $S^h \subset L^2(\Omega)$ be one of well-known plate element spaces with respect to $\mathcal{T}_h$ whose members vanish at the boundary nodes (cf. [**C**]), and S_j^h be a subspace of S^h whose members vanish at all nodes that are not interior to Ω_j. Then from (2.6) we have

$$(2.7)\qquad S^h = \sum_{j=1}^J S_j^h.$$

Define the following discrete bilinear form on S^h:

$$a^h(v,w) = \sum_{K\in\mathcal{T}_h} a(v,w)_K, \quad \forall v,w \in S^h. \tag{2.8}$$

Then the finite element method for problem (2.1) is defined as follows: find $u^h \in S^h$ such that

$$a^h(u^h, v) = (f, v)_\Omega, \quad \forall v \in S^h. \tag{2.9}$$

Let $\{\phi_j^h\}_{j=1}^n$ denote the nodal basis of the finite element space S^h and $\{p_j^h\}_{j=1}^n$ denote the nodal set corresponding to the nodal parameters (a node is counted k times if there are k nodal parameters attached to it). Then (2.9) gives the following linear system:

$$A\xi = b, \tag{2.10}$$

where

$$A = \left[a(\phi_i^h, \phi_j^h)\right]_{n\times n}, \quad \xi = \left(u^h(p_1^h), \dots, u^h(p_n^h)\right)^t, \quad b = \left((f,\phi_1^h)_\Omega, \dots, (f,\phi_n^h)_\Omega\right)^t.$$

It is well known that the above linear system is ill-conditioned since the condition number $\mathrm{cond}(A) = O(\underline{h}^{-4})$, this implies that the classical iterative methods such as the Jacobi, Gauss-Seidel and the Conjugate Gradient method are not efficient for the system. To overcome the difficulty, one popular approach, which is also the approach in this paper, is to find a good preconditioner for the system and then to solve the preconditioned system. Several methods are available for obtaining good preconditioner (cf. [**Y,X**]). The goal of this paper is to construct a good preconditioner for system (2.10) based on two-level overlapping domain decomposition methods using unstructured meshes and nonmatching coarse grids, which we proceed as follows.

Let $\mathcal{T}_H$ denote an *approximate* coarse mesh for Ω with the mesh size H. Where *approximate* means that the union of all elements in $\mathcal{T}_H$, denoted by Ω^H, is required to cover a significant part of Ω but not necessarily identical with Ω. Also, we assume that $\mathcal{T}_H$ is shape regular but not necessarily quasi-uniform. Let $S^H \subset L^2(\Omega^H)$ be a finite dimensional space associated with $\mathcal{T}_H$; $\{\phi_j^H\}_{j=1}^m$ and $\{p_j^H\}_{j=1}^m$ denote the nodal basis of S^H and the nodal set of the mesh $\mathcal{T}_H$.

For practical consideration and technical simplicity, we make the following two weak assumptions on $\mathcal{T}_H$:

(1) No coarse grid element lies completely outside of Ω, that is,

$$K \cap \Omega \neq \emptyset, \quad \forall K \in \mathcal{T}_H. \tag{2.11.i}$$

(2) The coarse mesh covers a significant part of the fine mesh, more precisely,

$$\Omega \setminus \Omega^H \subset \bigcup_{p_j^H} B_{p_j^H}(\mathrm{diam}\, O_j^H), \tag{2.11.ii}$$

where $O_j^H \equiv \mathrm{support}(\phi_j^H)$, and $B_p(r)$ denotes the ball with radius r and center p.

We conclude this section by recalling the following property of the standard interpolation operator π_h associated with $\mathcal{T}_h$: suppose that $W^{s,p}(K) \hookrightarrow C^1(K) \cap W^{\ell,q}(K)$, then

$$\|v - \pi_h v\|_{W^{\ell,q}(K)} \leq C h_K^{s-\ell} h_K^{2(1/q-1/p)} |v|_{W^{s,p}(K)}, \quad \forall v \in W^{s,p}(K),\ K \in \mathcal{T}_h; \tag{2.12}$$

and introducing some notations to be used in the following sections:

$$a^H(v,w) = \sum_{K \in \mathcal{T}_H} a(v,w)_K, \quad \forall v,\ w \in S^H, \tag{2.13.i}$$

$$h_i = \max_{\substack{K \subset \Omega_i \\ K \in \mathcal{T}_h}} h_K, \quad H_i = \max_{\substack{K \cap \Omega_i \neq \emptyset \\ K \in \mathcal{T}_H}} H_K, \tag{2.13.ii}$$

$$\Omega_i^H = \bigcup_{\substack{K \in \mathcal{T}_H \\ K \cap \Omega_i \neq \emptyset}} K, \quad \mathcal{T}_i = \{K \in \mathcal{T}_h; K \cap \Omega_i \neq \emptyset\}. \tag{2.13.iii}$$

3. Two-level Overlapping Schwarz Methods

In this section, based on the finite element spaces S^h, S_j^h and S^H defined in the previous section, we will present and analyze two-level overlapping domain decomposition methods for solving linear system (2.10). The methods to be described below are constructed based on the well-known approach and idea as employed in [**X**] and [**Z**] for two-level nested grids and spaces, and in [**T,S**] and [**F**] for two-level nested grids but nonnested spaces, that is, to use a coarse to fine grid prolongation operator in conjunction with local and coarse grid solvers. However, the use of unstructured meshes and, in particular, the use of nonmatching coarse grids cause some new difficulties both in the derivation and analysis of the methods.

Since we allow nonmatching coarse grids, clearly, $S^H \not\subset S^h$, in fact, even in the case of matching grids, most of well-known plate elements result in nonnested multi-level finite element spaces (cf. [**F**]). To overcome the nonnestedness between S^h and S^H and pass information from the coarse grid to the fine grid, we introduce a prolongation operator $I_H^h : S^H \to S^h$, and assume that I_H^h satisfies the following properties:

$$|v - I_H^h v|_{H^\ell(\Omega_j)} \leq C h_j^{2-\ell} |v|_{H^2(\Omega_j^H)}, \quad \forall v \in S^H;\ \ell = 0,1,2;\ j = 1,2,\dots,J, \tag{H.1}$$

where and throughout this paper C denotes a generic positive constant. We also define the restriction operator from fine to coarse grid as the adjoint operator of I_H^h with respect to $(\cdot,\cdot)_{\Omega^H}$ and $(\cdot,\cdot)_{\Omega^h}$, that is,

$$(I_h^H v, w)_{\Omega^H} = (v, I_H^h w)_{\Omega^h}, \quad \forall v \in S^h,\ w \in S^H. \tag{3.1}$$

As in [**J**] and [**X**], we define the following operators:

$$A_j : S_j^h \to S_j^h,\ j = 1,2,\dots,J,$$
$$(A_j v, w)_\Omega = a^h(v,w), \quad \forall v,\ w \in S_j^h. \tag{3.2}$$

$$Q_j : S^h \to S_j^h,\ j = 1,2,\dots,J,$$
$$(Q_j v, w)_\Omega = (v,w)_\Omega, \quad \forall v \in S^h,\ w \in S_j^h. \tag{3.3}$$

$$P_j : S^h \to S_j^h,\ j = 1,2,\dots,J,$$
$$(P_j v, w)_\Omega = a^h(v,w), \quad \forall v \in S^h,\ w \in S_j^h. \tag{3.4}$$

$$A^H: S^H \to S^H,$$

$$(A^H v, w)_{\Omega^H} = a^H(v, w), \quad \forall v,\ w \in S^H. \tag{3.5}$$

$$P_h^H: S^h \to S^H,$$

$$a^H(P_h^H v, w) = a^h(v, I_H^h w), \quad \forall v \in S^h,\ w \in S^H. \tag{3.6}$$

Also, let A^h denote the operator induced by the stiffness matrix A, that is,

$$(A^h v, w)_\Omega = a^h(v, w), \quad \forall v,\ w \in S^h. \tag{3.7}$$

Then, it is easy to check that

$$A_j P_j = Q_j A^h,\ j = 1, 2, \ldots, J, \tag{3.8}$$

$$A^H P_h^H = I_h^H A^h. \tag{3.9}$$

Finally, let R^H be some inexact coarse grid solver (an approximation of A_H^{-1}), which is symmetric positive definite with respect to $(\cdot,\cdot)_{\Omega^H}$; and R_j is an inexact local solver on Ω_j (approximation of A_j^{-1}, $j = 1, 2 \ldots J$), which is symmetric positive definite with respect to $(\cdot,\cdot)_\Omega$.

We now introduce the following two-level additive Schwarz preconditioning operator $B^h: S_h \to S_h$ for the operator A^h:

$$B^h = I_H^h R^H I_h^H + \sum_{j=1}^J R_j Q_j. \tag{3.10}$$

Hence the conditioned equation of (2.9) is

$$B^h A^h u = B^h f. \tag{3.11}$$

Let $T^h = B^h A^h$. Using the identities (3.8) and (3.9) we can rewrite T^h as

$$T^h = I_H^h R^H A^H P_h^H + \sum_{j=1}^J R_j A_j P_j \equiv \sum_{j=0}^J T_j, \tag{3.12}$$

where

$$T_0 = I_H^h R^H A^H P_h^H, \quad T_j = R_j A_j P_j,\ j = 1, 2, \ldots, J. \tag{3.13}$$

It is easy to see that $R^H A^H$ is symmetric positive definite with respect to $a^H(\cdot,\cdot)$; and $R_j A_j$ is symmetric positive definite with respect to $a(\cdot,\cdot)_{\Omega_j}$. Then T^h is symmetric positive definite with respect to $a^h(\cdot,\cdot)$. Hence

$$\operatorname{cond}(T^h) = \frac{\lambda_{\max}(T^h)}{\lambda_{\min}(T^h)}, \tag{3.14}$$

where $\lambda_{\max}(T^h)$ and $\lambda_{\min}(T^h)$ are the largest and smallest eigenvalue of T^h, respectively.

Next, we are going to show that T^h, the preconditioned operator, has an h-independent condition number, hence, (3.11) is well conditioned. We will proceed by deriving an upper bound for $\lambda_{\max}(T^h)$ and a lower bound for $\lambda_{\min}(T^h)$.

LEMMA 3.1. *There holds the following estimate:*

$$\lambda_{\max}(T^h) \le C\widetilde{\rho}(M+1), \tag{3.15}$$

where $\widetilde{\rho} = \max\{\lambda_{\max}(R^H A^H), \lambda_{\max}(R_1 A_1), \ldots, \lambda_{\max}(R_J A_J)\}$.

For a proof of the above Lemma, we refer to [**F**]. We also remark that the proofs of the upper bound estimate for $\lambda_{\max}(T^h)$ are same for the case of structured meshes with matching coarse grids and for the case of unstructured meshes with nonmatching coarse grids.

On the other hand, the lower bound estimate for $\lambda_{\min}(T^h)$ is much complicated even in the case structured meshes with matching coarse grids (cf. [**D,F,X**]). To derive the estimate, the idea we use here is to introduce another fine to coarse intergrid operator. It is important to recognize that the introduction of this additional intergrid operator is just for the technical analysis, and this intergrid operator has no effect on the implementation of overlapping Schwarz methods.

Suppose there is an intergrid operator $L_h^H : S^h \to S^H$ which satisfies

$$\text{(H.2)} \quad |v - L_h^H v|_{H^\ell(\Omega_j)} \le C H_j^{2-\ell} |v|_{H^2(\Omega_j)}, \quad \forall v \in S_j^h;\ \ell = 0,1,2;\ j = 1,2,\ldots,J.$$

Then we have the following partition lemma, which leads to an lower bound estimate for $\lambda_{\min}(T^h)$.

LEMMA 3.2. *For any* $v \in S^h$, *there exist* $v_0 \in S^H$, $v_j \in S_j$, $j = 1,2,\ldots,J$, *such that*

$$v = I_H^h v_0 + \sum_{j=1}^J v_j, \tag{3.16}$$

$$a^H(v_0, v_0) + \sum_{j=1}^J a^h(v_j, v_j) \le CM\left[1 + \max_j (H_j/\delta_j)^4\right] a^h(v, v). \tag{3.17}$$

PROOF. For any $v \in S^h$ we set

$$v_0 = L_h^H v \in S^H, \quad w = v - I_H^h v_0 \in S^h, \tag{3.18.i}$$

$$v_j = \pi_h(\varphi_j w), \quad j = 1,2,\ldots,J. \tag{3.18.ii}$$

Clearly, $v_j \in S_j$. Furthermore, by (2.6.i) we have

$$\sum_{j=1}^J v_j = \sum_{j=1}^J \pi_h(\varphi_j w) = \pi_h w = w = v - I_H^h v_0,$$

So (3.16) holds. Next, by (H.1), (H.2) and (2.13.i) we get

$$\begin{aligned} a^H(v_0, v_0) &\le C|v_0|^2_{H^2(\mathcal{T}_H)} = C|L_h^H v|^2_{H^2(\mathcal{T}_H)} \\ &\le C|v|^2_{H^2(\mathcal{T}_h)} \le C a^h(v, v), \end{aligned} \tag{3.19}$$

$$\begin{aligned} &\|w\|_{L^2(\Omega_j)} + H_j |w|_{H^1(\mathcal{T}_j)} \\ &= \left\|v - I_H^h(L_h^H v)\right\|_{L^2(\Omega_j)} + H_j \left|v - I_H^h(L_h^H v)\right|_{H^1(\mathcal{T}_j)} \\ &\le \|v - L_h^H v\|_{L^2(\Omega_j)} + \left\|L_h^H v - I_H^h(L_h^H v)\right\|_{L^2(\Omega_j)} \end{aligned} \tag{3.20}$$

$$
\begin{aligned}
&+ H_j\big(|v - L_h^H v|_{H^1(\mathcal{T}_j)} + \big|L_h^H v - I_H^h(L_h^H v)\big|_{H^1(\mathcal{T}_j)}\big)\\
&\le CH_j^2\big[|v|_{H^2(\mathcal{T}_j)} + |L_h^H v|_{H^2(\mathcal{T}_j)}\big]\\
&\le CH_j^2|v|_{H^2(\Omega_j)}.
\end{aligned}
$$

Let v^I be the linear or bilinear interpolation of v associated with $\mathcal{T}_h$, then

$$
\begin{aligned}
\|\varphi_j - \varphi_j^I\|_{L^2(K)} + h_j\|\nabla(\varphi_j - \varphi_j^I)\|_{L^2(K)} + h_j^2\|\nabla^2(\varphi_j - \varphi_j^I)\|_{L^\infty(K)}&\\
\le Ch_j^2\|\nabla^2\varphi_j\|_{L^\infty(K)},\quad \forall K\in\mathcal{T}_j;\ j=1,2,\dots,J.&
\end{aligned}
\tag{3.21}
$$

Thus for any $K\in\mathcal{T}_j$,

$$
|v_j|_{H^2(K)} = \big|\pi_h(\varphi_j w)\big|_{H^2(K)} \le \big|\pi_h(\varphi_j^I w)\big|_{H^2(K)} + \big|\pi_h\big((\varphi_j - \varphi_j^I)w\big)\big|_{H^2(K)}.
\tag{3.22}
$$

Therefore,

$$
\begin{aligned}
\big|\pi_h(\varphi_j^I w)\big|_{H^2(K)} &\le \big|\pi_h(\varphi_j^I w) - \varphi_j^I w\big|_{H^2(K)} + |\varphi_j^I w|_{H^2(K)}\\
&\le Ch_j|\varphi_j^I w|_{H^3(K)} + \|\varphi_j^I\|_{L^\infty(K)}|w|_{H^2(K)}\\
&\quad + \|\nabla\varphi_j^I\|_{L^\infty(K)}|w|_{H^1(K)}\\
&\le \|\nabla\varphi_j^I\|_{L^\infty(K)}|w|_{H^1(K)} + \|\nabla^2\varphi_j^I\|_{L^\infty(K)}\|w\|_{L^2(K)}\\
&\le C\big(\|\varphi_j^I\|_{L^\infty(K)}|w|_{H^2(K)} + \|\nabla\varphi_j^I\|_{L^\infty(K)}|w|_{H^1(K)}\\
&\quad + \|\nabla^2\varphi_j^I\|_{L^\infty(K)}\|w\|_{L^2(K)}\big)\\
&\le C\left[|w|_{H^2(K)} + \frac{1}{\delta_j}|w|_{H^1(K)} + \left(\frac{1}{\delta_j}\right)^2\|w\|_{L^2(\Omega)}\right],
\end{aligned}
\tag{3.23}
$$

and

$$
\begin{aligned}
\big|\pi_h\big((\varphi_j - \varphi_j^I)w\big)\big|_{H^2(K)} &\le Ch_j^{-2}\big|\pi_h\big((\varphi_j - \varphi_j^I)w\big)\big|_{L^2(K)}\\
&\le h_j^{-2}\big[|\varphi_j - \varphi_j^I|_{L^\infty(K)} + h_j\big|\nabla(\varphi_j - \varphi_j^I)\big|_{L^\infty(K)}\\
&\quad + h_j^2\big|\nabla^2(\varphi_j - \varphi_j^I)\big|_{L^\infty(K)}\big]|w|_{L^2(K)}\\
&\le C\left(\frac{1}{\delta_j}\right)^2|w|_{L^2(K)}.
\end{aligned}
\tag{3.24}
$$

Combining (3.21)–(3.24)

$$
\begin{aligned}
a^h(v_j, v_j) &\le C|v_j|^2_{H^2(\mathcal{T}_h)} = C|v_j|^2_{H^2(\mathcal{T}_j)}\\
&\le C\left[|w|^2_{H^2(\mathcal{T}_j)} + \left(\frac{1}{\delta_j}\right)^2|w|^2_{H^1(\mathcal{T}_j)} + \left(\frac{1}{\delta_j}\right)^4\|w\|^2_{L^2(\Omega_j)}\right].
\end{aligned}
\tag{3.25}
$$

Summing up the above inequality over j from 1 to J we obtain

$$
\begin{aligned}
\sum_{j=1}^J a^h(v_j, v_j) &\le C\sum_{j=1}^J\Bigg[|w|^2_{H^2(\mathcal{T}_j)} + \left(\frac{1}{\delta_j}\right)^2|w|^2_{H^1(\mathcal{T}_j)}\\
&\qquad [t] + \left(\frac{1}{\delta_j}\right)^4\|w\|^2_{L^2(\Omega_j)}\Bigg]\\
&\le C\sum_{j=1}^J\left[1 + \left(\frac{H_j}{\delta_j}\right)^2 + \left(\frac{H_j}{\delta_j}\right)^4\right]|v|^2_{H^2(\Omega_j)}
\end{aligned}
\tag{3.26}
$$

$$\leq CM\left[1 + \max_j \left(\frac{H_j}{\delta_j}\right)^4\right] |v|^2_{H^2(\mathcal{T}_h)}$$

$$\leq CM\left[1 + \max_j \left(\frac{H_j}{\delta_j}\right)^4\right] a^h(v, v).$$

Finally, the lemma follows from adding (3.26) to (3.19). □

Now using Lemma 3.2 and a standard argument (cf. Lemma 3.16 of [**X**]) we have

LEMMA 3.3. *There holds the following estimate*:

$$\lambda_{\min}(T) \geq \frac{C\underset{\sim}{\rho}}{M[1 + (\max_j (H_j/\delta_j))^4]}, \tag{3.27}$$

where $\underset{\sim}{\rho} = \min\{\lambda_{\min}(R^H A^H), \lambda_{\min}(R_1 A_1), \ldots, \lambda_{\min}(R_J A_J)\}$.

Therefore, from Lemmas 3.1 and 3.2 we get the following main theorem of this paper.

THEOREM 3.1. *Under the assumptions* (H.1) *and* (H.2), *the preconditioned operator* $T^h = B^h A^h$ *defined by* (3.12) *satisfies*

$$\operatorname{cond}(T^h) \leq \frac{C\widetilde{\rho}}{\underset{\sim}{\rho}} M^2 \left[1 + \left(\max_j (H_j/\delta_j)\right)^4\right]. \tag{3.28}$$

REMARK 3.1. Consider the following multiplicative Schwarz method as proposed in [**J**]: choose $u^0 \in S^h$ and define $u^k \in S^h$ recursively by

$$\begin{cases} w_{-1} = u^k, \\ w_j = w_{j-1} + T_j(u - w_{j-1}), \quad j = 0, 1, 2, \ldots, J, \\ u^{k+1} = w_J. \end{cases} \tag{3.29}$$

Then,

$$u - u^{k+1} = E^h(u - u^k), \quad E^h = (I - T_J)\ldots(I - T_1)(I - T_0). \tag{3.30}$$

We can show that the iteration matrix E^h satisfies the estimate:

$$\|E^h\|^2 \leq 1 - \frac{\underset{\sim}{\rho}(2 - \widetilde{\rho})}{CM^3[1 + \max_j (H_j/\delta_j)^4]}, \tag{3.31}$$

where $\|\cdot\|$ stands for operator norm with respect to energy norm induced by $a^h(\cdot, \cdot)$.

4. Application to the Morley Nonconforming Plate Element

In this section, we will apply the overlapping Schwarz methods developed in the previous two sections to the Morley nonconforming plate element (cf. [**C,M**]). This will be done by explicitly constructing the coarse space S^H, the prolongation operator I_H^h and the intergrid operator L_h^H, and showing that they satisfy assumptions (H.1) and (H.2).

For the Morley element, $\mathcal{T}_h$ is necessary to be a triangulation of Ω_j. Let S^h be the Morley finite element space associated with $\mathcal{T}_h$. It is well known that the standard interpolation operator π_h on S^h satisfies the approximation property (2.12). Let $\mathcal{T}_H$ is an *approximate* triangulation of $\vee$ which satisfies the two conditions listed in Section 2. We then define S^H as a reduced Bell finite element space associated with $\mathcal{T}_H$ (cf. [**B**,**C**]), more precisely, the element degrees of the freedom are

$$D^\alpha v(a^i), \quad |\alpha| \leq 1, \ i = 1,2,3, \tag{4.1.i}$$

$$D^\alpha v(a^i) = 0, \quad |\alpha| = 2, \ i = 1,2,3, \tag{4.1.ii}$$

where $\{a^i\}_{i=1}^3$ denote the three vertices of a triangle $K \in \mathcal{T}_H$. Hence, $S^H \subset C^1(\overline{\Omega^H}) \subset W^{2,r}(\Omega^H)(r \geq 3)$ and each $v \in S^H$ is piecewise polynomial of degree five and its all second order derivatives take value zero at all vertices of $\mathcal{T}_H$.

To define I_H^h, for each function in $v \in S^H$ we first need to give an extension of v on the portion of $\vee$ where it does not overlap with Ω^H. Let $\overline{v}$ denote an extension of v such that $\overline{v}|_{\Omega^H} = v$, and

$$\|\overline{v}\|_{H^2(\vee)} \leq C\|v\|_{H^2(\Omega^H)}.$$

From [**E**] we know that such an extension does exist. Next, we define $I_H^h \colon S^H \to S^h$ by

$$I_H^h v = \pi_h \overline{v}.$$

The usefulness of this choice is demonstrated by the following lemma.

LEMMA 4.1. *I_H^h satisfies the assumption* (H.1), *that is,*

$$|v - I_H^h v|_{H^\ell(\Omega_j)} \leq C h_j^{2-\ell} |v|_{H^2(\Omega_j^H)}, \quad \forall v \in S^H, \ \ell = 0,1,2, \ j = 1,2,\ldots,J, \tag{4.2}$$

where C is a positive constant which only depends on σ_0 in (2.4).

PROOF. For notation brevity, we abuse the notations by not distinguish v and $\overline{v}$ in the proof. Notice that $S^H \subset C^1(\overline{\Omega^H}) \subset W^{2,r}(\Omega^H)(r \geq 3)$, from [**C**] we have for any $v \in S^H$ and $K \in \mathcal{T}_h$

$$|v - I_H^h v|_{H^\ell(K)} = |v - \pi_h v|_{H^\ell(K)} \leq C h_K^{2-\ell} h_K^{1-2/r} |v|_{W^{2,r}(K)}. \tag{4.3}$$

By using the Cauchy inequality and the inverse inequality, we have for any $\tau \in \mathcal{T}_H$ and $\tau \subset \Omega_j^H$

$$\begin{aligned} \sum_{\substack{K\in\Omega_j \\ K\cap\tau\neq\emptyset}} |v - I_H^h v|^2_{H^\ell(K)} &\leq C h_j^{2(2-\ell)} \sum_{\substack{K\in\Omega_j \\ K\cap\tau\neq\emptyset}} h_K^{2(1/2-1/r)} |v|^2_{W^{2,r}(K)} \\ &\leq C h_j^{2(2-\ell)} \Bigg(\sum_{\substack{K\in\Omega_j \\ K\cap\tau\neq\emptyset}} h_K^2 \Bigg)^{1-2/r} \Bigg(\sum_{\substack{K\in\Omega_j \\ K\cap\tau\neq\emptyset}} |v|^r_{W^{2,r}(K)} \Bigg)^{2/r} \end{aligned} \tag{4.4}$$

$$\leq Ch_j^{2(2-\ell)} H_\tau^{2(1-2/r)} \Big(\sum_{\substack{\tau' \in \mathcal{T}_H \\ \tau' \cap \tau \neq \emptyset \\ \tau' \cap \Omega_j \neq \emptyset}} |v|_{W^{2,r}(\tau')}^r \Big)^{2/r}$$

$$\leq Ch_j^{2(2-\ell)} \sum_{\substack{\tau' \in \mathcal{T}_H \\ \tau' \cap \tau \neq \emptyset \\ \tau' \cap \Omega_j \neq \emptyset}} |v|_{H^2(\tau')}^2.$$

Summing up the above inequality over $\tau \subset \Omega_j^H$ we get

$$|v - I_H^h v|_{H^\ell(\Omega_j)}^2 \leq Ch_j^{2(2-\ell)} |v|_{H^2(\Omega_j^H)}^2, \quad \ell = 0, 1, 2.$$

which gives (4.2). The proof is completed. □

Finally, we need to define the intergrid operator $L_h^H : S^h \to S^H$ which satisfies the assumption (H.2). As the matter of fact, such an operator had already been constructed in [**F**] as some composite operator. In the following we will only recall the definition of it and refer to [**F**] for the verification of its properties.

Let W^h denote the Argyris finite element space associated with $\mathcal{T}_h$, define $\alpha_h : S^h \to W^h$ such that for any $v \in S^h$, $\alpha_h v$ satisfies

$$\text{(4.5)} \qquad \begin{cases} \alpha_h v(p^h) = v(p^h), & \\ \frac{\partial(\alpha_h v)}{\partial x_i}(p^h) = M\big(\frac{\partial v}{\partial x_i}(p^h)\big), & i = 1, 2, \\ \frac{\partial(\alpha_h v)}{\partial n}(m^h) = \frac{\partial v}{\partial n}(m^h), & \\ \frac{\partial^2(\alpha_h v)}{\partial x_i x_j}(p^h) = 0 \text{ or } M\big(\frac{\partial^2 v}{\partial x_i \partial x_j}(p^h)\big), & i, j = 1, 2, \end{cases}$$

where p^h and m^h run over all the nodal points and all the midpoints of the sides of $\mathcal{T}_h$ respectively, and $M(\cdot)$ is the averaging operator.

Let π_H be the standard interpolation operator on S^H. We define the intergrid operator L_h^H by

$$\text{(4.6)} \qquad L_h^H v = \pi_H(\overline{\alpha_h v}),$$

where $\overline{\alpha_h v}$ defined on Ω^H is the extension of $\alpha_h v$.

LEMMA 4.2. *The operator L_h^H satisfies the assumption* (H.2).

For a proof of Lemma 4.2, we refer to [**F**] and [**S**].

We now conclude the section by stating the following main result of this section.

THEOREM 4.1. *Let S^H and I_H^h be chosen as above, then the conclusion of Theorem* 3.1 *holds for the Morley plate element.*

REMARK 4.1. For the conforming plate elements, it can be shown that Theorem 3.1 holds if both I_H^h and L_h^H are chosen as the interpolation operators on S^h and S^H, respectively.

References

[R] R. Bank and J. Xu, *An algorithm for coarsening unstructured meshes*, Numer. Math. **73** (1996), 1–36.

[B] K. Bell, *A refined triangular plate bending element*, Internat. J. Numer. Methods Engrg. **1** (1969), 101–122.

[J] J. H. Bramble, J. E. Pasciak, J. Wang, and J. Xu, *Convergence estimates for product iterative methods with applications to domain decomposition*, Math. Comp. **57** (1991), 1–21.

[P] J. H. Bramble, J. E. Pasciak, and J. Xu, *Parallel multilevel preconditioners*, Math. Comp. **55** (1990), 1–22.

[S] S. C. Brenner, *Two-level additive Schwarz preconditioners for nonconforming plate elements*, Numer. Math. **72** (1996), 419–447.

[T] T. Chan, B. Smith, and J. Zou, *Overlapping Schwarz methods on unstructured meshes using nonmatching coarse grids*, Numer. Math. **73** (1996), 149–167.

[C] P. G. Ciarlet, *The finite element method for elliptic problems*, North-Holland, Amsterdam, 1978.

[F] Q. Deng and X. Feng, *Two-level Schwarz methods for conforming and nonconforming finite element approximations of plate bending problems*, Mathematical Report 94-01, The Univ. Tennessee, 1994.

[D] M. Dryja and O. B. Widlund, *An additive variant of the Schwarz alternating method for the case of many subregions*, Tech. Rep. **339**, Courant Institute (1989).

[G] P. Grisvard, *Singularities in boundary value problems*, Masson and Springer-Verlag, Paris, New York, 1992.

[L] P. L. Lions, *On the Schwartz alternating method.*, I. First International Symposium on Domain Decomposition Method for Partial Differential Equations (R. Glowinski *et al.*, eds.), SIAM, Philadelphia, PA, 1990, pp. 1–42.

[M] L. S. D. Morley, *The triangular equilibrium problem in the solution of plate bending problems*, Aerosp. Quart. **19** (1968), 149–169.

[E] E. M. Stein, *Singular integrals and differentiability properties of functions*, Princeton Univ. Press, Princeton, New Jersey, 1970.

[X] J. Xu, *Iterative methods by space decomposition and subspace correction*, SIAM Rev. **34** (1992), 581–613.

[Y] Y. Saad, *Iterative methods for sparse linear systems*, PWS Publ. Company, Boston, 1996.

[Z] X. Zhang, *Domain decomposition algorithms for the biharmonic Dirichlet problem*, Fifth International Symposium on Domain Decomposition Methods for Partial Differential Equations (Norfolk, VA, 1991) (D. E. Keyes, T. F. Chan, *et al.*, eds.), SIAM, Philadelphia, PA, 1992, pp. 119–126.

Department of Mathematics, University of Tennessee, Knoxville, TN 37996, USA

E-mail address: deng@cs.utk.edu

Department of Mathematics, University of Tennessee, Knoxville, TN 37996, USA

E-mail address: xfeng@math.utk.edu

Centre de Recherches Mathématiques
CRM Proceedings and Lecture Notes
Volume **21**, 1999

Large Elasto-Plastic Deformations of Thin Shells with Application to the Stamping of Anisotropic Sheet Metal Parts

J.-C. Gelin and L. Boubakar

ABSTRACT. The developments presented in the paper concern the formulation of a fully nonlinear shell theory including both membrane and bending components in the field of large elasto-plastic deformations for the modelling and simulation of sheet metal stamping processes. The paper focus on some numerical aspects of the intrinsic complexity of these problems. The first subject treated is the formulation of a fully non linear three node shell element including both membrane and bending components in the field of large deformations. The second part of the paper concerns the proper description of the anisotropic behaviour of the thin shells subjected to large elasto-plastic deformations. A general kinematic evolution of the anisotropy directions is proposed based on the restriction to the tangent plane of the polar decomposition of the deformation gradient. The stress calculation algorithm uses a particular implementation of the elastic predictor-plastic corrector method. The last part of the paper presents a numerical example that illustrates the proposed developments in the field of the simulation of the sheet metal forming processes.

1. Introduction

The possibility to use general shell elements for the simulation of sheet metal forming processes has been investigated in numerous papers [**17, 36, 37**] that emphasize the possibility of standard shell elements derived from the classical ones used in structural analysis problems. These elements [**4, 14, 15, 27**] are generally not well adapted to extensive nonlinear calculations involving nonlinear kinematics and nonlinear material behaviour in the large strains range.

The possibility to use a C^0 three node shell element with linear interpolation of the position and displacement have been investigated in References [**7**] and [**8**]. The point of departure of the element developed is the so called degenerated solid approach [**1**] combined with an assumed strain method [**39**] for the out of plane shear strain components. This formulation does not refer to special assumption as Kirchhoff-Love one and is applicable in the linearized geometric case or in the fully nonlinear geometric are. Furthermore, it is shown in [**16**] that such an element can

---

1991 *Mathematics Subject Classification.* Primary: 73G20; Secondary: 73K15.
This is the final form of the paper.

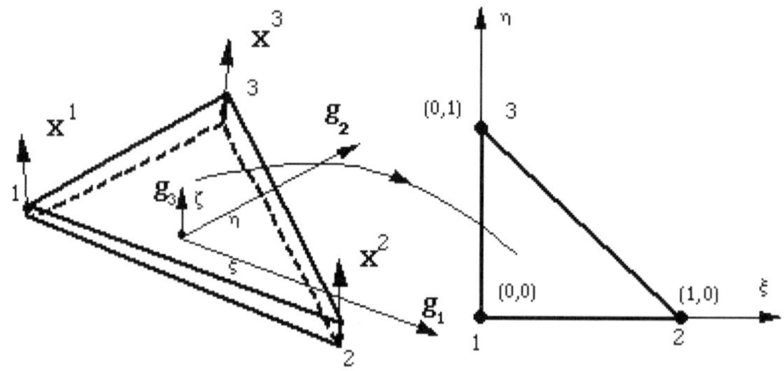

FIGURE 1. Representation of the geometry of the shell.

be used with one set of integration points in the mid plane that render the element very attractive for transient explicit dynamics solution procedures comparatively to well known four node shell elements [**5,6**] with hourglass control. It has been shown that the C^0 triangular element exhibits no spurious modes in constraint situations [**16**].

The second difficulty emphasized in the paper concerns the proper description of the anisotropic behaviour of the sheet metal during forming. It seems in the literature that the contributions in that field for sheet forming applications [**10,40**] are devoted to the implementation of general yield function capabilities, but are not very careful with the description of change in the anisotropy directions. The paper focus on a slightly different aspect in the sense that the quadratic Hill's yield criterion is considered. In fact the problem considered is relative to the proper description of the change in the principal anisotropy directions during the deformation of the sheet metal and a general kinematic description is defined based on a particular choice for the material rotation. Furthermore the case of the thin shell kinematics is studied in the sense that the kinematic is locally restrains to a plane one. The stress calculation algorithm is based on a particular implementation of the elastic predictor-plastic corrector method.The proposed stress calculation algorithm takes into account of the mixed interpolation used in the shell element formulation and of the zero normal stress assumption.

The last part of the paper is concerned with numerical examples that prove the validity of the approach proposed.

NOTATIONS. Tensors will be denoted by bold-face characters in direct notation. Their juxtaposition implies the usual summation operation over two adjacent indices, while double dots and the symbol $\otimes$ indicate the summation product and the tensor product. A superposed dot indicates the rate, a superposed T the transpose, a superposed S the symmetric part and a superposed A the skew-symmetric part.

2. Element Kinematics

Even in analyses where the thickness is very small and where the transverse shear can be neglected, the C^0 shell elements based on Mindlin kinematics appear to be more satisfactory than C^1 Kirchhoff elements. The proposed triangular shell element is included in the so-called continuum based element family and is built on C^0 isoparametric kinematics introduced by Ahmad [**1**]. Many authors have developed shell elements from such kinematics especially four nodes [**14, 15, 39**] or higher order elements that are proved to be among the most effective ones.

2.1. Interpolation of the geometry. In order to insure the continuity of the geometrical description, pseudonormal vectors $\mathbf{X}^i(\mathbf{X}^i = 1)$ are defined at each node as vectors that joint the top and the bottom of the shell, Fig. 1.

The classical Lagrange linear interpolation functions for the natural material coordinates ξ and η in the reference element are used. The third material coordinate in the thickness ζ is equal to 1 at the top of the shell and is equal to -1 at the bottom.

Position of a point M is defined as the sum of its position along the pseudonormal $\mathbf{X}$ and of the position $\overline{\mathbf{x}}(\xi,\eta)$ of the associated point of the mid-surface:

$$\mathbf{x}(\xi,\eta,\zeta) = \overline{\mathbf{x}}(\xi,\eta) + \zeta\frac{h}{2}\mathbf{X}(\xi,\eta). \tag{1}$$

The interpolation of the position vector $\mathbf{x}$ is given by:

$$\mathbf{x}(\xi,\eta,\zeta) = \sum_{i=1}^{n} N^i(\xi,\eta)\overline{\mathbf{x}}^i + \zeta\sum_{i=1}^{n} N^i(\xi,\eta)\frac{h^i}{2}\mathbf{X}^i, \tag{2}$$

where h^i is the thickness at node i

2.2. Displacement interpolation. The nodal displacement at each node is defined as (Fig. 2):

$$\mathbf{u}(M) = \overline{\mathbf{u}}(H) + \mathbf{u}_R(M), \tag{3}$$

where $\overline{\mathbf{u}}$ is the displacement of the associated point in the mid-surface and $\mathbf{u}_R$ the displacement given by the pseudonormal rotation. Vector $\mathbf{u}$ is interpolated from the nodal displacements $\overline{\mathbf{u}}^i$ and from $\mathbf{u}_R^i$, the displacement given by the rotation of the point M_0^i initially on the pseudonormal $\mathbf{X}_0^i$ at node i and at the same co-ordinate ζ as M:

$$\mathbf{u}(\xi,\eta,\zeta) = \sum_{i=1}^{n} N^i(\xi,\eta)\big[\overline{\mathbf{u}}^i + \mathbf{u}_R^i(\zeta)\big], \tag{4}$$

with

$$\mathbf{u}_R^i = \frac{h^i}{2}\zeta(\mathbf{X}^i - \mathbf{X}_0^i) = \frac{h^i}{2}\zeta(\mathbf{R}^i - \mathbf{I})\mathbf{X}_0^i, \tag{5}$$

where $\mathbf{R}^i$ is the orthogonal tensor which transforms $\mathbf{X}_0^i$ to $\mathbf{X}^i$ and $\mathbf{I}$ the second order identity tensor.

An orthogonal frame $(\mathbf{V}_{10}^i, \mathbf{V}_{20}^i, \mathbf{V}_{30}^i)$ is defined at each node with $\mathbf{V}_{30}^i = \mathbf{X}_{30}^i$. Assuming that the rotations in the loading step are small enough and denoting θ^i the rotation vector at node i:

$$\begin{aligned} \theta^i &= \theta_1^i\mathbf{V}_{10}^i + \theta_2^i\mathbf{V}_{20}^i, \\ (\mathbf{R}^i - \mathbf{I})\mathbf{X}_0^i &= \theta^i \wedge \mathbf{X}_0^i + \mathbf{O}\big[(\theta^i)^2\big], \end{aligned} \tag{6}$$

which gives the displacement interpolation:

$$\mathbf{u}(\xi,\eta,\zeta) = \sum_{i=1}^{n} N^i(\xi,\eta)\overline{\mathbf{u}}^i + \zeta\sum_{i=1}^{n} N^i(\xi,\eta)\frac{h^i}{2}(-\theta_1^i\mathbf{V}_{20}^i + \theta_2^i\mathbf{V}_{10}^i). \tag{7}$$

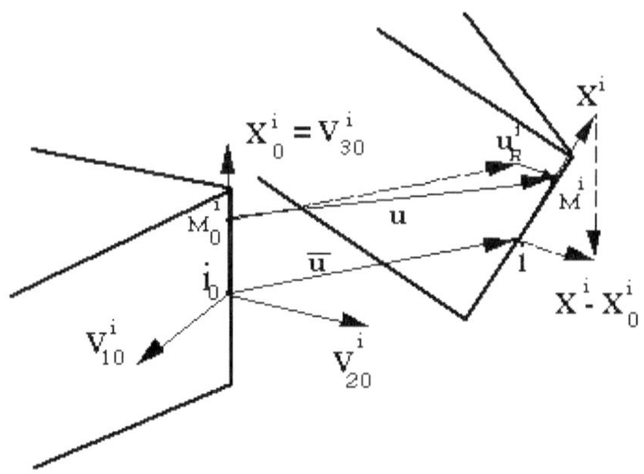

FIGURE 2. Description of the nodal kinematics.

The same interpolation functions N^i are used for the displacement and the position. When performing a nonlinear calculation, the coordinates at the end of the loading step are obtained by adding the element displacements to the original position, so the use of the same interpolation functions for both displacements and geometry is a consistent solution approach [**3**]. The shell element has five degrees of freedom per node. Three displacement components in the global coordinate system and two rotation components in $\mathbf{V}_{10}^i$ and $\mathbf{V}_{20}^i$.

Because achievement of an efficient element needs a modification of the interpolation of transverse shear components, the transverse notion leads to work in a coordinate system involving the plane tangent to the shell. The proposed formulation uses the material covariant frame (Fig. 2):

$$\mathbf{g}_1 = \frac{\partial \mathbf{x}}{\partial \xi}, \quad \mathbf{g}_2 = \frac{\partial \mathbf{x}}{\partial \eta}, \quad \mathbf{g}_3 = \frac{\partial \mathbf{x}}{\partial \zeta}. \tag{8}$$

The related contravariant frame $(\mathbf{g}^1, \mathbf{g}^2, \mathbf{g}^3)$ is defined by:

$$\mathbf{g}_k \mathbf{g}^1 = \delta_k^1. \tag{9}$$

Considering the state at the beginning of the load step as the initial configuration, the strain tensor

$$\varepsilon = \varepsilon_{ij} \mathbf{g}_0^i \otimes \mathbf{g}_0^j = \frac{1}{2}(\mathbf{g}_i \mathbf{g}_j - \mathbf{g}_{0i} \mathbf{g}_{0j}) \mathbf{g}_0^i \otimes \mathbf{g}_0^j, \tag{10}$$

where $\theta_1 = \xi$, $\theta_2 = \eta$ and $\theta_3 = \zeta$, has five significant components (ε_{33} associated to $\sigma^{33} = 0$, plane stress condition, does not appear in displacement calculation). It is yet a posteriori calculated using the behaviour law in order to update the thickness of the shell after each loading step. In plane strain components ε_{11}, ε_{22}, ε_{12} are classically derived from nodal displacements. Taking into account the displacement interpolation (7) and the derivatives of the interpolation functions N^i (ξ, η) and after straightforward calculations the in-plane membrane-bending interpolation matrix is expressed as:

$$\varepsilon_{\alpha\beta} = \mathbf{B}_{mb}(\mathbf{g}_{01}, \mathbf{g}_{02}, \mathbf{g}_{03}) \mathbf{u}_n^e. \tag{11}$$

2.3. Out of plane strain components interpolation. The previous approach could be used to interpolate the transverse shear strains. But the obtained C^0 elements based on the kinematics defined in the previous section are far too stiff. The Kirchhoff solution expected when the thickness of the shell is small imposes too many constraints per element in regard to its number of degrees of freedom. A direct interpolation of the transverse shear strain tensorial components is made from the values of those components at nodes:

$$\varepsilon_{\alpha 3} = \sum_{i=1}^{n} N^i(\xi, \eta)\varepsilon^i_{\alpha 3}, \tag{12}$$

where N^i are the linear shape functions. The main problem is now to get the nodal values of the out of plane shear strain components. In that goal, the out of plane shear strains in the direction of the side of the element are assumed to be constant. When making this assumption, only three conditions are imposed per element in order to get the Kirchhoff solution ($\varepsilon_{\alpha 3} = 0$) while the nullity of the shear components given by derivation of the displacements which are linear in ξ and η imposes six conditions per element. As a consequence the element built on this so called assumed strain method avoid shear locking. A material covariant frame related to the sides of the element $(\mathbf{f}^i_1, \mathbf{f}^i_2, \mathbf{f}^i_3)$ is introduced at each node i [**7, 8**]. Using the contravariant frame the strain tensor ε can be written:

$$\varepsilon = \varepsilon^*_{ij}\, \mathbf{f}^i_0 \otimes \mathbf{f}^j_0. \tag{13}$$

Along each side the transverse shear strain components $\varepsilon^*_{\alpha 3}$ are assumed to be constant and equal to the mean value that is the value at the mid-point of the side i.e.:

$$\varepsilon^{*i}_{\alpha 3} = \frac{1}{2}\left(\left(\frac{\partial \mathbf{u}}{\partial \xi}\right)_m (\mathbf{f}^i_{\alpha 0})_m + \left(\frac{\partial \mathbf{u}}{\partial r^i_\alpha}\right)_m (\mathbf{g}_{30})\right). \tag{14}$$

The different mean quantities $(\cdot)_m$ are calculated from the definition of $(\mathbf{f}^i_{10}, \mathbf{f}^i_{20})$ and the geometrical and kinematic interpolations (2) and (7). At each node, r^i_1 and r^i_2 are related to ξ and η co-ordinates and $(\mathbf{f}^i_{10}, \mathbf{f}^i_{20})$ vectors are related to $(\mathbf{g}^i_{10}, \mathbf{g}^i_{20})$ vectors. After straightforward calculations, the shear strain interpolation is expressed as:

$$\varepsilon_{\alpha 3} = \overline{\mathbf{B}}_s(\mathbf{g}_{01}, \mathbf{g}_{02}, \mathbf{g}_{03}).\mathbf{u}^e_n. \tag{15}$$

3. Elastoplastic Constitutive Equations in the Shell Context

3.1. Kinematic aspects. Based on Mandel's approach [**28–30**], the following decomposition of the deformation gradient $\mathbf{F}$ is introduced:

$$\mathbf{F} = \mathbf{V}^e a \widehat{\mathbf{F}}^p. \tag{16}$$

The tensor $\widehat{\mathbf{F}}^p$ is the plastic deformation gradient from the initial configuration to a so-called isoclinic relaxed configuration C_a. In this configuration, a triad of director vectors attached to the material substructure preserves its initial orientation in reference to a fixed cartesian coordinates system. During the movement, these vectors are updated by the orthogonal tensor $\mathbf{a}(\mathbf{a}\mathbf{a}^T = \mathbf{I}; \det \mathbf{a} = \mathbf{1})$. A new relaxed configuration is then obtained [**24**] from which a pure elastic deformation $\mathbf{V}^e$ is applied to complete the transformation.

Assuming that elastic strains are small ($\mathbf{V}^e = \mathbf{I}+\varepsilon^e; \|\varepsilon^e\| \ll 1$), the velocity gradient $\mathbf{L}$ is given by:

$$\mathbf{L} = \dot{\mathbf{F}}\mathbf{F}^{-1} = \dot{\mathbf{a}}\mathbf{a}^T + D_a\varepsilon^e + \mathbf{a}(\dot{\widehat{\mathbf{F}}}^p\widehat{\mathbf{F}}^{p-1})\mathbf{a}^T, \tag{17}$$

where $D_a\varepsilon^e$ is the rotational derivative of ε^e with respect to the rotation $\mathbf{a}$. This relation leads to the following decomposition of the strain rate tensor $\mathbf{D}$ and material spin tensor $\mathbf{W}$:

$$\begin{aligned} \mathbf{D} = \mathbf{L}^S &= D_a\varepsilon^e + \mathbf{a}(\dot{\widehat{\mathbf{F}}}^p\widehat{\mathbf{F}}^{p-1})^S\mathbf{a}^T = \mathbf{D}^e + \mathbf{D}^p, \\ \mathbf{W} = \mathbf{L}^A &= \dot{\mathbf{a}}\mathbf{a}^T + \mathbf{a}(\dot{\widehat{\mathbf{F}}}^p\widehat{\mathbf{F}}^{p-1})^A\mathbf{a}^T = \dot{\mathbf{a}}\mathbf{a}^T + \mathbf{W}^p. \end{aligned} \tag{18}$$

The first decomposition represents an extension of the well known additive decomposition of the strain rates for linear theory; it permits a thermodynamic formulation of the behaviour. The second relation is a particularity of finite strains that is essential to study the anisotropic media, it gives the evolution law of director vectors by coupling their spin $\dot{\mathbf{a}}\mathbf{a}^T$ to the material spin tensor $\mathbf{W}$ and the plastic spin $\mathbf{W}^p$. In the case of orthotropic materials, the anisotropic directions represent a preferential director frame and the rotation $\mathbf{a}$ defines its evolution.

3.2. Formulation of the behaviour. To describe the anisotropic plastic behaviour of the metals, Hill's orthotropic criterion [**18**] remains among the anisotropic criterion [**2, 19**] the most currently used. In deep drawing it permits a satisfactory representation of different kinds of deformations. Associated to an isotropic hardening, the yield function in the orthotropy axes is given by:

$$f(\sigma_a, K) = \frac{1}{2}\sigma_a : \mathbf{H} : \sigma_a - \frac{1}{2}K^2(X). \tag{19}$$

The fourth order tensor $\mathbf{H}$ is the Hill's orthotropic tensor; its components are classically expressed using the coefficients F, G, H, L, M and N. The function K is the flow stress depending on the scalar value X. The evolution laws are in this case:

$$\dot{\mathbf{d}}_a^p = \dot{\lambda}\mathbf{H} : \sigma_a, \quad \dot{X} = \dot{\lambda}\sqrt{\sigma_a : \mathbf{H} : \sigma_a}. \tag{20}$$

To use these relations we must define during the transformation the isoclinic configuration or the evolution law of the orthotropy frame. As it was introduced previously, we can use the relation (18) which gives a differential equation for the rotation $\mathbf{a}$ of anisotropy directions:

$$\dot{\mathbf{a}} = (\mathbf{W} - \mathbf{W}^p)\mathbf{a}, \quad \mathbf{a}(0) = \mathbf{I}. \tag{21}$$

Within a macroscopic approach, Dafalias [**11**] and Loret [**25**] have use the representation theory of isotropic tensorial functions to evaluate the plastic spin $\mathbf{W}^p$. Nevertheless, this method introduce new scalar parameters which remains difficult to identify experimentally. Another approach consists on a choice for $\mathbf{a}$ using the kinematics of the movement [**13, 23**]. In particular, the proper rotation $\mathbf{R}$ obtained from the polar decomposition of the deformation gradient $\mathbf{F}(\mathbf{F} = \mathbf{RU})$ and the rotation $\mathbf{Q}$ obtained from the material spin $\mathbf{W}$ can be regarded as two extreme choices. In the first case $\mathbf{W}^p = \mathbf{R}[\dot{\mathbf{U}}\mathbf{U}^{-1}]^A\mathbf{R}^T$ and in the second $\mathbf{W}^p = 0$. The effect of these rotations on the response of an orthotropic material has been studied for a rigid plastic behaviour [**13**] and when accounting of elastic behaviour [**9**]. In the case of simple shear test with plane stress condition, the use of $\mathbf{Q}$ leads

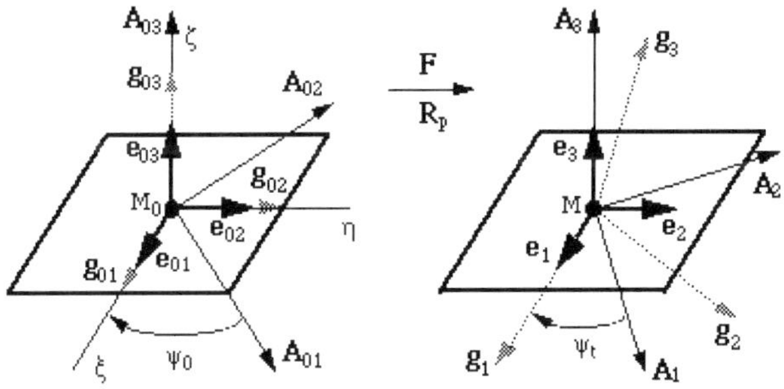

FIGURE 3. Orthotropy directions updating.

to an oscillatory response. These oscillations disappear when using **R**. Moreover, in this case, the normal stress evolution is in agreement with the experimental response obtained in a torsion test with fixed extremities [**31,32**]. This comparison has been carried out by Dafalias [**11**] within the plastic spin approach.

3.3. Anisotropy directions update. In a shell finite element model for sheet metal forming analysis, it is interesting to keep two of the orthotropy directions $(\mathbf{A}_1, \mathbf{A}_2)$ in the mid-plane of the the shell $(\mathbf{g}_1, \mathbf{g}_2)$ (relation (8)) to simply differentiate the mechanical behaviour through the thickness (Fig. 3). During the movement an orthonormal frame $(\mathbf{e}_1, \mathbf{e}_2, \mathbf{e}_3)$ is built from the same material direction $\mathbf{g}_1$:

$$\mathbf{e}_1 = \frac{\mathbf{g}_1}{\|\mathbf{g}_1\|}, \quad \mathbf{e}_2 = \mathbf{e}_3 \wedge \mathbf{e}_1, \quad \mathbf{e}_3 = \frac{\mathbf{g}_1 \wedge \mathbf{g}_2}{\|\mathbf{g}_1 \wedge \mathbf{g}_2\|}. \tag{22}$$

If we consider the restriction $\mathbf{F}_p$ of the deformation gradient $\mathbf{F}$ to the mid-plane

$$\mathbf{F}_p = \mathbf{g}_1 \otimes \mathbf{g}_0^1 + \mathbf{g}_2 \otimes \mathbf{g}_0^2, \quad \mathbf{g}_{0i}\mathbf{g}_0^j = \delta_{ij}. \tag{23}$$

Thus, when using the proper rotation $\mathbf{R}_p$ obtained from the polar decomposition of $\mathbf{F}_p$, we preserve strictly the orthotropy direction $\mathbf{A}_3$ normal to the mid-plane $(\mathbf{g}_1, \mathbf{g}_2)$:

$$\mathbf{R}_p = \cos\varphi \mathbf{e}_1 \otimes \mathbf{e}_{01} + \sin\varphi \mathbf{e}_2 \otimes \mathbf{e}_{01} - \sin\varphi \mathbf{e}_1 \otimes \mathbf{e}_{02} + \cos\varphi \mathbf{e}_2 \otimes \mathbf{e}_{02}. \tag{24}$$

The angle is given by [**9**]:

$$\varphi = \operatorname{arctg}\left(\frac{F_{p12}}{F_{p11} + F_{p22}}\right), \tag{25}$$

with:

$$F_{p11} = (\mathbf{g}_1\mathbf{e}_1)(\mathbf{g}_0^1\mathbf{e}_{01}), \quad F_{p22} = (\mathbf{g}_2\mathbf{e}_2)(\mathbf{g}_0^2\mathbf{e}_{02}),$$
$$F_{p12} = (\mathbf{g}_1\mathbf{e}_1)(\mathbf{g}_0^1\mathbf{e}_{02}) + (\mathbf{g}_2\mathbf{e}_1)(\mathbf{g}_0^2\mathbf{e}_{02}). \tag{26}$$

The current orientation ψ_t of anisotropy directions with respect to the orthonormal frame $(\mathbf{e}_1, \mathbf{e}_2, \mathbf{e}_3)$ is simply given in using the initial orientation ψ_0 and the φ angle by the following relation: $\psi_t = \psi_0 - \varphi$.

The updating obtained by $\mathbf{R}_p$ is identical to those obtained by $\mathbf{R}$ if the nodal normals of the shell elements presented previously are perpendicular to $(\mathbf{g}_1, \mathbf{g}_2)$. Practical cases are closed of this case.

3.4. Incremental description of the behaviour. Within an incremental method, the state of the material is defined from the previous convergent state and the incremental displacement $\mathbf{u}$. Integrating equations (20) over the time step Δt (subscripts 0 and 1 correspond to values at the beginning and at the end of the loading step under consideration), we obtain the following incremental laws:

$$(\sigma_a)_1 = (\sigma_a)_0 + \mathbf{C}^e : [\Delta \mathbf{d}_a - \Delta \mathbf{d}_a^p], \qquad (27)$$
$$(\mathbf{X}_{ai})_1 = (\mathbf{X}_{ai})_0 + \Delta \mathbf{X}_{ai},$$

with:

$$\Delta \mathbf{d}_a = \int_{t_0}^{t_1} \dot{\mathbf{d}}_a \, dt, \quad \Delta \mathbf{d}_a^p = \int_{t_0}^{t_1} \dot{\mathbf{d}}_a^p \, dt, \quad \Delta X_{ai} = \int_{t_0}^{t_1} \dot{\mathbf{X}}_{ai} \, dt. \qquad (28)$$

To model the behaviour, these laws must be consistent and objective and lead to accurate and stable solutions, [**20, 21, 35, 38**].

The incremental laws in the isoclinic configuration are close to those of the linear theory. That permits to calculate for each displacement increment $\mathbf{u}$ the state of the material in both steps: first an elastic prediction step then a possible plastic correction one. At the time of the prediction the state is completely defined by:

$$(\sigma_a)_1^{\text{trial}} = (\sigma_a)_0 + \mathbf{C}^e : \Delta \mathbf{d}_a, \quad (\mathbf{X}_{ai})_1 = (\mathbf{X}_{ai})_0. \qquad (29)$$

Using the generalized mid-point rule (unconditionally stable for $\alpha \geq 1/2$) we obtain the following form for the strain increment $\Delta \mathbf{d}_a$:

$$\Delta \mathbf{d}_a = \int_{t_0}^{t_1} \dot{\mathbf{d}}_a \, dt = \mathbf{R}_\alpha^T [\Delta t \mathbf{D}_\alpha] \mathbf{R}_\alpha. \qquad (30)$$

The subscript $\alpha\big(\alpha \in [0,1]\big)$ means that the quantities are calculated in reference to an intermediate configuration C_α. To evaluate these quantities we must choose an interpolation of the geometry over the increment. In the case of linear interpolation, $\Delta t D_\alpha$ can be approximated by [**20, 21, 35**]:

$$\Delta t \mathbf{D}_\alpha = \frac{1}{2}\left(\frac{\partial \Delta \mathbf{u}}{\partial \mathbf{x}_\alpha} + \left[\frac{\partial \Delta \mathbf{u}}{\partial \mathbf{x}_\alpha} \right]^T \right), \qquad (31)$$

where $\mathbf{x}_\alpha = (1-\alpha)\mathbf{x}_0 + \alpha \mathbf{x}_1$ is the vector position of a material point in C_α. If the medium transformation is described in reference to a frame attached to a material direction as it was done in the previous section, the incremental objectivity is insured for any α value. In this case, the classical choice $\alpha = 1/2$ only permits a second order accuracy of the scheme. The components of $\Delta t \mathbf{D}_{1/2}$ in reference to the contravariant convective basis $\mathbf{g}_{1/2}^i \otimes \mathbf{g}_{1/2}^j \big(\mathbf{g}_{(1/2)i} = 1/2(\mathbf{g}_{0i} + \mathbf{g}_{1i})\big)$ are [**9**]:

$$\Delta t \mathbf{D}_{1/2} = \frac{1}{2}\left(\frac{\partial \mathbf{u}}{\partial \theta_i} \mathbf{g}_{(1/2)j} + \frac{\partial \mathbf{u}}{\partial \theta_j} \mathbf{g}_{(1/2)i} \right) \mathbf{g}_{1/2}^j \otimes \mathbf{g}_{1/2}^j, \qquad (32)$$

where $\theta_1 = \xi$, $\theta_2 = \eta$ and $\theta_3 = \zeta$. These components, in the form of (10), are interpolated within the three node shell element using the interpolation matrix for (11) and (15) for quantities in the intermediate configuration $C_{1/2}$ [**21**] i.e. and $\mathbf{B}_{mb}(\mathbf{g}_{(1/2)1}, \mathbf{g}_{(1/2)2}, \mathbf{g}_{(1/2)3})$ and $\overline{\mathbf{B}}_s(\mathbf{g}_{(1/2)1}, \mathbf{g}_{(1/2)2}, \mathbf{g}_{(1/2)3})$.

To obtain the strain increment $\Delta \mathbf{d}_a$, the tensor is rotated by $\mathbf{R}_{1/2}$. To simply make this transport, we use the fact that the components of $\Delta t \mathbf{D}_{1/2}$ with respect to an orthonormal frame $(\mathbf{e}_{(1/2)1}, \mathbf{e}_{(1/2)2}, \mathbf{e}_{(1/2)3})$ in the intermediate configuration are also those of $\Delta \mathbf{d}_a$ with respect to the frame $\mathbf{e}_{0i} = \mathbf{R}^T_{1/2}\mathbf{e}_{(1/2)i}$ in the isoclinic configuration. In particular, we use the anisotropy frame$\mathbf{A}_{(1/2)i} = \mathbf{R}_{1/2}\mathbf{A}_{0i}$ (Fig. 3):

$$(33) \qquad \Delta \mathbf{d}_a = (\Delta t \mathbf{D}_{1/2})_{ij}(\mathbf{g}^i_{1/2}\mathbf{A}_{(1/2)k})(\mathbf{g}^j_{1/2}\mathbf{A}_{(1/2)1})\mathbf{A}_{0k} \otimes \mathbf{A}_{01}.$$

From (27) and (31) the elastic trial stress $(\sigma_a)_1^{\text{trial}}$ is calculated. If the condition $f \geq 0$ is confirmed with this stress state, a correction is made using:

$$(34) \qquad (\sigma_a)_1 = (\sigma_a)_1^{\text{trial}} - \mathbf{C}^e : \Delta \mathbf{d}^p_a, \quad (\mathbf{X}_{ai})_1 = (\mathbf{X}_{ai})_0 + \Delta \mathbf{X}_{ai}.$$

The generalized mid-point rule leads to the following general expressions for $\Delta \mathbf{d}^p_a$ and $\Delta \mathbf{X}_{ai}$:

$$(35) \qquad \Delta \mathbf{d}^p_a = \int_{t_0}^{t_1} \dot{\mathbf{d}}^p_a \, dt = \Delta\lambda \mathbf{d}_\beta(\sigma_a, \mathbf{X}_{ai}),$$
$$\Delta \mathbf{X}_{ai} = \int_{t_0}^{t_1} \dot{\mathbf{X}}_{ai} \, dt = \Delta\lambda \mathbf{X}_\beta(\sigma_a, \mathbf{X}_{ai}),$$

with:

$$(36) \quad \mathbf{d}_\beta(\sigma_a, \mathbf{X}_{ai}) = \mathbf{d}\big[(1-\beta)(\sigma_a)_0 + \beta(\sigma_a)_1, (1-\beta)(\mathbf{X}_{ai})_0 + \beta(\mathbf{X}_{ai})_1\big],$$
$$X_\beta(\sigma_a, \mathbf{X}_{ai}) = \mathbf{X}\big[(1-\beta)(\sigma_a)_0 + \beta(\sigma_a)_1, (1-\beta)(\mathbf{X}_{ai})_0 + \beta(\mathbf{X}_{ai})_1\big]\beta \in [0,1].$$

The trial stress $(\sigma_a)_1^{\text{trial}}$ is projected on the yield surface at the end of the loading step in the direction of the normal $\mathbf{n}_\beta$ to $(\sigma_a)_\beta = (1-\beta)(\sigma_a)_0 + \beta(\sigma_a)_1$. The plastic multiplier $\Delta\lambda$ is obtained from the yield condition $f = 0$ at t_1.

Contrary to the generalized trapezoidal rule whose stability depends on the form of the yield function, the generalized mid-point rule is unconditionally stable for $\beta \geq 1/2$ [**34**]. In the case of Hill's orthotropic criterion associated to an isotropic hardening, the correction stage is made as (we adopt a matrix notation; $\mathbf{I}_4$ is the fourth order identity tensor):

$$(37) \qquad (\sigma_a)_1 = \mathbf{G}^{-1}\mathbf{P}, \quad X_1 = X_0 + \Delta\lambda\sqrt{(\sigma_a)^T_\beta \mathbf{H}(\sigma_a)_\beta},$$

where:

$$(38) \qquad \mathbf{G} = \mathbf{I}_4 + \beta\Delta\lambda\mathbf{C}^e\mathbf{H}, \quad \mathbf{P} = (\sigma_a)_1^{\text{trial}} - (1-\beta)\Delta\lambda\mathbf{C}^e\mathbf{H}(\sigma_a)_0.$$

The loading condition $f = 0$ at the end of the increment leads to a nonlinear equation solved by a local Newton method. At the $(i+1)$-th iteration, $\Delta\lambda$ is given by $\Delta\lambda_{i+1} = \Delta\lambda_i - f(\Delta\lambda_i)(\partial f/\partial\Delta\lambda)_i^{-1}$ with:

$$(39) \quad \frac{\partial f}{\partial \Delta\lambda} = \big[-\beta\mathbf{G}^{-1}(\mathbf{C}^e\mathbf{H})\mathbf{G}^{-1}\mathbf{P} - (1-\beta)\mathbf{G}^{-1}\mathbf{C}^e\mathbf{H}(\sigma_a)_0\big]^T$$
$$\mathbf{H}(\sigma_a)_1 - K(X_1)\frac{\partial K(X_1)}{\partial X_1}\left[\frac{\partial X_1}{\partial \Delta\lambda}\right].$$

For $\beta = 1$ we find the implicit Euler backward method. The algorithm, in this case, is closed from the one proposed by De borst and Feenestra [**12**] in small strain hypothesis. Compared to classical schemes, the implicit Euler backward method is recommended for its accuracy [**12, 33**].

In the case of finite deformations and especially finite rotations involving the rotation of anisotropy directions, an accuracy analysis for the family of algorithms (37), (38) and (39) has been performed in [**9**]. The response of an orthotropic material in simple shear has been studied for two hardening laws and three initial

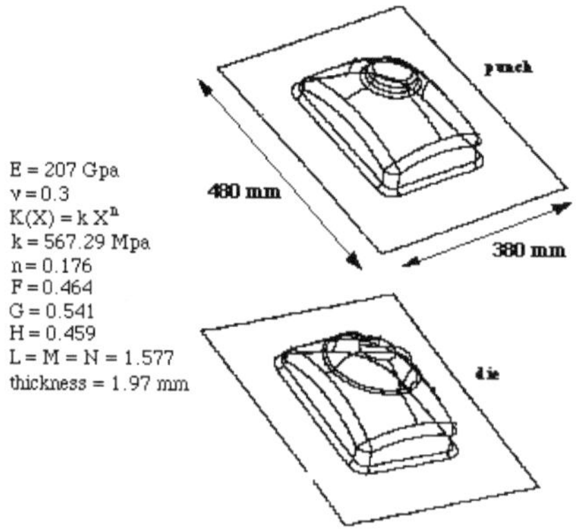

FIGURE 4. Geometry, tool definition and material properties.

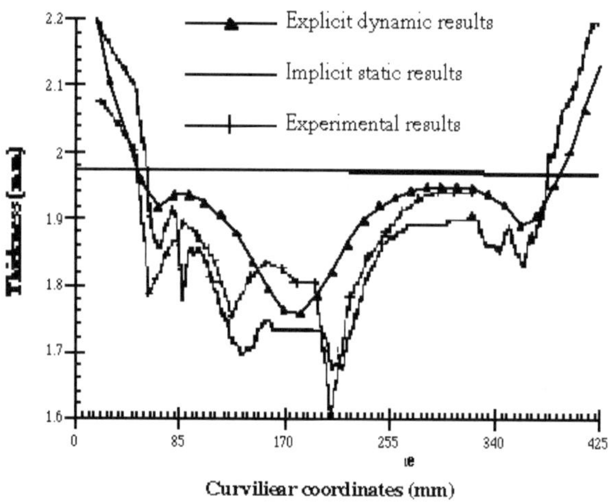

FIGURE 5. Thickness distribution along a cross section.

orientations of anisotropy directions. Still there, the choice $\beta = 1$ appears the most interesting; it permits a good accuracy with a very simple state calculation algorithm.

The Cauchy stresses σ in the current configuration are obtained directly from the stresses σ_a in the isoclinic configuration. We use the fact that the components of σ_a with respect to the initial orthotropy frame are also those of σ with respect to the updated orthotropy frame in the current configuration:

$$\sigma_a = \sigma_{ij} \mathbf{A}_{0i} \otimes \mathbf{A}_{0j} \iff \sigma = \sigma_{ij} \mathbf{A}_i \otimes \mathbf{A}_j. \tag{40}$$

4. Numerical Example

We treat the simulation of the dashpot cup used in Renault Twingo car. In this example, the sheet-metal is plane and bowed along x-axis by 1° and along y-axis by 5°. Geometry, tool definition and material properties are given in Fig. 4. A mesh containing 2100 three node shell elements is used in the finite element

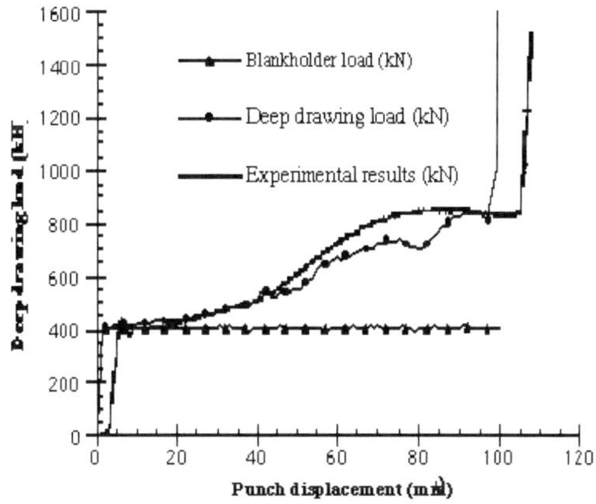

FIGURE 6. Punch force and blankholder force versus punch displacement.

approximation. The results concerning thickness distribution along section S corresponding to a punch displacement equal 95 mm is compared to the experiment (Fig. 5). The punch force vs punch travel is also presented and compared to the experiment (Fig. 7). The sheet-metal is orthotropic and the initial orientation of anisotropy directions is equal 90°. The friction coefficient is equal 0.15 and the blankholder force is equal 400 kN.

Still there, the numerical results are in agreement with experimental ones. That shows the efficiency and the accuracy of the proposed approach in industrial cases. In particular, Fig. 6 shows that the blankholder load remains unchanged and equal 400 kN during the forming process.

5. Conclusion

Within a shell finite element modelling, a numerical scheme for the state and internal variables calculation in finite strains is presented in the case of Hill's orthotropic plasticity. Derived from a behaviour modelling in a rotating frame, the incremental laws are a closed form of those in small strain theory, the predictor-corrector scheme is therefore applied. The proposed algorithm takes into account of the formulation of the mixed interpolation used in a degenerated three node shell element formulation in order to avoid shear locking and of the zero normal stress assumption. In that aim, the anisotropy direction updating is done using the restriction of the proper rotation to the mid-plane of the shell. The accuracy and the efficiency of the proposed approach are shown through two examples including real forming process parameters. The obtained results are in good agreement with the experimental ones.

References

1. S. Ahmad, B. M. Irons, and O. C. Zienkiewiecz, *Analysis of thick and thin shell structures by curved finite elements*, Internat. J. Numer. Methods Engrg. **2** (1971), 419–451.
2. F. Barlat, D. J. Lege, and J. C. Brem, *A six-component yield function for anisotropic materials*, Internat. J. Plasticity **7** (1991), 693–712.
3. K. J. Bathe, *Finite element procedures in engineering analysis*, Prentice Hall, 1982.
4. J. L. Batoz, K. J. Bathe, and L. Ho, *Three node triangular plate bending elements*, Internat. J. Numer. Methods Engrg. **15** (1980), 1771–1812.

5. T. Belytschko and C. S. Tsay, *A stabilization procedure for the quadrilateral plate element with one point quadrature*, Internat. J. Numer. Methods Engrg. **1** (1983), 405–417.
6. T. Belytschko, J. L. Lin, and C. S. Tsay, ———, *Explicit algorithms for nonlinear dynamics of shells*, Comput. Methods Appl. Mech. Engrg. **43** (1984), 251–276.
7. P. Boisse, J. L. Daniel, and J. C. Gelin, *A simple isoparametric three node shell finite element*, Comput. & Structures **44-6** (1992), 1263–1273.
8. ———, *A C0 three-node shell element for nonlinear structural analysis*, Internat. J. Numer. Methods Engrg. **37** (1994), 2339–2364.
9. M. L. Boubakar, *Contribution à la simulation numérique de l'emboutissage des tôles, prise en compte du comportement élastolplastique anisotrope*, Ph.D thesis, Univ. Besançon, 1994.
10. K. Chung and K. Shah, *Forming of planar anisotropic metals*, Internat. J. Plasticity **4** (1992), 453–476.
11. Y. F. Dafalias and M. M. Rashid, *The effect of plastic spin on anisotropic material behavior*, Internat. J. Plasticity **5** (1989), 227–246.
12. R. De Borst and P. H. Fenestra, *Studies in anisotropic plasticity with reference to the Hill criterion*, Internat. J. Numer. Methods Engrg. **29** (1990), 315–336.
13. A. Dogui and F. Sidoroff, *Quelques remarques sur la plasticité anisotrope en grandes dformations*, C. R. Acad. Sci. Paris Sér. II **299** (1984), no. 18, 1225–1228.
14. N. Dvorkin and K. J. Bathe, *A continuum mechanics based four node shell elements for general non linear analysis*, Engrg. Comput. **1** (1984), 77–78.
15. ———, *A formulation of general shell elements—the use of mixed interpolation of tensorial components*, Internat. J. Numer. Methods Engrg. **22** (1986), 697–722.
16. J. C. Gelin, L. Boulmane, and P. Boisse, *Quasi-static implicit and transient explicit analysis of sheet metal forming using a C three node sheet element*, J. Mater. Process. Tech. **50** (1994), no. 1–4, 54–69.
17. J. C. Gelin and J. L. Daniel, *A finite element simulation of sheet metal forming processes using a general nonflat shell element*, Proc. NUMIFORM 89 (E. G. Thompson, ed), Balkema, Rotterdam, 1989, pp. 389–394.
18. R. Hill, *The mathematical theory of plasticity*, The Oxford Engineering Science Series, 1950.
19. ———, *Theoretical plasticity of textured aggregates*, Math. Proc. Cambridge Philos. Soc. **85** (1979), 179–261.
20. T. J. R. Hughes, *Numerical implementation of constitutive models: Rate independent deviatoric plasticity*, Theoretical Foundation for Large Scale Computation for Non-linear Material Behavior (S. Nemat-Nasser, R. J. Asaro and G. A. Hgemier, eds.), Martinus Nijhoff Publ., Dordrecht, 1984, pp. 29-57.
21. T. J. R. Hughes and J. Winget, *Finite rotation effects in numerical integration of rate constitutive equations arising in large deformation analysis*, Internat. J. Numer. Methods Engrg. **15** (1980), 1862–1867.
22. S. W. Key and Z. E. Beisinger, *The transient dynamic analysis of the shells by the element methods*, Proceedings of the 3rd Conference on Matrix Methods in Structural Mechanics (Ohio, 1971), Wright-Patterson Air Force Base, 1971, pp. 142–153.
23. P. Ladevèze, *Sur la théorie de la plasticité en grandes déformations*, Rapport interne **9**, LMT-ENSET-Cachan, 1980.
24. E. H. Lee, *Elastic-plastic deformation at finite strains*, J. Appl. Mech. **36** (1969), 1–6.
25. B. Loret, *On the effects of plastic rotations in the finite deformation of anisotropic elasto-plastic materials*, Mech. Materials **2** (1983), 287–304.
26. B. Loret, F. Hammoum, and Y. S. Dafalias, *A note on the accuracy of stress-point algorithms for anisotropic elastic-plastic plastic solids*, Comput. Methods Appl. Mech. Engrg. **98** (1992), 399–409.
27. R. H. Macneal, *Derivation of element stiffness matrices by assumed strain distributions*, J. Nuclear Engrg. & Design. **70** (1982), 3–12.
28. J. Mandel, *Plasticité et viscoplasticité*, Cours C.I.S.M 97, Udine, Springer, New York, 1971.
29. ———, *Thermodynamics and plasticity*, Proc. Symp. Foundations of Continuum Thermodynamics (Lisbon, 19??) (D. Domingos, ed.), MacMillan, 1973, pp. 283–298.
30. ———, *Equations constitutives et directeurs dans les milieux plastiques et viscoplastiques*, Internat. J. Solids Structures **9** (1973), 725–740.
31. F. Montheillet, M. Cohen, and J. J. Jonas, *Axial stresses and texture development during the torsion testing of Al, Cu and a-Fe*, Acta Metal. **32** (1984), no. 11, 2077–2089.

32. ———, *Relation between axial stresses and texture development during torsion testing: A simplified theory*, Acta Metal. **33** (1985), no. 4, 705–717.
33. J. C. Nagtegaal and F. E. Veldpaus, *On the implementation of finite strain plasticity equations in a numerical model*, Numerical Analysis of Forming Processes (J. F. T. Pittman, O. C. Zienkiewicz, R. D. Wood, and J. M. Alexander, eds.), Pineridge Press, Swanrea, 1984.
34. M. Ortiz and E. P. Popov, *Accuracy and stability of integration algorithms for elastoplastic constitutive relations*, Internat. J. Numer. Methods Engrg. **21** (1985), 1561–1576.
35. P. M. Pinsky, M. Ortiz, and K. S. Pister, *Numerical integration of rate constitutive equations in finite deformation analysis*, Comput. Methods Appl. Mech. Engrg. **40** (1983), 137–158.
36. N. Rebelo, J. C. Nagtegaal, and H. D. Hibbit, *Finite element analysis of sheet forming processes*, Internat. J. Numer. Methods Engrg. **30** (1990), 1739–1758.
37. J. M. Roelandt, X. J. Liu, and J. L. Batoz, *Axisymmetric and general shell elements for large transformations*, Large Plastic Deformations (C. Teodosiu, J. Raphanel, and F. Sidoroff, eds.), Balkema, Rotterdam, 1993, pp. 457–463.
38. R. Rubinstein and S. N. Atluri, *Objectivity of incremental constitutive relations over finite time steps in computational finite deformation analyses*, Comput. Methods Appl. Mech. Engrg. **36** (1983), 277–290.
39. G. Stanley, *Continuum based shell elements*, Ph.D thesis, Stanford Univ., 1985.
40. D. Zhou and R. H. Wagoner, *A numerical method for introducing an arbitrary yield function into rigid-viscoplastic* FEM *programs*, Internat. J. Numer. Methods Engrg. **37** (1994), 3467–3488.

Laboratoire de Mécanique Appliquée, Université de Franche-Comté (Besançon), 25030 Besançon Cedex, France

E-mail address, Gelin: `jean-claude.gelin@univ-Fcomte.fr`

E-mail address, Boubakar: `lamine.boubakar@univ-Fcomte.fr`

Centre de Recherches Mathématiques
CRM Proceedings and Lecture Notes
Volume **21**, 1999

Justification of Two-Dimensional Linear Shell Models by the Use of Γ-Convergence Theory

Karine Genevey

ABSTRACT. We give a new justification of the membrane and flexural shell models in linearized elasticity. Under appropriate assumptions on the data, we prove that the sequence of energy functionals associated with the scaled problem Γ-converges to the energy functional associated with a two-dimensional model. This two-dimensional model is membrane-dominated or flexural, depending on the geometric and kinematic conditions. Then, a classical argument allows to prove the convergence result of the displacements.

Introduction

The deformations of an elastic body submitted to forces are governed by *three-dimensional* mathematical equations. This means that the unknown, which is the vector of the components of the displacement, depends on three variables. However, when the elastic body is "thin" in one dimension, for instance when it is a plate or a shell, one can use two-dimensional shell models, such as those of Naghdi, Koiter, Budiansky-Sanders. Thus, the important point is to explain *which* model is the "good" one in a given situation and *why*. Hence, an important aspect of the mathematical analysis in elasticity consists in studying the validity of the two-dimensional equations to describe the physical behavior of a three-dimensional body. This is what we call the *justification of the model.*

For *linearly elastic shells*, [**S**] pointed out the importance of the geometry of the shell: depending on the geometric and kinematic conditions, the *formal asymptotic analysis* leads to identify one of *two distinct models*: the "membrane" model or the "flexural" model. Thus, it is not possible to derive these two models simultaneously for shells, unlike the case of plates. For other works on this subject, see [**D1, D2, CaS, MS**].

Essentially, a two-dimensional model is considered justified when one can prove convergence of the three-dimensional unknown to the leading term of the asymptotic expansion, as the thickness ε of the shell goes to zero. For nonlinear membranes,

1991 *Mathematics Subject Classification.* Primary: 73C02; Secondary: 35Q72.

This work is part of Human Capital and Mobility Program "Shells: Mathematical Modeling and Analysis, Scientific Computing" of the Commission of the European Communities (Contract ERBCHRXCT940536).

This is the final form of the paper.

such results were obtained by [**LR**] using Γ-convergence and following the approach of [**ABP**]. In the linear case, the articles of [**CL2, CLM**] give the complete justification of the membrane and flexural models.

Here, we give another method to obtain these results using Γ-convergence theory. A similar approach was done for linearly elastic plates by [**BCGR**].

We study separately the membrane case and the flexural case. First, we recall the main notations about the geometry of the shell, and we make appropriate scalings, so that the problem can be posed over a domain independent of ε. Next, we prove the Γ-convergence of the energy functionals associated with the scaled three-dimensional problem to a functional corresponding to a variational problem posed over a two-dimensional domain. We then deduce the weak convergence of the displacements, the strong convergence being obtained as in [**CL2, CLM**].

1. The Three-Dimensional Shell Problem in Linearized Elasticity

We begin with geometric preliminaries. Throughout this work, Greek indices and exponents (except ε) belong to the set $\{1,2\}$, Latin indices and exponents (except when used to index sequences) take their values in the set $\{1,2,3\}$, and we use the summation convention on repeated indices and exponents.

Let ω be a bounded, open and connected subset of $\mathbf{R}^2$, with a Lipschitz-continuous boundary γ. We note $y = (y_\alpha)$ a generic point of $\overline{\omega}$, and $\partial_\alpha := \partial/\partial y_\alpha$ the partial derivatives. Let $\varphi : \overline{\omega} \to \mathbf{R}^3$ be an injective mapping, at least of class $\mathcal{C}^3$. We assume that the two vectors

$$\mathbf{a}_\alpha(y) := \partial_\alpha \varphi(y)$$

are linearly independent at all points $y \in \overline{\omega}$. They form the *covariant basis* of the tangent plane to the surface $S = \varphi(\overline{\omega})$ at the point $\varphi(y)$; the two vectors $\big(\mathbf{a}^\alpha(y)\big)$ defined by

$$\mathbf{a}^\alpha(y) \cdot \mathbf{a}_\beta(y) = \delta^\alpha_\beta$$

constitute the *contravariant basis* at this same point $\varphi(y)$. We also define the vector

$$\mathbf{a}_3 = \mathbf{a}^3 := \frac{\mathbf{a}_1 \times \mathbf{a}_2}{|\mathbf{a}_1 \times \mathbf{a}_2|}.$$

The covariant and contravariant components of the *metric tensor* are given by

$$a_{\alpha\beta} := \mathbf{a}_\alpha \cdot \mathbf{a}_\beta, \quad a^{\alpha\beta} := \mathbf{a}^\alpha \cdot \mathbf{a}^\beta,$$

the covariant and mixed components of the *curvature tensor* are

$$b_{\alpha\beta} := \mathbf{a}^3 \cdot \partial_\beta \mathbf{a}_\alpha, \quad b^\beta_\alpha := a^{\beta\sigma} b_{\sigma\alpha},$$

and the symbols of Christoffel of the surface S are defined by

$$\Gamma^\sigma_{\alpha\beta} := \mathbf{a}^\sigma \cdot \partial_\beta \mathbf{a}_\alpha.$$

Since the metric tensor is symmetric and definite positive at all points of $\overline{\omega}$, there exists a constant a_0 such that:

$$a(y) := \det\big(a_{\alpha\beta}(y)\big) \geq a_0 > 0 \quad \text{for all } y \in \overline{\omega}.$$

The *area element* along S is $\sqrt{a}\,dy$.

For each $\varepsilon > 0$, we define the sets

$$\Omega^\varepsilon := \omega \times \left]-\varepsilon, \varepsilon\right[\quad \text{and} \quad \Gamma^\varepsilon := \gamma \times [-\varepsilon, \varepsilon].$$

We let $x^\varepsilon = (x_i^\varepsilon)$ denote a generic point in the set Ω^ε, and $\partial_i^\varepsilon = \partial/\partial x_i^\varepsilon$, so that we have $x_\alpha^\varepsilon = y_\alpha$ and $\partial_\alpha^\varepsilon = \partial_\alpha$. Let the mapping $\Phi : \overline{\Omega}^\varepsilon \to \mathbf{R}^3$ be defined by

$$\Phi(x^\varepsilon) := \varphi(y) + x_3^\varepsilon \mathbf{a}^3(y) \quad \text{for all } x^\varepsilon = (y, x_3^\varepsilon) \in \overline{\Omega}^\varepsilon.$$

The three vectors

$$\mathbf{g}_i^\varepsilon(x^\varepsilon) := \partial_i^\varepsilon \Phi(x^\varepsilon)$$

form the *covariant basis* at the point $\Phi(x^\varepsilon)$, with which we associate the *contravariant basis* formed by the vectors $\mathbf{g}^{i,\varepsilon}(x^\varepsilon)$ given by

$$\mathbf{g}^{i,\varepsilon}(x^\varepsilon) \cdot \mathbf{g}_j^\varepsilon(x^\varepsilon) = \delta_j^i.$$

The covariant and contravariant components of the *metric tensor*, and the Christoffel symbols are

$$g_{ij}^\varepsilon := \mathbf{g}_i^\varepsilon \cdot \mathbf{g}_j^\varepsilon, \quad g^{ij,\varepsilon} := \mathbf{g}^{i,\varepsilon} \cdot \mathbf{g}^{j,\varepsilon}, \quad \Gamma_{ij}^{p,\varepsilon} := \mathbf{g}^{p,\varepsilon} \cdot \partial_i^\varepsilon \mathbf{g}_j^\varepsilon.$$

The volume element in the set $\Phi(\overline{\Omega}^\varepsilon)$ is $\sqrt{g^\varepsilon}\, dx^\varepsilon$, where

$$g^\varepsilon := \det(g_{ij}^\varepsilon).$$

For all $\varepsilon > 0$, the set $\Phi(\overline{\Omega}^\varepsilon)$ is the *reference configuration* of an *elastic shell* with *middle surface* S and *thickness* $2\varepsilon > 0$. We assume that the elastic material constituting the shell is *homogeneous* and *isotropic*, and that the reference configuration is a *natural state*. The Lamé constants of the material are $\lambda^\varepsilon > 0$ and $\mu^\varepsilon > 0$. We assume that the shell is *clamped* on a portion $\Gamma_0^\varepsilon := \gamma_0 \times [-\varepsilon, \varepsilon]$ of its lateral surface, where $\gamma_0 \subset \gamma$ satisfies mes $\gamma_0 > 0$, and that it is submitted to volume forces with density $f^{i,\varepsilon}\mathbf{g}^{i,\varepsilon}$, where $f^{i,\varepsilon} \in L^2(\Omega)$.

The covariant components $u_i^\varepsilon : \overline{\Omega}^\varepsilon \to \mathbf{R}$ of the displacement $u_i^\varepsilon(x^\varepsilon)\mathbf{g}^{i,\varepsilon}(x^\varepsilon)$ of the points of the shell are the unknowns of the *three-dimensional shell problem* in linearized elasticity. This problem can be written as

$$\mathbf{u}^\varepsilon \in \mathbf{V}(\Omega^\varepsilon) := \{\mathbf{v}^\varepsilon = (v_i^\varepsilon) \in \mathbf{H}^1(\Omega^\varepsilon);\ \mathbf{v}^\varepsilon = \mathbf{0} \text{ on } \Gamma_0^\varepsilon\}, \tag{1.1}$$

$$\int_{\Omega^\varepsilon} A^{ijk\ell,\varepsilon} e_{k\|\ell}^\varepsilon(\mathbf{u}^\varepsilon) e_{i\|j}^\varepsilon(\mathbf{v}^\varepsilon) \sqrt{g^\varepsilon}\, dx^\varepsilon = \int_{\Omega^\varepsilon} f^{i,\varepsilon} v_i^\varepsilon \sqrt{g^\varepsilon}\, dx^\varepsilon \quad \text{for all } \mathbf{v}^\varepsilon \in \mathbf{V}(\Omega^\varepsilon), \tag{1.2}$$

where the covariant components of the *three-dimensional elasticity tensor* are defined by

$$A^{ijk\ell,\varepsilon} := \lambda^\varepsilon g^{ij,\varepsilon} g^{k\ell,\varepsilon} + \mu^\varepsilon (g^{ik,\varepsilon} g^{j\ell,\varepsilon} + g^{i\ell,\varepsilon} g^{jk,\varepsilon}), \tag{1.3}$$

and the covariant components of the *linearized strain tensor* are

$$e_{i\|j}^\varepsilon(\mathbf{v}^\varepsilon) := \frac{1}{2}(\partial_i^\varepsilon v_j^\varepsilon + \partial_j^\varepsilon v_i^\varepsilon) - \Gamma_{ij}^{p,\varepsilon} v_p^\varepsilon. \tag{1.4}$$

The variational problem (1.1)–(1.2) has one and only one solution for each $\varepsilon > 0$; this follows from the $\mathbf{V}(\Omega^\varepsilon)$-ellipticity of the bilinear form appearing in (1.2), see [**C**].

From now on, we will *not* work with the variational formulation (1.1)–(1.2) (which is the approach of [**CL2,CLM**]) but with the *formulation in terms of energy*. Indeed, the vector field $\mathbf{u}^\varepsilon = (u_i^\varepsilon)$ can also de defined as the unique solution of the following *minimization problem*:

$$\mathbf{u}^\varepsilon \in \mathbf{V}(\Omega^\varepsilon) \quad \text{and} \quad J^\varepsilon(\mathbf{u}^\varepsilon) = \inf_{\mathbf{v}^\varepsilon \in \mathbf{V}(\Omega^\varepsilon)} J^\varepsilon(\mathbf{v}^\varepsilon), \tag{1.5}$$

where

$$J^\varepsilon(\mathbf{v}^\varepsilon) := \frac{1}{2}\int_{\Omega^\varepsilon} A^{ijk\ell,\varepsilon} e^\varepsilon_{k\|\ell}(\mathbf{v}^\varepsilon) e^\varepsilon_{i\|j}(\mathbf{v}^\varepsilon)\sqrt{g^\varepsilon}\,dx^\varepsilon - \int_{\Omega^\varepsilon} f^{i,\varepsilon} v^\varepsilon_i \sqrt{g^\varepsilon}\,dx^\varepsilon. \tag{1.6}$$

Of course, formulations (1.1)–(1.2) and (1.5)–(1.6) are equivalent.

2. Justification of the Two-Dimensional Membrane Shell Model

In this section, we assume that the shell is *clamped* along its whole lateral face $\Phi(\Gamma^\varepsilon)$, i.e., $\gamma = \gamma_0$, and the displacements vanish there. We also assume that the middle surface of the shell is uniformly elliptic in the sense that there exists a constant $b > 0$ such that

$$\left|b_{\alpha\beta}(y)\xi^\alpha\xi^\beta\right| \geq b\,\xi^\alpha\xi^\alpha \tag{2.1}$$

for all $y \in \overline{\omega}$ and $(\xi^\alpha) \in \mathbf{R}^2$. For instance, a portion of a sphere and a portion of an ellipsoid are uniformly elliptic surfaces. Inequality (2.1) means that the principal radii of curvature $R_1(y)$ and $R_2(y)$ have the same sign at all points $\varphi(y) \in S$ and that there exists a constant $\rho > 0$ such that

$$\rho^{-1} \leq \left|R_\alpha(y)\right| \leq \rho, \quad \alpha = 1, 2, \qquad \text{for all } y \in \overline{\omega}.$$

2.1. The scaled three-dimensional problem. We first set the three-dimensional problem (1.1)–(1.2) over a domain *independent* of ε. Next, we make the appropriate *scalings* on the components of the displacement, and the assumptions on the data. We let

$$\Omega := \omega \times\,]-1, 1[, \quad \Gamma := \gamma \times [-1, 1],$$

and with any point $x^\varepsilon = (x^\varepsilon_i) \in \overline{\Omega}^\varepsilon$, we associate the point $x = (x_i) \in \overline{\Omega}$ defined by $x_\alpha = x^\varepsilon_\alpha (= y_\alpha)$ and $x_3 = (1/\varepsilon)x^\varepsilon_3$.

Then, for all $x^\varepsilon \in \overline{\Omega}^\varepsilon$, the *scaled displacement* $\mathbf{u}(\varepsilon) = \big(u_i(\varepsilon)\big)$ is defined by

$$u_i(\varepsilon)(x) = u^\varepsilon_i(x^\varepsilon), \tag{2.2}$$

and with the vector field $\mathbf{v}^\varepsilon = (v^\varepsilon_i) \in \mathbf{V}(\Omega^\varepsilon)$ we associate the *scaled vector field*

$$\mathbf{v} = \big(v_i(x)\big) = \big(v^\varepsilon_i(x^\varepsilon)\big) \quad \text{for all } x^\varepsilon \in \overline{\Omega}^\varepsilon.$$

We next assume that there exist constants $\lambda > 0$ and $\mu > 0$ independent of ε, and there exist functions $f^i \in L^2(\Omega)$ independent of ε such that

$$\lambda^\varepsilon = \lambda, \quad \mu^\varepsilon = \mu, \tag{2.3}$$

$$f^{i,\varepsilon}(x^\varepsilon) = f^i(x) \quad \text{for all } x \in \Omega. \tag{2.4}$$

The choice of these scalings is of prime importance in order to find the "right" two-dimensional model; in particular, one has to make *different* assumptions in the flexural case.

Finally, with the functions $\Gamma^{p,\varepsilon}_{ij}, g^\varepsilon, A^{ijk\ell,\varepsilon} : \overline{\Omega}^\varepsilon \to \mathbf{R}$, we associate the functions $\Gamma^p_{ij}(\varepsilon)$, $g(\varepsilon)$, $A^{ijk\ell}(\varepsilon) : \overline{\Omega}^\varepsilon \to \mathbf{R}$ defined by (cf. [**CL2**]), for all $x \in \overline{\Omega}$,

$$\Gamma^p_{ij}(\varepsilon)(x) := \Gamma^{p,\varepsilon}_{ij}(x^\varepsilon), \quad g(\varepsilon)(x) := g^\varepsilon(x^\varepsilon), \quad A^{ijk\ell}(\varepsilon)(x) := A^{ijk\ell,\varepsilon}(x^\varepsilon). \tag{2.5}$$

In addition, for any vector field $\mathbf{v} = (v_i) \in \mathbf{H}^1(\Omega)$, we define the symmetric tensor $\big(e_{i\|j}(\varepsilon)(\mathbf{v})\big) \in \mathbf{L}^2(\Omega)$ given by

$$e_{\alpha\|\beta}(\varepsilon)(\mathbf{v}) := \frac{1}{2}(\partial_\alpha v_\beta + \partial_\beta v_\alpha) - \Gamma^p_{\alpha\beta}(\varepsilon) v_p, \tag{2.6}$$

$$e_{\alpha\|3}(\varepsilon)(\mathbf{v}) := \frac{1}{2}\left(\partial_\alpha v_3 + \frac{1}{\varepsilon}\partial_3 v_\alpha\right) - \Gamma^\sigma_{\alpha 3}(\varepsilon) v_\sigma, \tag{2.7}$$

$$e_{3\|3}(\varepsilon)(\mathbf{v}) := \frac{1}{\varepsilon}\partial_3 v_3. \tag{2.8}$$

One can easily check that $e^\varepsilon_{i\|j}(\mathbf{v}^\varepsilon) = e_{i\|j}(\varepsilon)(\mathbf{v})$.

We thus have the following result:

THEOREM 2.1. *The scaled unknown* $\mathbf{u}(\varepsilon) = \big(u_i(\varepsilon)\big)$ *defined in* (2.2) *satisfies*

$$\mathbf{u}(\varepsilon) \in \mathbf{V}(\Omega) := \big\{\mathbf{v} = (v_i) \in H^1(\Omega); \mathbf{v} = \mathbf{0} \text{ on } \Gamma\big\}, \tag{2.9}$$

$$J(\varepsilon)\big(\mathbf{u}(\varepsilon)\big) = \inf_{\mathbf{v}\in\mathbf{V}(\Omega)} J(\varepsilon)(\mathbf{v}), \tag{2.10}$$

where

$$J(\varepsilon)(\mathbf{v}) = \frac{1}{2}\int_\Omega A^{ijk\ell}(\varepsilon) e_{k\|\ell}(\varepsilon)(\mathbf{v}) e_{i\|j}(\varepsilon)(\mathbf{v})\sqrt{g(\varepsilon)}\,dx - L(\varepsilon)(\mathbf{v}), \tag{2.11}$$

and

$$L(\varepsilon)(\mathbf{v}) = \int_\Omega f^i v_i \sqrt{g(\varepsilon)}\,dx. \tag{2.12}$$

REMARK 2.2. We give another expression for the functional $J(\varepsilon)$. Using (1.3) and (2.5) we can write

$$\begin{aligned} J(\varepsilon)(\mathbf{v}) = {} & \frac{1}{2}\int_\Omega \left\{ \frac{2\lambda\mu}{\lambda+2\mu} g^{\alpha\beta}(\varepsilon) g^{\sigma\tau}(\varepsilon) \right\} e_{\alpha\|\beta}(\varepsilon)(\mathbf{v}) e_{\sigma\|\tau}(\varepsilon)(\mathbf{v})\sqrt{g(\varepsilon)}\,dx \\ & + \frac{1}{2}\int_\Omega \left\{ 2\mu g^{\alpha\sigma}(\varepsilon) g^{\beta\tau}(\varepsilon) \right\} e_{\alpha\|\beta}(\varepsilon)(\mathbf{v}) e_{\sigma\|\tau}(\varepsilon)(\mathbf{v})\sqrt{g(\varepsilon)}\,dx - L(\varepsilon)(\mathbf{v}) \\ & + \frac{1}{2}\int_\Omega \left\{ (\lambda+2\mu)\left[\frac{\lambda}{\lambda+2\mu} g^{\alpha\beta}(\varepsilon) e_{\alpha\|\beta}(\varepsilon)(\mathbf{v}) \right.\right. \\ & \qquad \left.\left. + g^{33}(\varepsilon) e_{3\|3}(\varepsilon)(\mathbf{v}) \right]^2 \right\} \sqrt{g(\varepsilon)}\,dx \\ & + \frac{1}{2}\int_\Omega \left\{ 4\mu g^{\alpha\sigma}(\varepsilon) g^{33}(\varepsilon) e_{\alpha\|3}(\varepsilon)(\mathbf{v}) e_{\sigma\|3}(\varepsilon)(\mathbf{v}) \right\} \sqrt{g(\varepsilon)}\,dx. \end{aligned} \tag{2.13}$$

Note that a similar expression was used for the scaled functional of the three-dimensional *plate* problem to study with Γ-convergence theory the asymptotic behavior of the three-dimensional unknown (cf. [**BCGR**]).

2.2. A Korn's inequality. We recall a fundamental result of [**CL2**] which plays a crucial rôle in their proof as well as in the one we give in Section 2.4. It is a *Korn's inequality for uniformly elliptic surfaces*, which will allow us to prove a priori estimates for $\big(\mathbf{u}(\varepsilon)\big)_{\varepsilon>0}$.

From now on, we will suppose that one of the following assumptions holds:

$$\varphi \text{ is analytic in an open set containing } \overline{\omega} \text{ and } \gamma \text{ is of class } \mathcal{C}^3 \tag{2.14}$$

or

$$\varphi \in \mathcal{C}^5(\overline{\omega}; \mathbf{R}^3) \quad \text{and} \quad \gamma \text{ is of class } \mathcal{C}^4. \tag{2.15}$$

These assumptions were stated in [**CS**] and [**CL1**], respectively, to prove existence theorems for the two-dimensional linear membrane shell problem.

The following theorem, whose proof is given in [**CL2**], states a generalized Korn's inequality.

THEOREM 2.3. *Assume that S is elliptic, and that either* (2.14) *or* (2.15) *holds. Then, there exist constants $\varepsilon_0 > 0$ and $C > 0$ such that, for all $0 < \varepsilon \leq \varepsilon_0$ and for all $\mathbf{v} = (v_i) \in \mathbf{V}(\Omega)$*

$$\Big\{\sum_\alpha \|v_\alpha\|^2_{1,\Omega} + \|v_3\|^2_{0,\Omega}\Big\}^{1/2} \leq C\Big\{\sum_{ij} \big\|e_{ij}(\varepsilon)(\mathbf{v})\big\|^2_{0,\Omega}\Big\}^{1/2} \tag{2.16}$$

where the functions $e_{i\|j}(\varepsilon)(\mathbf{v})$ and the space $\mathbf{V}(\Omega)$ are defined as in(2.6)–(2.9).

REMARK 2.4. Inequality (2.16) involves the $H^1(\Omega)$ norms of the horizontal components (v_α), but only the $L^2(\Omega)$ norm of the vertical component v_3.

REMARK 2.5. For any $\eta = (\eta_i) \in H^1_0(\omega) \times H^1_0(\omega) \times L^2(\omega)$, let us define the functions

$$\gamma_{\alpha\beta}(\eta) := \frac{1}{2}(\partial_\alpha \eta_\beta + \partial_\beta \eta_\alpha) - \Gamma^\sigma_{\alpha\beta}\eta_\sigma - b_{\alpha\beta}\eta_3.$$

A sufficient condition to establish (2.16) is that there exists a constant $c > 0$ such that

$$\Big\{\sum_\alpha \|\eta_\alpha\|^2_{1,\omega} + \|\eta_3\|^2_{0,\omega}\Big\}^{1/2} \leq c\Big\{\sum_{\alpha,\beta} \big\|\gamma_{\alpha\beta}(\eta)\big\|^2_{0,\omega}\Big\}^{1/2}$$

for all $\eta \in H^1_0(\omega) \times H^1_0(\omega) \times L^2(\omega)$. One can show that the assumptions of Theorem 2.3 imply this condition (cf. [**CS**] and [**CL1**]).

REMARK 2.6. When there is no geometric condition on S and if we assume that the shell is only partially clamped, then the constant in inequality (2.16) depends on ε. The geometric and kinematic conditions we made are essential to establish the Korn's inequality of Theorem 2.3.

2.3. Review on Γ-convergence. In this section, we recall the definitions and the main properties on Γ-convergence theory that we will use in the proofs of Section 2.4. For a general survey, see [**Da**] and [**A**].

DEFINITION 2.7. Let $\mathcal{V}$ be a reflexive Banach space, and $\big(J^*(\varepsilon)\big)_{\varepsilon>0}$ a sequence of functionals $J^*(\varepsilon) : \mathcal{V} \to \mathbf{R} \cup \{+\infty\}$, where ε is a parameter going to zero. We denote by $\rightharpoonup$ the weak convergence in $\mathcal{V}$. We say that the functional $J^* : \mathcal{V} \to \mathbf{R}$ is the Γ-limit of the functionals $J^*(\varepsilon)$ if the following properties hold:

(i) If $\big(\mathbf{v}(\varepsilon)\big)_{\varepsilon>0}$ is a sequence in $\mathcal{V}$, weakly convergent in $\mathcal{V}$, then

$$\mathbf{v}(\varepsilon) \rightharpoonup \mathbf{v} \in \mathcal{V} \implies J^*(\mathbf{v}) \leq \liminf_{\varepsilon\to 0} J^*(\varepsilon)\big(\mathbf{v}(\varepsilon)\big). \tag{2.17}$$

(ii) For any $\mathbf{v} \in \mathcal{V}$, there exists a sequence $\big(\mathbf{v}(\varepsilon)\big)$ in $\mathcal{V}$ such that

$$\mathbf{v}(\varepsilon) \rightharpoonup_{\varepsilon\to 0} \mathbf{v} \quad \text{and} \quad J^*(\varepsilon)\big(\mathbf{v}(\varepsilon)\big) \to J^*(\mathbf{v}). \tag{2.18}$$

REMARK 2.8. When the Γ-limit exists, it is unique. This is an easy consequence of Definition 2.7.

REMARK 2.9. We gave the definition using the weak topology of $\mathcal{V}$. This is not the most general setting: the Γ-limit depends on the topology with which $\mathcal{V}$ is equipped.

THEOREM 2.10. *Assume that the sequence $\big(J^*(\varepsilon)\big)_{\varepsilon>0}$ Γ-converges to J^*, and assume that there exists a compact subset $\mathcal{U}$ of $\mathcal{V}$ independent of ε such that, for all $\varepsilon > 0$, there exists $\mathbf{u}(\varepsilon)$ satisfying*

$$\mathbf{u}(\varepsilon) \in \mathcal{U} \quad \text{and} \quad J^*(\varepsilon)\big(\mathbf{u}(\varepsilon)\big) = \inf_{\mathbf{v}\in\mathcal{V}} J^*(\varepsilon)(\mathbf{v}).$$

Then there exists $\mathbf{u} \in \mathcal{U}$ such that

$$\mathbf{u}(\varepsilon) \rightharpoonup_{\varepsilon\to 0} \mathbf{u} \quad \text{and} \quad J^*(\mathbf{u}) = \inf_{\mathbf{v}\in\mathcal{V}} J^*(\mathbf{v}).$$

In addition, one has

$$J^*(\varepsilon)\big(\mathbf{u}(\varepsilon)\big) \to_{\varepsilon\to 0} J^*(\mathbf{u}).$$

2.4. Convergence theorems. We first establish the Γ-convergence of a sequence of energy functionals by identifying its Γ-limit.

Let $\mathcal{V}$ be the space defined by

$$\mathcal{V} := \big\{\mathbf{v} = (v_i); v_\alpha \in H_0^1(\Omega), v_3 \in L^2(\omega)\big\}. \tag{2.19}$$

We extend the energies of the three-dimensional problem to $\mathcal{V}$ by letting, for any $\mathbf{v} \in \mathcal{V}$,

$$J^*(\varepsilon)(\mathbf{v}) := \begin{cases} J(\varepsilon)(\mathbf{v}) & \text{if } \mathbf{v} \in \mathbf{V}(\Omega), \\ +\infty & \text{otherwise,} \end{cases} \tag{2.20}$$

where the space $\mathbf{V}(\Omega)$ and the functional $J(\varepsilon)$ are defined as in (2.9) and (2.11), respectively. Let J^* be the functional given by

$$J^*(\mathbf{v}) := \begin{cases} J(\mathbf{v}) & \text{if } \mathbf{v} \text{ does not depend on } x_3, \\ +\infty & \text{otherwise,} \end{cases} \tag{2.21}$$

where

$$J(\mathbf{v}) = \frac{1}{4}\int_\Omega a^{\alpha\beta\sigma\tau}\gamma_{\sigma\tau}(\mathbf{v})\gamma_{\alpha\beta}(\mathbf{v})\sqrt{a}\,dx - L(\mathbf{v}), \tag{2.22}$$

$$L(\mathbf{v}) = \int_\Omega f^i v_i \sqrt{a}\,dx, \tag{2.23}$$

the functions $a^{\alpha\beta\rho\sigma}$ are the contravariant components of the elasticity tensor of S, defined by

$$a^{\alpha\beta\sigma\tau} := \frac{4\lambda\mu}{\lambda+2\mu}a^{\alpha\beta}a^{\sigma\tau} + 2\mu(a^{\alpha\sigma}a^{\beta\tau} + a^{\alpha\tau}a^{\beta\sigma}), \tag{2.24}$$

and the functions $\gamma_{\alpha\beta}(\cdot)$ are the covariant components of the linearized change of metric tensor given by

$$\gamma_{\alpha\beta}(\mathbf{v}) := \frac{1}{2}(\partial_\alpha v_\beta + \partial_\beta v_\alpha) - \Gamma^\sigma_{\alpha\beta}v_\sigma - b_{\alpha\beta}v_3 \text{ for all } \mathbf{v} \in \mathcal{V}. \tag{2.25}$$

The main result is the following:

THEOREM 2.11. *Let the space $\mathcal{V}$ be equipped with the weak topology of $H^1(\Omega) \times H^1(\Omega) \times L^2(\Omega)$. Then, for this topology, the functional J^* is the Γ-limit of the functionals $J^*(\varepsilon)$.*

PROOF. It suffices to check that the functional J^* satisfies properties (i) and (ii) of Definition 2.7.

(i) We first show that, for any $\mathbf{v} \in \mathcal{V}$,

$$\mathbf{v}(\varepsilon) \rightharpoonup \mathbf{v} \text{ in } \mathcal{V} \implies J^*(\mathbf{v}) \leq \liminf_{\varepsilon \to 0} J^*(\varepsilon)\big(\mathbf{v}(\varepsilon)\big). \tag{2.26}$$

FIRST CASE. The weak limit $\mathbf{v}$ depends on x_3. By definition of J^*, we have $J^*(\mathbf{v}) = +\infty$. It then suffices to show that $\liminf_{\varepsilon\to 0} J^*(\varepsilon)\big(\mathbf{v}(\varepsilon)\big) = +\infty$. Assume that this is false, i.e., assume that $\liminf_{\varepsilon\to 0} J^*(\varepsilon)\big(\mathbf{v}(\varepsilon)\big) < +\infty$. Then one can show that there exist a constant $c_1 > 0$ and a subsequence, still denoted by ε, such that $J^*(\varepsilon)\big(\mathbf{v}(\varepsilon)\big) \geq c_1$. Thus $\mathbf{v}(\varepsilon) \in H^1(\Omega)$ and $J^*(\varepsilon)\big(\mathbf{v}(\varepsilon)\big) = J(\varepsilon)\big(\mathbf{v}(\varepsilon)\big)$. It follows that there exists a constant $c_2 > 0$ such that

$$\int_\Omega A^{ijk\ell}(\varepsilon) e_{k\|\ell}(\varepsilon)\big(\mathbf{v}(\varepsilon)\big) e_{i\|j}(\varepsilon)\big(\mathbf{v}(\varepsilon)\big)\sqrt{g(\varepsilon)}\, dx \leq c_2.$$

Using the generalized Korn's inequality (2.16), we deduce that $\big\|e_{i\|j}(\varepsilon_k)\big(\mathbf{v}(\varepsilon_k)\big)\big\|_{0,\Omega}$, $\big\|v_\alpha(\varepsilon_k)\big\|_{1,\Omega}$ and $\big\|v_3(\varepsilon_k)\big\|_{0,\Omega}$ are bounded independently of ε. Consequently, there exists a subsequence, still denoted $\big(\mathbf{v}(\varepsilon)\big)_{\varepsilon>0}$ for convenience, and there exist functions $e_{i\|j} \in L^2(\Omega)$ such that

$$e_{i\|j}(\varepsilon_k)\big(\mathbf{v}(\varepsilon_k)\big) \rightharpoonup e_{i\|j} \text{ in } L^2(\Omega), \tag{2.27}$$

$$v_\alpha(\varepsilon_k) \rightharpoonup v_\alpha \text{ in } H^1(\Omega) \quad \text{and} \quad v_\alpha(\varepsilon_k) \to v_\alpha \text{ in } L^2(\Omega), \tag{2.28}$$

$$v_3(\varepsilon_k) \rightharpoonup v_3 \text{ in } L^2(\Omega), \tag{2.29}$$

where the function $\mathbf{v} = (v_i)$ is the weak limit in (2.26). It follows from the convergences (2.27)–(2.29) that the functions v_i are independent of x_3 (see Theorem 5.1 of [**CL2**]), which is a contradiction . Thus $\liminf_{\varepsilon\to 0} J^*(\varepsilon)\big(\mathbf{v}(\varepsilon)\big) = +\infty$.

SECOND CASE. The weak limit $\mathbf{v}$ does not depend on x_3. If

$$\liminf_{\varepsilon \to 0} J^*(\varepsilon)\big(\mathbf{v}(\varepsilon)\big) = +\infty,$$

then (2.26) always holds. Assume that $\liminf_{\varepsilon\to 0} J^*(\varepsilon)\big(\mathbf{v}(\varepsilon)\big) < +\infty$. As we did in the first case, it follows that, for a subsequence still denoted by ε, we have $\mathbf{v}(\varepsilon) \in \mathbf{H}^1(\Omega)$, and there exist functions $e_{i\|j} \in L^2(\Omega)$ such that the convergences (2.27)–(2.29) hold. Using the convergences $\Gamma^\sigma_{\alpha\beta}(\varepsilon) \to \Gamma^\sigma_{\alpha\beta}$, $\Gamma^3_{\alpha\beta}(\varepsilon) \to b_{\alpha\beta}$ in $\mathcal{C}^0(\overline{\Omega})$ (see Lemma 3.1 of [**CL2**]), we get

$$e_{\alpha\|\beta} = \gamma_{\alpha\beta}(\mathbf{v}). \tag{2.30}$$

The expression (2.13) of $J(\varepsilon)$, combined with the positive definiteness of the tensor $\big(g^{ij}(\varepsilon)\big)$, implies

$$\begin{aligned} J(\varepsilon)(\mathbf{v}) \geq \frac{1}{2}\int_\Omega &\Big\{\frac{2\lambda\mu}{\lambda + 2\mu} g^{\alpha\beta}(\varepsilon) g^{\sigma\tau}(\varepsilon)\Big\} e_{\alpha\|\beta}(\varepsilon)(\mathbf{v}) e_{\sigma\|\tau}(\varepsilon)(\mathbf{v})\sqrt{g(\varepsilon)}\, dx \\ &+ \frac{1}{2}\int_\Omega \Big\{2\mu g^{\alpha\sigma}(\varepsilon) g^{\beta\tau}(\varepsilon)\Big\} e_{\alpha\|\beta}(\varepsilon)(\mathbf{v}) e_{\sigma\|\tau}(\varepsilon)(\mathbf{v})\sqrt{g(\varepsilon)}\, dx - L(\varepsilon)(\mathbf{v}). \end{aligned}$$

With the convergences $g^{\alpha\beta}(\varepsilon) \to a^{\alpha\beta}$ and $g(\varepsilon) \to a$ in $\mathcal{C}^0(\overline{\Omega})$ this inequality becomes

$$\lim_{\varepsilon \to 0} J(\varepsilon)\big(\mathbf{v}(\varepsilon)\big) \geq \frac{1}{4}\int_\Omega a^{\alpha\beta\sigma\tau}\gamma_{\alpha\beta}(\mathbf{v})\gamma_{\sigma\tau}(\mathbf{v})\sqrt{a}\, dx - L(\mathbf{v}).$$

This means that, for any convergent subsequence $J(\varepsilon)\big(\mathbf{v}(\varepsilon)\big)$, we have

$$\lim_{\varepsilon \to 0} J(\varepsilon)\big(\mathbf{v}(\varepsilon)\big) \geq J(\mathbf{v}).$$

Then proposition (2.26) is satisfied.

(ii) Now we show that, for any $\mathbf{v} \in \mathcal{V}$, there exists a sequence $\big(\mathbf{v}(\varepsilon)\big)$ in $\mathcal{V}$ such that

$$\big(\mathbf{v}(\varepsilon)\big) \rightharpoonup \mathbf{v} \text{ in } \mathcal{V} \quad \text{and} \quad J^*(\varepsilon)\big(\mathbf{v}(\varepsilon)\big) \to J^*(v). \tag{2.31}$$

FIRST CASE. The weak limit $\mathbf{v}$ depends on x_3. Let $\big(\mathbf{v}(\varepsilon)\big)_{\varepsilon>0}$ be the sequence defined by $\mathbf{v}(\varepsilon) = \mathbf{v}$ for all $\varepsilon > 0$. Then it follows from the first part of the proof that

$$J^*(\mathbf{v}) = \liminf_{\varepsilon\to 0} J^*(\varepsilon)\big(\mathbf{v}(\varepsilon)\big).$$

On the other hand, since $\mathbf{v}$ depends on x_3, definition (2.21) of the functional J^* implies $J^*(\mathbf{v}) = +\infty$. Thus

$$J^*(\mathbf{v}) = \liminf_{\varepsilon\to 0} J^*(\varepsilon)\big(\mathbf{v}(\varepsilon)\big),$$

and property (2.31) is verified in this case.

SECOND CASE. The weak limit $\mathbf{v}$ does not depend on x_3. We consider a sequence $\big(\mathbf{v}(\varepsilon)\big)_{\varepsilon>0}$ in the space $\mathbf{V}(\Omega)$ where, for all $\varepsilon > 0$, the function $\mathbf{v}(\varepsilon)$ is the unique solution of the following minimization problem:

$$\mathbf{v}(\varepsilon) \in \mathbf{V}(\Omega) \quad \text{and} \quad J_{\mathbf{v}}(\varepsilon)\big(\mathbf{v}(\varepsilon)\big) = \inf_{\mathbf{w}\in\mathbf{V}(\Omega)} J_{\mathbf{v}}(\varepsilon)(\mathbf{w}),$$

where, for all $\mathbf{u}, \mathbf{w} \in \mathbf{V}(\Omega)$,

$$J_{\mathbf{v}}(\varepsilon)(\mathbf{w}) := \frac{1}{2} B(\varepsilon)(\mathbf{w}, \mathbf{w}) - \tilde{L}(\varepsilon)(\mathbf{v}, \mathbf{w}),$$

the bilinear form $B(\varepsilon)$ being defined on $V(\Omega) \times V(\Omega)$ by

$$B(\varepsilon)(\mathbf{u}, \mathbf{w}) := \int_\Omega A^{ijk\ell}(\varepsilon) e_{k\|\ell}(\varepsilon)(\mathbf{u}) e_{i\|j}(\varepsilon)(\mathbf{w}) \sqrt{g(\varepsilon)}\, dx,$$

and the linear form $\tilde{L}(\varepsilon)$ being defined on $\mathbf{V}(\Omega)$ by

$$\tilde{L}(\varepsilon)(\mathbf{v}, \mathbf{w}) = \frac{1}{2} \int_\Omega a^{\alpha\beta\sigma\tau} \gamma_{\sigma\tau}(\mathbf{v}) e_{\alpha\|\beta}(\varepsilon)(\mathbf{w}) \sqrt{a}\, dx.$$

We recall that the function $\mathbf{v}$ is fixed.

Equivalently, the function $\mathbf{v}(\varepsilon)$ solves the following variational problem

$$B(\varepsilon)\big(\mathbf{v}(\varepsilon), \mathbf{w}\big) = \tilde{L}(\varepsilon)(\mathbf{v}, \mathbf{w}) \quad \text{for all } \mathbf{w} \in \mathbf{V}(\Omega). \tag{2.32}$$

Since $J_{\mathbf{v}}(\varepsilon)\big(\mathbf{v}(\varepsilon)\big) \le J_{\mathbf{v}}(\varepsilon)(\mathbf{0}) = 0$, we deduce that there exists a constant $C(\mathbf{v})$ depending on $\mathbf{v}$, such that

$$\sum_{i,j} \big\| e_{i\|j}(\varepsilon)\big(\mathbf{v}(\varepsilon)\big) \big\|^2_{0,\Omega} \le \mathcal{C}(\mathbf{v}).$$

This, together with Korn's inequality (2.16), then implies that $\big\|e_{i\|j}(\varepsilon)\big(\mathbf{v}(\varepsilon)\big)\big\|_{0,\Omega}$, $\big\|v_\alpha(\varepsilon)\big\|_{1,\Omega}$ and $\big\|v_3(\varepsilon)\big\|_{0,\Omega}$ are bounded independently of ε. Consequently, there exists a subsequence, still denoted $(\mathbf{v}(\varepsilon))$, and there exist functions $e_{i\|j} \in L^2(\Omega)$, $\tilde{v}_\alpha \in H_0^1(\Omega)$ and $\tilde{v}_3 \in L^2(\Omega)$ such that

$$e_{i\|j}(\varepsilon)(\mathbf{v}(\varepsilon)) \rightharpoonup e_{i\|j} \text{ in } L^2(\Omega), \tag{2.33}$$

$$v_\alpha(\varepsilon) \rightharpoonup \tilde{v}_\alpha \text{ in } H^1(\Omega) \quad \text{and} \quad v_\alpha(\varepsilon) \to \tilde{v}_\alpha \text{ in } L^2(\Omega), \tag{2.34}$$

$$v_3(\varepsilon) \rightharpoonup \tilde{v}_3 \text{ in } L^2(\Omega). \tag{2.35}$$

These convergences imply that (see Step (i)) the functions $\tilde{v}_i$ are independent of x_3 and that

$$e_{\alpha\|\beta} = \gamma_{\alpha\beta}(\tilde{\mathbf{v}}). \tag{2.36}$$

We fix $\mathbf{w} = (w_i)$ in the space $\mathbf{V}(\Omega)$, we multiply equation (2.32) by ε, and we let ε tend to 0. Using the weak convergences (2.33)–(2.35), we get

$$\int_\Omega \{2\mu a^{\alpha\sigma} e_{\sigma\|3} \partial_3 w_\alpha + [\lambda a^{\sigma\tau} e_{\sigma\|\tau} + (\lambda + 2\mu) e_{3\|3}] \partial_3 w_3\} \sqrt{a}\, dx = 0. \tag{2.37}$$

By letting $\mathbf{w}$ vary in $\mathbf{V}(\Omega)$, we obtain

$$e_{3\|3} = -\frac{\lambda}{\lambda + 2\mu} a^{\alpha\beta} e_{\alpha\|\beta}, \tag{2.38}$$

$$e_{\alpha\|3} = 0. \tag{2.39}$$

Next, in equation (2.32), we choose $\mathbf{w} \in \mathbf{V}(\Omega)$ such that $\partial_3 \mathbf{w} = 0$, and we let ε tend to 0. We find that the function $\tilde{\mathbf{v}}$ satisfies

$$\int_\Omega \left\{ \frac{2\lambda\mu}{\lambda + 2\mu} a^{\alpha\beta} a^{\sigma\tau} + \mu(a^{\alpha\sigma} a^{\beta\tau} + a^{\alpha\tau} a^{\beta\sigma}) \right\} \gamma_{\sigma\tau}(\tilde{\mathbf{v}}) \gamma_{\alpha\beta}(\mathbf{w}) \sqrt{a}\, dx$$
$$= \frac{1}{2} \int_\Omega a^{\alpha\sigma\beta\tau} \gamma_{\sigma\tau}(\mathbf{v}) \gamma_{\alpha\beta}(\mathbf{w}) \sqrt{a}\, dx.$$

Since the functions $\tilde{\mathbf{v}}$ and $\mathbf{w}$ do not depend on x_3, this equation can be written as

$$B(\overline{\tilde{\mathbf{v}}}, \overline{\mathbf{w}}) = B(\overline{\mathbf{v}}, \overline{\mathbf{w}}) \text{ for } \overline{\mathbf{w}} \in \mathbf{V}_M(\omega), \tag{2.40}$$

where the space $\mathbf{V}_M$ is defined by

$$\mathbf{V}_M(\omega) = H_0^1(\omega) \times H_0^1(\omega) \times L^2(\omega), \tag{2.41}$$

the bilinear form B is given by

$$B(\overline{\mathbf{w}}^1, \overline{\mathbf{w}}^2) = \int_\omega a^{\alpha\beta\sigma\tau} \gamma_{\sigma\tau}(\overline{\mathbf{w}}^1) \gamma_{\alpha\beta}(\overline{\mathbf{w}}^2) \sqrt{a}\, dy \quad \text{for } (\overline{\mathbf{w}}^1, \overline{\mathbf{w}}^2) \in \mathbf{V}_M(\omega) \times \mathbf{V}_M(\omega), \tag{2.42}$$

and where, if $\mathbf{w}$ is a function defined almost everywhere on Ω, we denote by $\overline{\mathbf{w}} = (\overline{w}_i)$ the function defined almost everywhere on ω by

$$\overline{\mathbf{w}}(y) := \frac{1}{2} \int_{-1}^{1} \mathbf{w}(y, x_3)\, dx_3. \tag{2.43}$$

In particular, if $\mathbf{w}$ does not depend on x_3, then

$$\mathbf{w}(y, x_3) = \overline{\mathbf{w}}(y).$$

Under either one of assumptions (2.14) and (2.15), it was shown by [**CS**] and [**CL1**] that the bilinear form B is coercive on the space $\mathbf{V}_M(\omega)$. Thus, equation (2.40) implies

$$\tilde{\mathbf{v}} = \mathbf{v}.$$

Therefore, since $\mathbf{v}$ is unique, the whole sequence $\mathbf{v}(\varepsilon)$ weakly converges to $\mathbf{v}$ in $\mathbf{H}^1(\Omega)$.

Finally, we have

$$\begin{aligned} J(\varepsilon)\big(\mathbf{v}(\varepsilon)\big) &= \frac{1}{2}B(\varepsilon)\big(\mathbf{v}(\varepsilon), \mathbf{v}(\varepsilon)\big) - L(\varepsilon)\big(\mathbf{v}(\varepsilon)\big) \\ &= \frac{1}{2}\tilde{L}(\varepsilon)\big(\mathbf{v}, \mathbf{v}(\varepsilon)\big) - L(\varepsilon)\big(\mathbf{v}(\varepsilon)\big) \\ &= \frac{1}{4}\int_\Omega a^{\alpha\beta\sigma\tau}\gamma_{\sigma\tau}(\mathbf{v})e_{\alpha\|\beta}\big(\mathbf{v}(\varepsilon)\big)\sqrt{a}\,dx - L(\varepsilon)\big(\mathbf{v}(\varepsilon)\big), \\ &\to \frac{1}{4}\int_\Omega a^{\alpha\beta\sigma\tau}\gamma_{\sigma\tau}(\mathbf{v})\gamma_{\alpha\beta}(\mathbf{v})\sqrt{a}\,dx - L(\mathbf{v}) = J(\mathbf{v}), \end{aligned}$$

which we may write as

$$J^*(\varepsilon)\big(\mathbf{v}(\varepsilon)\big) \to J^*(\mathbf{v}). \quad \square$$

Now, we deduce a convergence result for the displacements:

THEOREM 2.12. *The sequence* $\big(\mathbf{u}(\varepsilon)\big)_{\varepsilon>0}$ *where, for all* $\varepsilon > 0$, $\mathbf{u}(\varepsilon)$ *is the solution of the scaled minimization problem* (2.9)–(2.10) *satisfies*

$$u_\alpha(\varepsilon) \to u_\alpha \text{ in } H^1(\Omega), \quad u_3(\varepsilon) \to u_3 \text{ in } L^2(\Omega),$$

where the function $\mathbf{u}$ *is independent of* x_3 *and* $\mathbf{u}$ *is the solution of the minimization problem*

$$J^*(\mathbf{u}) = \inf_{\mathbf{v}\in\mathcal{V}} J^*(\mathbf{v}),$$

the functional J^* *being defined in* (2.21). *The function* $\overline{\mathbf{u}}$ *is the solution of the two-dimensional membrane shell problem in linearized elasticity.*

PROOF. We infer from Korn's inequality (2.16) that $\|u_\alpha(\varepsilon)\|_{0,\Omega}$ and $\|u_3(\varepsilon)\|_{0,\Omega}$ are bounded independently of ε. Thus the vector fields $\mathbf{u}(\varepsilon)$ belong to a compact set of $\mathcal{V}$ for the weak topology of $H^1(\Omega) \times H^1(\Omega) \times L^2(\Omega)$. Moreover, standard estimates show that any weak limit of $\mathbf{u}(\varepsilon)$ is independent of x_3. Theorem 2.9 implies that there exist a function $\mathbf{u} \in \mathcal{V}$ and a subsequence $\big(\mathbf{u}(\varepsilon_k)\big)$ such that $\mathbf{u}(\varepsilon_k) \rightharpoonup \mathbf{u}$ in $\mathcal{V}$ and $\mathbf{u}$ satisfies $J^*(\mathbf{u}) = \inf_{\mathbf{v}\in\mathcal{V}} J^*(\mathbf{v})$. Thus, the function $\mathbf{u}$ is unique and the whole sequence $\big(\mathbf{u}(\varepsilon)\big)$ converges to $\mathbf{u}$ in $\mathcal{V}$.

The strong convergences are obtained as in [**CL2**]. $\square$

3. Justification of the Two-Dimensional Flexural Model

The process is the same as in the membrane case. There is no geometric assumption on the middle surface S and the shell is only *partially* clamped on $\Gamma_0^\varepsilon := \gamma_0 \times [-\varepsilon, \varepsilon]$. The "right" choice of scalings is:

$$u_i(\varepsilon)(x) = u_i^\varepsilon(x^\varepsilon) \quad \text{for all } x \in \Omega, \tag{3.1}$$

and the forces must be of order 2 in ε:

$$f^{i,\varepsilon}(x^\varepsilon) = \varepsilon^2 f^i(x) \quad \text{for all } x \in \Omega. \tag{3.2}$$

The energy functional corresponding to the scaled three-dimensional problem is

$$J(\varepsilon)(\mathbf{v}) = \frac{1}{2\varepsilon}\Big\{\int_\Omega A^{ijk\ell}(\varepsilon)e_{k\|\ell}(\varepsilon)(\mathbf{v})e_{i\|j}(\varepsilon)(\mathbf{v})\Big\}\sqrt{g(\varepsilon)}\,dx - L(\varepsilon)(\mathbf{v}); \tag{3.3}$$

see (2.5)–(2.8) and (2.12) for the definitions of the functions $A^{ijk\ell}(\varepsilon)$, $e_{i\|j}(\varepsilon)$ and $L(\varepsilon)$.

Let $\mathbf{V}_F(\omega)$ be the space defined by:

$$\mathbf{V}_F(\omega) := \{\eta = (\eta_i) \in H^1(\omega) \times H^1(\omega) \times H^2(\omega), \\ \gamma_{\alpha\beta}(\eta) = 0 \text{ in } \omega, \eta_i = \partial_\nu \eta_3 = 0 \text{ on } \gamma_0\}, \tag{3.4}$$

where the functions $\gamma_{\alpha\beta}$ are defined as in (2.25). The covariant components of the change of curvature tensor are

$$\rho_{\alpha\beta}(\eta) := \partial_{\alpha\beta}\eta_3 - \Gamma^\sigma_{\alpha\beta}\partial_\sigma\eta_3 + b^\sigma_\beta(\partial_\alpha\eta_\sigma - \Gamma^\tau_{\alpha\sigma}\eta_\tau) \\ + b^\sigma_\alpha(\partial_\beta\eta_\sigma - \Gamma^\tau_{\beta\sigma}\eta_\tau) + b^\sigma_\alpha|_\beta\eta_\sigma - c_{\alpha\beta}\eta_3,$$

where

$$b^\sigma_\beta|_\alpha := \partial_\alpha b^\alpha_\beta + \Gamma^\sigma_{\alpha\tau}b^\tau_\beta - \Gamma^\tau_{\beta\alpha}b^\sigma_\tau, \quad c_{\alpha\beta} := b^\sigma_\alpha b_{\sigma\beta}.$$

For any $\mathbf{v} \in \mathbf{V}(\Omega) := \{\mathbf{v} = (v_i) \in H^1(\Omega); \mathbf{v} = 0 \text{ sur } \Gamma_0\}$, the functional $J^* : \mathbf{V}(\Omega) \to \mathbf{R}$ is defined in the following way:

$$J^*(\mathbf{v}) = \begin{cases} J(\mathbf{v}) & \text{if } \mathbf{v} \text{ is independent of } x_3 \quad \text{and} \quad \overline{\mathbf{v}} \in \mathbf{V}_F(\omega) \\ +\infty & \text{otherwise,} \end{cases} \tag{3.5}$$

where

$$J(\mathbf{v}) := \bar{J}(\overline{\mathbf{v}}) = \frac{1}{6}\int_\omega a^{\alpha\beta\sigma\tau}\rho_{\sigma\tau}(\overline{\mathbf{v}})\rho_{\alpha\beta}(\overline{\mathbf{v}})\sqrt{a}\,dy - L(\overline{\mathbf{v}}), \tag{3.6}$$

$$L(\overline{\mathbf{v}}) := \int_\omega \Big(\int_{-1}^1 f^i dx_3\Big)\overline{\mathbf{v}}_i\sqrt{a}\,dy, \tag{3.7}$$

and $\overline{\mathbf{v}}$ is the average associated with $\mathbf{v}$ (see (2.43)).

We use the following *Korn's inequality* (see [**CLM**]):

$$\|\mathbf{v}\|_{1,\Omega} \le \frac{C}{\varepsilon}\Big\{\sum_{i,j}\|e_{i\|j}(\varepsilon)(\mathbf{v})\|^2_{0,\Omega}\Big\}^{1/2} \quad \text{for all } \mathbf{v} \in \mathbf{V}(\Omega), \tag{3.8}$$

to prove the Γ-convergence of the functionals $J(\varepsilon)$.

THEOREM 3.1. *The functional J^* is the Γ-limit of functionals $J(\varepsilon)$ in $\mathbf{V}(\Omega)$ equipped with the weak topology of $H^1(\Omega) \times H^1(\Omega) \times H^1(\Omega)$.*

For a proof, see [**G**].

As in the membrane case, we deduce from this result a convergence theorem for the displacements. We call $\mathbf{u}(\varepsilon)$ the solution of the scaled minimization problem:

$$\mathbf{v}(\varepsilon) \in \mathbf{V}(\Omega) \quad \text{and} \quad J_\mathbf{v}(\varepsilon)\big(\mathbf{v}(\varepsilon)\big) = \inf_{\mathbf{w}\in\mathbf{V}(\Omega)} J_\mathbf{v}(\varepsilon)(\mathbf{w}). \tag{3.9}$$

THEOREM 3.2. *The sequence $\big(\mathbf{u}(\varepsilon)\big)_{\varepsilon>0}$ satisfies*

$$u_i(\varepsilon) \to u_i \text{ in } H^1(\Omega), \tag{3.10}$$

where the function $\mathbf{u}$ is independent of x_3, $\overline{\mathbf{u}} = \int_{-1}^1 \mathbf{u}(y, x_z)\,dx_z \in \mathbf{V}_F(\omega)$ and $\mathbf{u}$ is the solution of the minimization problem

$$J^*(\mathbf{u}) = \inf_{\mathbf{v}\in\mathbf{V}(\Omega)} J^*(\mathbf{v}), \tag{3.11}$$

where J^* is defined by (3.5). The function $\overline{\mathbf{u}}$ is the solution of the two-dimensional flexural shell problem in linearized elasticity.

REMARK 3.3. A similar analysis can be made in the case of linearly elastic plates. With the right scalings, it leads to identify separately the membrane model and the flexural model.

References

[ABP] E. Acerbi, G. Buttazzo, and D. Percivale, *A variational definition of the strain energy for an elastic string*, J. Elasticity **25**, (1991), 137–148.

[A] H. Attouch, *Variational convergence for functions and operators*, Pitman, Boston, 1984.

[BCGR] F. Bourquin, P. G. Ciarlet, G. Geymonat, and A. Raoult, *Γ-convergence et analyse asymptotique des plaques minces*, C. R. Acad. Sci. Paris, Sér. I **315** (1992), 1017–1024.

[CaS] D. Caillerie and E. Sanchez-Palencia, *Elastic thin shells: Asymptotic theory in the anisotropic and heterogeneous cases*, Math. Models Methods Appl. Sci. **5** (1996), 473–496.

[C] P. G. Ciarlet, *Mathematical elasticity.* II. *Theory of plates*, Studies in Mathematics and its Applications, vol. 27, North-Holland Publ. Co., Amsterdam, 1997.

[CL1] P. G. Ciarlet and V. Lods, *On the ellipticity of linear membrane shell equations*, J. Math. Pures Appl. **75** (1996), 107–124.

[CL2] ———, *Asymptotic analysis of linearly elastic shells.* I. *Justification of membrane shell equations*, Arch. Rational Mech. Anal. **136** (1996), no. 2, 119–161.

[CLM] P. G. Ciarlet, V. Lods, and B. Miara, *Asymptotic analysis of linearly elastic shells.* II. *Justification of flexural shell equations*, Arch. Rational Mech. Anal. **136** (1996), no. 2, 163–190.

[CS] P. G. Ciarlet and E. Sanchez-Palencia, *An existence and uniqueness theorem for the two-dimensional linear membrane shell equations*, J. Math. Pures Appl. **75** (1996), 51–67.

[Da] G. Dal Maso, *An introduction to Γ-convergence*, Nonlinear Differential Equations and Their Applications, Birkäuser, Basel, 1993.

[D1] P. Destuynder, *Sur une Justification des modèles de plaques et de coques par les méthodes asymptotiques*, Doctoral Dissertation, Univ. Pierre et Marie Curie, Paris, 1980.

[D2] ———, *A classification of thin shell theories*, Acta Appl. Math. **4** (1985), 15–63.

[G] K. Genevey, *Sur quelques questions liées à la théorie mathématique des coques minces*, Ph.D., Univ. Pierre et Marie Curie, 1998.

[LR] H. Le Dret and A. Raoult, *The membrane shell model in nonlinear elasticity: A variational asymptotic derivation*, J. Nonlinear Sci. **6** (1996), 59–84.

[MS] B. Miara and E. Sanchez-Palencia *Asymptotic analysis of linearly elastic shells*, Asymptotic Anal. **12** (1996), 41–54.

[S] E. Sanchez-Palencia, *Passages à la limite de l'élasticité tri-dimensionnelle à la théorie asymptotique des coques minces*, C. R. Acad. Sci. Paris, Sér. II **311** (1990), 909–916.

ESPARTERIA 6, 2-1, 08003 BARCELONE, SPAIN
E-mail address: genevey@ann.jussieu.fr

Centre de Recherches Mathématiques
CRM Proceedings and Lecture Notes
Volume **21**, 1999

Numerical Simulation of High-Rate Stamping of Tubes and Sheets

S. F. Golovashchenko and N. M. Bessonov

Abstract. A numerical finite difference procedure with natural approximation of derivatives is applied to high-rate forming of tubes and sheets. The workhardening and strain-rate hardening of the material was taken into account using experimental data on the expansion of rings subjected to pressure pulses. Numerical examples are also given together with experimental data on final deflections of sheets and tubes as a function of pressure pulse amplitude. Axial carrying capacity of tubular nodes with circular grooves, filled with tube material, was predicted.

1. Introduction

High-rate pulsed metal forming processes can be based on short duration high-voltage discharge of electric capacitors through a coil, or through interval between electrodes in a chamber filled with water or using the energy of chemical explosion [**1**]. Velocity of the deformed material in these processes is usually around 100–200 m/sec and the duration of the process is $\approx 10^{-4}$ sec. That means that inertia forces are very important in the mechanics of these processes. Several practical problems need to be solved to make the design better: calculation of pressure pulse amplitude needed to shape the blank, prediction of blank fracture possibility, and reduction blank with die impact velocity.

2. Theory

The numerical simulation which we present in this paper is based on the finite difference method with natural approximation of derivatives [**5**].

The numerical procedure which eliminates the need of global stiffness matrix calculations and allows use of orthogonal tensor theory instead of general tensor theory is developed. This procedure takes into account effects of blank stretching, bending and shearing.

1991 *Mathematics Subject Classification.* Primary: 73E50; Secondary: 73C05.

This paper was written while the authors were visiting Ford Motor Company Scientific Research Laboratory, 20000 Rotunda Dr., P.O.Box 2053, MD3182, Room 2621C, SRL, Dearborn, MI 48121-2053, USA.

This is the final form of the paper.

The initial set of equations includes solid mechanics equation of motion

$$\operatorname{Div}\sigma = \rho\frac{d\vec{\mathbf{v}}}{dt}, \tag{2.1}$$

where σ is a stress tensor, $\vec{\mathbf{v}}$ is a velocity vector, ρ is a material density.

Hook's law for plane stress state in local n, τ, b coordinate system was employed, where normal to middle surface stress is neglected

$$\dot{\sigma}_\tau = \frac{E}{1-\mu^2}(\dot{\varepsilon}_\tau + \mu\dot{\varepsilon}_b), \quad \dot{\sigma}_b = \frac{E}{1-\mu^2}(\dot{\varepsilon}_b + \mu\dot{\varepsilon}_\tau), \tag{2.2}$$

$$\dot{\tau}_{bn} = \frac{E}{2(1+\mu)}\dot{\gamma}_{bn}, \quad \dot{\tau}_{\tau n} = \frac{E}{2(1+\mu)}\dot{\gamma}_{\tau n}, \quad \dot{\tau}_{b\tau} = \frac{E}{2(1+\mu)}\dot{\gamma}_{b\tau}, \tag{2.3}$$

where $\dot{\sigma}_i$, $\dot{\tau}_{ij}$ and $\dot{\varepsilon}_i$, $\dot{\gamma}_{ij}$ are stress and strain tensor component time derivatives.

The Von Mises yield criterion in the following form is used

$$\sqrt{\sigma_\tau^2 + \sigma_b^2 - \sigma_\tau\sigma_b + 3(\tau_{bn}^2 + \tau_{\tau n}^2 + \tau_{b\tau}^2)} \le \sigma_s, \tag{2.4}$$

where σ_s is a current yield stress depending upon plastic strain and strain-rate intensities. The velocity field along the blank thickness was assumed linear, i.e. the initial normal to the middle surface remained a straight line during the deformation process [**3**]

$$\vec{\mathbf{v}} = \vec{\mathbf{v}}^0 + \vec{\omega} \times \vec{\mathbf{r}}_k, \tag{2.5}$$

where $\vec{\mathbf{v}}^0$ is a velocity vector of the middle surface node, $\vec{\omega}$ is an angular velocity vector of the normal to the middle surface rotation, $\vec{\mathbf{r}}_k$ is a radius vector along the normal.

The middle surface of the blank being deformed was discretisized by plane triangles. Every node of this grid was surrounded by a final volume V of the blank material. Every final volume consists of parts of several elements volumes V_i based on middle surface grid triangles. It corresponds to the part of the middle surface around the grid node under consideration, which is bound by the contour L as is shown in Fig. 1. The solid mechanics equation of motion was integrated over the nodal volume

$$\int_V \rho\frac{d}{dt}\begin{bmatrix} v_x \\ v_y \\ v_z \end{bmatrix} dV = \sum_i [\alpha]_i \int_{V_i} \begin{bmatrix} \frac{\partial\sigma_n}{\partial n} + \frac{\partial\tau_{n\tau}}{\partial\tau} + \frac{\partial\tau_{nb}}{\partial b} \\ \frac{\partial\tau_{n\tau}}{\partial n} + \frac{\partial\sigma_\tau}{\partial\tau} + \frac{\partial\tau_{b\tau}}{\partial b} \\ \frac{\partial\tau_{nb}}{\partial n} + \frac{\partial\tau_{b\tau}}{\partial\tau} + \frac{\partial\sigma_b}{\partial b} \end{bmatrix} dV. \tag{2.6}$$

Node motion was considered in a stable global coordinate system x, y, z. Stress tensor divergence was taken in a local basis n, τ, b for each middle surface triangle, where all the shell theory hypotheses were assumed to be valid. The transformation from global to local coordinates system was realized using the direction cosine matrix $[\alpha]_i$ for each V_i. This integration resulted in equation of motion for the nodes:

$$M\frac{d}{dt}\begin{bmatrix} v_x \\ v_y \\ v_z \end{bmatrix} = \sum_i [\alpha]_i \begin{bmatrix} p_n S_i + \int_{L_i} Q_\tau\, db + \int_{L_i} Q_b\, d\tau \\ p_\tau S_i + \int_{L_i} N_\tau\, db + \int_{L_i} N_{b\tau}\, d\tau \\ p_b S_i + \int_{L_i} N_{b\tau}\, db + \int_{L_i} N_b\, d\tau \end{bmatrix}, \tag{2.7}$$

where

$$M = \sum_i \rho V_i, \quad N_{ij} = \int_{-h/2}^{h/2} \tau_{ij}\, dn,$$

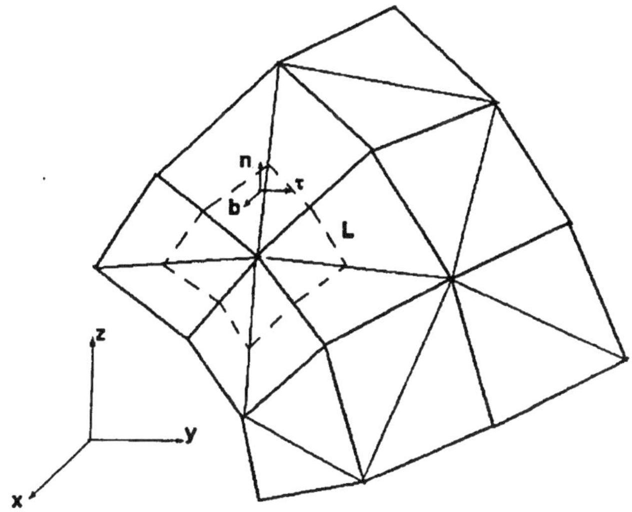

FIGURE 1. Nonorthogonal grid.

and h is the blank thickness for a given triangle. The same approach was used to obtain the equation of rotation of the middle surface normal. The solid mechanics equation of motion was multiplied by the radius-vector

$$\vec{\mathbf{r}} \times \rho \frac{d\vec{\mathbf{v}}}{dt} = \vec{\mathbf{r}} \times \operatorname{Div} \sigma. \tag{2.8}$$

Its integration for each part of nodal volume gave the equation of normal rotation.

$$\begin{bmatrix} J_{xx} & J_{xy} & J_{xz} \\ J_{yx} & J_{yy} & J_{yz} \\ J_{zx} & J_{zy} & J_{zz} \end{bmatrix} \frac{d}{dt} \begin{bmatrix} \omega_x \\ \omega_y \\ \omega_z \end{bmatrix} = \sum_i [\alpha]_i \begin{bmatrix} 0 \\ \int_{L_i} M_\tau \, db + \int_{L_i} M_{\tau b} \, d\tau - \int_{S_i} Q_\tau \, dS \\ \int_{L_i} M_{\tau b} \, db + \int_{L_i} M_b \, d\tau - \int_{S_i} Q_b \, dS \end{bmatrix} dV. \tag{2.9}$$

where

$$M_{ij} = \int_{-h/2}^{h/2} \sigma_{ij} \, n \, dn.$$

The moment of inertia tensor J components in the global set of coordinates were calculated for each volume part using the local components of this tensor in the local basis of each triangle

$$J_{nn} = J_{n\tau} = J_{nb} = J_{\tau n} = J_{bn} = J_{b\tau} = J_{\tau b} = 0, \tag{2.10}$$

$$J_{\tau\tau} = J_{bb} = \int_{-h/2}^{h/2} s\rho n^2 \, dn = \rho s \frac{h^3}{12}. \tag{2.11}$$

The components of the global coordinates J_{kl} were calculated through the local coordinates components for each V_i

$$J_{kl} = \sum_{a=1}^{3} \sum_{c=1}^{3} \alpha_{ka} \alpha_{lc} J_{ac},$$

and accumulated

$$J = \sum_i J_i.$$

An explicit numerical procedure was involved to integrate the node equation of motion and normal equation of rotation. Strain-rate tensor components were calculated in local basis for each triangle as derivatives of the velocity vector projections through their nodal values

$$\vec{\mathbf{v}} = v_n^0 \vec{\mathbf{e}}_n + (v_\tau^0 + \omega_b n)\vec{\mathbf{e}}_\tau + (v_b^0 - \omega_\tau n)\vec{\mathbf{e}}_b, \tag{2.12}$$

$$\begin{gathered}
\dot{\varepsilon}_\tau = \frac{\partial v_\tau^0}{\partial \tau} + n\frac{\partial \omega_b}{\partial \tau}, \quad \dot{\varepsilon}_b = \frac{\partial v_b^0}{\partial b} - n\frac{\partial \omega_\tau}{\partial b}, \\
\dot{\gamma}_{b\tau} = \frac{\partial v_\tau^0}{\partial b} + \frac{\partial v_b^0}{\partial \tau} + n\left(\frac{\partial \omega_b}{\partial b} - \frac{\partial \omega_\tau}{\partial \tau}\right), \\
\dot{\gamma}_{bn} = \left(\frac{\partial v_n^0}{\partial b} - \omega_\tau\right) f(n), \quad \dot{\gamma}_{\tau n} = \left(\frac{\partial v_n^0}{\partial \tau} - \omega_b,\right) f(n), \\
f(n) = 1 - \left(\frac{2n}{h}\right)^2.
\end{gathered} \tag{2.13}$$

The local basis was chosen for each triangle in a following manner

$$\vec{\mathbf{e}}_n = \frac{\overrightarrow{12} \times \overrightarrow{23}}{|\overrightarrow{12} \times \overrightarrow{23}|}, \quad \vec{\mathbf{e}}_\tau = \frac{\overrightarrow{12}}{|\overrightarrow{12}|}, \quad \vec{\mathbf{e}}_b = \vec{\mathbf{e}}_n \times \vec{\mathbf{e}}_\tau, \tag{2.14}$$

where 1, 2, 3 are middle surface triangle vertices.

The same approach was used for axisymmetrical analysis. The middle surface generator was discretisized by segments. To surround each node with a final volume, each segment was divided into two equal parts. Volume integration of the solid mechanics equation of motion results in the following node equation of motion

$$M\frac{d}{dt}\begin{bmatrix} v_r \\ v_z \end{bmatrix} = \sum_i [\alpha']_i \begin{bmatrix} \int_{T_i} p_n r \, d\tau + (Q_\tau r)_i - \int_{T_i} N_\theta \sin\varphi \, d\tau \\ \int_{T_i} p_\tau r \, d\tau + (N_\tau r)_i - \int_{T_i} N_\theta \cos\varphi \, d\tau \end{bmatrix}, \tag{2.15}$$

where p_n, p_τ are external pressure components, r, θ, z are cylindrical coordinates and φ is an angle between axis n and r. The normal to the middle surface equation of rotation was obtained in the same way as equation (2.8). In this case, the rotation of the normal takes place in the radial cross-section plane and the inertia moment becomes a scalar value.

$$M\frac{h^2}{12}\frac{d\omega}{dt} = \sum_{i=1}^{2} r\left(M_{\tau_i} - \int_{T_i} Q_\tau \, d\tau\right) - \int_{T_i} M_\theta \cos\varphi \, d\tau. \tag{2.16}$$

The strain-rate tensor components for the middle surface deformation were obtained similar to (2.11)

$$\dot{\varepsilon}_\tau = \frac{\partial v_\tau}{\partial \tau}, \quad \dot{\varepsilon}_\varphi = \frac{\dot{r}}{r}, \quad \dot{\gamma}_{n\tau} = \frac{\partial v_n}{\partial \tau} + \frac{\partial v_\tau}{\partial n}, \quad \dot{\varepsilon}_n = -\frac{\partial v_\tau^0}{\partial \tau} - \frac{\dot{r}^0}{r}. \tag{2.17}$$

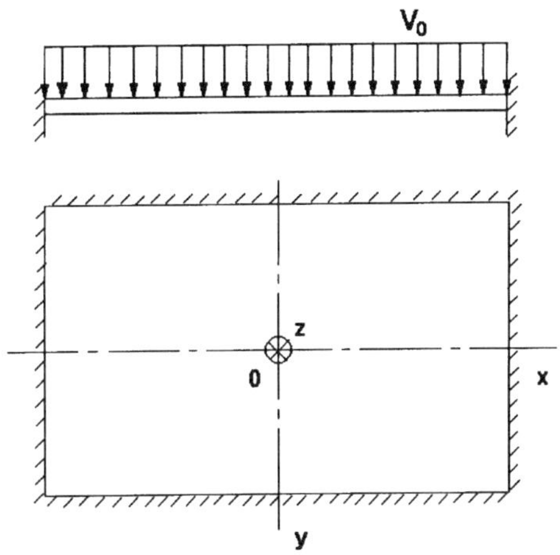

FIGURE 2. The scheme of clamped plate loading.

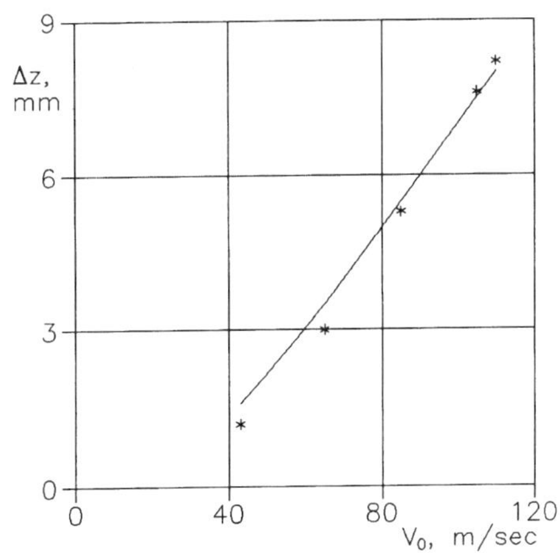

FIGURE 3. The influence of the plate initial velocity on its maximum final deflection.

3. Numerical Examples and Experimental Validation

The earlier discussed model was employed to analyse deformation of clamped rectangular plates loaded impulsively as shown in Fig. 2. Plate dimensions and material properties were taken from the experimental paper [**2**]. According to this paper, pulsed load was assumed as uniform transverse initial velocity. Numerical results in comparison with experimental data on maximum final deflection of the plate vs its initial velocity are shown in Fig. 3 for Al 6061. The final shape of the plate, obtained both theoretically and experimentally is shown for different cross-sections in Fig. 4.

The significant feature of tube high-rate forming or tubular nodes assembling processes analysis is a high strain-rate value. It can reach 10000 $\sec^{-1}$ or more. In order to take into account material strain-rate sensitivity, the experimental device

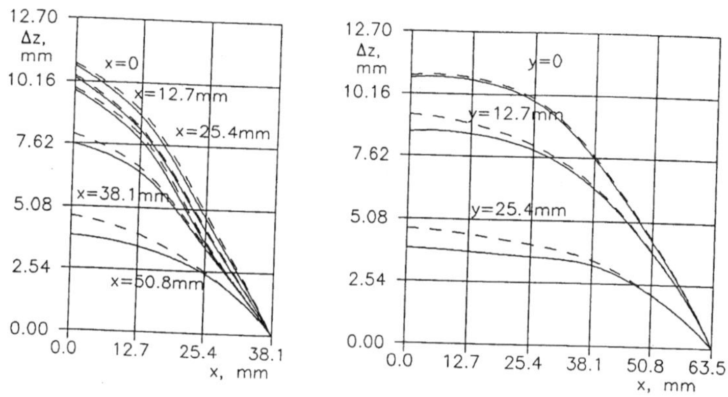

FIGURE 4. The shape of the deformed rectangular plate in different cross-sections.

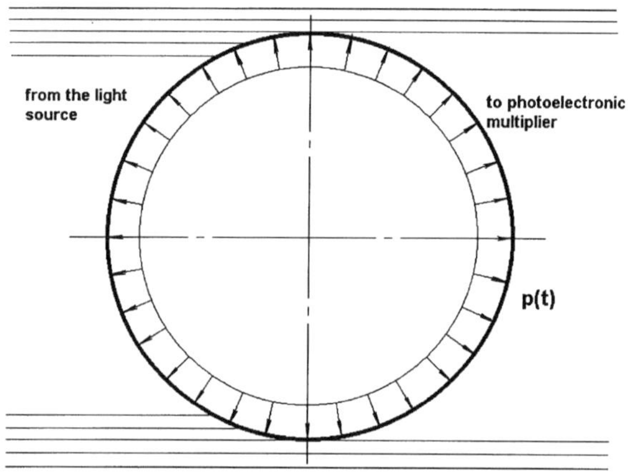

FIGURE 5. Schematic of ring high-rate widening registration.

for the ring pulsed electromagnetic uniform expansion was designed. The stress state of the short thin walled ring being expanded is close to uniaxial. The displacement of the ring vs time was measured using an optical system. It included the source of light, which beams were shaded by the ring during its movement as shown in Fig. 5. The light intensity was measured from the other side of the ring by means of photo multiplier through a system of lens. The electromagnetic pressure pulse parameters were registered using an inductive gage, installed in the gap between the electromagnetic machine coil and the ring being deformed. Such kind of experimental results give us two important opportunities: to obtain yield stress function vs strain and strain-rate and to check the data obtained in some other tests to increase reliability. Some experiments were done for annealed aluminum, which is a very strain-rate sensitive material. Its mechanical properties were taken from experimental data [**4**] and used in the ring expansion calculation. Numerical results were compared with the experimental curve of the ring radial displacement vs time in Fig. 6.

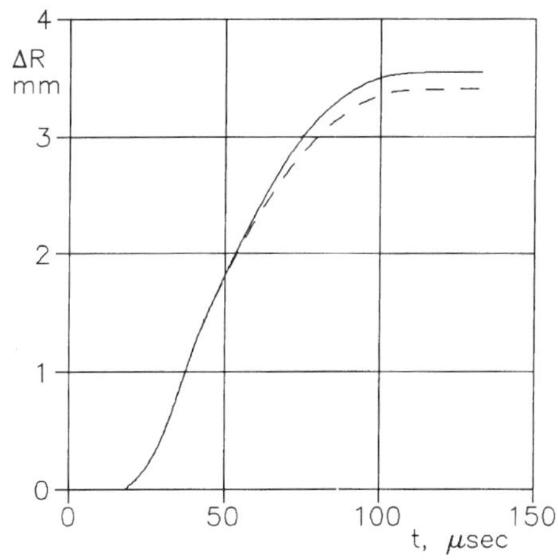

FIGURE 6. Ring radial displacement vs time theoretical (solid line) and experimental (broken line) graphs.

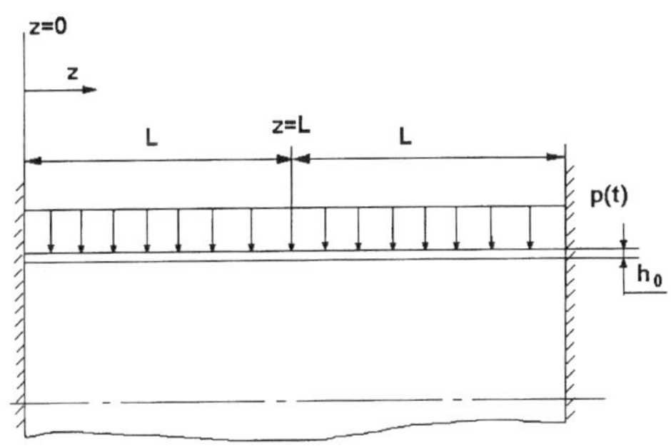

FIGURE 7. Schematic of the tube reduction on a mandrel with circular groove rectangular profile.

The axisymmetrical model was applied to tube forming analysis. The first example is a process of the copper tube reduction into circular grooves with rectangular profile. The schematic of the process is shown in Fig. 7. The ends of the tube were assumed as clamped. The electromagnetic pressure pulse was approximated as the following function of time t:

$$p(t) = p_0 e^{-\beta t} \sin^2 \omega t, \tag{3.1}$$

where $\omega = 63800\,1/\text{sec}$, $\beta = 25900\,1/\text{sec}$. Experimental and theoretical shapes of the tube generator for different pulse amplitudes are shown in Fig. 8. Numerical results and experimental data for tube maximum final deflection varying the length of the groove or thickness of the tube are given in Fig. 9. Analysis of the axial strength of tubular nodes, obtained by preliminary reduction of tubes into circular grooves was done applying stepwise axial load. The plastic strain distribution from the reduction process was taken into account in this analysis. To damp the vibrations the artificial viscosity was involved in calculations. The practical application was done for high-strength steel tubes with internal diameter 24 mm, thickness 1mm and circular groove length $L = 5$ mm. The groove depth was varied from 0.3 to

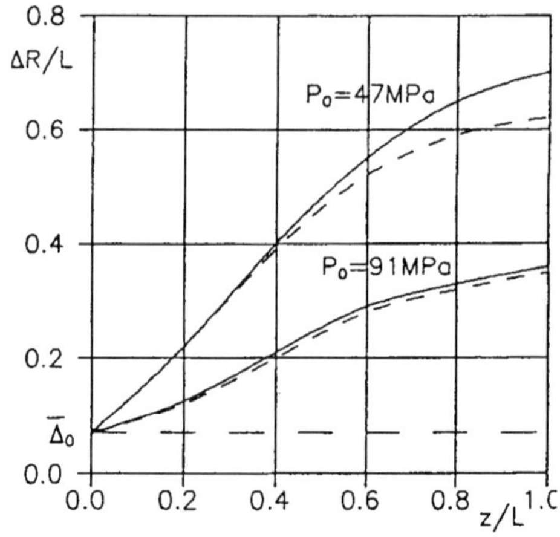

FIGURE 8. Theoretical (solid lines) and experimental (broken lines) final shapes of the tube generator for the different pressure pulse amplitudes.

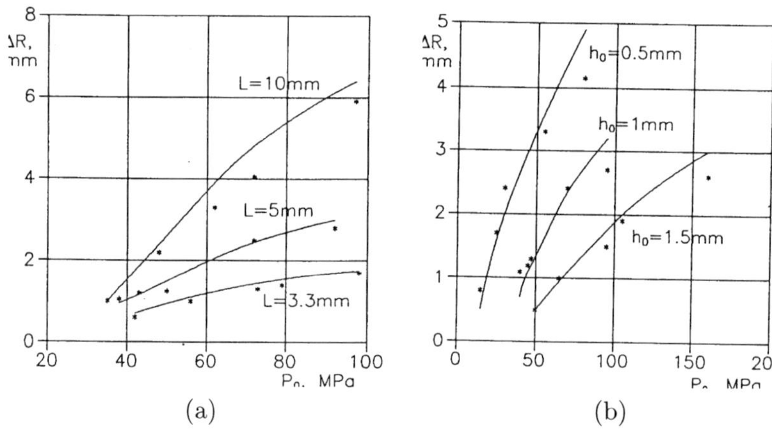

FIGURE 9. Theoretical predictions (solid lines) and experimental data (stars) of tube maximum final deflection depending on pressure pulse amplitude for different groove length (a) and tube thicknesses (b).

1.8 mm. In the case of tensile axial load the carrying capacity of the node grew as the groove depth was increased both in the theoretical analysis and experiments. Compressive loading showed different results. For small depth of the groove, the carrying capacity increased with its growth. In this case, the node was losing its carrying capacity being expanded along the distant from the axial load edge of the groove. For some depth value, the mechanism was changing—the tube was losing its stability and circular fold was growing from the close to the applied force edge of the groove. Further increasing of the groove depth resulted in decreasing of the node carrying capacity.

4. Conclusions

A simple numerical procedure based on the finite difference method with natural approximation of derivatives was applied to solid mechanics equations of motion solution employing Timoshenko theory of shells hypotheses. Experimental validation of the numerical results for the transverse loading of rectangular plates and axisymmetrical high-rate forming of tubes showed reasonable agreement for engineering practice accuracy. It was able to predict axial compressive force for preliminary deformed tube to lose its axial stability and to press out of the circular groove.

References

1. E. J. Bruno (ed.), *High velocity forming of metals*, American Society of Tool and Manufacturing Engineers, Dearborn, Michigan, 1996.
2. N. Jones, T. O. Uran, and S. A. Tekin, *The dynamic plastic behavior of fully clamped rectangular plates*, Internat. J. Solids Structures **6** (1970), no. 12, 1499–1512.
3. S. P. Timoshenko and J. N. Goodier, *Theory of elasticity*, McGraw-Hill, New York, 1970.
4. R. A.Vasin, V. S. Lenskiy, and E. V. Lenskiy, *Dynamic relationships between stresses and strains*, Problems of Dynamics of Elasto-Plastic Continuum, Mir, Moscow, 1975, pp. 7-38.
5. M. L. Wilkins, *Calculation of elasto-plastic flow*, Methods of Computational Physics (B. Adler, S. Fernback and M. Rotenberg, eds.) (New York, 1964), vol.3, Academic Press, New York, 1964.

BAUMAN MOSCOW STATE TECHNICAL UNIV., DIVISION MT-6, 2ND BAUMANSKAYA STREET 5, MOSCOW, 107005, RUSSIA.
E-mail address: sgolovas@bmstu.ru

INSTITUTE OF THE PROBLEMS OF MECHANICAL ENGINEERING, ACAD. OF SCI. OF RUSSIA, BOLSHOJ PROSPECT, 61, ST.PETERSBURG, 199178, RUSSIA
E-mail address: Bess_@robotek.ru

Centre de Recherches Mathématiques
CRM Proceedings and Lecture Notes
Volume **21**, 1999

An Intrinsic Form of the Stress Resultant Geometrically Exact Shell Theory

Adnan Ibrahimbegović

ABSTRACT. A nonconventional derivation of the stress resultant shell theory is presented that crucially relies on the exact shell balance equations; Hence the resulting shell theory is referred to as the geometrically exact. An intrinsic, vector-form of the shell strain measures is derived making use of the virtual work principle, and the various component forms are proposed. A very useful by-product in terms of a linear shell theory is obtained by linearizing the present nonlinear shell theory about the reference configuration.

1. Introduction

Last several years have marked an intense development in the computational treatment of shell theory (e.g., see [**8, 9, 13, 16, 18, 37, 38, 33, 41**], among others), clarifying the issues relating to the choice of the nonlocking finite element interpolations and consistent linearization procedure. The classical theoretical foundations of the shell theory (e.g., see [**7, 27**] or [**28**]) have been extended in this development, through a novel capability to account for finite rotations (rotations unrestricted in size).

The main thrust of this work is directed towards establishing the link between the classical and the novel developments. The most important among the results obtained concern an intrinsic, vector-form of the strain measures for the nonlinear shell theory and a novel form of the linear shell theory.

The main idea exploited in this work is to first find a set of stress resultants, which satisfy the shell balance equations and then deduce the corresponding strain measures through the applications of the virtual work principle. Since the shell balance equations are derived by taking into account the true shell geometry (e.g., see [**1, 32**]), the shell theory is referred to as *geometrically exact*. This idea is similar to the one first proposed by [**31**] for derivation of a geometrically exact theory of two-dimensional beams. In that respect the present work can be considered a proper extension to shells, with the major difficulty being related to the presence of the three-dimensional finite rotations.

The proposed procedure stands in sharp contrast with a more traditional one which invariably starts with the considerations of the shell surface kinematics in

---

1991 *Mathematics Subject Classification.* Primary: 73K15; Secondary: 73G05.
This is the final form of the paper.

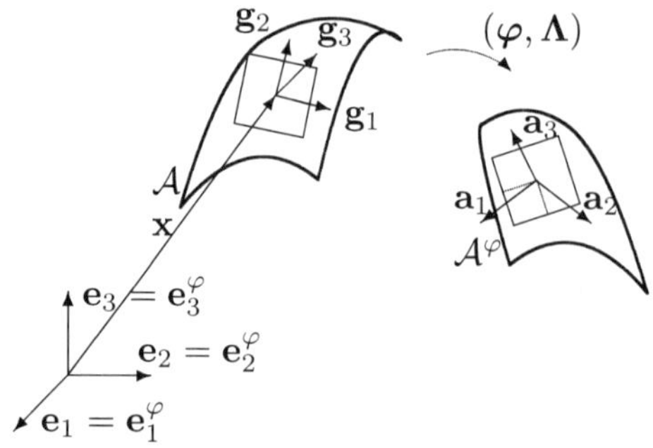

FIGURE 1. Shell reference and current configuration with the base vectors.

order to obtain the corresponding strain measures and then deduce the stress resultants by integrating through the thickness. Such an approach is utilized not only in the classical expositions on the subject (e.g., see [**27**]) but also in more recent developments (e.g., see [**36**]).

Despite the fact that the present development is quite different from the classical shell theory derivations, we can show that the component form of the proposed intrinsic strain measures in the convected coordinate system, remarkably enough, corresponds to the classical result. However, another component form of the intrinsic strain measures, obtained by using the rotated coordinate system, can also be interesting for practical applications in that it offers a natural framework for handling so-called drilling rotation component (e.g., see [**16**]).

A very interesting by-product of the present development in the nonlinear shell theory is a corresponding form of the geometrically linear shallow shell theory. The latter is obtained simply by performing the consistent linearization of the presented nonlinear theory at the shell reference configuration, which can be assumed as shallow in order to simplify the subsequent finite element implementation. It is shown that this non-conventional approach not only clarifies often confusing previous attempts to derive the linear shell theory of this kind, but also leads to a novel result on proper displacement-rotation coupling due to the shallow shell geometry.

2. Shell Theory for Finite Rotations

2.1. Shell kinematics: Virtual displacement and rotation. We consider a shell as a two-dimensional surface embedded in the Euclidean space (see Fig. 1).

The position vector for a point in the shell reference configuration is

$$\mathbf{x} = x_i \mathbf{e}_i, \tag{2.1}$$

where x_i are Euclidean coordinates and $\mathbf{e}_i$ are the base vectors (see Fig. 1). We further consider that a vector, so-called director, is attached to each point in the shell. It proves convenient for treatment of finite rotations to construct an orthonormal triad $\mathbf{g}_i$ with the director being its member $\mathbf{g}_3$. Without loss of generality one can assume that the director $\mathbf{g}_3$ is initially along the normal to the shell undeformed surface, thus base vectors $\mathbf{g}_1$ and $\mathbf{g}_2$ are in the shell tangent plane. The coordinates of vectors in such a frame are called the normal coordinates (e.g., see [**28**]). A convenient way to set the base vectors $\mathbf{g}_i$ is simply by rotating the global

Euclidean base vectors $\mathbf{e}_i$ as

$$\mathbf{g}_i = \mathbf{\Lambda}_0 \mathbf{e}_i, \tag{2.2}$$

where $\mathbf{\Lambda}_0$ is an orthogonal tensor, a known function of the reference configuration.

With the preceding notation in hand a placement of the shell in $\Re^3$ is assumed to be defined as the set

$$\mathcal{S} := \{\mathbf{x}^* \in \Re^3 \mid \mathbf{x}^* = \mathbf{x} + \zeta \mathbf{g}_3; \zeta \in [h^-, h^+]\}, \tag{2.3}$$

where $h = h^+ - h^-$ is the thickness of the shell.

Each material point in the shell reference configuration is assigned a new position vector in the deformed configuration

$$\mathbf{x}^\varphi = \boldsymbol{\varphi}(\mathbf{x}) = \varphi_i \mathbf{e}_i^\varphi, \tag{2.4}$$

where $\mathbf{e}_i^\varphi$ are the base vectors of the Euclidean coordinate system used for the current configuration. Although one typically chooses two Euclidean systems for shell reference and current configurations to coincide, i.e. $\mathbf{e}_i^\varphi = \delta_{ij}\mathbf{e}_j$ with δ_{ij} as the Kronecker symbol, in the foregoing we keep the different symbols for the two sets of the base vectors in order to provide a clear distinction between the material and spatial objects.

We rely on the generalized Kirchhoff hypothesis along with the classical assumption of inextensibility of the director to describe the motion of the director as

$$\mathbf{a}_i = \Lambda \mathbf{g}_i. \tag{2.5}$$

Therefore, by the virtue of (2.4) and (2.5), the placement of the *deformed* shell in $\Re^3$ is described as

$$\mathcal{S}^\varphi := \{\mathbf{x}^{\varphi *} \in \Re^3 \mid \mathbf{x}^{\varphi *} = \boldsymbol{\varphi} + \zeta \mathbf{a}_3; \zeta \in [h^-, h^+]\}. \tag{2.6}$$

We consider that the transformation in (2.5) producing a new position of the director, is isometric and that the director $\mathbf{a}_3$ has no other restrictions except for being a unit vector, which is in sharp contrast with the classical Kirchhoff-Love kinematic hypothesis where the director remains normal to the shell deformed surface. The main advantage of the former kinematic hypothesis with respect to the latter is in separating the displacement and the rotation field in the shell configuration space given as

$$\mathcal{C} := \{\boldsymbol{\phi} = (\boldsymbol{\varphi}, \mathbf{\Lambda}) \mid \boldsymbol{\phi} : \mathcal{A} \mapsto \Re^3 \times \mathrm{SO}(3)\}, \tag{2.7}$$

where SO(3) is the group of the proper orthogonal transformations

$$\mathrm{SO}(3) := \{\mathbf{\Lambda} \colon \Re^3 \mapsto \Re^3 \mid \mathbf{\Lambda}^T = \mathbf{\Lambda}^{-1}, \det \mathbf{\Lambda} = +1\}. \tag{2.8}$$

Virtual rotations, $\delta\mathbf{W}$, are defined as the infinitesimal rotations superposed on the shell deformed configuration

$$\delta \mathbf{W} = \frac{d}{d\varepsilon}\bigg|_{\varepsilon=0} [\mathbf{\Lambda}_\varepsilon]. \tag{2.9}$$

In constructing the admissible variations of finite rotations, $\mathbf{\Lambda}_\varepsilon$, one should take into account the proper rotation sequence in the multiplicative update (e.g., see [**2**]) and the Rodrigues formula (e.g., see [**12**]) for exponential mapping of infinitesimal rotations, which yields

$$\mathbf{\Lambda}_\varepsilon = \exp(\varepsilon \delta \mathbf{W}) \mathbf{\Lambda}. \tag{2.10}$$

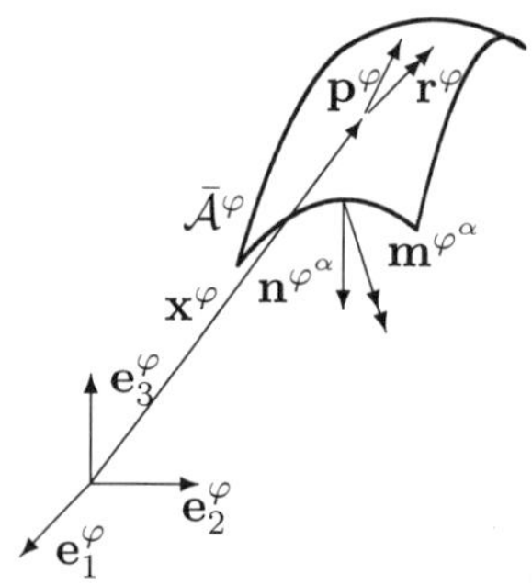

FIGURE 2. Stress resultants and couples.

The skew-symmetric tensor of infinitesimal rotations $\delta\mathbf{W}$ can also be represented by its axial vector $\delta\mathbf{w}$,

$$\delta\mathbf{W}\mathbf{b} = \delta\mathbf{w} \times \mathbf{b}, \quad \forall \mathbf{b} \in \Re^3. \tag{2.11}$$

The virtual displacements, $\delta\boldsymbol{\varphi}$, are defined in a standard manner as the small displacements superposed onto the deformed configuration with

$$\delta\boldsymbol{\varphi} = \frac{d}{d\varepsilon}\bigg|_{\varepsilon=0} [\boldsymbol{\varphi}_\varepsilon]; \quad \boldsymbol{\varphi}_\varepsilon = \boldsymbol{\varphi} + \varepsilon\delta\boldsymbol{\varphi}. \tag{2.12}$$

In accordance with (2.8), and presented considerations the tangent configuration space of the shell can be defined as

$$TC := \big\{\delta\boldsymbol{\phi} = (\delta\boldsymbol{\varphi}, \delta\mathbf{w}) \mid \delta\boldsymbol{\phi}\colon \mathcal{A} \mapsto \Re^3 \times T_\Lambda \, \mathrm{SO}(3)\big\}. \tag{2.13}$$

2.2. Geometrically exact local balance equations. Consider a connected region in the shell reference (undeformed) configuration $\bar{\mathcal{A}} \subset \mathcal{A}$, with a (piecewise) smooth boundary $\partial\bar{\mathcal{A}}$ and the unit normal vector $\boldsymbol{\upsilon} = \upsilon_\alpha \mathbf{g}^\alpha$. In the current (deformed) configuration, the corresponding shell mid-surface, its boundary and the unit normal vector are denoted, respectively, with $\bar{\mathcal{A}}^\varphi$, $\partial\bar{\mathcal{A}}^\varphi$ and $\boldsymbol{\upsilon}^\varphi$. The stress resultants $\mathbf{n}^{\varphi^\alpha}$ and couples $\mathbf{m}^{\varphi^\alpha}$ are acting in the deformed configuration along with the external force $\mathbf{p}^\varphi$ and external couple $\mathbf{r}^\varphi$ (see Fig. 2).

Following Budiansky [**7**], we introduce the change of coordinates with

$$\mathbf{p} = \mathbf{p}^\varphi \frac{d\mathcal{A}^\varphi}{d\mathcal{A}}, \quad \mathbf{r} = \mathbf{r}^\varphi \frac{d\mathcal{A}^\varphi}{d\mathcal{A}}, \quad \mathbf{n}^\alpha = \mathbf{n}^{\varphi\alpha} \frac{\upsilon_\alpha^\varphi ds^\varphi}{\upsilon_\alpha ds}, \quad \mathbf{m}^\alpha = \mathbf{m}^{\varphi\alpha} \frac{\upsilon_\alpha^\varphi ds^\varphi}{\upsilon_\alpha ds}. \tag{2.14}$$

In the above, $\mathbf{n}^\alpha$ and $\mathbf{m}^\alpha$ are stress resultants and couples still acting in the current configuration but parameterized in the reference configuration. (The transformation (2.14) is a shell-like equivalent to the transformation from Cauchy to Piola-Kirchhoff stress vector.) Integral form of the balance equations for linear and angular momentum balance can then be written as

$$\int_{\partial\bar{\mathcal{A}}} \mathbf{n}^\alpha \upsilon_\alpha \, ds + \int_{\bar{\mathcal{A}}} \mathbf{p} \, d\mathcal{A} = \mathbf{0}, \tag{2.15}$$

and

$$\int_{\partial\bar{\mathcal{A}}} (\boldsymbol{\varphi} \times \mathbf{n}^\alpha + \mathbf{m}^\alpha)\upsilon_\alpha \, ds + \int_{\bar{\mathcal{A}}} (\boldsymbol{\varphi} \times \mathbf{p} + \mathbf{m}) \, d\mathcal{A} = \mathbf{0}. \tag{2.16}$$

Using the standard tools, the divergence theorem for the surfaces and the localization theorem, we get the *local* form of shell balance equations as

$$\frac{1}{j}(j\mathbf{n}^\alpha)_{,\alpha} + \mathbf{p} = \mathbf{0}, \tag{2.17}$$

$$\boldsymbol{\varphi}_{,\alpha} \times \mathbf{n}^\alpha + \frac{1}{j}(j\mathbf{m}^\alpha)_{,\alpha} + \mathbf{r} = \mathbf{0}, \tag{2.18}$$

where j is the surface Jacobian. Note that the balance equations are derived with no approximation with respect to the shell geometry—hence they are often referred to as the *geometrically exact* (e.g., see [**1**]). Contrary to the classical expositions on the subject (e.g., see [**27**], and references therein), there is no explicit appearance of the Christoffel symbols in the local shell balance equations (2.17) and (2.18); they are eliminated by making use of the identity $\Gamma^\beta_{\beta\alpha} = j^{-1} j_{,\alpha}$ (see [**34**]).

Selecting the set of normal coordinates and the local Cartesian frames, the shell balance equation can further be simplified, and the partial derivatives with respect to the surface coordinates s_α can be easily constructed by making use of the isoparametric representation of the shell surface as given with:

PROPOSITION 2.1. *Partial derivatives with respect to normal coordinates*

$$\begin{pmatrix} \frac{\partial}{\partial s_1} \\ \frac{\partial}{\partial s_2} \end{pmatrix} = \begin{bmatrix} \mathbf{x}_{,\xi_1} \cdot \mathbf{g}_1 & \mathbf{x}_{,\xi_1} \cdot \mathbf{g}_2 \\ \mathbf{x}_{,\xi_2} \cdot \mathbf{g}_1 & \mathbf{x}_{,\xi_2} \cdot \mathbf{g}_2 \end{bmatrix}^{-1} \begin{pmatrix} \frac{\partial}{\partial \xi_1} \\ \frac{\partial}{\partial \xi_2} \end{pmatrix}. \tag{2.19}$$

PROOF. Exploiting underlying structure of the Euclidean space, the isoparametric surface representation, and the chain rule we can write

$$\left\langle \frac{\partial}{\partial \xi_1}, \frac{\partial}{\partial \xi_2} \right\rangle^T = [\mathbf{x}_{,\xi_1}, \mathbf{x}_{,\xi_2}]^T \left\langle \frac{\partial}{\partial x_1}, \frac{\partial}{\partial x_2}, \frac{\partial}{\partial x_3} \right\rangle^T,$$

where ξ_α are natural coordinates (e.g., obtained by the finite element representation of the shell surface), and x_i are Euclidean coordinates. In addition, the partial derivatives with respect to the chosen set of normal coordinates can be obtained by making use of the local Cartesian base vectors (see Fig. 1)

$$\left\langle \frac{\partial}{\partial x_1}, \frac{\partial}{\partial x_2}, \frac{\partial}{\partial x_3} \right\rangle^T = \left([\mathbf{g}_1, \mathbf{g}_2, \mathbf{g}_3]^T\right)^{-1} \left\langle \frac{\partial}{\partial s_1}, \frac{\partial}{\partial s_2}, \frac{\partial}{\partial s_3} \right\rangle^T,$$

and, since

$$\left([\mathbf{g}_1, \mathbf{g}_2, \mathbf{g}_3]^T\right)^{-1} = [\mathbf{g}_1, \mathbf{g}_2, \mathbf{g}_3] =: \boldsymbol{\Lambda}_0,$$

by combining the last two expressions we can obtain the result stated in (2.19). □

2.3. Strain measures. Having defined the balance equations with the stress resultants and couples, we would like to get the corresponding energy-conjugate strain measures. The crucial idea used to derive the strain measures goes back to Reissner and co-workers (e.g., see [**31, 32, 35**]) and exploits the virtual work principle. Namely, by the virtue of the fact that the stress resultants $\mathbf{n}^\alpha$ and couples $\mathbf{m}^\alpha$ satisfy the shell balance equations, they also satisfy the virtual work principle for the shell. It holds

$$\begin{aligned} \int_{\mathcal{A}} (\delta\boldsymbol{\varphi} \cdot \mathbf{p} + \delta\mathbf{w} \cdot \mathbf{r})\, d\mathcal{A} + \int_{\partial_n \mathcal{A}} \delta\boldsymbol{\varphi} \cdot \overline{\mathbf{n}}_\upsilon \, ds + \int_{\partial_m \mathcal{A}} \delta\mathbf{w} \cdot \overline{\mathbf{m}}_\upsilon \, ds \\ = \int_{\mathcal{A}} (\delta\boldsymbol{\varepsilon}_\alpha \cdot \mathbf{n}^\alpha + \delta\boldsymbol{\kappa}_\alpha \cdot \mathbf{m}^\alpha)\, d\mathcal{A}, \end{aligned} \tag{2.20}$$

where $\delta\boldsymbol{\varepsilon}_\alpha$ and $\delta\boldsymbol{\kappa}_\alpha$ are virtual strain measures, and $\overline{\mathbf{n}}_\upsilon$ and $\overline{\mathbf{m}}_\upsilon$ are prescribed traction forces and couples. Making use of the local balance equations to express the external force and couple

$$\mathbf{p} = -\mathbf{n}^\alpha_{,\alpha}, \tag{2.21}$$

$$\mathbf{r} = -\boldsymbol{\varphi}_{,\alpha} \times \mathbf{n}^\alpha - \mathbf{m}^\alpha_{,\alpha}, \tag{2.22}$$

and the two-dimensional form of the Cauchy principle to express the traction forces

$$\overline{\mathbf{n}}_\upsilon = \mathbf{n}^\alpha \upsilon_\alpha, \quad \overline{\mathbf{m}}_\upsilon = \mathbf{m}^\alpha \upsilon_\alpha, \tag{2.23}$$

the left hand side of (2.20) can be transformed and the virtual work equations can be rewritten as

$$\begin{aligned} \int_{\mathcal{A}} \{(\delta\boldsymbol{\varphi}_{,\alpha} - \delta\mathbf{w} \times \boldsymbol{\varphi}_{,\alpha}) \cdot \mathbf{n}^\alpha + \delta\mathbf{w}_{,\alpha} \cdot \mathbf{m}^\alpha\} \, d\mathcal{A} \\ = \int_{\mathcal{A}} (\delta\boldsymbol{\varepsilon}_\alpha \cdot \mathbf{n}^\alpha + \delta\boldsymbol{\kappa}_\alpha \cdot \mathbf{m}^\alpha) \, d\mathcal{A}. \end{aligned} \tag{2.24}$$

Considering $\mathbf{n}^\alpha$ and $\mathbf{m}^\alpha$ to be arbitrary functions, from (2.24) we get

$$\delta\boldsymbol{\varepsilon}_\alpha = \delta\boldsymbol{\varphi}_{,\alpha} - \delta\mathbf{w} \times \boldsymbol{\varphi}_{,\alpha}, \tag{2.25}$$

and

$$\delta\boldsymbol{\kappa}_\alpha = \delta\mathbf{w}_{,\alpha}, \tag{2.26}$$

as the explicit forms of the virtual strain measures in terms of the virtual displacements and rotations. We can make the forms of two expressions for the virtual strains somewhat more similar. First, by using the results in (2.5) and (2.10), we can compute the variation of the base vectors $\mathbf{a}_i$ as

$$\begin{aligned} \delta\mathbf{a}_i &= \delta\mathbf{W}\mathbf{a}_i \\ &= \delta\mathbf{w} \times \mathbf{a}_i. \end{aligned} \tag{2.27}$$

Second, by taking the derivative of the base vectors $\mathbf{a}_i$ with respect to the shell surface coordinates (see Proposition 2.1), we get

$$\begin{aligned} \mathbf{a}_{i,\alpha} &= \boldsymbol{\Lambda}_{,\alpha}\boldsymbol{\Lambda}^T\mathbf{a}_i + \boldsymbol{\Lambda}\boldsymbol{\Lambda}_{0,\alpha}\boldsymbol{\Lambda}_0^T\boldsymbol{\Lambda}^T\mathbf{a}_i \\ &= [\boldsymbol{\Omega}_\alpha + \boldsymbol{\Lambda}\boldsymbol{\Omega}_{0\alpha}\boldsymbol{\Lambda}^T]\mathbf{a}_i. \end{aligned} \tag{2.28}$$

Since $\boldsymbol{\Omega}_\alpha$ and $\boldsymbol{\Omega}_{0\alpha}$ are skew-symmetric tensors, the last expression can also be rewritten in terms of the corresponding axial vectors $\boldsymbol{\omega}_\alpha$ and $\boldsymbol{\omega}_{0\alpha}$ as

$$\begin{aligned} \mathbf{a}_{i,\alpha} &= (\boldsymbol{\omega}_\alpha + \boldsymbol{\Lambda}\boldsymbol{\omega}_{0\alpha}) \times \mathbf{a}_i \\ &= \bar{\boldsymbol{\omega}}_\alpha \times \mathbf{a}_i. \end{aligned} \tag{2.29}$$

Finally, by taking into account that the derivation and variation being linear operators commute, from (2.27) and (2.29) we can get

$$\delta\bar{\boldsymbol{\omega}}_\alpha \times \mathbf{a}_i + (\bar{\boldsymbol{\omega}}_\alpha \times \delta\mathbf{w}) \times \mathbf{a}_i = \delta\mathbf{w}_{,\alpha} \times \mathbf{a}_i, \tag{2.30}$$

which compared to the result in (2.26) leads to an alternative expression for the bending strain measure

$$\delta\boldsymbol{\kappa}_\alpha = \delta\bar{\boldsymbol{\omega}}_\alpha - \delta\mathbf{w} \times \bar{\boldsymbol{\omega}}_\alpha. \tag{2.31}$$

By exploiting result in (2.27), the virtual strain measures in (2.25) can be rewritten in equivalent form as

$$\delta\boldsymbol{\varepsilon}_\alpha = \delta\boldsymbol{\varphi}_{,\alpha} - \delta\mathbf{w} \times \boldsymbol{\varphi}_{,\alpha} - (\delta\mathbf{a}_\alpha - \delta\mathbf{w} \times \mathbf{a}_\alpha). \tag{2.32}$$

Similarly, making use of (2.29) the virtual bending strains in (2.31) can be rewritten as

$$\delta\boldsymbol{\kappa}_\alpha = \delta\boldsymbol{\omega}_\alpha - \delta\mathbf{w} \times \boldsymbol{\omega}_\alpha. \tag{2.33}$$

As shown shortly afterwards, these two modifications of the strain measures amount to the elimination of the rigid body modes.

The last two expressions for the virtual strains $\delta\boldsymbol{\varepsilon}_\alpha$ and $\delta\boldsymbol{\kappa}_\alpha$ can formally be written as

$$\delta\boldsymbol{\varepsilon}_\alpha = \boldsymbol{\Lambda}\frac{d}{d\varepsilon}\bigg|_{\varepsilon=0}\left[\boldsymbol{\Lambda}_\varepsilon^T(\boldsymbol{\varphi}_{\varepsilon,\alpha} - \mathbf{a}_{\alpha\,\varepsilon})\right], \tag{2.34}$$

and

$$\delta\boldsymbol{\kappa}_\alpha = \boldsymbol{\Lambda}\frac{d}{d\varepsilon}\bigg|_{\varepsilon=0}\left[\boldsymbol{\Lambda}_\varepsilon^T(\boldsymbol{\omega}_{\alpha\,\varepsilon})\right], \tag{2.35}$$

which corresponds to a special form of the Lie derivative (e.g., see [**25**]) of the spatial vector field[1], where the pull-back and push-forward in this case are performed by the two-point tensor $\boldsymbol{\Lambda}$. From (2.34) and (2.35) it follows immediately that

PROPOSITION 2.2. *Shell strain measures (objective spatial vector fields)*

$$\boldsymbol{\varepsilon}_\alpha = \boldsymbol{\varphi}_{,\alpha} - \mathbf{a}_\alpha, \tag{2.36}$$

and

$$\boldsymbol{\kappa}_\alpha = \boldsymbol{\omega}_\alpha. \tag{2.37}$$

PROOF. We first show that the expression in (2.36) and (2.37) are indeed the strain measures. Let the shell motion be a rigid body motion of the form

$$\boldsymbol{\varphi}^{\mathrm{rbm}} = \mathbf{Q}\mathbf{x} + \mathbf{c}, \quad \boldsymbol{\Lambda}^{\mathrm{rbm}} = \mathbf{Q},$$

where $\mathbf{Q}$ is a (constant) orthogonal tensor representing the rigid body rotation, while $\mathbf{c}$ is a (constant) vector of rigid body translation. It is easy to verify that both strain measure are indeed equal to zero, i.e.

$$\begin{aligned}\boldsymbol{\varepsilon}_\alpha^{\mathrm{rbm}} &= \mathbf{Q}\mathbf{x}_{,\alpha} - \mathbf{a}_\alpha \\ &= \mathbf{Q}(\mathbf{g}_\alpha - \mathbf{g}_\alpha) = \mathbf{0},\end{aligned}$$

and

$$\boldsymbol{\Omega}_\alpha^{\mathrm{rbm}} = \mathbf{Q}_{,\alpha}\mathbf{Q}^T = \mathbf{0} \implies \boldsymbol{\kappa}_\alpha^{\mathrm{rbm}} = \boldsymbol{\omega}_\alpha^{\mathrm{rbm}} = \mathbf{0}.$$

We can also show that the proposed strain measures are objective spatial vector fields. Consider the superposed rigid body motion introduced at the deformed configuration, $\boldsymbol{\phi} = (\boldsymbol{\varphi}, \boldsymbol{\Lambda})$, to be of the form

$$\boldsymbol{\varphi}^{\mathrm{srbm}} = \mathbf{Q}\boldsymbol{\varphi} + \mathbf{c}, \quad \boldsymbol{\Lambda}^{\mathrm{srbm}} = \mathbf{Q}\boldsymbol{\Lambda},$$

The corresponding strain measures can be computed as

$$\begin{aligned}\boldsymbol{\varepsilon}_\alpha^{\mathrm{srbm}} &= \boldsymbol{\varphi}_{,\alpha}^{\mathrm{srbm}} - \mathbf{a}_\alpha^{\mathrm{srbm}} \\ &= \mathbf{Q}(\boldsymbol{\varphi}_{,\alpha} - \mathbf{a}_\alpha) \\ &= \mathbf{Q}\boldsymbol{\varepsilon}_\alpha,\end{aligned}$$

[1]Using the Lie derivative formalism in order to obtain the virtual strain measures, the real strain measures are pulled-back to the reference configuration, their admissible variations are obtained and the result is pushed-forward onto the current configuration.

which is the proper form of the transformation for an objective spatial vector field (e.g., see [**14**]).

Similarly, we compute skew-symmetric tensor form of the bending strain measures

$$\begin{aligned}\boldsymbol{\Omega}_\alpha^{\text{srbm}} &= (\mathbf{Q}\boldsymbol{\Lambda})_{,\alpha}(\mathbf{Q}\boldsymbol{\Lambda})^T \\ &= \mathbf{Q}\boldsymbol{\Lambda}_{,\alpha}\boldsymbol{\Lambda}^T\mathbf{Q}^T \\ &= \mathbf{Q}\boldsymbol{\Omega}_\alpha\mathbf{Q}^T,\end{aligned}$$

from where, by using the result on similarity transformation of the corresponding axial vector (e.g., see [**12**]), we further obtain

$$\begin{aligned}\boldsymbol{\kappa}_\alpha^{\text{srbm}} &= \boldsymbol{\omega}_\alpha^{\text{srbm}} \\ &= \mathbf{Q}\boldsymbol{\omega}_\alpha \\ &= \mathbf{Q}\boldsymbol{\kappa}_\alpha,\end{aligned} \tag{2.38}$$

which concludes the proof. □

2.4. Component form of strain measures. In this section we take a closer look at different component forms of the strain measures in (2.36) and (2.37). If the strain measures are pictured in the local Cartesian basis $\mathbf{a}_i$ we obtain

$$\begin{aligned}\varepsilon_{\alpha\beta} &= \mathbf{a}_\alpha \cdot \boldsymbol{\varphi}_{,\beta} - \mathbf{a}_\alpha \cdot \mathbf{a}_\beta \\ &= \mathbf{a}_\alpha \cdot \boldsymbol{\varphi}_{,\beta} - \mathbf{g}_\alpha \cdot \mathbf{g}_\beta,\end{aligned} \tag{2.39}$$

$$\begin{aligned}\gamma_\alpha &= \mathbf{a}_3 \cdot \boldsymbol{\varphi}_{,\alpha} - \mathbf{a}_3 \cdot \mathbf{a}_\alpha \\ &= \mathbf{a}_3 \cdot \boldsymbol{\varphi}_{,\alpha} - \mathbf{g}_3 \cdot \mathbf{g}_\alpha,\end{aligned} \tag{2.40}$$

and

$$\begin{aligned}\kappa_{\alpha\beta} &= \mathbf{a}_\alpha \cdot \boldsymbol{\omega}_\beta \\ &= \mathbf{a}_\alpha \cdot (\bar{\boldsymbol{\omega}}_\beta - \boldsymbol{\Lambda}\boldsymbol{\omega}_{0\beta}) \\ &= \mathbf{a}_\alpha \cdot \bar{\boldsymbol{\omega}}_\beta - \mathbf{g}_\alpha \cdot \boldsymbol{\omega}_{0\beta},\end{aligned} \tag{2.41}$$

for membrane, shear and bending strain measure components, respectively.

In shell theory, it proves convenient to change the numbering of the bending strain components by introducing

$$\hat{\kappa}_{\alpha\gamma} = e_{\alpha\beta}\kappa_{\beta\gamma}, \tag{2.42}$$

where $e_{\alpha\beta}$ is a two-dimensional alternating tensor. With this notation and (2.29) on hand, we can rewrite the expression in (2.41) for bending strain measures as

$$\hat{\kappa}_{\alpha\beta} = \mathbf{a}_\alpha \cdot \mathbf{a}_{3,\beta} - \mathbf{g}_\alpha \cdot \mathbf{g}_{3,\beta}. \tag{2.43}$$

The graphical representation of these strain measures is given in Fig. 3.

The choice of the local Cartesian coordinates with the base vector $\mathbf{a}_i$ stands in sharp contrast with the typically used convected coordinates of the classical shell theory (e.g., see [**27**]), with base vectors $\boldsymbol{\varphi}_{,\alpha}$ and $\mathbf{a}_3$. (Only in the absence of the shear deformation that these two sets of base vectors become the same.)

It is interesting to note that the component form of the strain measures for the classical shell theory (e.g., see [**7**] or [**27**]) can be recovered as the consistent

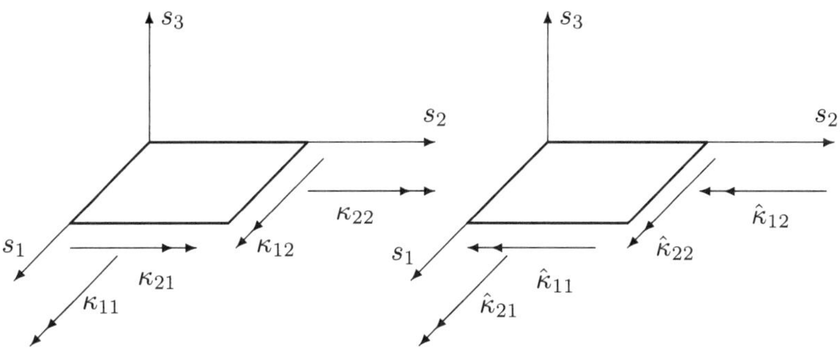

FIGURE 3. Sign convention of shell bending deformation components.

projection of the vector-forms in (2.36) and (2.37) to the convected base vectors. Hence, we obtain

(2.44) $$\tilde{\varepsilon}_{\alpha\beta} = \boldsymbol{\varphi}_{,\alpha} \cdot \boldsymbol{\varphi}_{,\beta} - \mathbf{g}_\alpha \cdot \mathbf{g}_\beta,$$

(2.45) $$\tilde{\gamma}_\alpha = \boldsymbol{\varphi}_{,\alpha} \cdot \mathbf{a}_3 - \mathbf{g}_\alpha \cdot \mathbf{g}_3,$$

and

(2.46) $$\tilde{\kappa}_{\alpha\beta} = \boldsymbol{\varphi}_{,\alpha} \cdot \mathbf{a}_{3,\beta} - \mathbf{g}_\alpha \cdot \mathbf{g}_{3,\beta},$$

as the classical membrane, shear and bending strain measures.

We find the last observation quite remarkable, since the procedure used herein to obtain the strain measures of the classical shell theory is completely different from the usual one. Namely, in most classical (e.g., see [**7, 27**]) and more recent expositions (e.g., see [**28, 36**]), one starts directly with a consideration of kinematics (bending and stretching) of the shell surface to obtain the strain measures. One so obtains in particular the membrane and bending strain measures precisely as those in (2.44) and (2.46), given, respectively, as the difference of the first and second fundamental forms of the shell surface in the deformed and undeformed configuration.

Therefore, the vector forms established in (2.36) and (2.37) represent intrinsic strain measures valid in both shell theories. The only difference stems from the choice of the base vectors: rotated basis for the former and convected basis for the latter shell theory. Interestingly enough, the shear strain components are the same in both shell theories (see expressions (2.40) and (2.45)), since the unit vector $\mathbf{a}_3$ is an element of both vector bases.

2.5. Finite element implementation aspects. Choosing the appropriate finite interpolations for shells is far from trivial. Having established early on (e.g., see [**42**]) that the standard isoparametric interpolations lead to strong locking problem, numerous methods have been proposed as a cure (e.g., see [**29**] and reference therein).

The main idea we would like to communicate in this section is that the performance of a suitable finite element interpolation can be further enhanced when coupled with a rigorous theoretical formulation. The model problem chosen for consideration is the assumed shear strain method (e.g., see [**11, 15, 22**] or [**5**]), which is meant to alleviate the shear locking problems. It was shown in [**17**] that using the standard assumed shear strain field with the presented shell theory and the careful treatment of the numerical implementation details, we can improve upon

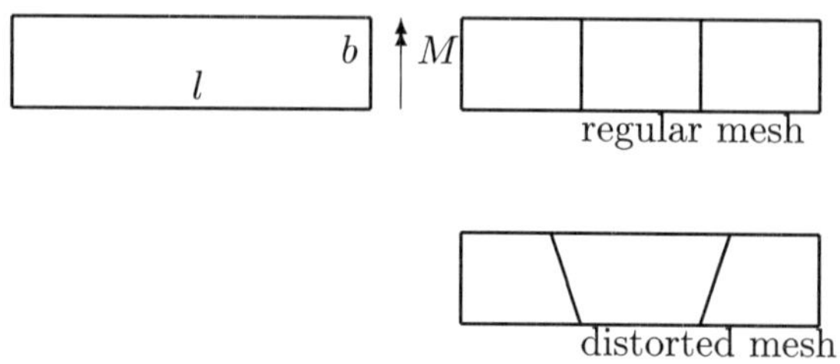

FIGURE 4. Cantilever beam with free-end moment. Regular and distorted three-element mesh.

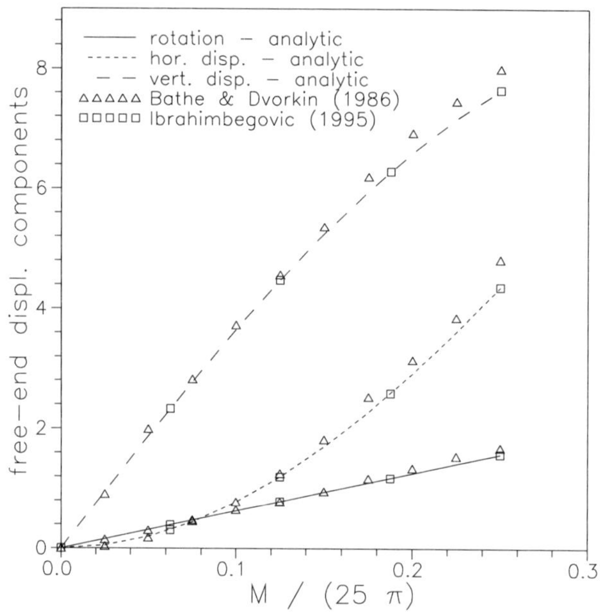

FIGURE 5. Free-end displacement components and free-end rotations for the cantilever beam obtained with 3-element regular mesh.

the previously reported results. An illustrative numerical result, which corroborates this statement, concerns the bending test problem (see Fig. 4) proposed by Dvorkin and Bathe [**11**] in order to display an excellent performance of their 4-node shell element. It is readily apparent from Fig. 5 and Table 1 that with the proposed theory these already excellent results can be improved even further.

TABLE 1. Free-end displ. for distorted mesh.

Result / Disp. Comp.	Hor.	Ver.	Rot.
Dvorkin and Bathe [**11**]	0.813	3.849	37.8^o
Ibrahimbegović [**17**]	1.154	4.287	43.4^o
Analytic	1.196	4.475	45.0^o

3. Geometrically Linear Shell theory

Early works on nonlinear analysis of shells were mainly directed towards extending the formulation obtained in the linear analysis; typical developments of this kind are: the updated Lagrangian formulation (e.g., see [**4**] and references therein), the corotational formulation (e.g., see [**40**]) and the natural approach of Argyris and co-workers (see [**3**]). We note in passing that most of these developments are inherently restricted to low-order shell finite element interpolations (e.g., either 3-node or 4-node shell element), and that it is not at all obvious how to generalize them to higher-order elements. (It should be noted in that respect, that the direct development of the nonlinear shell theory presented in the previous sections, by no means suffers from similar deficiencies.)

In this section we describe a theoretical development based on the idea which is entirely opposite to the one just described. Namely, having developed a nonlinear shell theory we then proceed to obtain its geometrically linear version. This is carried out simply by performing the consistent linearization of the nonlinear shell theory at the reference configuration.

In that respect we first note that the consistent linearization of the finite rotation tensor $\mathbf{\Lambda}$ leads to a skew-symmetric tensor $\mathbf{\Theta}$ of infinitesimal rotations, i.e.

$$\mathbf{\Lambda}(\boldsymbol{\theta}) \overset{\text{lin}}{\mapsto} \mathbf{I} + \mathbf{\Theta}. \tag{3.1}$$

The linearized bending strain measures can then be computed as

$$\begin{aligned} \boldsymbol{\Omega}_\alpha &= \mathbf{\Lambda}_{,\alpha}\mathbf{\Lambda}^T \overset{\text{lin}}{\mapsto} \mathbf{\Theta}_{,\alpha}(\mathbf{I}-\mathbf{\Theta}) \approx \mathbf{\Theta}_{,\alpha}, \\ \boldsymbol{\omega}_\alpha &\overset{\text{lin}}{\mapsto} \boldsymbol{\theta}_{,\alpha} = \boldsymbol{\kappa}^{\text{lin}}. \end{aligned} \tag{3.2}$$

The further linearization of the remaining strain measures is influenced by the chosen shallow shell representation of the shell surface geometry[2].

In the shallow shell approximation, we can assume that the shell is obtained from a corresponding plate by a local isometric transformation, $\mathbf{\Lambda}_0$. Within the order of approximation used for shallow shells (see [**20, 24**]) we can write

$$\mathbf{\Lambda}_0 \overset{\text{lin}}{\mapsto} \mathbf{I} + \mathbf{\Theta}_0, \quad \mathbf{\Theta}_0 = \begin{bmatrix} 0 & 0 & f_{,1} \\ 0 & 0 & f_{,2} \\ -f_{,1} & -f_{,2} & 0 \end{bmatrix}, \tag{3.3}$$

where $x_3 = f(x_1, x_2)$ represents a functional form of the shallow shell reference configuration (see Fig. 6).

The linearized form of the base vectors can than be readily computed as

$$\mathbf{a}_i^{lin} = \mathbf{\Lambda}^{lin}\mathbf{\Lambda}_0^{lin}\mathbf{e}_i. \tag{3.4}$$

Similarly, the linearized form of the position vector can be obtained as

$$\boldsymbol{\varphi} \overset{\text{lin}}{\mapsto} \begin{pmatrix} x_1 + u_1 \\ x_2 + u_2 \\ f + u_3 \end{pmatrix}, \tag{3.5}$$

[2] In the finite element setting, the shallow shell approximation of the reference geometry is introduced over each shell element, which, along with the use of the element local Cartesian coordinates, ensures the convergence to the deep shell solutions (see [**21**])

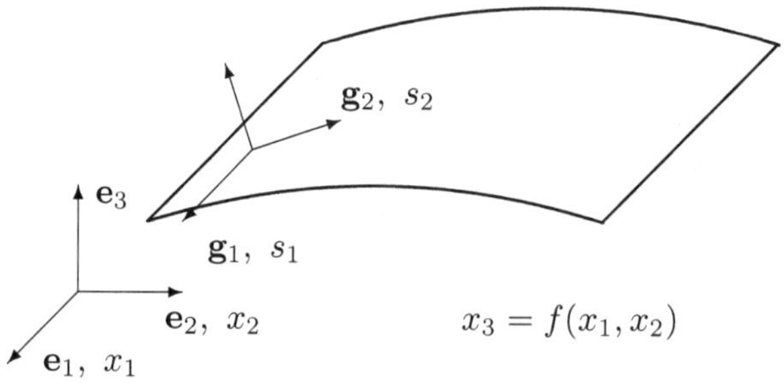

FIGURE 6. Shallow shell approximation of the reference configuration.

which, along with the expression in (3.4), leads to the linearized vector form of the strain measures as

$$\varepsilon_1^{\text{lin}} = \begin{pmatrix} u_{1,1} - \theta_2 f_{,1} \\ u_{2,1} - \theta_3 + \theta_1 f_{,1} \\ u_{3,1} + \theta_2 \end{pmatrix}, \quad \varepsilon_2^{\text{lin}} = \begin{pmatrix} u_{1,2} + \theta_3 - \theta_2 f_{,2} \\ u_{2,2} + \theta_1 f_{,2} \\ u_{3,2} - \theta_1 \end{pmatrix}. \tag{3.6}$$

Renumbering the curvature components as in (2.42) and changing the sign convention for the rotation components to

$$\hat{\theta}_\alpha = e_{\alpha\beta}\theta_\beta, \quad \psi = \theta_3, \tag{3.7}$$

will not affect the linearized bending strains in (3.2), whereas the remaining strain measures (3.6) can be rewritten as

$$\varepsilon_1^{\text{lin}} = \begin{pmatrix} u_{1,1} + \hat{\theta}_1 f_{,1} \\ u_{2,1} + \hat{\theta}_2 f_{,1} - \psi \\ u_{3,1} - \hat{\theta}_1 \end{pmatrix}, \quad \varepsilon_2^{\text{lin}} = \begin{pmatrix} u_{1,2} + \hat{\theta}_1 f_{,2} + \psi \\ u_{2,2} + \hat{\theta}_2 f_{,2} \\ u_{3,2} - \hat{\theta}_2 \end{pmatrix}. \tag{3.8}$$

With these results on hand, the symmetric parts of the membrane and bending strain measures can be written, respectively, as

$$\varepsilon_{(\alpha\beta)}^{\text{lin}} = \frac{1}{2}(u_{\alpha,\beta} + u_{\beta,\alpha}) + \frac{1}{2}(\hat{\theta}_\alpha f_{,\beta} + \hat{\theta}_\beta f_{,\alpha}), \tag{3.9}$$

and

$$\hat{\kappa}_{(\alpha\beta)}^{\text{lin}} = \frac{1}{2}(\hat{\theta}_{\alpha,\beta} + \hat{\theta}_{\beta,\alpha}). \tag{3.10}$$

Similarly, the shear strains are given as

$$\gamma_\alpha^{\text{lin}} = u_{3,\alpha} - \hat{\theta}_\alpha. \tag{3.11}$$

Note that both the bending and the shear strain measures are those of the Reissner [**30**]—Mindlin [**26**] plate theory. Hence, when searching for an optimal finite element interpolation scheme one can directly benefit from already accumulated, very rich experience regarding this problem (e.g., see [**22, 15**], or [**5**]).

The expression for the membrane strains is different from the one usually encountered in a number of previous works (e.g., see [**6, 10**]) in that the rotation field $\hat{\theta}_\alpha$ rather than the derivatives of the lateral displacement are coupled with the in-plane displacements through the shallow shell approximation of the warped element. The letter simplifies the construction of nonlocking finite element interpolations, since it implies that the problems of membrane and shear locking would appear uncoupled. This enables an excellent performance of the shell element (see [**19**]) over the complete set of the standard test problems for linear analysis of shells (e.g., see [**23**] or [**6**]).

Acknowledgements

This work was supported by "Pôle modélisation numérique de Picardie".

4. Conclusions

One of the major breakthroughs of the recent research work on nonlinear shells theory concerns the capability to account for finite rotations. This work bears upon the theoretical foundations of such a shell theory, its geometrically linear version and its relationship with the classical shell theory. Some of the most important findings in that respect, can be summarized as follows:

(i) An intrinsic, vector-form of shell strain measures is identified and different component forms are obtained including the classical results on the membrane and bending strain measures, which make use of the first and second fundamental forms of the shell surface. This finding is even more remarkable considering that the proposed procedure to derive the shell theory departs from exact shell balance equations rather than the usual considerations of the shell kinematics.

(ii) Consistent linearization of the proposed nonlinear shell theory is used to obtain the corresponding linear shell theory. It is shown that in such a way we not only eliminate a number of often confusing hypotheses (as alluded to in [**39**]), but also obtain a novel form of the shell theory which is more suitable for the finite element implementation.

(iii) The issues of the theoretical formulation and finite element implementation are tightly coupled in shells. It is thus not surprising that, as shown in this work, the performance of some well-known methods for constructing the nonlocking finite element interpolations can be enhanced when used with the presented theoretical framework.

References

1. S. S. Antman, *Nonlinear problems in elasticity*, Springer, Berlin, 1995.
2. J. H. Argyris, *An excursion into large rotations*, ENOMECH 81, Part I (Stuttgart, 1981), Comput. Methods Appl. Mech. Engrg. **32** (1982), 85–155.
3. J. H. Argyris, H. Balmer, J. St. Doltsinis, P. C. Dunne, M. Haase, G. A. Malejannakis, H. P. Mlejnek, and D. W. Scharp, *Finite element method—The natural approach*, Comput. Methods Appl. Mech. Engrg. **17–18** (1979), 1–106.
4. K. J. Bathe, *Finite element procedures*, Prentice Hall Englewood Cliffs, N.J., 1995.
5. K. J. Bathe and E. N. Dvorkin, *A four-node plate bending element based on Mindlin-Reissner plate theory and a mixed interpolation*, Internat. J. Numer. Methods Engrg. **21** (1985), 367–383.
6. T. Belytschko and I. Leviathan, *Projection schemes for one-point quadrature shell elements*, Comput. Methods Appl. Mech. Engrg. **115** (1994), 277–286.
7. B. Budiansky, *Notes on nonlinear shell theory*, J. Appl. Mech. **35** (1968), 393–401.
8. N. Buechter and E. Ramm, *Shell theory versus degeneration—A comparison in large rotation finite element analysis*, Internat. J. Numer. Methods Engrg. **34** (1992), 39–59.
9. J. Chroscielewski, J. Makowski, and H. Stumpf, *Genuinely resultant shell finite elements for geometric and material nonlinearity*, Internat. J. Numer Methods Engrg. **35** (1992), 63–94.

10. M. A. Crisfield, *Nonlinear finite element analysis of solids and structures. I. Essentials*, John Wiley, London, 1991.
11. E. N. Dvorkin and K. J. Bathe, *A continuum mechanics based four-node shell element for general nonlinear analysis*, Engrg. Comput. **1** (1984), 77–88.
12. H. Goldstein, *Classical mechanics*, Addison-Wesley, Reading, 1980.
13. F. Gruttmann, E. Stein, and P. Wriggers, *Theory and numerics of thin elastic shells with finite rotations*, Ing.. Archives **59** (1989), 54–67.
14. M. E. Gurtin, *An introduction to continuum mechanics*, Academic Press, New York, 1981.
15. T. J. R. Hughes and T. E. Tezduyar, *Finite elements based upon mindlin plate theory with particular reference to the four-node bilinear isoparametric element*, J. Appl. Mech. **46** (1981), 587–596.
16. A. Ibrahimbegović, *Stress resultant geometrically nonlinear shell theory with drilling rotations. I. A consistent formulation*, Comput. Methods Appl. Mech. Engrg. **118** (1994), 265–284.
17. ______, *On assumed shear strain in finite rotation shell analysis*, Engrg. Comput. **12** (1995), 425–438.
18. A. Ibrahimbegović and F. Frey, *Stress resultant geometrically nonlinear shell theory with drilling rotations. II. Computational aspects*, Comput. Methods Appl. Mech. Engrg. **118** (1994), 285–308.
19. ______, *Stress resultant geometrically nonlinear shell theory with drilling rotations. III. Linearized kinematics*, Internat. J. Numer. Methods Engrg. **37** (1994), 3659–3683.
20. A. Ibrahimbegović, F. Frey, G. Fonder, and Ch. Massonnet, *A variational formulation for shallow shells finite elements in the* 1990'*s.* (E. Onate *et al.*, eds.), Springer-Verlag, Berlin, 1991, pp. 68–80.
21. S. Idelsohn, *On the use of deep, shallow or flat shell finite elements for the analysis of thin shell structures*, Comput. Methods Appl. Mech. Engrg. **26** (1981), 321–330.
22. R. MacNeal, *Derivation of element stiffness matrices by assumed strain distribution*, Nuclear Engrg. Des. **70** (1983), 3–12.
23. R. MacNeal and R. L. Harder, *A proposed standard set of problems to test finite element accuracy*, Finite Elem. Anal. Des. **1** (1985), 3–20.
24. K. Marguerre, *Zur Theorie der Gekrümmten Platte Grosser Formånderung*, Proc. 5th Internat. Conf. Appl. Mech., 1938, pp. 93-101.
25. J. E. Marsden and T. J. R. Hughes, *Mathematical foundations of elasticity Prentice-Hall*, Englewoods Cliffs, N.J., 1983.
26. R. D. Mindlin, *Influence of rotatory inertia and shear in flexural motion of isotropic elastic plates*, J. Appl. Mech. **18** (1951), 31–38.
27. P. M. Naghdi, *The theory on shells and plates encyclopedia of physics*, (S. Flugge, ed.), Springer-Verlag, Berlin, 1972.
28. F. I. Niordson, *Shell theory*, North-Holland, Amsterdam, 1985.
29. A. K. Noor, T. Belytschko, and J. C. Simo, *Analytical and computational models for shells*, ASME Publication CED-vol. 3, 1989.
30. E. Reissner, *The effect of transverse shear deformation on the bending of elastic plates*, J. Appl. Mech. **13** (1946), 69–76.
31. ______, *On one-dimensional finite strain beam theory: The plane problem*, J. Appl. Math. Phys. **23** (1972), 795–804.
32. ______, *Linear and nonlinear theory of shells*, Thin Shell Structures: Theory, Experiment and Design (Y. C. Fung and E. E. Sechler, eds.), Princeton-Hall, N.J., 1974, pp. 29–44.
33. C. Sansour and H. Bufler, *An exact finite rotation shell theory, its mixed variational formulation and its finite element implementation*, Internat. J. Numer. Methods Engrg. **34** (1992), 73–115.
34. J. G. Simmonds, *A brief on tensor analysis*, Springer-Verlag, Berlin, 1982.
35. J. G. Simmonds and D. A. Danielson, *Nonlinear shell theory with finite rotation and stress-function vectors*, J. Appl. Mech. **39** (1972), 1085–1090.
36. J. C. Simo and D. D. Fox, *On a stress resultants geometrically exact shell model.I. Formulation and optimal parameterization*, Comput. Methods Appl. Mech. Engrg. **72** (1989), 267–304.

37. J. C. Simo, D. D. Fox, and M. S. Rifai, *On a stress resultants geometrically exact shell model. II. The linear theory; computational aspects*, Comput. Methods Appl. Mech. Engrg. **73** (1989), no. 1, 53–92.
38. ———, *On a stress resultants geometrically exact shell model. III. Computational aspects of the nonlinear theory*, Comput. Methods Appl. Mech. Engrg. **79** (1990), no. 1, 117–164.
39. A. Tessler, *A C^o anisoparametric three-node shell element*, Comput. Methods Appl. Mech. Engrg. **78** (1990), 89–103.
40. G. Wempner, *Finite elements, finite rotations and small strains of flexible shells*, Internat. J. Solids Structures **5** (1969), 117–153.
41. P. Wrigger and F. Gruttmann, *Thin shells with finite rotations in biot stresses: Theory and FE implementation*, Internat. J. Numer. Methods Engrg. **36** (1993), 2049–2072.
42. O. C. Zienkiewicz, J. Too, and R. L. Taylor, *Reduced integration technique in general analysis of plates and shells*, Internat. J. Numer. Methods Engrg. **3** (1971), 275–290.

COMPIÈGNE UNIVERSITY OF TECHNOLOGY, DEPARTMENT OF MECHANICAL ENGINEERING, DIVISION M.N.M., LAB. G2MS, CNRS URA 1505, 60206 COMPIÈGNE, FRANCE
E-mail address: Adnan@mx.Ibrahimbegovic@utc.fr

Centre de Recherches Mathématiques
CRM Proceedings and Lecture Notes
Volume **21**, 1999

Numerical Inf-Sup Analysis of MITC Plate Bending Elements

Alexander Iosilevich, Klaus-Jürgen Bathe, and Franco Brezzi

ABSTRACT. This paper focuses on the numerical evaluation of the inf-sup condition for MITC plate bending elements. Since no analytical results exist whether these elements satisfy this fundamental condition for stability and optimality of a mixed finite element formulation, we develop a numerical procedure which allows to calculate the inf-sup value, and apply this procedure to two members of the family. We find that the elements pass the tests, which underlines their reliability for use in engineering practice.

1. Introduction

Many plate bending finite elements based on the Reissner-Mindlin theory have been proposed since the first application of finite element analysis to plate bending problems.

Using the Reissner-Mindlin model in the displacement-based procedure leads to the use of simple isoparametric interpolations, as employed in the two-dimensional linear elastic analysis, for the transverse displacement and section rotations. However, this approach results into a well-known difficulty: when the plate thickness becomes relatively small, the finite element discretization stiffens, resulting in almost zero displacements. This numerical phenomenon is referred to as (*shear*) *locking* in the literature [**1**].

Many elements have been designed to overcome this difficulty; however, almost all approaches lack a satisfactory mathematical analysis and were proposed solely on the basis of numerical experiments.

The MITC plate bending elements, however, show superior performance and these elements are supported by a strong mathematical foundation. The properties of the mixed-interpolated elements were deduced from very powerful elements used in incompressible elasticity [**2, 3**], and a mathematical basis was further provided in [**6**] and [**8**].

Nonetheless, still, we have no analytical results whether the elements satisfy the inf-sup condition [**5**], the crucial criterion for stability and optimality of mixed finite element formulations. Moreover, even if these results existed, it would be

1991 *Mathematics Subject Classification.* Primary: 73K10, 73V05; Secondary: 65N30, 65N12.

This is the final form of the paper.

almost impossible to extend the analysis to general discretizations with distorted elements.

In this paper we present a numerical test to identify whether the inf-sup condition is satisfied. Although we concentrate on the analysis of the MITC element family, the proposed procedure is quite general and can be applied to other formulations as well.

Our findings suggest that the elements of the MITC family indeed satisfy the inf-sup condition, and thus are optimal and reliable for practical use.

2. Problem Statement

2.1. Reissner-Mindlin plate bending model. In the plate bending model we consider a two-dimensional boundary-value problem, whose solution approximates the exact solution of a three-dimensional linear elasticity problem for a body

$$\mathcal{B} = \left\{ \underline{x} \in \mathbb{R}^3 \mid (x_1, x_2) \in \Omega \subset \mathbb{R}^2,\ x_3 \in \left]-\frac{t}{2}, \frac{t}{2}\right[, t \ll \operatorname{diam}(\Omega) \right\}$$

under the action of a transverse load $\underline{p} = \big(0, 0, f(x_1, x_2)\big)$.

The Reissner-Mindlin plate bending model is based on the assumption that a straight fibre normal to the undeformed plate mid-surface Ω remains straight but not necessarily normal during deformation. The equations governing the model are [**1**]:

1. Displacements

$$u_1 = -x_3\beta_1(x_1, x_2); \quad u_2 = -x_3\beta_2(x_1, x_2); \quad u_3 = w(x_1, x_2);$$

2. Strains (linear part of the strain tensor)

$$\varepsilon_{11} = -x_3 \frac{\partial \beta_1}{\partial x_1}; \quad \varepsilon_{22} = -x_3 \frac{\partial \beta_2}{\partial x_2}; \quad \varepsilon_{12} = -\frac{1}{2} x_3 \left(\frac{\partial \beta_1}{\partial x_2} + \frac{\partial \beta_2}{\partial x_1} \right);$$

$$\varepsilon_{13} = \frac{1}{2}\left(\frac{\partial w}{\partial x_1} - \beta_1 \right); \quad \varepsilon_{23} = \frac{1}{2}\left(\frac{\partial w}{\partial x_2} - \beta_2 \right);$$

3. Stresses (linear elastic isotropic material assumption)

$$\sigma_{11} = \frac{E}{1-\nu^2}(\varepsilon_{11} + \nu\varepsilon_{22}); \quad \sigma_{22} = \frac{E}{1-\nu^2}(\varepsilon_{22} + \nu\varepsilon_{11}); \quad \sigma_{33} = 0;$$

$$\sigma_{12} = 2\mu\varepsilon_{12}; \quad \sigma_{13} = 2\mu\varepsilon_{13}; \quad \sigma_{23} = 2\mu\varepsilon_{23},$$

where E is the Young's modulus, ν is the Poisson's ratio, and $\mu = E/\big[2(1+\nu)\big]$ stands for the shear modulus.

4. The variational form as a starting point for the finite element solution is based upon the principle of minimum potential energy [**1**]. Without loss of generality we consider the case of homogeneous boundary conditions, i.e., a plate clamped along the boundary $\partial\Omega$. Thus, we look for $\underline{u} = (\underline{\beta}, w) \in V = B \times W$, $B = \big[H_0^1(\Omega)\big]^2$, $W = H_0^1(\Omega)$, such that

$$a(\underline{\beta}, \underline{\eta}) + \frac{\mu k}{t^2}(\nabla w - \underline{\beta}, \nabla \xi - \underline{\eta}) = \frac{1}{t^3}(f, \xi) \quad \forall \underline{v} = (\underline{\eta}, \xi) \in V, \tag{2.1}$$

where

$$a(\underline{\eta}_1, \underline{\eta}_2) = \frac{E}{12(1-\nu^2)} \int_\Omega \big[(1-\nu)\underline{\underline{\varepsilon}}(\underline{\eta}_1) : \underline{\underline{\varepsilon}}(\underline{\eta}_2) + \nu(\nabla \cdot \underline{\eta}_1)(\nabla \cdot \underline{\eta}_2)\big]\, d\Omega$$

is a symmetric bilinear form; $\underline{\varepsilon}(\eta) = (\partial \eta_i / \partial x_j)\underline{e}_i\underline{e}_j$, $i, j = 1, 2$, is a linear strain operator; $\nabla = (\partial / \partial x_i)\underline{e}_i$, $i = 1, 2$; $(\cdot,\cdot)$ is the $L^2(\Omega)$-inner product; t is the plate thickness; and k is the shear correction factor, usually $k = 5/6$.

Introducing the shear term

$$\underline{\gamma} = \frac{\mu k}{t^2}(\nabla w - \underline{\beta}) \tag{2.2}$$

and rescaling the applied transverse load to $g = f/t^3$, we can equivalently rewrite (2.1) as:

Find $\underline{u} = (\underline{\beta}, w) \in V$ *and* $\underline{\gamma} \in G$, *such that*:

$$\begin{cases} a(\underline{\beta}, \underline{\eta}) + (\underline{\gamma}, \nabla \xi - \underline{\eta}) = (g, \xi) & \forall \underline{v} = (\underline{\eta}, \xi) \in V, \\ \left(\frac{t^2}{\mu k}\underline{\gamma}, \underline{\zeta}\right) = (\nabla w - \underline{\beta}, \underline{\zeta}) & \forall \underline{\zeta} \in G, \end{cases} \tag{2.3}$$

where G is a functional space that should be chosen appropriately.

For any finite thickness we have $\underline{\gamma} \in L^2(\Omega)$. However, as $t \to 0$ the corresponding norm $\|\underline{\gamma}\|_{L^2(\Omega)}$ is no longer bounded from above, and in the limit case $t = 0$ the appropriate space for the shear term becomes [**2**]:

$$\begin{gathered} \Gamma' = H^{-1}(\text{div}; \Omega) = \{\underline{\zeta}, \underline{\zeta} \in [H^{-1}(\Omega)]^2, \nabla \cdot \underline{\zeta} \in H^{-1}(\Omega)\}, \\ \|\underline{\zeta}\|^2_{H^{-1}(\text{div};\Omega)} = \|\underline{\zeta}\|^2_{H^{-1}(\Omega)} + \|\nabla \cdot \underline{\zeta}\|^2_{H^{-1}(\Omega)}. \end{gathered} \tag{2.4}$$

Thus, we can rewrite (2.3) for the case of zero thickness as the following *limit problem*:

Find $\underline{u} = (\underline{\beta}, w) \in V$ *and* $\underline{\gamma} \in G = \Gamma'$, *such that*:

$$\begin{cases} a(\underline{\beta}, \underline{\eta}) + (\underline{\gamma}, \nabla \xi - \underline{\eta}) = (g, \xi) & \forall \underline{v} = (\underline{\eta}, \xi) \in V, \\ (\nabla w - \underline{\beta}, \underline{\zeta}) = 0 & \forall \underline{\zeta} \in \Gamma'. \end{cases} \tag{2.5}$$

2.2. Finite element approximation. We specify finite-dimensional spaces for the transverse displacement and rotations, $W_h \subset W$ and $B_h \subset B$, restricting our analysis to the use of conforming approximations. Using these spaces with equation (2.1) we arrive at the *discrete variational problem*:

Find $\underline{u}_h = (\underline{\beta}_h, w_h) \in V_h = B_h \times W_h$, *such that*

$$a(\underline{\beta}_h, \underline{\eta}_h) + \frac{\mu k}{t^2}(\nabla w_h - \underline{\beta}_h, \nabla \xi_h - \underline{\eta}_h) = (g, \xi_h) \quad \forall \underline{v}_h = (\underline{\eta}_h, \xi_h) \in V_h. \tag{2.6}$$

The difficulty to obtain good convergence with this variational formulation to the exact solution lies in that the space V_h must have a sufficient number of elements which are able to satisfy the Kirchhoff constraint $\nabla w_h = \underline{\beta}_h$. This condition of zero shear strain is a serious restriction and when not satisfied introduces spurious shear strain energy in the numerical solution. As the plate thickness decreases, this strain energy error becomes large and leads to much too small displacements. This purely numerical phenomenon is known as (*shear*) *locking*.

The general approach to remedy this situation is to introduce a *reduction operator* R_h into the variational formulation, so that (2.6) becomes:

Find $\underline{u}_h = (\underline{\beta}_h, w_h) \in V_h = B_h \times W_h$, *such that* $\forall \underline{v}_h = (\underline{\eta}_h, \xi_h) \in V_h$

$$a(\underline{\beta}_h, \underline{\eta}_h) + \frac{\mu k}{t^2}(R_h(\nabla w_h - \underline{\beta}_h), R_h(\nabla \xi_h - \underline{\eta}_h)) = (g, \xi_h). \tag{2.7}$$

If $R_h = I$ (identity operator), we have the *displacement-based approach*; other possible choices include *reduced/selective integration*, and *mixed* and *hybrid formulations* [**1**].

To proceed with the mixed interpolation, we introduce a discrete version of (2.2),

$$\underline{\gamma}_h = \frac{\mu k}{t^2} R_h(\nabla w_h - \underline{\beta}_h),$$

and rewrite (2.7) to obtain an equivalent discrete variational formulation:

Find $\underline{u}_h = (\underline{\beta}_h, w_h) \in V_h$ *and* $\underline{\gamma}_h \in G_h = R_h(\nabla W_h \oplus B_h)$, *such that:*

$$\text{(2.8)} \qquad \begin{cases} a(\underline{\beta}_h, \underline{\eta}_h) + (\underline{\gamma}_h, R_h(\nabla \xi_h - \underline{\eta}_h)) = (g, \xi_h) & \forall \underline{v}_h = (\underline{\eta}_h, \xi_h) \in V_h, \\ \left(\frac{t^2}{\mu k}\underline{\gamma}_h, \underline{\zeta}_h\right) = (R_h(\nabla w_h - \underline{\beta}_h), \underline{\zeta}_h) & \forall \underline{\zeta}_h \in G_h. \end{cases}$$

In the limit case $t = 0$, (2.8) becomes:

$$\text{(2.9)} \qquad \begin{cases} a(\underline{\beta}_h, \underline{\eta}_h) + (\underline{\gamma}_h, R_h(\nabla \xi_h - \underline{\eta}_h)) = (g, \xi_h) & \forall \underline{v}_h = (\underline{\eta}_h, \xi_h) \in V_h, \\ (R_h(\nabla w_h - \underline{\beta}_h), \underline{\zeta}_h) = 0 & \forall \underline{\zeta}_h \in G_h. \end{cases}$$

Since the limit problem (2.9) can serve as the most severe test whether the given choice of the discrete spaces B_h and W_h and the reduction operator R_h exhibit the locking behavior, we will further concentrate on the analysis of the system (2.9).

2.3. The ellipticity and the inf-sup conditions. The variational problem (2.9) represents a typical case of the well-studied saddle-point problem (see e.g. [**5, 7**]), and the corresponding conditions for solvability and stability of the system are given by:

(i) The *ellipticity condition*: $\exists \alpha > 0$, such that $\forall \underline{v}_h = (\underline{\eta}_h, \xi_h) \in V_{0h}$

$$\text{(2.10)} \qquad a(\underline{\eta}_h, \underline{\eta}_h) \geq \alpha \|\underline{v}_h\|_V^2 = \alpha\left[\|\underline{\eta}_h\|^2_{H^1(\Omega)} + \|\xi_h\|^2_{H^1(\Omega)}\right],$$

where

$$V_{0h} = \left\{\underline{v}_h = (\underline{\eta}_h, \xi_h) \in V_h \;\middle|\; (\underline{\zeta}_h, R_h(\nabla \xi_h - \underline{\eta}_h)) = 0 \;\forall \underline{\zeta}_h \in G_h\right\};$$

(ii) The inf-sup *condition*: $\exists k_0 > 0$, such that

$$\text{(2.11)} \qquad \inf_{\underline{\zeta}_h \in G_h} \sup_{\underline{v}_h = (\underline{\eta}_h, \xi_h) \in V_h} \frac{(R_h(\nabla \xi_h - \underline{\eta}_h), \underline{\zeta}_h)}{\|\underline{\zeta}_h\|_{\Gamma'} \|\underline{v}_h\|_V} \geq k_0.$$

Moreover, if (2.10) and (2.11) are satisfied, and R_h is a good enough approximation of the identity operator, then the element is optimally convergent (when used in uniform meshes) in the corresponding natural error norms [**7**], i.e.,

$$\text{(2.12)} \qquad \|\underline{v} - \underline{v}_h\|_V + \|\underline{\gamma} - \underline{\gamma}_h\|_{\Gamma'} \leq ch^m,$$

where h is a characteristic mesh size, and m is the order of the interpolation space V_h.

3. MITC Elements

The MITC elements were originally designed based on results available for incompressible elasticity (see [**2, 3, 6**]).

The following "algorithm" is used to construct the elements:

1. Start with a pair of interpolation spaces (B_h, Q_h) that is known to work for the Stokes problem, i.e., which satisfies the inf-sup condition

$$\inf_{q_h \in Q_h} \sup_{\underline{\eta}_h \in B_h} \frac{(q_h, \nabla \cdot \underline{\eta}_h)}{\|\underline{\eta}_h\|_B \|q_h\}_Q} \geq k_0 > 0. \tag{3.1}$$

   B_h will serve as a space for rotations, while $Q_h \subset L^2(\Omega)$ is an auxiliary space, which never appears in actual calculations.

2. Find another space $G_h = \Gamma_h$ as a discrete subspace of the functional space $\Gamma = \left[H^{-1}(\mathrm{div}; \Omega)\right]' = H_0(\mathrm{rot}; \Omega)$,

$$H_0(\mathrm{rot}; \Omega) = \left\{ \underline{\zeta} \mid \underline{\zeta} \in \left[L^2(\Omega)\right]^2, \mathrm{rot}\, \underline{\zeta} \in L^2(\Omega), \underline{\zeta} \cdot \underline{\tau} = 0 \text{ on } \partial\Omega \right\},$$

$$\|\underline{\zeta}\|^2_{H_0(\mathrm{rot};\Omega)} = \|\underline{\zeta}\|^2_{L^2(\Omega)} + \|\mathrm{rot}\, \underline{\zeta}\|^2_{L^2(\Omega)},$$

   where $\mathrm{rot}\, \underline{\zeta} = -(\partial \zeta_1 / \partial x_2) + (\partial \zeta_2 / \partial x_1)$, and a reduction operator R_h, such that the following diagram commutes:

$$\begin{array}{ccc} B_h & \xrightarrow{\ \mathrm{rot}\ } & L^2(\Omega) \\ {\scriptstyle R_h}\big\downarrow & & \big\downarrow{\scriptstyle P_h} \\ \Gamma_h & \xrightarrow[\ \mathrm{rot}\ ]{} & Q_h \end{array}$$

   where P_h is the $L^2(\Omega)$-projection operator.

3. Choose a space for the transverse displacement W_h to satisfy

$$\nabla W_h = \{\underline{\zeta}_h \in \Gamma_h, \mathrm{rot}\, \underline{\zeta}_h = 0\}.$$

Elements designed according to this algorithm can be shown to satisfy the ellipticity condition (2.10) and a weak form of the inf-sup condition (2.11) [**7**],

$$\inf_{\underline{\zeta}_h \in G_h} \sup_{\underline{v}_h = (\underline{\eta}_h, \xi_h) \in V_h} \frac{\left(R_h(\nabla \xi_h - \underline{\eta}_h), \underline{\zeta}_h\right)}{\|\underline{\zeta}_h\|_{\Gamma'_h} \|\underline{v}_h\|_V} \geq \beta > 0, \tag{3.2}$$

where

$$\|\underline{\zeta}_h\|_{\Gamma'_h} = \sup_{\underline{\chi}_h \in \Gamma_h} \frac{(\underline{\zeta}_h, \underline{\chi}_h)}{\|\underline{\chi}_h\|_\Gamma}, \tag{3.3}$$

with β being a constant independent of t and h, and be optimally convergent in the associated norms (when used in uniform meshes), namely

$$\|\underline{v} - \underline{v}_h\|_V + \|\underline{\gamma} - \underline{\gamma}_h\|_{\Gamma'_h} \leq ch^m. \tag{3.4}$$

We shall now refer to condition (3.2) as the inf-sup *condition in the* Γ'_h*-norm.*

REMARK 3.1. The MITC4 element is based on the well-known $Q_1 - P_0$ pair for incompressible analysis, which indeed does not satisfy the inf-sup condition (3.1). Moreover, no analytical results exist whether the element satisfies (3.2) and/or the error bound of type (3.4) is applicable.

In our analysis we will consider the MITC4 element as a special case, and the MITC9 element as a typical representative of the family. We define all spaces and operators over a reference 2×2 element K; generalizations are straightforwardly obtained via covariant transformations.

For a summary of the elements and numerical results we refer to [**1, 4**]; more on error and convergence analysis can be found in [**8, 10, 11**].

For the mixed-interpolated four-node element we use

$$\begin{aligned} W_h &= \{\xi_h \in H_0^1(\Omega), \xi_h|_K \in Q_1(K)\}, \\ B_h &= \{\underline{\eta}_h \in [H_0^1(\Omega)]^2, \underline{\eta}_h|_K \in [Q_1(K)]^2\}, \\ Q_h &= \{q_h : q_h|_K \in \mathbb{R}\}, \\ \Gamma_h &= \{\underline{\zeta} \in H_0(\mathrm{rot};\Omega), \underline{\zeta}|_K \in \{\mathrm{span}(1,s), \mathrm{span}(1,r)\}\}. \end{aligned}$$

The corresponding spaces for the MITC9 elements are defined as

$$\begin{aligned} W_h &= \{\xi_h \in H_0^1(\Omega), \xi_h|_K \in Q_2'(K)\}, \\ B_h &= \{\underline{\eta}_h \in [H_0^1(\Omega)]^2, \underline{\eta}_h|_K \in [Q_2(K)]^2\}, \\ Q_h &= \{q_h : q_h|_K \in P_1(K)\}, \\ \Gamma_h &= \{\underline{\zeta} \in H_0(\mathrm{rot};\Omega), \underline{\zeta}|_K \in \{\mathrm{span}(1,r,s,rs,s^2), \mathrm{span}(1,r,s,rs,r^2)\}\}, \end{aligned}$$

where Q_2' is the second-order Serendipity space [**1**].

4. Numerical Analysis

In this section we develop a test methodology, which allows to numerically check whether the inf-sup condition (2.11) holds for a particular finite element discretization. In the development we follow the guidelines of the numerical inf-sup test proposed by Chapelle and Bathe [**9**] for the problems of incompressible elasticity.

We start by observing that in order to prove the inf-sup condition (2.11) for a given discretization, it is sufficient to show that:

(a) The inf-sup condition in the Γ_h'-norm (3.2) is satisfied; and
(b) The following norm equivalence hold

$$\|\underline{\zeta}_h\|_{\Gamma_h'} \geq \rho \|\underline{\zeta}_h\|_{\Gamma'} \quad \forall \underline{\zeta}_h \in \Gamma_h, \tag{4.1}$$

with a constant ρ independent of the mesh size h.

Correspondingly, we develop two tests, which allow to check whether conditions (a) and (b) are satisfied for a given problem, and we apply these tests to the MITC4 and MITC9 elements, and to the displacement-based quadrilateral elements QUAD4 and QUAD9.

4.1. inf-sup test in the Γ_h'-norm. For the case of a bounded domain Ω, we have (by the Poincaré-Friedrichs inequality [**1**]) the following norm for the space of rotations and transverse displacement:

$$\begin{aligned} \|\underline{v}_h\|_V^2 \simeq |\underline{v}_h|_{H^1(\Omega)}^2 &= |\underline{\eta}_h|_{H^1(\Omega)}^2 + |\xi_h|_{H^1(\Omega)}^2 \\ &= L^2 \sum_{i,j=1}^{2} \left\| \frac{\partial \eta_{h,i}}{\partial x_j} \right\|_{L^2(\Omega)}^2 + \sum_{i=1}^{2} \left\| \frac{\partial \xi_h}{\partial x_i} \right\|_{L^2(\Omega)}^2 \\ &= \mathbf{V}^T \mathbf{S} \mathbf{V}, \end{aligned} \tag{4.2}$$

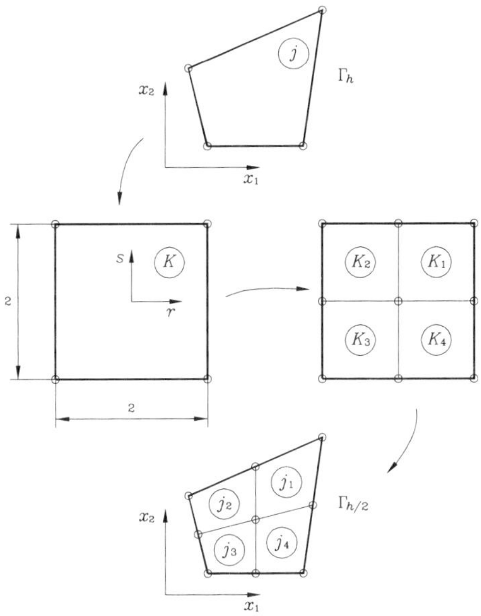

FIGURE 1. $\Gamma_h \rightarrow \Gamma_{h/2}$ mapping through a covariant transformation for a four-node element.

where L is a characteristic dimension of the plate, $\mathbf{V} \in \mathbb{R}^n$ is a vector of nodal point displacements, and $\mathbf{S} \in \mathbb{R}^n \times \mathbb{R}^n$ is a positive definite norm matrix.

The shear term is calculated from the nodal point displacements $\mathbf{U}$ using $\underline{\zeta}_h = \mathbf{B}_\gamma \mathbf{U}$. Defining $\mathbf{G} = (\mathbf{B}_\gamma^T, \mathbf{B}_\gamma)$, $\mathbf{G} \in \mathbb{R}^n \times \mathbb{R}^n$, we have $\big(\underline{\zeta}_h, R_h(\nabla \xi_h - \underline{\eta}_h)\big) = \mathbf{U}^T\mathbf{G}\mathbf{V}$. To avoid trivial solutions we require that $\mathbf{V}, \mathbf{U} \in \mathcal{D} = \mathbb{R}^n \setminus \ker(\mathbf{G})$, where $\ker(\mathbf{G})$ stands for the null space of matrix $\mathbf{G}$.

To calculate $\|\underline{\zeta}_h\|_{\Gamma_h'}$, we will use the norm duality

$$\|\underline{\zeta}_h\|_{\Gamma_h'} = \sup_{\underline{\chi}_h \in \Gamma_h} \frac{(\underline{\zeta}_h, \underline{\chi}_h)}{\|\underline{\chi}_h\|_\Gamma}. \tag{4.3}$$

Also, we use

$$\begin{aligned}\|\underline{\chi}_h\|_\Gamma^2 = \|\underline{\chi}_h\|^2_{H_0(\mathrm{rot};\Omega)} &= \|\underline{\chi}_h\|^2_{L^2(\Omega)} + L^2\|\operatorname{rot}\underline{\chi}_h\|^2_{L^2(\Omega)}\\ &= \mathbf{W}^T(\mathbf{G}+\mathbf{Q})\mathbf{W} = \mathbf{W}^T\mathbf{D}\mathbf{W},\end{aligned}$$

where $\mathbf{Q}$ is a symmetric positive semidefinite matrix. Thus we can rewrite (4.3) as

$$\|\underline{\zeta}_h\|_{\Gamma_h'} = \sup_{\mathbf{W}\in\mathcal{D}} \frac{\mathbf{U}^T\mathbf{G}\mathbf{W}}{\sqrt{\mathbf{W}^T\mathbf{D}\mathbf{W}}}. \tag{4.4}$$

Of course, the condition $\mathbf{W} \in \mathcal{D}$ is sufficient to guarantee that the seminorm in the denominator of (4.4) can be used as a norm[1]. Summarizing, the inequality (3.2) can be written in matrix form as

$$\inf_{\mathbf{U}\in\mathcal{D}} \sup_{\mathbf{V}\in\mathcal{D}} \frac{\mathbf{U}^T\mathbf{G}\mathbf{V}}{\sup_{\mathbf{W}\in\mathcal{D}}(\mathbf{U}^T\mathbf{G}\mathbf{W}/\sqrt{\mathbf{W}^T\mathbf{D}\mathbf{W}})\sqrt{\mathbf{V}^T\mathbf{S}\mathbf{V}}} \geq \beta. \tag{4.5}$$

[1]Indeed, we have $\mathbf{W}^T\mathbf{G}\mathbf{W} > 0\ \forall \mathbf{W} \in \mathcal{D}$, and $\mathbf{V}^T\mathbf{Q}\mathbf{V} \geq 0\ \forall \mathbf{V} \in \mathbb{R}^n$, thus $\mathbf{W}^T\mathbf{D}\mathbf{W} > 0$ $\forall \mathbf{W} \in \mathcal{D}$.

We show in Appendix A that a lower bound for β can be calculated as

$$\beta \geq \beta^* = \sqrt{\delta_{\min}/\sigma_{\max}}, \tag{4.6}$$

where $\delta_{\min}$ stands for the smallest nonzero eigenvalue of the matrix $\mathbf{D}$, and $\sigma_{\max}$ is the largest eigenvalue of $\mathbf{S}$.

To test an element, we calculate β^* for a sequence of meshes obtained by successive refinements and plot $\lg(\beta^*)$ vs. $\lg(1/N)$, where N stands for the square root of the total number of elements. If the value β^* converges to zero, the element is failing our inf-sup test to identify whether the inf-sup condition (3.2) is satisfied. Conversely, if β^* stops decreasing as the number of elements is increased, the test is passed.

4.2. From Γ'_h to Γ'. To approach the continuous space Γ', we perform a series of successive refinements, simply subdividing each element j, $j = 1, \ldots, N^2$ into four "subelements" (see Fig. 1).

This allows us to obtain a sequence of meshes, such that the spaces for shears satisfy $\Gamma_h \subseteq \Gamma_{h/2} \subseteq \cdots \subseteq \Gamma_{h/n} \subset \Gamma$, and, increasing the refinement, we can approach the continuous space Γ as closely as we want by our finite element discretization. Moreover, for $\underline{\zeta}_h \in \Gamma_h$, we have

$$\|\underline{\zeta}_h\|_{\Gamma'_{h/n}} = \sup_{\underline{\chi}_h \in \Gamma_{h/n}} \frac{(\underline{\zeta}_h, \underline{\chi}_h)}{\|\underline{\chi}_h\|_{\Gamma_{h/n}}} \xrightarrow[n\to\infty]{} \|\underline{\zeta}_h\|_{\Gamma'}. \tag{4.7}$$

Therefore, considering

$$\theta_n(h) = \inf_{\underline{\zeta}_h \in \Gamma_h} \frac{\|\underline{\zeta}_h\|_{\Gamma'_h}}{\|\underline{\zeta}_h\|_{\Gamma'_{h/n}}}, \tag{4.8}$$

we can find an estimate (for large n) for the constant

$$\theta_\infty(h) = \inf_{\underline{\zeta}_h \in \Gamma_h} \frac{\|\underline{\zeta}_h\|_{\Gamma'_h}}{\|\underline{\zeta}_h\|_{\Gamma'}}.$$

If the $\theta_\infty(h)$ does not converge to zero as $h \to 0$, that is,

$$\theta_\infty(h) \geq \overline{\theta} > 0, \tag{4.9}$$

with $\overline{\theta}$ independent of h, we can conclude that the Γ'_h and Γ' norms, for the finite element spaces under analysis, are in fact equivalent.

REMARK 4.1. In some cases it is possible to show that there exists a constant $\Theta^* > 0$ independent of h and the plate thickness, such that

$$\|\underline{\zeta}_h\|_{\Gamma'_{h/2}} \geq \Theta^* \|\underline{\zeta}_h\|_{\Gamma'}.$$

Then (4.9) will also be satisfied provided $\theta_2(h)$ is bounded from below by a positive constant independent of h. Appendix C presents such a proof for the MITC9 element.

Expressing (4.8) in terms of dual norms (4.3) and (4.7), we obtain

$$\theta_n(h) = \inf_{\underline{\zeta}_h \in \Gamma_h} \frac{\sup_{\underline{\chi}_h \in \Gamma_h} [(\underline{\zeta}_h, \underline{\chi}_h)/\|\underline{\chi}_h\|_{\Gamma_h}]}{\sup_{\underline{\kappa}_h \in \Gamma_{h/n}} [(\underline{\zeta}_h, \underline{\kappa}_h)/\|\underline{\kappa}_h\|_{\Gamma_{h/n}}]}. \tag{4.10}$$

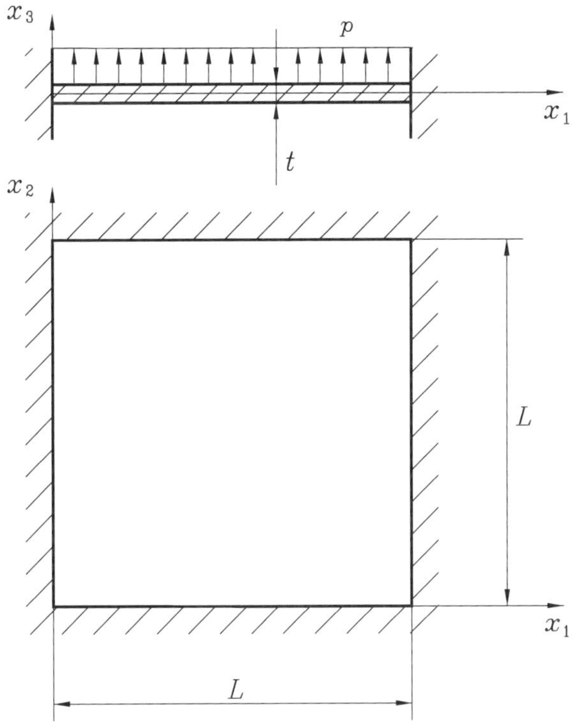

FIGURE 2. Clamped plate.

To represent equation (4.10) in matrix form, we use

$$\underline{\zeta}_h = \overline{\mathbf{B}}_\gamma \mathbf{U}; \quad \underline{\chi}_h = \overline{\mathbf{B}}_\gamma \overline{\mathbf{V}}; \quad \mathbf{U}, \overline{\mathbf{V}} \in \mathbb{R}^{\overline{m}};$$
$$\|\underline{\chi}_h\|_{\Gamma_h} = \sqrt{\overline{\mathbf{V}}^T \overline{\mathbf{D}}\, \overline{\mathbf{V}}}; \quad \overline{\mathbf{G}} = (\overline{\mathbf{B}}_\gamma^T, \overline{\mathbf{B}}_\gamma); \quad \overline{\mathbf{D}}, \overline{\mathbf{G}} \in \mathbb{R}^{\overline{m}} \times \mathbb{R}^{\overline{m}};$$
$$\underline{\kappa}_h = \mathbf{B}_\gamma \mathbf{V}; \quad \mathbf{V} \in \mathbb{R}^m; \quad \|\underline{\kappa}_h\|_{\Gamma_{h/n}} = \sqrt{\mathbf{V}^T \mathbf{D} \mathbf{V}};$$
$$\mathbf{D} \in \mathbb{R}^m \times \mathbb{R}^m; \quad \mathbf{G} = (\mathbf{B}_\gamma^T, \overline{\mathbf{B}}_\gamma); \qquad \mathbf{G} \in \mathbb{R}^m \times \mathbb{R}^{\overline{m}};$$
$$\overline{\mathcal{D}} = \mathbb{R}^{\overline{m}} \setminus \ker(\overline{\mathbf{G}}); \quad \mathcal{D} = \mathbb{R}^m \setminus \ker(\mathbf{D}),$$

where $\overline{m}$ and m are the dimensions of the nodal displacement vectors corresponding to the spaces Γ_h and $\Gamma_{h/n}$, respectively. Substituting these expressions into equation (4.10), we obtain

$$\theta_n(h) = \inf_{\mathbf{U} \in \overline{\mathcal{D}}} \frac{\sup_{\overline{\mathbf{V}} \in \overline{\mathcal{D}}} [\overline{\mathbf{V}}^T \overline{\mathbf{G}} \mathbf{U} / \sqrt{\overline{\mathbf{V}}^T \overline{\mathbf{D}}\, \overline{\mathbf{V}}}]}{\sup_{\mathbf{V} \in \mathcal{D}} [\mathbf{V}^T \mathbf{G} \mathbf{U} / \sqrt{\mathbf{V}^T \mathbf{D} \mathbf{V}}]}. \tag{4.11}$$

In Appendix B we demonstrate that the following lower bound for θ_n can be calculated:

$$\theta_n(h) \geq \theta_n^*(h) = \overline{\gamma}_{\min} \sqrt{\frac{\delta_{\min}}{\overline{\delta}_{\max} \gamma_{\max}}}, \tag{4.12}$$

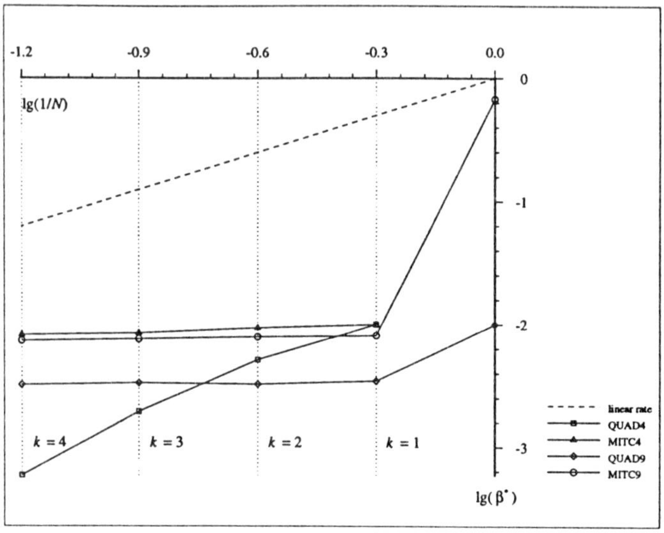

FIGURE 3. Inf-sup test in the Γ'_h-norm, uniform meshes. Values k refer to the test point numbering used in the θ_n-test. We used 1×1, 2×2, 4×4, 8×8, and 16×16 meshes.

where $\overline{\gamma}_{\min}$ is the smallest nonzero eigenvalue of $\overline{\mathbf{G}}$; $\delta_{\min}$ is the smallest nonzero eigenvalue of $\mathbf{D}$; $\overline{\delta}_{\max}$ stands for the largest eigenvalue of $\overline{\mathbf{D}}$; and $\gamma_{\max}$ is calculated as the largest eigenvalue of matrix $\mathbf{G}^T\mathbf{G}$.

Therefore, the following test procedure can be employed: for every point, $k = 1$, $2, \ldots$, on the graph for the inf-sup test in the Γ'_h-norm, we calculate a series of values θ_n^*, $n = 1, 2, 4, \ldots, M$, for some M, and plot $\lg(\theta_n^*)$ vs. $\lg(1/n)$. If, keeping either n or k fixed, the $\theta_n^*(h)$ converges to zero, the element fails the test. On the other hand, if the $\theta_n^*(h)$ does not converge to zero for any n or k fixed, the test is passed. This scheme is referred to as the *θ_n-test*.

4.3. Test results. We apply the developed test methodology to a square plate with all edges clamped, loaded by a uniformly distributed load p (see Fig. 2).

The geometry and material characteristics are $L = 100$ mm, $E = 2 \cdot 10^5$ MPa, $\nu = 0.3$, $k = 5/6$.

First we consider a sequence of uniform meshes of square elements obtained by equal subdivisions of the discretized domain $\Omega = \{(x_1, x_2) \in]0; L[^2\}$. Fig. 3 demonstrates the behavior of β^* for the elements of the MITC family and the four- and nine-node displacement-based elements QUAD4 and QUAD9. Clearly, the β^* value of the QUAD4 element converges to zero, predicting that the element will lock in the analysis. Other elements do not exhibit this trend, so we conclude that the QUAD9 and the two MITC elements have passed the inf-sup test in the Γ'_h-norm.

Fig. 4 presents the results of the θ_n-test. For all elements but the QUAD4 element we can conclude that the $\theta_n^*(h)$ value is bounded from below and above zero as $h \to 0$ and $n \to \infty$. Therefore we predict that the norm equivalence (4.1) holds for the QUAD9 element and the MITC elements with a constant ρ independent of the mesh size h.

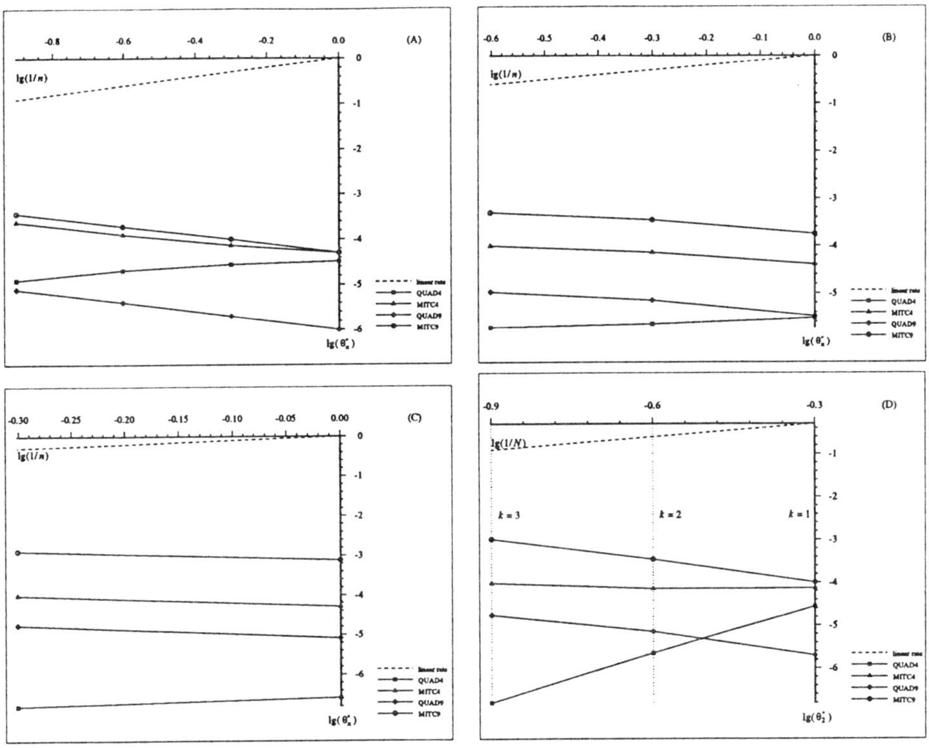

FIGURE 4. (A) θ_n-test of the clamped plate problem, uniform meshes. Starting mesh is the 2×2 mesh ($k = 1$). (B) θ_n-test of quadrilateral plate bending elements (uniform meshes with square elements). Starting mesh is the 4×4 mesh ($k = 2$). (C) θ_n-test of quadrilateral plate bending elements (uniform meshes with square elements). Starting mesh is the 8×8 mesh ($k = 3$). (D) θ_n-test of quadrilateral plate bending elements (uniform meshes with square elements). $\theta_2^*(h)$ values corresponding to $k = 1, 2, 3$.

Fig. 5 shows the calculated transverse displacement for the case $t/L = 1/10,000$, $p = 1.454 \cdot 10^7$ MPa (the corresponding Kirchhoff theory solution is $w_{\max} = 1$ mm).

We now turn our attention to the analysis of slightly distorted meshes (Fig. 6).

Fig. 7 displays the results of the inf-sup test in the Γ_h'-norm. In this case both displacement-based elements give β^* values converging to zero. The MITC elements, however, pass the test, showing robustness to the mesh distortions used in this test.

The numerical results for the transverse displacement for the $t/L = 1/10,000$ case are presented in Fig. 8.

5. Conclusions

In this paper we developed a numerical test methodology, which allows to check whether the inf-sup condition is satisfied for a particular finite element scheme. The proposed procedure can be applied to new and existing finite element formulations for plate bending problems.

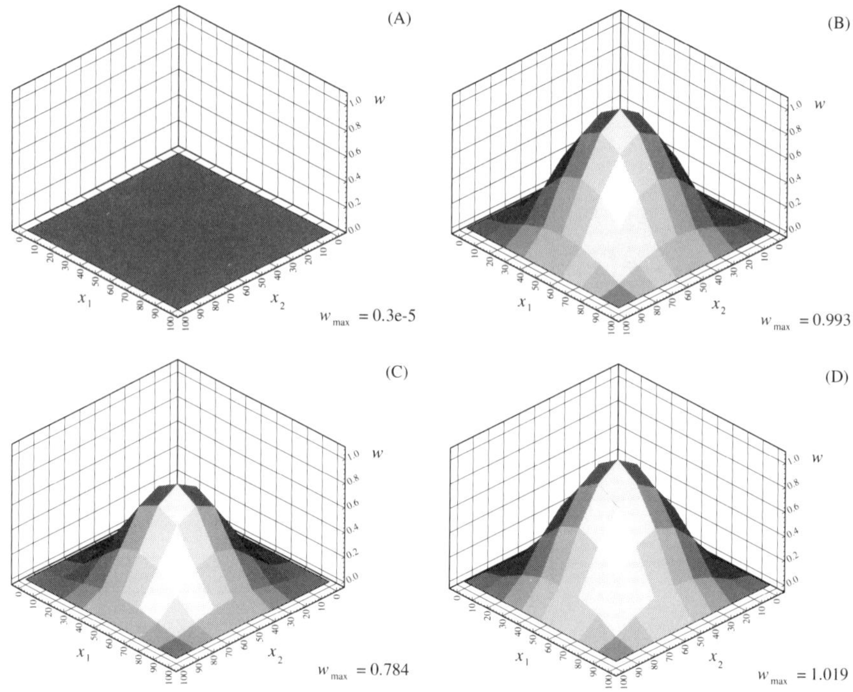

FIGURE 5. Finite element solution for the transverse displacement w [mm]; 8×8 uniform meshes for four-node elements, 4×4 uniform meshes for nine-node elements. (A) QUAD4 element; (B) MITC4 element; (C) QUAD9 element; (D) MITC9 element.

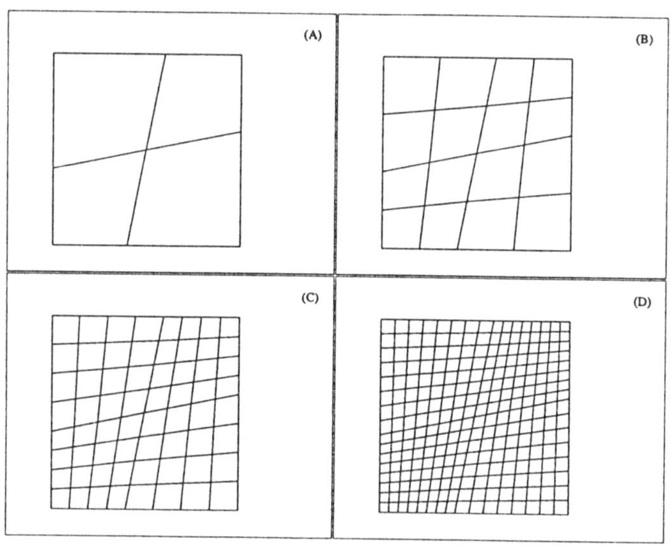

FIGURE 6. Distorted meshes used in the tests. (A) 2×2 mesh; (B) 4×4 mesh; (C) 8×8 mesh; and (D) 16×16 mesh.

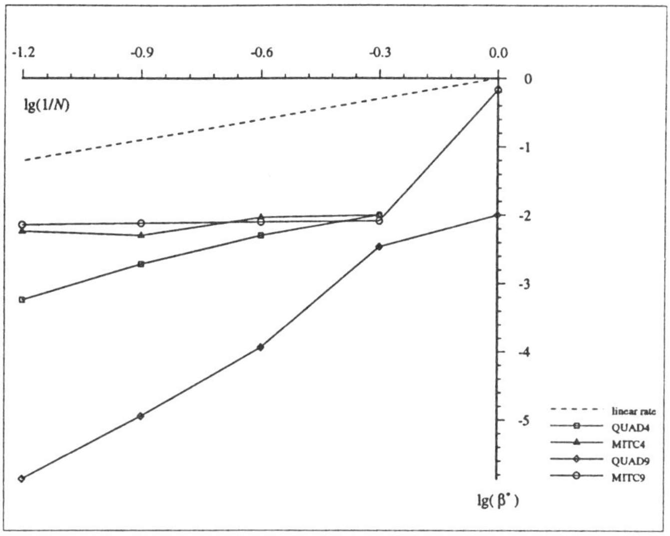

FIGURE 7. Inf-sup test in the Γ'_h-norm, distorted meshes.

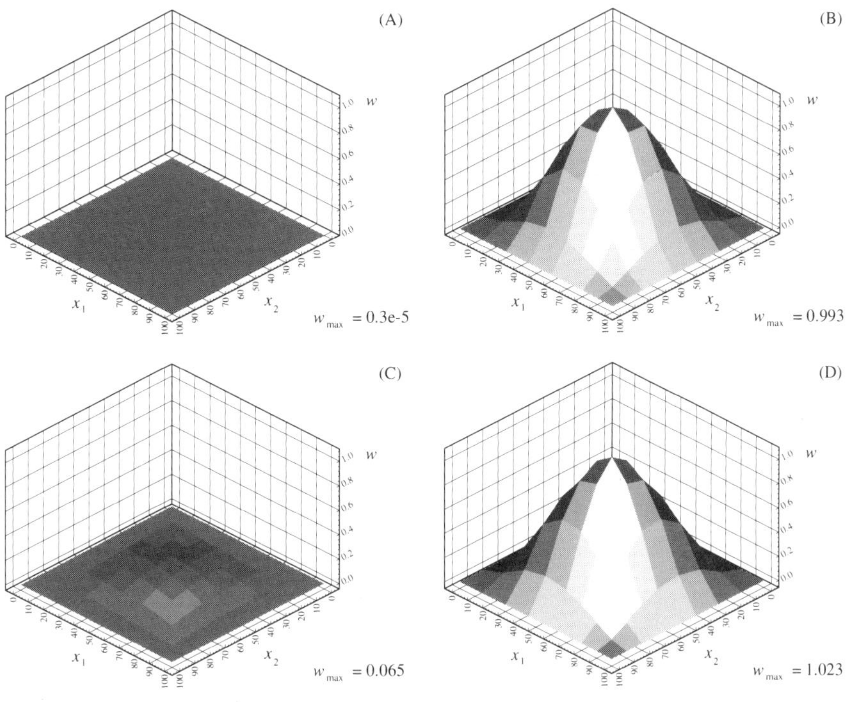

FIGURE 8. Finite element solution for the transverse displacement w [mm]; 8×8 distorted meshes for four-node elements, 4×4 distorted meshes for nine-node elements. (A) QUAD4 element; (B) MITC4 element; (C) QUAD9 element; (D) MITC9 element.

Summarizing our results on the analysis of the tested elements, we report:

1. Displacement-based elements, of course, fail to satisfy the inf-sup condition, and exhibit locking. While the four-node element failed the test using uniform and distorted meshes, the QUAD9 element passed the test with uniform meshes, but failed to work on reasonably distorted meshes. This result explicitly emphasizes the importance of using distorted discretizations in the test.
2. The elements of the MITC family (including the MITC4 element) passed the test, including the test using a sequence of reasonably distorted meshes. Thus we predict that for uniform meshes the elements satisfy the inf-sup condition (2.11) and have the optimal error bounds (2.12). The results on the distorted meshes are also promising, but as only a specific sequence was considered, a general conclusion requires further studies.

Appendix A. Derivation of the inf-sup Test in the Γ_h'-norm

Using the Rayleigh quotient [**1**], we can identify the following upper bound for $\|\underline{\zeta}_h\|_{\Gamma_h'}$:

$$\|\underline{\zeta}_h\|_{\Gamma_h'} = \sup_{\mathbf{W}\in\mathcal{D}} \frac{\mathbf{U}^T\mathbf{G}\mathbf{W}}{\sqrt{\mathbf{W}^T\mathbf{D}\mathbf{W}}} \le \frac{1}{\sqrt{\delta_{\min}}} \sup_{\mathbf{W}\in\mathcal{D}} \frac{\mathbf{U}^T\mathbf{G}\mathbf{W}}{\|\mathbf{W}\|_2} = \frac{1}{\sqrt{\delta_{\min}}}\|\mathbf{G}\mathbf{U}\|_2, \tag{A.1}$$

where we made the second transformation invoking the Cauchy-Schwarz inequality for real numbers with $\mathbf{W} = \mathbf{U}$.

Substituting (A.1) into (4.5), we obtain

$$\begin{aligned}
\inf_{\mathbf{U}\in\mathcal{D}} \sup_{\mathbf{V}\in\mathcal{D}} \frac{\mathbf{U}^T\mathbf{G}\mathbf{V}}{\sup_{\mathbf{W}\in\mathcal{D}} \frac{\mathbf{U}^T\mathbf{G}\mathbf{W}}{\sqrt{\mathbf{W}^T\mathbf{D}\mathbf{W}}}\sqrt{\mathbf{V}^T\mathbf{S}\mathbf{V}}} &\ge \sqrt{\delta_{\min}} \inf_{\mathbf{U}\in\mathcal{D}} \sup_{\mathbf{V}\in\mathcal{D}} \frac{\mathbf{U}^T\mathbf{G}\mathbf{V}}{\|\mathbf{G}\mathbf{U}\|_2\sqrt{\mathbf{V}^T\mathbf{S}\mathbf{V}}} \\
&\ge \sqrt{\frac{\delta_{\min}}{\sigma_{\max}}} \inf_{\mathbf{U}\in\mathcal{D}} \sup_{\mathbf{V}\in\mathcal{D}} \frac{\mathbf{U}^T\mathbf{G}\mathbf{V}}{\|\mathbf{G}\mathbf{U}\|_2\|\mathbf{V}\|_2} \\
&= \sqrt{\frac{\delta_{\min}}{\sigma_{\max}}} \inf_{\mathbf{U}\in\mathcal{D}} \frac{\|\mathbf{G}\mathbf{U}\|_2}{\|\mathbf{G}\mathbf{U}\|_2} = \sqrt{\frac{\delta_{\min}}{\sigma_{\max}}},
\end{aligned}$$

where the first inequality follows from the Rayleigh quotient. Then we applied the Cauchy-Schwarz inequality with $\mathbf{V} = \mathbf{U}$, and invoked the condition $\mathbf{U} \in \mathcal{D}$, to cancel out the infimum expression.

Appendix B. Derivation of the θ_n-test

Following the procedure used to derive (4.6) in Appendix A, we obtain the following bounds for the expressions in the numerator and denominator of (4.11):

$$\sup_{\overline{\mathbf{V}}\in\overline{\mathcal{D}}} \frac{\overline{\mathbf{V}}^T\overline{\mathbf{G}}\mathbf{U}}{\sqrt{\overline{\mathbf{V}}^T\overline{\mathbf{D}}\,\overline{\mathbf{V}}}} \ge \frac{1}{\sqrt{\overline{\delta}_{\max}}} \sup_{\overline{\mathbf{V}}\in\overline{\mathcal{D}}} \frac{\overline{\mathbf{V}}^T\overline{\mathbf{G}}\mathbf{U}}{\|\overline{\mathbf{V}}\|_2} = \frac{1}{\sqrt{\overline{\delta}_{\max}}}\|\overline{\mathbf{G}}\mathbf{U}\|_2 \ge \frac{\overline{\gamma}_{\min}}{\sqrt{\overline{\delta}_{\max}}}\|\mathbf{U}\|_2;$$

$$\sup_{\mathbf{V}\in\mathcal{D}} \frac{\mathbf{V}^T\mathbf{G}\mathbf{U}}{\sqrt{\mathbf{V}^T\mathbf{D}\mathbf{V}}} \le \frac{1}{\sqrt{\delta_{\min}}} \sup_{\mathbf{V}\in\mathcal{D}} \frac{\mathbf{V}^T\mathbf{G}\mathbf{U}}{\|\mathbf{V}\|_2} = \frac{1}{\sqrt{\delta_{\min}}}\sqrt{\mathbf{U}^T\mathbf{G}^T\mathbf{G}\mathbf{U}} \le \sqrt{\frac{\gamma_{\max}}{\delta_{\min}}}\|\mathbf{U}\|_2.$$

Substituting these expressions into equation (4.11), and invoking the condition $\mathbf{U} \in \mathcal{D}$ to cancel $\|\mathbf{U}\|_2$ terms, one can easily obtain (4.12).

Appendix C.

In this appendix we want to show that, at least in some simple special cases, one can easily prove that

$$\text{(C.1)} \qquad \sup_{\underline{\delta}_h \in \Gamma_{h/2}} \frac{(\underline{\delta}_h, \underline{\gamma}_h)}{\|\underline{\delta}_h\|_\Gamma} \geq \Theta^* \sup_{\underline{\delta} \in \Gamma} \frac{(\underline{\delta}, \underline{\gamma}_h)}{\|\underline{\delta}\|_\Gamma} = \Theta^* \|\underline{\gamma}_h\|_{\Gamma'}$$

(with Θ^* a constant independent of $\underline{\gamma}_h$ and h), for all $\underline{\gamma}_h \in \Gamma_h$, so that the fulfillment of the θ_2-test is sufficient to have the desired estimate.

As an example, we will indicate a proof of (C.1) in a very special case, in which Ω is a square divided into equal subsquares, and we are using MITC9 elements for the solution.

It is clear that (C.1) will follow if we can build a linear operator $\Pi\colon H_0(\text{rot};\Omega) \to \Gamma_{h/2}$, such that

$$\text{(C.2)} \qquad (\underline{\delta} - \Pi\underline{\delta}, \underline{\gamma}_h) = 0 \quad \forall \underline{\delta} \in \Gamma, \quad \forall \underline{\gamma}_h \in \Gamma_h,$$

$$\text{(C.3)} \qquad \|\Pi\underline{\delta}\|_\Gamma \leq \theta_0 \|\underline{\delta}\|_\Gamma \quad \forall \underline{\delta} \in \Gamma,$$

with θ_0 independent of $\underline{\delta}$ and h. Indeed, from (C.2) and (C.3) we have

$$\text{(C.4)} \qquad \begin{aligned} \sup_{\underline{\delta}_h \in \Gamma_{h/2}} \frac{(\underline{\delta}_h, \underline{\gamma}_h)}{\|\underline{\delta}_h\|_\Gamma} &\geq \sup_{\underline{\delta} \in \Gamma} \frac{(\Pi\underline{\delta}, \underline{\gamma}_h)}{\|\Pi\underline{\delta}\|_\Gamma} = \sup_{\underline{\delta} \in \Gamma} \frac{(\underline{\delta}, \underline{\gamma}_h)}{\|\Pi\underline{\delta}\|_\Gamma} \\ &\geq \frac{1}{\theta_0} \sup_{\underline{\delta} \in \Gamma} \frac{(\underline{\delta}, \underline{\gamma}_h)}{\|\underline{\delta}\|_\Gamma} = \frac{1}{\theta_0} \|\underline{\gamma}_h\|_{\Gamma'}, \end{aligned}$$

where the first inequality holds since, a priori, the set spanned by $\Pi\underline{\delta}$ when $\underline{\delta}$ varies in $H_0(\text{rot};\Omega)$ is contained in $\Gamma_{h/2}$; the first equality follows from (C.2), and the second inequality follows from (C.3). Then (C.4) (first term compared with the last) easily implies (C.1) with $\Theta^* = 1/\theta_0$.

It remains, therefore, to prove (C.2) and (C.3). As the vectors $\underline{\delta}$ in $\Gamma = H_0(\text{rot};\Omega)$ can be written as **[7]**

$$\text{(C.5)} \qquad \underline{\delta} = \underline{\eta} - \nabla\chi, \quad \underline{\eta} \in \left[H_0^1(\Omega)\right]^2, \chi \in H_0^1(\Omega),$$

with

$$\text{(C.6)} \qquad \|\underline{\delta}\|^2_{H_0(\text{rot};\Omega)} \sim \|\underline{\eta}\|^2_{H^1(\Omega)} + \|\chi\|^2_{H^1(\Omega)},$$

we can construct two operators Π_1 and Π_2, such that $(\underline{\eta}, \chi) \to (\Pi_1\underline{\eta}, \Pi_2\chi)$ maps $\left[H_0^1(\Omega)\right]^2 \times H_0^1(\Omega)$ into the MITC9 spaces $(\Gamma_{h/2}, W_{h/2})$, and then set

$$\text{(C.7)} \qquad \Pi(\underline{\eta} - \nabla\chi) := \Pi_1\underline{\eta} - \nabla(\Pi_2\chi) \in \Gamma_{h/2},$$

provided that we have

$$\text{(C.8)} \qquad (\nabla(\chi - \Pi_2\chi), \underline{\gamma}_h) = 0 \quad \forall \underline{\gamma}_h \in \Gamma_h,$$

$$\text{(C.9)} \qquad \|\Pi_2\chi\|_{H^1(\Omega)} \leq \theta_2 \|\chi\|_{H^1(\Omega)},$$

and

$$\text{(C.10)} \qquad (\underline{\eta} - \Pi_1\underline{\eta}, \underline{\gamma}_h) = 0 \quad \forall \underline{\gamma}_h \in \Gamma_h,$$

$$\text{(C.11)} \qquad \|\Pi_1\underline{\eta}\|_\Gamma \leq \theta_1 \|\underline{\eta}\|_{H^1(\Omega)}.$$

Let us start by constructing Π_1. We shall do so using a grid of macroelements of size h for the $h/2$ grid (see Fig. 1). We shall present the construction of the first

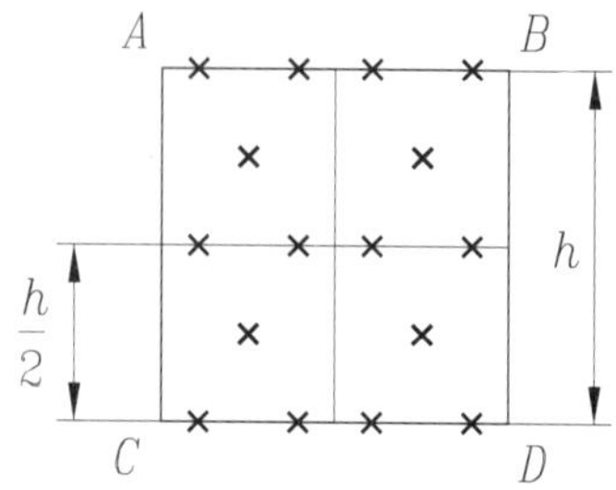

FIGURE 9. A 2×2 grid of MITC9 elements used for construction of $\underline{\eta}_1^I$.

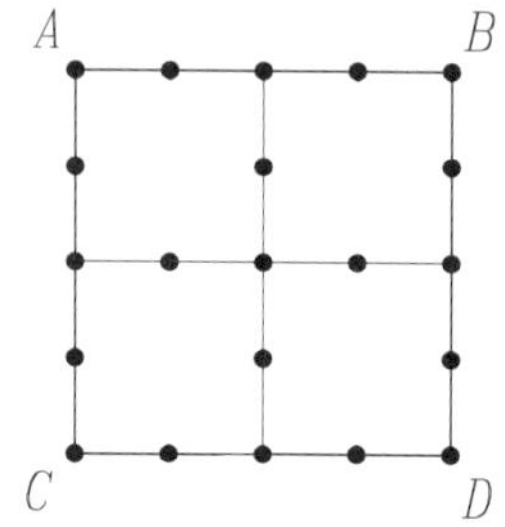

FIGURE 10. A 2×2 grid of MITC9 elements used for construction of χ^I.

component η_1^I of $\underline{\eta}^I = \Pi_1 \underline{\eta}$ since the second component η_2^I can be constructed using an identical procedure.

To construct η_1^I on the 2×2 grid of Fig. 9, we need to assign its value at 16 points (8 on the boundary, 8 in the interior).

We first set the 8 values of $(\Pi_1 \underline{\eta})$ on $\overline{AB}$ and on $\overline{CD}$ by requiring

$$\int_A^B (\eta_1 - \eta_1^I) q(x)\, dx = \int_C^D (\eta_1 - \eta_1^I) q(x)\, dx = 0, \tag{C.12}$$

for all functions $q(x)$ that are piecewise linear on the $h/2$-grid and discontinuous (there are four independent such q on $\overline{AB}$ and four on $\overline{CD}$). The remaining 8 degrees of freedom can then be fixed by requiring that

$$\int_{K_h} (\eta_1 - \eta_1^I) q(x, y)\, dx\, dy = 0, \tag{C.13}$$

for every $q \in \text{span}\{1, x, y, xy, y^2, x^2, xy^2, x^3\}$, where K_h is the square $ABCD$. It is not difficult to check that the 16 conditions (C.12)–(C.13) determine η_1^I in a unique way. However, (C.13) (and its companion for η_2^I) imply

$$\int_{K_h} (\underline{\eta} - \underline{\eta}^I) \cdot \underline{\gamma}_h = 0 \quad \forall \underline{\gamma}_h \in \Gamma_h, \tag{C.14}$$

so that (C.10) surely holds true. The verification of (C.11) is somewhat technical but standard.

We turn now to the construction of Π_2. Still reasoning on a square of the h-grid as before, we see (Fig. 10) that we have to assign the value of $\chi^I = \Pi_2 \chi$ at 21 points.

We then assign as value of χ^I at each vertex of the h-grid (A, B, C, and D in Fig. 10) the mean value of χ on the squares of the h-grid that share that vertex (this will use also the values of χ on the three other squares that are not shown in Fig. 10). Then we assign the values at the nodes on each edge of the h-grid by requiring, for instance, for the edge AB, that

$$\int_A^B (\chi - \chi^I)(x) q(x)\, dx = 0, \tag{C.15}$$

for all $q(x)$ polynomial of degree ≤ 2 on the edge AB. It is clear that, by this procedure, we have prescribed, in a unique way, the value of χ^I on the boundary of the elements of the h-grid. It remains to assign the value of χ^I at the five internal nodes of Fig. 10. For this we can require that

$$\int_{K_h} (\chi - \chi^I) q(x, y)\, dx\, dy = 0, \tag{C.16}$$

for all $q(x, y) = \text{span}\{1, x, y, x^2, y^2\}$. It is not difficult to check that (C.16) determines the values of χ^I at the internal nodes in a unique way. Moreover, for every $\underline{\gamma}_h$ in Γ_h we have

$$\begin{aligned} \int_{K_h} \nabla(\chi - \chi^I) \cdot \underline{\gamma}_h\, dx\, dy = \int_{\partial K_h} &(\chi - \chi_h) \underline{\gamma}_h \cdot \underline{n}\, ds \\ &- \int_{K_h} (\chi - \chi_h) \nabla \cdot \underline{\gamma}_h\, dx\, dy. \end{aligned} \tag{C.17}$$

The first (boundary) integral on the right-hand side of (C.17) vanishes (because of (C.15)), since $\underline{\gamma}_h \cdot \underline{n}$ is a polynomial of degree ≤ 2 on every edge of K_h, while the second integral vanishes (because of (C.16)), since $\nabla \cdot \underline{\gamma}_h$ is a polynomial of degree ≤ 1 in K_h. Hence (C.17) becomes

$$\int_{K_h} \nabla(\chi - \chi^I) \cdot \underline{\gamma}_h\, dx\, dy = 0 \quad \forall \underline{\gamma}_h \in \Gamma_h,$$

and (C.8) holds true. As before, the proof of (C.9) is technical but standard.

References

1. K. J. Bathe, *Finite element procedures*, Prentice Hall, 1996.
2. K. J. Bathe and F. Brezzi, *On the convergence of a four-node plate bending element based on Mindlin/Reissner plate theory and a mixed interpolation*, The Mathematics of Finite Elements and Applications V (Uxbridge, 1984), Academic Press, London-New York, 1985, 491–503.
3. ———, *A simplified analysis of two plate bending elements—the* MITC4 *and* MITC9 *elements*, Numerical Methods in Engineering: Theory and Applications, University College (Swansea, UK, 1998), Martinus Nijhoff Publ. (Kluwer), 1987.
4. K. J. Bathe, M. L. Bucalem, and F. Brezzi, *Displacement and stress convergence of our* MITC *plate bending elements*, Engrg. Comput. **7** (1990), no. 4, 291–302.
5. F. Brezzi and K. J. Bathe, *A discourse on the stability conditions for mixed finite element formulations*, Comput. Methods Appl. Mech. Engrg. **82** (1990), no. 1-3, 27–57.
6. F. Brezzi, K. J. Bathe, and M. Fortin, *Mixed-interpolated elements for Reissner-Mindlin plates*, Internat. J. Numer. Methods. Engrg. **28** (1989), no. 8, 1787–1801 .
7. F. Brezzi and M. Fortin, *Mixed and hybrid finite element methods*, Springer Ser. Comput. Math., vol. 15, Springer-Verlag, New York, 1991.
8. F. Brezzi, M. Fortin, and R. Stenberg, *Error analysis of mixed-interpolated elements for Reissner-Mindlin plates*, Math. Models Methods Appl. Sci. **1** (1991), no. 2, 125–151.
9. D. Chapelle and K. J. Bathe, *The* inf-sup *test*, Comput. & Structures **47** (1993), no. 4-5, 537–545.

10. R. Stenberg and M. Suri, *An hp error analysis of* MITC *plate elements*, SIAM J. Numer. Anal. **34** (1997), no. 2, 544–568.
11. J. Pitkäranta and M. Suri, *Design principles and error analysis for reduced-shear plate-bending finite elements*, Numer. Math. **75** (1996), no. 2, 223–266.

Department of Mechanical Engineering, Massachusetts Institute of Technology, Cambridge, MA 02139, U.S.A.

Istituto di Analisi Numerica del Consiglio Nazionale delle Ricerche, 27100 Pavia, Italy

E-mail address: brezzi@dragon.ian.pv.ncv.it

Centre de Recherches Mathématiques
CRM Proceedings and Lecture Notes
Volume **21**, 1999

Optimisation des plaques et coques

E. Pagnacco et J. E. Souza de Cursi

Résumé. Nous étudions un problème d'optimisation topologique : la détermination d'une coque ou plaque de masse minimum, avec des contraintes d'encombrement et de défaillance de type Von Mises. La topologie est caractérisée par une carte d'épaisseurs inconnue ρ : les trous correspondent à $\rho = 0$. Un de nos résultats fondamentaux établit que ρ est solution d'une équation de type **algébrique** : il en résulte une méthode numérique qui est testée en quelques situations académiques.

Introduction

L'optimisation topologique est un domaine particulièrement important pour l'industrie automobile : dans la conception de pièces destinées aux véhicules, il est usuel de chercher à déterminer des formes ou paramètres en tenant compte à la fois de contraintes géométriques et mécaniques. Par exemple, on cherche à définir automatiquement des formes capables de transmettre des efforts entre deux positions données, avec des contraintes d'encombrement et de résistance mécanique (défaillance plastique). Pour des raisons de coût et de consommation, il faut chercher les formes conduisant aux pièces de masse minimum : on doit en quelque sorte canaliser le flux de contraintes en employant le minimum de matière possible.

Ce type de problème a été souvent abordé en caractérisant la pièce par un squelette connu ou inconnu à habiller avec des sections droites aux inerties ou surfaces inconnues. Une variante de cette méthode consiste à déterminer une courbe moyenne et des sections droites supportées par cette même courbe.

Ces approches présentent l'inconvénient de faire des hypothèses préalables sur la connectivité, *i.e.*, sur la topologie de la pièce : par exemple, la dernière méthode ne peut conduire qu'à une forme tubulaire et la première qu'à un assemblage de poutres. Ainsi, l'évolution des besoins de l'industrie a conduit à la recherche de méthodologies où aucune hypothèse *a priori* n'est introduite sur la connectivité de la pièce. Cela nécessite une nouvelle démarche, où la détermination de la topologie de la pièce, c'est-à-dire les zones sans matière (les trous), est un préalable à l'optimisation de paramètres ou de forme définissant cette même topologie : on

1991 *Mathematics Subject Classification.* Primary : 73K15, 73K40, 65C20; Secondary : 65K10, 73K50.

Le premier auteur est financé par le groupe automobile P.S.A.

This is the final form of the paper.

commence par déterminer, même de façon un peu grossière, la forme générale de la pièce, l'emplacement et la forme des trous, puis on raffine la forme ainsi trouvée. On est alors conduit à aborder la conception dans l'ordre suivant :

1. Détermination de la topologie optimale : dans cette étape, aucune hypothèse n'est faite sur la forme ou la connectivité de la pièce. L'objectif est de définir l'allure de la pièce, c'est-à-dire sa morphologie.
2. Optimisation des paramètres définissant la forme de la pièce : on introduit ici une représentation paramétrée de la forme trouvée et on détermine la valeur optimale des paramètres.

La résolution de la première de ces étapes a conduit tout d'abord à caractériser la topologie par une «densité de matière»inconnue à déterminer où les trous correspondent aux zones à «densité»nulle. Des recherches en ce domaine ont mené à une première méthodologie de résolution s'appuyant sur les méthodes d'homogénéisation (cf., par exemple, [**1, 5, 3, 7**]). Cependant, cette approche conduit à des problèmes mathématiques avec des difficultés profondes non entièrement résolues à ce jour (existence des solutions, connexion entre le problème relaxé (homogénéisé) et le problème original). De plus, leur analyse numérique pose également beaucoup de problèmes : instabilité numérique ; forte dépendance de la finesse du maillage, du type d'élément utilisé et de la conception initiale. Enfin, ces méthodes modifient les propriétés mécaniques du matériau de conception et conduisent à des matériaux homogénéisés qui peuvent ne pas représenter correctement le matériau réel. Une méthodologie alternative est issue de l'optimisation combinatoire : en caractérisant la présence de matière ou non dans un élément d'une discrétisation par une variable logique, des algorithmes tels le recuit simulé ou les méthodes génétiques peuvent être envisagées. Cependant ces méthodes ne sont vraiment efficaces que lorsque la discrétisation est grossière : elles sont souvent très coûteuses ou en échec devant un grand nombre de variables, ce qui est le cas usuel dans la pratique industrielle.

Ce travail développe une méthodologie nouvelle pour cette même étape 1, faisant suite aux travaux précédents concernant des pièces soumises à des champs de contraintes planes [**8, 9**]. Ici, la pièce est considérée comme une plaque ou coque et son comportement est élastique linéaire : comme dans le cas des pièces soumises à des contraintes planes, il est possible de caractériser leur topologie à l'aide d'une carte d'épaisseurs : les trous correspondent aux zones d'épaisseur nulle. Un résultat fondamental établit que la carte optimale est caractérisée par une équation de type algébrique, ce qui permet de surmonter certaines des difficultés citées, notamment celles liées au grand nombre de variables et à la rapidité : les procédures de résolution numérique seront celles liées aux systèmes d'équations algébriques.

1. Description de la topologie

Ce travail envisage des pièces de type plaque ou coque (figure 1) : pour de telles pièces, il est usuel d'introduire la surface moyenne $S \subset R^3$, supposée régulière. S est décrite à l'aide de coordonnées curvilignes $a = (a_1, a_2) \in \Omega \subset R^2$:

$$S = x(\Omega) = \{x = (x_1, x_2, x_3) \in R^3 \mid x = x(a), a \in \Omega\}.$$

On note $e_1 = \partial x/\partial a_1$, $e_2 = \partial x/\partial a_2$ et $e_3 = (e_1 \wedge e_2)/|e_1 \wedge e_2|$. La région Q occupée par la pièce est définie par une carte d'épaisseurs $\rho\colon \Omega \to R$ associant à

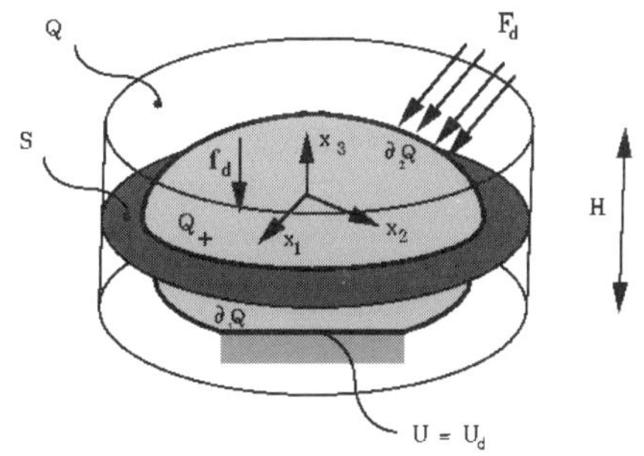

FIGURE 1. Situation considérée.

chaque particule de S l'épaisseur de la coque en ce point :

$$Q = \left\{ x + re_3 \in R^3 \;\middle|\; x \in S, -\frac{\rho(x)}{2} < r < \frac{\rho(x)}{2} \right\}.$$

Le paramètre ρ définit alors la topologie de la pièce : les trous correspondent aux régions où $\rho = 0$. On peut ainsi décomposer Q en deux sous-régions $Q_0 = \{x \in Q \mid \rho = 0\}$ et $Q_+ = \{x \in Q \mid \rho > 0\}$ telles que $Q_+ \cup Q_0 = Q$. Les trous correspondent à Q_0 et la pièce à Q_+. Dans ce qui suit, nous nous intéresserons aussi à $Q(\alpha+) = \{x \in Q \mid \rho \geq \alpha\}$. Les contraintes d'encombrement sont traitées de façon implicite par la donnée de la surface S et d'une épaisseur maximum $H > 0$ (Les résultats s'étendent aisément au cas où $H : \Omega \to R$ est une fonction). La partie de S correspondant aux trous est $S_0 = \{x \in S \mid \rho(x) = 0\}$ et la partie de S appartenant réellement à la pièce est $S_+ = \{x \in S \mid \rho(x) > 0\}$. Pour déterminer la topologie de la pièce, nous devons déterminer S_0 ou, ce qui est équivalent son complémentaire S_+ : ces deux ensembles sont déterminés par la carte d'épaisseurs ρ. Ainsi, pour la situation considérée, la détermination de ρ entraîne celle de la topologie : nous étudions dans ce qui suit le problème de la détermination de la carte d'épaisseurs optimale.

En outre, nous supposons qu'il existe une épaisseur minimale $\rho_{\min}$ constituant un seuil de matière en dessous duquel apparaît un trou. Il s'agit donc d'une borne inférieure pour ρ (figure 1), ce qui revient à supposer que $Q_+ = Q(\rho_{\min}+)$.

2. Formulation de l'équilibre d'une coque mince

L'objectif de ce travail n'est pas la construction d'une théorie des coques : nous ne reviendrons pas ici sur les aspects théoriques d'une telle construction et renvoyons à la littérature existante le lecteur intéressé par ces développements [**10**].

La structure occupe la région Q (figure 1). On note le champ de déplacement U et le champ de contrainte σ. Sur une partie du bord $\partial_1 Q$, on suppose que le champ de déplacement est imposé : $U = U_d$. Sur la partie complémentaire : $\partial_2 Q = \partial Q - \partial_1 Q$, on impose une densité surfacique d'efforts F_d. D'autre part, Q est soumis à une densité volumique d'efforts f_d.

En statique, les données U_d, F_d, f_d, et donc les inconnues U et σ, sont indépendantes du temps. Le problème de l'équilibre peut alors être formulé de la façon suivante :

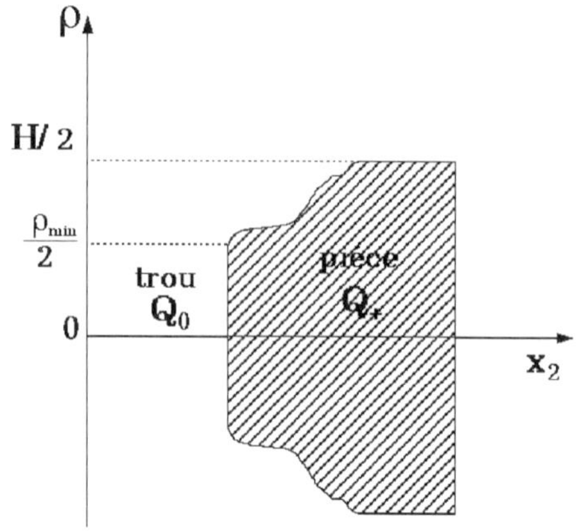

FIGURE 2. Coupe dans l'épaisseur au voisinage d'un trou.

Trouver un champ de déplacement tels que :

– U et un champ de contrainte σ,

– U vérifie les liaisons cinématiques : $U = U_d$ (+ régularité),

– σ vérifie les équations d'équilibre :

$$\forall U^* \text{ (régulier) tel que } U^* = 0 \text{ sur } \partial_1 Q,$$
$$\int_Q \mathrm{Tr}\big(\sigma\varepsilon(U^*)\big)\,dV = \int_Q f\cdot U^*\,dV + \int_{\partial_2 Q} F_d\cdot U^*\,dA,$$

– σ et $\varepsilon(U)$ vérifient la relation de comportement élastique :

$$\sigma = H\varepsilon(U) = \mathrm{grad}\Big(\frac{1}{2}\varepsilon(U)H\varepsilon(U)\Big),$$

où H désigne le tenseur symétrique de Hooke du matériau. Ici, pour un matériau élastique isotrope, on a : $H\varepsilon(U) = \lambda\mathrm{Tr}\big(\varepsilon(U)\big)Id + 2\mu\varepsilon(U)$ où λ et μ sont les coefficients de Lamé et on utilise la définition du tenseur de déformations en petites perturbations : $\varepsilon(U) = \frac{1}{2}\big(\mathrm{grad}(U) + \mathrm{grad}^T(U)\big)$.

Dans le cas des coques minces (Kirchhoff-Love), $U = U(a)$ et nous pouvons intégrer l'équation d'équilibre suivant l'épaisseur avec $\Gamma = \partial_2 Q$ dans le repère local en découplant les déformations dues aux effets de membrane et de flexion ($\varepsilon = e + x_3\chi$) et en négligeant les effets de cisaillement transversal :

$$\int_\Omega \Big(\rho e(U^*)He + \frac{\rho^3}{12}\chi(U^*)H\chi\Big)\,d\Omega = \int_\Omega \rho f\cdot U^*\,d\Omega + \int_\Gamma F_d\cdot U^*\,d\Gamma, \quad \forall U^*.$$

Cette formulation suppose donc qu'il est possible de donner un sens au produit $\rho\sigma$ dans Q tout entier et non pas seulement dans Q_+ : cette fonction est prolongée par zéro sur Q_0.

En terme d'énergie, le problème revient alors à chercher la solution en déplacement $\mathrm{U_{ex}}$ qui minimise l'énergie potentielle sur l'ensemble des champs de déplacement cinématiquement admissibles (C.A.), c'est à dire les champs qui vérifient les liaisons cinématiques : $E_p(U_{\text{ex}}) = \text{Min}_{U C.A.} E_p(U)$ avec :

$$E_p(U) = \frac{1}{2}\int_\Omega \Big(\rho(eHe) + \frac{\rho^3}{12}(\chi H \chi)\Big)\, d\Omega - \Big(\int_\Omega \rho f \cdot U\, d\Omega + \int_{\partial_2\Omega} F_d \cdot U\, d\Gamma\Big),$$

et la contrainte solution est obtenue via la loi de comportement.

On définit alors la contrainte équivalente au sens de Mises : $\sigma_{\text{eq}} = 3/2\text{Tr}\big(\sigma^D \cdot \sigma^D)\big)^{1/2}$ où σ^D représente le déviateur des contraintes.

3. Le problème d'optimisation

La formulation du problème est la suivante :

$$\begin{cases} \text{Min}_\rho\big(M(\rho) = \mu \int_\Omega \rho\, d\Omega\big), \\ \sigma_{\text{eq}} \leq \bar{\sigma}, \\ \rho = 0 \text{ ou } \rho_{\min} \leq \rho \leq H. \end{cases}$$

Le paramètre d'optimisation est la carte d'épaisseur ρ. La fonction objectif représente la masse de la structure. La première contrainte représente des restrictions aux comportement structural ($\bar{\sigma}$ est le seuil de plasticité) et la seconde définit les limitations sur les valeurs du paramètre de conception

Les épaisseurs admissibles sont celles respectant l'épaisseur maximum et conduisant à des champs de contraintes en domaine élastique :

$$C = \big\{\rho \colon \Omega \to R,\ 0 \leq \rho \leq H \text{ et } \sigma(\rho) \in \Sigma_E\big\}.$$

On a $C = K(0+) \cap K(H-) \cap K_E$ où $K(\varepsilon+) = \{\rho \colon \Omega \to R, \rho \geq \varepsilon\}$, $K(\eta-) = \{\rho \colon \Omega \to R, \rho \leq \eta\}$, $K_E = \big\{\rho \colon \Omega \to R, \sigma_{\text{eq}}(\rho) \leq \bar{\sigma}\big\}$. L'épaisseur optimale $\rho^\star$ vérifie : $\rho^\star \in C$ et $M(\rho^\star) = \inf_C M$. Dans ce qui suit, on note $\sigma^\star = \sigma(\rho^\star)$ et $\sigma^\star_{\text{eq}} = \sigma_{\text{eq}}(\sigma^\star)$ et on définit pour un nombre réel α, la projection orthogonale de α sur $[0, H]$:

$$\text{proj}_{[0,H]}(\alpha) = \begin{cases} 0 & \text{si } \alpha < 0, \\ \alpha & \text{si } \alpha \in [0, H], \\ H & \text{si } \alpha > H. \end{cases}$$

Nous avons alors le résultat suivant : Soit $h \colon [0,1] \to [0,1]$ une fonction telle que $0 \leq h(\alpha) < 1$, si $\alpha < 1$ et $h(1) = 1$. Si $\rho^\star\sigma^\star$ est assez régulier, alors

$$\rho^\star = g(\rho^\star) = \text{proj}_{[0,H]}\Big\{\rho^\star h\Big(\frac{\sigma^\star_{\text{eq}}}{\bar{\sigma}}\Big)\Big\}. \tag{1}$$

Nous avons donc $\rho^\star > 0 \implies \sigma^\star_{\text{eq}} = \bar{\sigma}$ de façon que ce résultat peut être interprété sous la forme d'une alternative : **ou bien** l'épaisseur est nulle ($\rho^\star = 0$) **ou bien** la contrainte de Von Mises est saturée ($\sigma^\star_{\text{eq}} = \bar{\sigma}$).

Nous retrouvons ainsi une méthode classique de la conception mécanique : la topologie optimale est celle où **la contrainte de Von Mises est saturée partout dans la pièce** (*fully stressed design*) [**6**].

Ce résultat peut aussi être interprété en termes de multiplicateurs de Lagrange : l'ensemble C est caractérisé par trois *contraintes* au sens de l'optimisation ($\rho \geq 0$, $\rho \leq H$ et $\sigma \leq \bar{\sigma}$). La dérivée de $\rho \to M(\rho)$ est constante et non nulle. Ainsi, pour l'optimum, au moins un des multiplicateurs de Lagrange associés aux contraintes doit être non nul et, par conséquent, au moins une des *contraintes* doit être saturée. D'autre part, si $\rho^\star = H$ et $\sigma^\star_{\text{eq}} < \bar{\sigma}$, il doit être possible de diminuer $\rho^\star$ jusqu'à ce que la contrainte de Von Mises soit saturée. Cependant, la démonstration n'utilise pas la notion de différentielle mais le Lemme de Borel-Cantelli [**8**].

REMARQUE 1. La fonction h est, en principe, arbitraire. On peut, par exemple, considérer : $h(\alpha) = h_1(\alpha) = \sqrt{\alpha}$ ou $h(\alpha) = h_2(\alpha) = \alpha$. On peut aussi considérer une fonction qui permet de prendre en compte la donnée d'une épaisseur minimale de matière (dans ce cas, $Q_+ = Q(\alpha_{\min}+)$:

$$h(\alpha) = h_3(\alpha) = \begin{cases} 0 & \text{si } \alpha < \alpha_{\min}, \\ \sqrt{\alpha} & \text{si } \alpha \geq \alpha_{\min}, \end{cases}$$

ainsi que des régularisations de h_3.

4. Résolution numérique

Le résultat précédent caractérise $\rho^\star$ comme étant un point fixe de g. Ainsi, $\rho^\star$ est solution d'une équation de type algébrique (Eq. 1), et donc la détermination de la topologie optimale peut être obtenue par les méthodes classiques de résolution des équations algébriques au lieu des algorithmes d'optimisation : par exemple, pour $\omega > 0$, on peut employer une méthode de relaxation [**4**] du type

$$\rho^{n+1} = \omega g(\rho^n) + (1 - \omega)\rho^n.$$

Les itérations sont arrêtées lorsque le nombre d'éléments où la condition de Von Mises est violée (soit $\sigma^\star_{\text{eq}} > \bar{\sigma}$) est inférieur à un t. Dans nos tests, nous utilisons $t = 1$.

Cette méthode ne nécessite pas le calcul des sensibilitées ce qui induit un volume de calcul minimal. On note que chaque itération se décompose en deux étapes : l'évaluation du champ de contraintes, nécessaire au calcul de g, puis la modification de la carte d'épaisseurs.

Pour la résolution numérique, la procédure employée reste la même que celle mise en oeuvre précédemment pour les champs de contraintes planes :

1. la région Ω est discrétisée à l'aide d'un maillage. Ce maillage est fixe au cours des itérations et sert à la discrétisation de la carte d'épaisseurs (approchée comme étant constante dans chaque élément), ainsi qu'à l'évaluation numérique du champ de contraintes dans la pièce par une méthode d'éléments finis [**2**] ;
2. d'une façon cohérente avec cette approche, les contraintes d'encombrement sont données par des valeurs maximales de l'épaisseur en chaque élément ;

On note que cette démarche introduit un seul maillage de la région considérée et autant de variables d'optimisation que d'éléments : on est ainsi conduit à traiter des problèmes d'optimisation avec un grand nombre de variables. Aucune stratégie de réduction de ce nombre de variables n'a été nécessaire, car la méthode a pu traiter correctement des discrétisations comptant quelques milliers d'éléments. Par ailleurs, rappelons que la détermination de cette carte d'épaisseur n'est qu'une première étape de recherche de la topologie optimale avant une étape d'optimisation de formes, donc elle ne requiert pas une précision prés importante et, par conséquent, une épaisseur constante par élément semble suffisante.

Dans la pratique, une des difficultés numériques à surmonter est le traitement des zones à épaisseur nulle : afin d'obtenir une méthode plus rapide, la prise en compte des trous dans la pièce doit être réalisée sans aucun remaillage. Nous avons mis en oeuvre deux procédures permettant d'aboutir à un tel résultat :

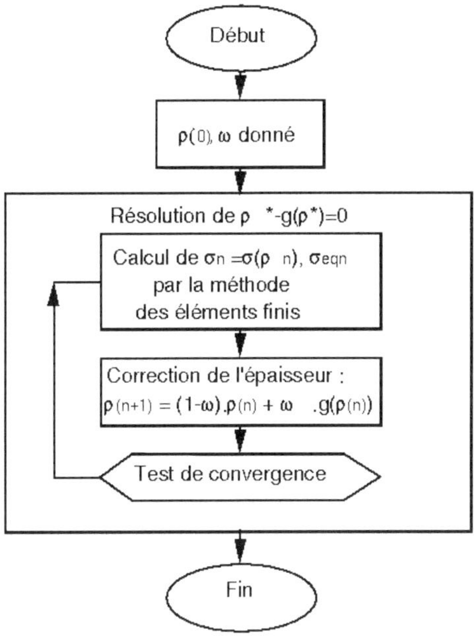

FIGURE 3. Algorithme de résolution.

1. La première consiste à poser que l'épaisseur d'un élément dans un trou n'est pas complètement nulle, mais simplement négligeable par rapport au seuil minimal de matière : on fixe une valeur $\varepsilon > 0$ comme « zéro » numérique et on associe cette épaisseur aux éléments correspondant aux trous. Dans ce cas, les contraintes dans ces éléments ne sont pas prises en compte dans les tests de convergence. Cependant, le produit $\rho\sigma$ dans ces mêmes éléments est contrôlé pour vérifier qu'il tend bien vers zéro.

2. La deuxième méthode consiste à annuler l'épaisseur des éléments dans un trou de façon à ce que la contribution de ces éléments à la rigidité globale de la structure soit bien nulle. Dans ce cas, le problème aux limites décrivant l'équilibre de la coque n'est plus coercif et la matrice du système linéaire résultant de l'approximation par éléments finis n'est pas inversible. Nous introduisons des conditions cinématiques de blocage complet des noeuds du maillage situés dans les trous, de manière à obtenir un système inversible.

Ces méthodes ont montré leur efficacité pratique et la recherche de topologie optimale a pu être effectuée sur de simples stations de travail en sessions interactives, même pour des problèmes comportant quelques milliers d'inconnues. De plus, l'allure globale de la pièce a été toujours obtenue en très peu d'itérations (ce nombre reste inférieur à une dizaine).

L'algorithme utilisé est alors présenté dans la figure 4. Le choix de l'épaisseur initiale de conception ρ_0 est guidé par des valeurs ayant un sens physiquement. Le paramètre ω introduit dans la formule de point fixe afin de corriger les épaisseurs est un facteur de relaxation qui doit prendre ses valeurs entre 0 et 1.

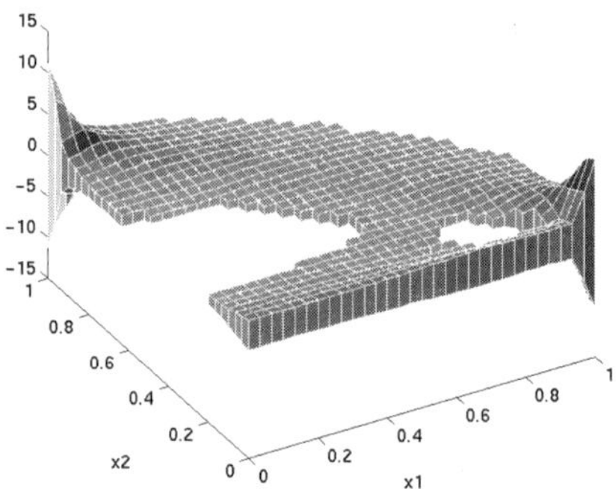

FIGURE 4. Exemple 1. La conception initiale et une solution possible.

5. Essais numériques

EXEMPLE 1. Cet exemple concerne une plaque : Ω est un carré de côté 1,

$$\Omega = \{(a_1, a_2) \mid 0 < a_i < 1, i = 1, 2\},$$

et la surface moyenne S est un plan,

$$S = \{x \in R^3 \mid (x_1, x_2) \in \Omega; x_3 = 0\}.$$

Dans ce cas, $x_1 = a_1$ et $x_2 = a_2$. Le coin inférieur gauche est complètement fixé dans ses déplacements et le côté gauche ne supporte aucun déplacement horizontal. La région est maillée par des éléments quadrilatéraux à quatre noeuds (Q4). La carte d'épaisseur ρ^* est approchée par une fonction constante sur chaque élément. Les champs de déplacements u* et de contraintes σ^* sont calculés par une méthode d'éléments finis de type P_1, c'est à dire que σ^* est approché par une fonction constante sur chaque élément. On applique une force concentrée dans le coin inférieur droit, dirigée vers le bas et de norme unitaire. Le matériau est caractérisé par : $E = 1$, $\nu = 0.3$, $\bar{\sigma} = 2$. Le maillage utilisé est de 20×20 éléments (figure 5). L'épaisseur initiale ρ_0 est constante et unitaire pour tous les éléments. Nous obtenons en 13 itérations la solution présentée dans la figure 5 pour un seuil de matière $\alpha_{\min}$ fixé à 0.8.

EXEMPLE 2. Cet exemple concerne une plaque carrée, analogue à celle de l'exemple précèdent, mais de côté 50mm. Deux de ses faces sont encastrées et une force concentrée est exercée dans le coin opposé, dirigée perpendiculairement à la surface moyenne et de valeur 400N (figure 5). Le matériau est caractérisé par : $E = 200000\text{N/mm}^2$, $\nu = 0.3$, $\bar{\sigma} = 200$Mpa. Le maillage utilisé est de 7x7 éléments. L'épaisseur initiale ρ_0 est constante de valeur 2mm pour tous les éléments. Nous obtenons en 10 itérations la solution présentée dans la figure 5 pour un seuil de matière fixé à 0.1mm.

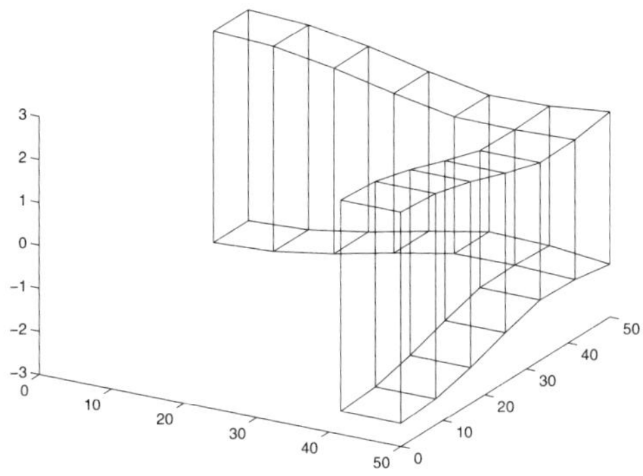

FIGURE 5. Exemple 2. La conception initiale et une solution possible.

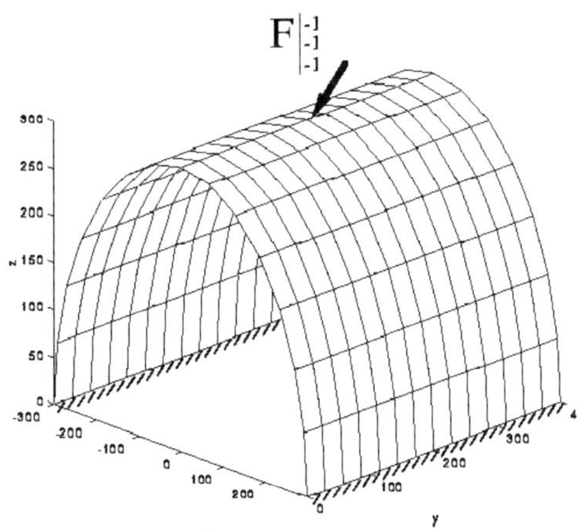

FIGURE 6. Exemple 3. La conception initiale.

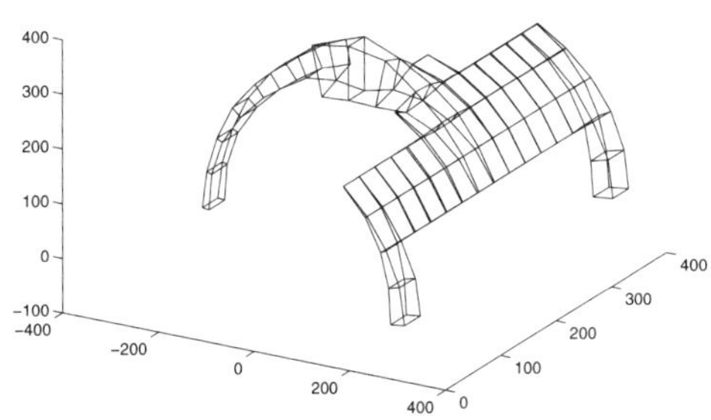

FIGURE 7. Exemple 3. Une solution possible.

EXEMPLE 3. Cet exemple concerne une coque cylindrique (arche) de rayon $R = 300$mm et de longueur $L = 400$mm (figure 5). On a

$$\Omega = \{(a_1, a_2) \mid 0 < a_1 < \pi,\ 0 < a_2 < L\},$$

et

$$S = \{x \in R^3 \mid x_1 = R\cos a_1, x_2 = a_2; x_3 = R\sin a_1, a \in \Omega\}.$$

Ses deux bords droits sont encastrés et une force concentrée oblique est exercée au centre de la région de conception de valeur 5000000N sur chaque composante. Le matériau est caractérisé par : $E = 200000\text{N/mm}^2$, $\nu = 0.3$, $\bar{\sigma} = 200$Mpa. Le maillage utilisé est de 14x14 éléments. L'épaisseur initiale ρ_0 est constante de valeur 30mm pour tous les éléments. Nous obtenons en 50 itérations la solution présentée dans la figure 5 pour un seuil de matière fixé à 0.1mm.

Conclusion

Nous avons exposé un problème de détermination de la topologie optimale, avec un critère de masse minimum, d'une coque élastique homogène soumise à des forces données. La topologie est caractérisée par une carte d'épaisseurs, dont la valeur optimale satisfait une équation algébrique, équivalente à un critère de *fully stressed design*, c'est à dire que la contrainte équivalente au sens de Von Mises doit être saturée partout dans la pièce. Cette propriété suggère une méthode numérique : la résolution de cette équation par des méthodes usuelles du domaine des équations algébriques. Des essais numériques ont été présentés en utilisant une méthode de relaxation : dans ce cas, ni le calcul des sensibilités ni des remaillages au cours des itérations ne sont nécessaires.
Dans ces essais, la méthode a été efficace et la résolution interactive a été possible même pour un grand nombre de variables de conception.

La comparaison entre cette méthode simple et d'autres stratégies de résolution des équations non linéaires (Newton-Raphson, par exemple) est un des axes du travail actuellement en cours, ainsi que la prise en compte d'autres limitations (des déplacements limités par exemple), et une extension aux problèmes de vibrations.

Références

1. M. P. Bendsoe, A. Diaz, and N. Kikuchi, *Topology and generalized layout optimization of structures*, Topology Optimization of Structures (Sesimbra, 1992), NATO Adv. Sci. Inst. Ser. E Appl. Sci., vol. 227, Kluwer Acad. Publ., Dordrecht, 1993, pp. 159–205.
2. G. Dhatt and G. Touzot, *Une présentation de la méthode des éléments finis*, Collection Université de Compiègne, Librarie Maloine, Paris ; Les Presses de l'Université Laval, Québec, 1981
3. A. Diaz and N. Kikuchi, *Solutions to shape and topology eigenvalue optimization problems using a homogenization method*, Internat. J. Numer. Methods Engrg. **35** (1992), no. 7, 1487–1502.
4. A. Gourdin and M. Boumahrat, *Méthodes numériques appliquées*, Lavoisier, Paris,1983.
5. C. S. Haber, C. S. Jog, and M. P. Bendsoe, *Topology design with optimized*, Self-Adaptative Materials, Internat. J. Numer. Meth. Engrg. **37** (1994), 1323–1350.
6. Raphaël T. Haftka and Zafer Gürdal, *Elements of structural optimization*, 3rd ed., Solid Mechanics and its Applications, vol. 11, Kluwer Academic Publ. Group, Dordrecht, 1992.
7. H. Rodrigues and P. Fernandes, *Topology optimization of linear elastic structures subjected to thermal loads*, Topology Optimization of Structures, Kluwer, Dordrecht.
8. J. E. Souza De Cursi, *Allégement d'une pièce elastique homogène soumises à des contraintres planes*, Institut de Mécanique de Rouen, Rapport de recherche no. 1/94, 1994.

9. ______, *Fast topological design in* 2D *elasticity*, Developments in Computational Techniques for Structural Engineering, BHV Topping, 1995, pp. 313–322.

10. S. Timoshenko and S. Woinowsky-Krieger, *Théorie des plaques et coques*, Dunod, 1961

LABORATOIRE DE MÉCANIQUE DE ROUEN (UPRESA CNRS 6104), INSTITUT NATIONAL DES SCIENCES APPLIQUÉES DE ROUEN, PLACE EMILE BLONDEL, BP 8, 76131 MONT SAINT-AIGNAN CEDEX, FRANCE.

E-mail address, E. Pagnacco: Emmanuel.Pagnacco@insa-rouen.fr
E-mail address, J. E. Souza de Cursi: souza@insa-rouen.fr

Centre de Recherches Mathématiques
CRM Proceedings and Lecture Notes
Volume **21**, 1999

Vibroacoustic Behavior of a Rectangular Box

M. A. Tournour and N. Atalla

ABSTRACT. A powerful method for the study of the vibroacoustic response of a rectangular box is presented. A free interface component mode synthesis is used for the modal analysis of the box. The structure modes are then coupled to the cavity modes using a modal superposition technique. For the radiation of the box, an indirect variational formulation was developed in terms of surface integral equations. A special integration scheme was used to handle the singularities of the *Green*'s function and its derivatives. In order to reduce significantly the computation time of the regular integrals, a *Gauss* quadrature with varying number of integration points was investigated. The techniques presented are validated and the results for a typical box problem are presented.

1. Introduction

A detailed study of the response of a rectangular box has been performed. The study involved state-of-the-art techniques such as component mode synthesis, modal superposition technique, variational boundary element approach and original integration schemes.

While performing the finite element vibration analysis, the memory requirement rapidly increases with the dimension or the complexity of the structure and with the extent of the analysis frequency range. Recourse is often taken to use component mode synthesis (CMS) approach which allows for a decomposition of the structure in several smaller parts. The method presented in Section 2.2 is based on the works done by MacNeal [**4**] and Rubin [**6**]. A further development of this method was given by Craig and Chang [**2**]. An incomplete modal basis of the plates forming the box is used. The truncated basis is completed with a residual flexibility obtained by a pseudo-static approximation. To correctly handle the rigid body modes, inertia relief modes are taken into account. A substantial advantage of the present method lies in the fact that all interface coordinates are eliminated from the final system of equations. The modal basis of the plates is obtained using a finite element method—bending, shear stress and in plane motion are considered.

The coupling of the box with its cavity is considered in Section 2.3. The system equations are written in terms of structure and cavity modes. In order to obtain a symmetric eigenproblem, a transformation based on the work done by Bouhioui [**1**]

1991 *Mathematics Subject Classification.* 00B25, 65-06.
This is the final form of the paper.

is applied to the system. The coupled structure-cavity modes are obtain by solving a small symmetric system.

The radiation analysis is performed using a variational boundary element approach described is Section 2.4. The formulation, originally presented by Hamdi [**3**], is discretized using the integration scheme developed by Wang and Atalla [**10**]. The main handicap of the formulation lies in the cost of the double surface integral computation. This problem is reduced using the selective integration criteria developed by Tournour [**8**].

Section 3 presents the validation of the techniques. A customized finite element program and a commercial package were used for the validation of the structural and coupled structure-cavity modal analysis. A typical box problem is considered. Since no results were found in the literature for the radiation of the box, the algorithm was validated using results for vibrating plates and pulsating bodies, immersed in air or water.

Section 4 uses the modal analysis performed in Section 3 to calculate the vibroacoustic response of the box. The mean square velocity is calculated using finite element method and component mode synthesis. For both cases the coupling with the cavity is considered.

2. Theory

2.1. The problem. The calculation of the vibroacoustic behavior of an elastic box immersed in light fluid is considered. The box is supposed to be free and is excited through a point load. The coupling with the interior fluid is considered but the effect of the exterior fluid is not accounted for. The calculated vibroacoustic indicators are the modes of the box with cavity effects, the quadratic velocity of the box and the radiated acoustic power.

2.2. Structural analysis. A free interface component mode synthesis was used. The method writes the displacement of each component i in terms of its modes:

(1)
$$\begin{aligned} \boldsymbol{z}_i &= \boldsymbol{\Phi}_i \boldsymbol{p}_i, \\ &= \boldsymbol{\Phi}_{i,k} \boldsymbol{p}_{i,k} + \boldsymbol{\Phi}_{i,d} \boldsymbol{p}_{i,d}, \end{aligned}$$

where $\boldsymbol{\Phi}_{i,d}$ are the dropped modes assessed with a residual flexibility and $\boldsymbol{\Phi}_{i,k}$ are the kept free modes obtained using a finite element method. The model includes bending, shear stress and in-plane motion effects.

2.2.1. *Residual flexibility.* For the dropped modes $\omega \ll \omega_{i,d}$ and a static approximation can be used. The contribution of the dropped modes is then assessed as:

(2)
$$\boldsymbol{\Phi}_{i,d} \boldsymbol{p}_{i,d} = \boldsymbol{G}_{i,d} \boldsymbol{f}_{i,\mathrm{junc}},$$

where $\boldsymbol{f}_{i,\mathrm{junc}}$ is the force at the junction and $\boldsymbol{G}_{i,d}$ is the residual flexibility, obtained using the relation:

(3)
$$\boldsymbol{G}_{i,d} = \boldsymbol{G}_i - \boldsymbol{\Phi}_{i,k} \boldsymbol{\Omega}_{i,k}^{-1} \boldsymbol{\Phi}_{i,k}^T,$$

where $\boldsymbol{G}_i$ is the component flexibility, $\boldsymbol{\Phi}_{i,k}\ \boldsymbol{\Omega}_{i,k}^{-1}\ \boldsymbol{\Phi}_{i,k}^T$ is the contribution of the kept elastic modes to the overall component flexibility and $\boldsymbol{\Omega}_{i,k}$ is a diagonal matrix containing the square of natural frequencies of the kept elastic modes.

Because of the free interfaces, the overall component flexibility $\boldsymbol{G}_i$ cannot be obtained by simply inverting the component stiffness matrix $\boldsymbol{K}_i$ since the later is singular. It is necessary to "remove" the rigid body motion by the use of a filter operator $\boldsymbol{A}_i$:

$$\boldsymbol{A}_i = \boldsymbol{I} - \boldsymbol{M}_i \boldsymbol{\Phi}_{i,r} \boldsymbol{\Phi}_{i,r}^T, \tag{4}$$

where $\boldsymbol{\Phi}_{i,r}$ are the component rigid body modes and $\boldsymbol{M}_i$ is the component mass matrix. A more complete description of this method is given by Craig and Chang [**2**]. The overall component flexibility is then given by :

$$\boldsymbol{G}_i = \boldsymbol{A}_i^T \boldsymbol{G}_{i,c} \boldsymbol{A}_i, \tag{5}$$

where $\boldsymbol{G}_{i,c}$ is obtained by complementing with zeros a reduced flexibility. The reduced flexibility is obtained by inverting the stiffness matrix which has been regularized by constraining arbitrary determinate degrees of freedom.

2.2.2. *Equation of motion of the components.* Using Eqs. (1) and (2), the equation of motion of each component can be written as:

$$\begin{aligned} \widetilde{\boldsymbol{M}}_i \ddot{\boldsymbol{Z}}_i + \widetilde{\boldsymbol{K}}_i \boldsymbol{Z}_i &= [\boldsymbol{\Phi}_{i,k}\ \boldsymbol{G}_{i,d}]^T [\boldsymbol{f}_{i,\mathrm{junc}}\ \boldsymbol{f}_{i,\mathrm{ext}}] \\ &= \boldsymbol{F}_{i,\mathrm{junc}} + \boldsymbol{F}_{i,\mathrm{ext}}, \end{aligned} \tag{6}$$

where $\boldsymbol{f}_{i,\mathrm{ext}}$ is the external force applied to the component and where

$$\widetilde{\boldsymbol{M}}_i = \begin{bmatrix} \boldsymbol{I} & \boldsymbol{0} \\ \boldsymbol{0} & \boldsymbol{G}_{i,d}^T \boldsymbol{M}_i \boldsymbol{G}_{i,d} \end{bmatrix}, \quad \widetilde{\boldsymbol{K}}_i = \begin{bmatrix} \boldsymbol{\Omega}_{i,k} & \boldsymbol{0} \\ \boldsymbol{0} & \boldsymbol{G}_{i,d} \end{bmatrix}, \quad \boldsymbol{Z}_i = \begin{bmatrix} \boldsymbol{p}_{i,k} \\ \boldsymbol{f}_{i,\mathrm{junc}} \end{bmatrix}, \tag{7}$$

2.2.3. *Assembly of the components.* Using the continuity of the displacements and forces, the assembly of the structure leads to the equation:

$$\boldsymbol{M}_{\mathrm{str}} \ddot{\boldsymbol{P}}_{\mathrm{str}} + \boldsymbol{K}_{\mathrm{str}} \boldsymbol{P}_{\mathrm{str}} = \boldsymbol{Q}_{\mathrm{str}}, \tag{8}$$

where

$$\boldsymbol{M}_{\mathrm{str}} = \sum_{i=1}^{n_{\mathrm{comp}}} \boldsymbol{D}_i^T \widetilde{\boldsymbol{M}}_i \boldsymbol{D}_i, \quad \boldsymbol{K}_{\mathrm{str}} = \sum_{i=1}^{n_{\mathrm{comp}}} \boldsymbol{D}_i^T \widetilde{\boldsymbol{M}}_i \boldsymbol{D}_i, \quad \boldsymbol{Q}_{\mathrm{str}} = \sum_{i=1}^{n_{\mathrm{comp}}} \boldsymbol{D}_i^T \boldsymbol{F}_{i,\mathrm{ext}}, \tag{9}$$

and

$$\boldsymbol{P}_{\mathrm{str}} = \{\boldsymbol{p}_{1,k}^T\ \ \boldsymbol{p}_{2,k}^T \cdots \boldsymbol{p}_{n_{\mathrm{comp}},k}^T\}^T, \quad \boldsymbol{D}_i = \begin{bmatrix} \boldsymbol{E}_i \\ -\boldsymbol{B}_i \boldsymbol{K}_{\mathrm{junc}} \end{bmatrix}. \tag{10}$$

The matrix $\boldsymbol{E}_i$ is an assembly matrix, $\boldsymbol{B}_i$ is a transformation matrix issued from the force continuity and which elements are either $-1, 0$ or 1. The matrix $\boldsymbol{K}_{\mathrm{junc}}$ is defined as follows:

$$\boldsymbol{K}_{\mathrm{junc}} = \boldsymbol{G}_{\mathrm{junc}}^{-1} \boldsymbol{\Phi}_{\mathrm{junc}}, \tag{11}$$

$$\boldsymbol{G}_{\mathrm{junc}} = \sum_{i=1}^{n_{\mathrm{comp}}} \boldsymbol{B}_i^T \boldsymbol{G}_{i,k}^{\mathrm{junc}} \boldsymbol{B}_i, \tag{12}$$

$$\boldsymbol{\Phi}_{\mathrm{junc}} = \sum_{i=1}^{n_{\mathrm{comp}}} \boldsymbol{B}_i^T \boldsymbol{\Phi}_{i,k}^{\mathrm{junc}} \boldsymbol{E}_i, \tag{13}$$

where the superscript junc stands for the coordinates at the junction.

2.3. Structure-cavity coupling. Depending on the dimensions of the box, the influence of the cavity is of great importance and cannot be neglected [**7**]. The rigid mode (constant pressure) of the cavity adds stiffness to the system and affects the modes of the box. The effect of the other cavity modes is more complicated and can add either mass or stiffness depending on their frequency.

To account for this effect the following technique is used. The linear acoustic equation of the fluid is coupled to the equation of motion of the structure using the continuity conditions at the fluid-structure interface

$$\sigma_{ij,j} + \rho_s \omega^2 u_i = 0 \quad \text{equation of motion of the structure,} \tag{14}$$

$$\Delta p + k^2 p = 0 \quad \text{Helmholtz equation,} \tag{15}$$

$$\frac{\partial p}{\partial n} = \rho_f \omega^2 u_n \quad \text{Euler equation,} \tag{16}$$

$$\sigma_{ij} n_j = -p n_i \quad \text{normal stress continuity,} \tag{17}$$

where σ represent the strain, u the displacement, p the fluid pressure, ρ_s and ρ_f the structure and fluid densities, ω and k the pulsation and the wavenumber and where n represent the normal pointing inside the cavity.

Equations (14) to (17) lead to the weak formulations:

$$\int_{\Omega_s} \sigma_{ij}(u)\varepsilon(\delta u)\, d\Omega_s - \omega^2 \int_{\Omega_s} (u, \delta u)\, d\Omega_s - \int_{\Sigma} p, \delta u_n \, d\Sigma = 0, \tag{18}$$

$$\frac{1}{\rho_f} \int_{\Omega_f} (\nabla p, \nabla \delta p)\, d\Omega_f - \frac{\omega^2}{\rho_f c^2} \int_{\Omega_f} (p, \delta p)\, d\Omega_f - \omega^2 \int_{\Sigma} p, \delta u_n \, d\Sigma = 0, \tag{19}$$

where δp and δu are any admissible variation of functions p and u. Applying the variational's principle and after discretisation, the weak formulations lead to the coupled system:

$$\begin{bmatrix} \boldsymbol{K} - \omega^2 \boldsymbol{M} & -\boldsymbol{C}^T \\ -\boldsymbol{C} & \frac{1}{\rho_f \omega^2}\boldsymbol{H} - \frac{1}{\rho_f c^2}\boldsymbol{Q} \end{bmatrix} \begin{Bmatrix} \boldsymbol{u} \\ \boldsymbol{p} \end{Bmatrix} = \begin{Bmatrix} 0 \\ 0 \end{Bmatrix}, \tag{20}$$

where $\boldsymbol{K}$ and $\boldsymbol{M}$ are the structure stiffness and mass matrices, $\boldsymbol{H}$ and $\boldsymbol{Q}$ are the inertia and compression fluid matrices and $\boldsymbol{C}$ is the coupling matrix.

The corresponding eigenvalue problem is not symmetric. Furthermore, because matrix $\boldsymbol{H}$ involves the pressure gradient, it is singular (presence of a rigid body mode). To eliminate these problems we use the following transformation. The unknown $\boldsymbol{u}$ and $\boldsymbol{p}$ are written in terms of modal coordinates, $\boldsymbol{u} = \boldsymbol{\Phi}_s \gamma$ and $\boldsymbol{p} = \boldsymbol{\Phi}_f \psi$. The modal matrices $\boldsymbol{\Phi}_s$ and $\boldsymbol{\Phi}_f$ are chosen such that:

$$\boldsymbol{\Phi}_s^T \boldsymbol{M} \boldsymbol{\Phi}_s = \boldsymbol{I}_{n\times n}, \qquad \boldsymbol{\Phi}_s^T \boldsymbol{K} \boldsymbol{\Phi}_s = \boldsymbol{\Lambda}_s, \tag{21}$$

$$\boldsymbol{\Phi}_f^T \boldsymbol{Q} \boldsymbol{\Phi}_f = \rho_f c^2 \begin{bmatrix} 1 & \\ & \boldsymbol{\Lambda}_f^{-1} \end{bmatrix}, \quad \boldsymbol{\Phi}_f^T \boldsymbol{H} \boldsymbol{\Phi}_f = \rho_f \begin{bmatrix} 0 & \boldsymbol{0} \\ \boldsymbol{0} & \boldsymbol{I}_{m\times m} \end{bmatrix}, \tag{22}$$

where $\boldsymbol{\Lambda}_s$ is a diagonal matrix containing the square natural frequencies of the structure and $\boldsymbol{\Lambda}_f$ is a diagonal matrix containing the square natural non zero frequencies of the fluid. Equation (20) can be rewritten as:

$$\begin{bmatrix} \boldsymbol{\Lambda_s} - \omega^2 \boldsymbol{I} & -\boldsymbol{C}_r^T & -\boldsymbol{C}_e^T \\ -\boldsymbol{C}_r & -1 & \boldsymbol{0} \\ -\boldsymbol{C}_e & \boldsymbol{0} & \frac{1}{\omega^2}\boldsymbol{I} - \boldsymbol{\Lambda}_f^{-1} \end{bmatrix} \begin{Bmatrix} \gamma \\ \psi_r \\ \psi_e \end{Bmatrix} = \begin{Bmatrix} 0 \\ 0 \\ 0 \end{Bmatrix}, \tag{23}$$

where $\boldsymbol{C}_r = \Phi_{f_r}^{T}\boldsymbol{C}\boldsymbol{\Phi}_s$ and $\boldsymbol{C}_e = \boldsymbol{\Phi}_{f_e}^{T}\boldsymbol{C}\boldsymbol{\Psi}_s$ and where the superscript "e" stands for elastic and "r" for rigid. Solving for ψ_r leads to the two transformations:

$$\left\{\begin{matrix}\gamma\\ \psi_r\\ \boldsymbol{\psi}_e\end{matrix}\right\} = \begin{bmatrix}\boldsymbol{I} & \boldsymbol{0}\\ -\boldsymbol{C}_r & \boldsymbol{0}\\ -\boldsymbol{\Lambda}_f\boldsymbol{C}_e & \boldsymbol{\Lambda}_f\end{bmatrix}\left\{\begin{matrix}\gamma\\ \frac{1}{\omega^2}\boldsymbol{\psi}_e\end{matrix}\right\} = \begin{bmatrix}\boldsymbol{I} & \boldsymbol{0}\\ -\boldsymbol{C}_r & \boldsymbol{0}\\ \boldsymbol{0} & \omega^2\boldsymbol{I}\end{bmatrix}\left\{\begin{matrix}\gamma\\ \frac{1}{\omega^2}\boldsymbol{\psi}_e\end{matrix}\right\}. \tag{24}$$

Transformations (24) are applied to Eq. (23) which gives the final set of equations:

$$\begin{bmatrix}\boldsymbol{\Lambda}_s - \omega^2\boldsymbol{I} + \boldsymbol{C}_r^T\boldsymbol{C}_r + \boldsymbol{C}_e^T\boldsymbol{\Lambda}_f\boldsymbol{C}_e & -\boldsymbol{C}_e^T\boldsymbol{\Lambda}_f\\ -\boldsymbol{C}_e\boldsymbol{\Lambda}_f & \boldsymbol{\Lambda}_f - \omega^2\boldsymbol{I}\end{bmatrix}\left\{\begin{matrix}\gamma\\ \frac{1}{\omega^2}\boldsymbol{\psi}_e\end{matrix}\right\} = \left\{\begin{matrix}\boldsymbol{0}\\ \boldsymbol{0}\end{matrix}\right\}. \tag{25}$$

Eq. (25) is a typical symmetric eigenproblem. The main advantage of this technique lies in the fact that (25) is compact and symmetric. On the other hand, one has to be really careful about the choice of the number of kept modes. Indeed because of the coupling between structural and cavity modes, higher cavity modes can effect lower structural modes.

2.4. Radiation analysis.

2.4.1. *Formulation.* With the intention of being concise, only light fluids will be considered. For the case of heavy fluids, a development similar to the one presented here is rather straightforward.

The radiation problem of the box immersed in a light fluid is governed by the following system:

$$\Delta p + k^2 p = 0 \qquad \text{Helmholtz equation,} \tag{26}$$

$$\frac{\partial p}{\partial n} = -j\rho_f\omega v_n \qquad \text{Euler equation,} \tag{27}$$

$$\lim_{r\to\infty}\left(\frac{\partial p}{\partial r} + jkp\right) = 0 \qquad \text{Sommerfeld's condition,} \tag{28}$$

where the notation of the previous section is used. The indirect boundary integral formulation of this system is given by [**3**]

$$p(M) = \int_S \mu(Q)\frac{\partial G(M,Q)}{\partial n_Q}\,dS_q, \tag{29}$$

where μ is the pressure jump $p^+ - p^-$ through the surface of the box, and where $G(M,Q)$ is the *Green*'s function defined as

$$G(M,Q) = \frac{e^{-jk\|\vec{r}_M - \vec{r}_Q\|}}{4\pi\|\vec{r}_M - \vec{r}_Q\|}. \tag{30}$$

To solve for the pressure, one must know the pressure jump μ at the surface of the structure. The pressure jump is related to the normal velocity by the relation:

$$-j\rho_f\omega v_n(M) = \frac{\partial}{\partial n_M}\int_{S_Q}\mu(Q)\frac{\partial G(M,Q)}{\partial n_Q}\,dS_Q, \tag{31}$$

which could be re-formulated using the following variational functional [**3**]:

$$\begin{aligned} J(\mu) = \frac{1}{2}\int_{S_M}\int_{S_Q} & G(M,Q)\big\{k^2\mu(M)\mu(Q)(\vec{n}_M\cdot\vec{n}_Q)\\ & -\big[\vec{n}_M\times\nabla_M\mu(M)\big]\cdot\big[\vec{n}_Q\times\nabla_Q\mu(Q)\big]\big\}\,dS_Q\,dS_M\\ & + \int_{S_M} j\rho_f\omega v_n(M)\mu(M)\,dS_M. \end{aligned} \tag{32}$$

The solution of Eq. (31) must make the functional $J(\mu)$ stationary:

$$\delta J(\mu) = 0. \tag{33}$$

2.4.2. *Numerical implementation.* Boundary elements were used to solve Eq. (33). A special 8-nodes parabolic element was developed to described the behavior of the surface of the box. Once numerically implemented, Eq. (33) gives:

$$[A]\{\mu\} = [B]\{V_n\}, \tag{34}$$

where V_n is the discretized normal velocity, and where the matrices $[A]$ and $[B]$ are given by:

$$[B] = -i\rho\omega \sum_{i=1}^{n_{\text{elem}}} [T_i]^T \iint_{-1}^{+1} [N_i(M)]^T [N_i(M)] |\vec{r}_{,\varsigma}| \, d\xi \, d\eta [T_i], \tag{35}$$

$$[A] = \sum_{i=1}^{n_{\text{elem}}} \sum_{j=1}^{n_{\text{elem}}} [T_i]^T [a_{ij}][T_j], \tag{36}$$

with

$$\begin{aligned}
[a_{ij}] = \iiiint_{-1}^{+1} G(k, M, Q)\{ & k^2 (\vec{r}_{,\varsigma}(M) \cdot \vec{r}_{,w}(Q)) [N_i(M)]^T [N_j(Q)] \\
& - (\vec{r}_{,\eta}(M) \cdot \vec{r}_{,v}(Q)) [N_{i,\xi}(M)]^T [N_{j,u}(Q)] \\
& + (\vec{r}_{,\eta}(M) \cdot \vec{r}_{,u}(Q)) [N_{i,\xi}(M)]^T [N_{j,v}(Q)] \\
& + (\vec{r}_{,\xi}(M) \cdot \vec{r}_{,v}(Q)) [N_{i,\eta}(M)]^T [N_{j,u}(Q)] \\
& - (\vec{r}_{,\xi}(M) \cdot \vec{r}_{,u}(Q)) [N_{i,\eta}(M)]^T [N_{j,v}(Q)] \} \, d\xi \, d\eta \, du \, dv,
\end{aligned} \tag{37}$$

(ξ, η, ζ) and (u, v, w) w represent the local co-ordinates of M and Q respectively, $[N_i]$ and $[T_i]$ are the interpolation and assembly matrices of element i, $\vec{r}$ is the position vector, the subscripts $_{,\xi}$ $_{,\eta}$ $_{,\zeta}$ $_{,u}$ $_{,v}$ and $_{,w}$ represent the partial derivatives with respect to the corresponding co-ordinate and where $\vec{r}_{,\zeta} = \vec{r}_{,\xi} \wedge \vec{r}_{,\eta}$.

It's easily understandable from Eq. (30) that the integrand defined in Eq. (37) is singular when $Q \to M$. Although the integral is regular it cannot be calculated with enough accuracy by a normal Gauss integration scheme. A special scheme for the integration of functions containing an $1/r$ singularity was used. This numerical integration scheme, defined by Wang and Atalla [**10**] includes the $1/r$ singularity in the weights:

$$\iiiint_{-1}^{+1} \frac{f(\xi, \eta, u, v)}{r} \, d\xi \, d\eta \, du \, dv = \sum_{i=1}^{m_\xi} \sum_{j=1}^{m_\eta} \sum_{k=1}^{m_u} \sum_{l=1}^{m_v} f(\xi_i, \eta_j, u_k, v_l) W_{ijkl}, \tag{38}$$

where ξ_i, η_j, u_k and v_l are the integration points and W_{ijkl} is the corresponding weight.

For the nonsingular integrands $(i \neq j)$, a Gauss integration scheme was used. In order to reduce the calculation time of the double surface integral, a varying number of integration points was used. Two criteria were investigated, one for the lowest frequencies and the other one for the higher frequencies. For the lowest frequencies, the total number of integration points is over estimated since the mesh was chosen in order to satisfy the convergence criteria for the highest frequency. It is thus possible to use a reduced integration scheme. For the higher frequencies, wavelength are small and the cross influence of two distant elements is not dominant.

TABLE 1. Test box modes.

Mode	structure only			structure-cavity		
	Nastran	FEM	CMS	Nastran	FEM	CMS
15-s	251.52	252.18	252.15	251.45	252.01	251.98
16-s	270.55	270.59	270.57	270.52	270.47	270.44
17-s	276.99	277.62	277.61	272.83	273.24	273.23
18-s	286.91	286.46	286.47	286.58	286.12	286.12
2-c	-	-	-	300.95	300.79	300.80
19-s	314.51	315.09	315.08	312.25	312.72	312.71
20-s	319.88	319.40	319.43	319.75	319.27	319.29
21-s	328.20	328.68	328.70	327.66	328.11	328.12
22-s	338.66	339.10	339.10	337.08	337.41	337.42
3-c	-	-	-	344.27	343.88	343.89
23-s	360.11	360.33	360.37	359.48	359.66	359.70
24-s	369.53	369.32	369.33	368.63	368.41	368.41
4-c	-	-	-	374.14	373.37	373.38

For such frequencies and distances, it is also possible to use a reduced integration scheme. A detailed parametric study yielded to the parameters defining these two criteria.

3. Validation

3.1. Structure modes. To check the accuracy of the component mode synthesis algorithm used, the modes obtained with the component mode synthesis were compared to the modes obtained with the same finite element approach used for the study of the components. The algorithm was customized for the box problem. The results were also compared to the results obtained by the Nastran© commercial package.

A test case of a steel $0.458 \times 0.499 \times 0.575\text{m}^3$ 0.005m thick rectangular box is considered. For the fem and cms study, a $8 \times 8 \times 8$ 9-node element mesh is used, while for the Nastran commercial package a $16 \times 20 \times 20$ 4-node element mesh is used. In the CMS study, sixteen elastic modes are kept for each plate. The results are presented in Table 3.1.

Excellent agreement is obtained. The difference observed between the CMS and the FEM methods is less than 0.5% for the first forty natural frequencies. The finite element method took 8mn 35sec CPU time whereas the component modal synthesis took 3mn 38sec. Although a powerful sparse lanczos solver is used, most of the CPU time (8mn 22sec) for the finite element method is spent to solve the resulting 7878-dof eigenproblem.

3.2. Coupled structure-cavity modes. To test the structure-cavity coupling algorithm presented in section a $0.35 \times 0.29\text{m}^2$ simply supported aluminium plate coupled to a 0.14m deep hard-walled cavity filled with air. The results are compared to an analytical calculation by Sgard [**7**]. Fig. (1) presents the mean square velocity of the plate calculated by the two approaches. Excellent agreement is observed. Fig. (1) shows also the important effects of the cavity coupling. One can clearly observe the effect of the cavity on the first mode and the appearance of a new mode around 480Hz.

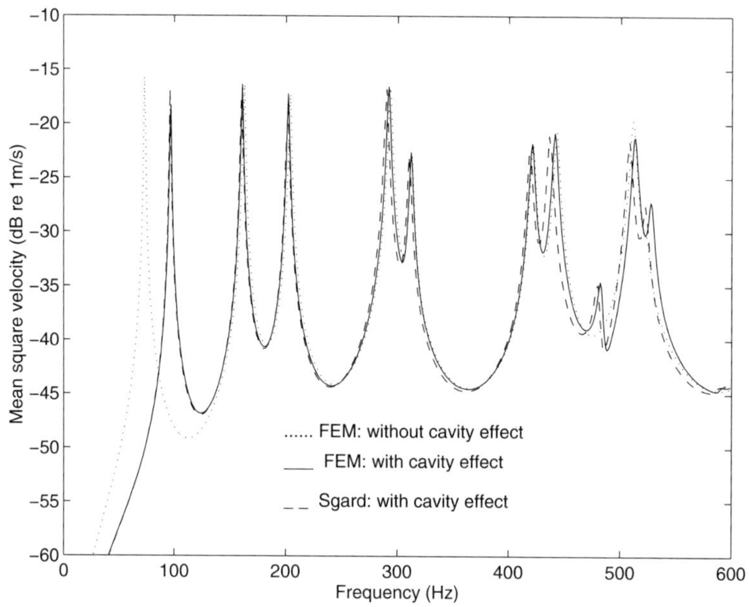

FIGURE 1. Response of a plate coupled with a cavity. Reprinted from [**9**] with permission.

Next, the modes with cavity effects of the box studied in Section 3.1 are calculated and compared to the results obtained by Nastran© commercial package and the customized finite element algorithm. The results for the box are presented in Table 3.1. Forty modes were kept for both the structure and the cavity. Once more, excellent agreement is found with the F.E. calculation.

3.3. Radiation power. Since no data on the radiation from a vibrating box are available in the literature, the validation of the presented algorithm is inferred from simple cases involving a plate and a cylinder. First, the radiation of a clamped square plate immersed in water is calculated. Comparison is made with the results obtained by Nelisse *et al.* [**5**] who use a Ritz method and a semi-analytical integration for the Green's Kernel. Fig. 2 compares Nelisse's results with the results obtained using the presented algorithm. Excellent agreement is observed.

Secondly, an open pulsating cylinder is considered. The results are compared to a semi-analytical integration scheme using 3-noded triangular elements [**3**]. The comparison is shown in Fig. 3. Once again, excellent agreement is found. Note that for these two cases, selective integration cuts computer time by a factor of 4 compared to a full 3×3 integration scheme.

4. Application to a Typical Box Problem

4.1. Mean square velocity. The box of Section 3 is excited through a point force at (0.229, 0.12475, 0.14375) from the center of the box. Fig. 4 presents the quadratic velocity of the box calculated using finite elements and the presented algorithm. Cavity coupling is accounted for both cases. Forty modes are kept for both the structure and the cavity. Excellent agreement is observed

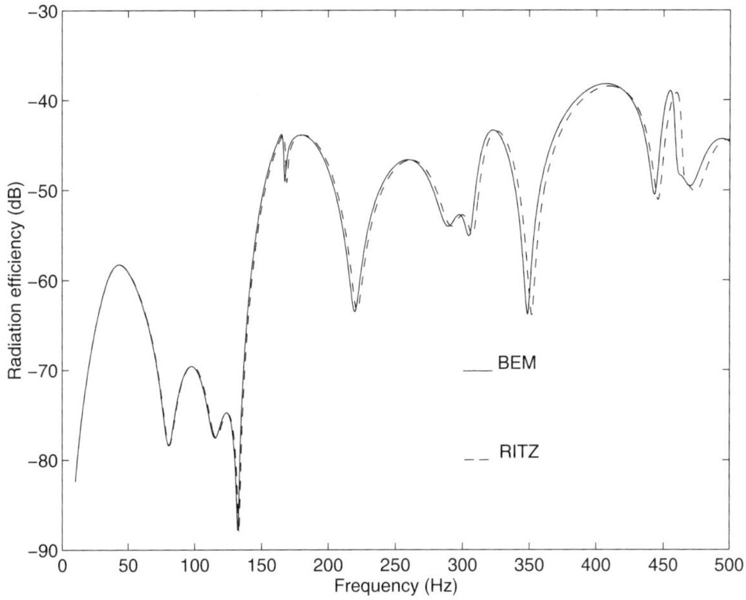

FIGURE 2. Response of a clamped square plate immersed in water. Reprinted from [**9**] with permission.

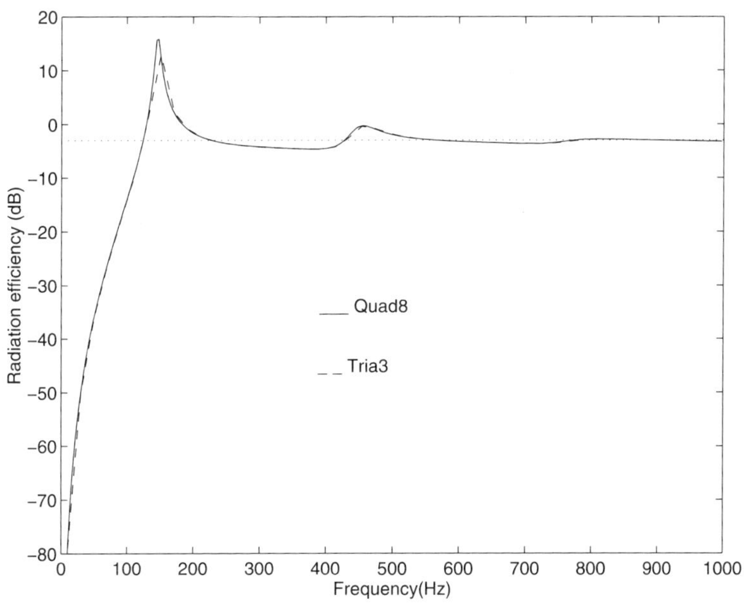

FIGURE 3. Radiation efficiency of a pulsating cylinder open at its extremities. Reprinted from [**9**] with permission.

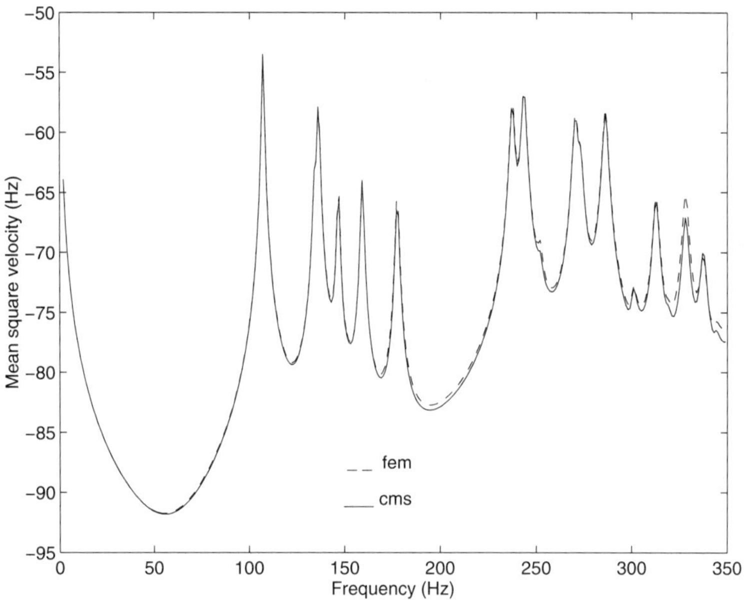

FIGURE 4. Response of the test box coupled with its cavity.

Note that the quadratic velocity peaks mainly at the box modes. Only cavity modes that are strongly coupled with the structure appear in the quadratic velocity. In the studied exemple only one cavity mode (see Table 3.1) is observed.

4.2. Radiation power. Fig. 5 represents the radiation power of the test box. A reduced $4 \times 4 \times 4$ 8-node element mesh was used together with the selective integration scheme. Unlike the quadratic velocity, the radiated power exhibits peaks at both the structure and the cavity modes.

5. Conclusion

State-of-the-art techniques for the study of the response of a rectangular box have been presented. They consist of a component mode synthesis method, a modal superposition technique for the coupling with a cavity, a variational boundary element approach and an original integration scheme.

The component modal synthesis method was compared to a finite element technique. The cms method gave good results in half time and with reduced memory requirement.

The coupling with the cavity was taken into account using a modal superposition technique. This technique is fast and allows a modal study of the coupled structure-cavity system. The results for a simply supported plate coupled to a cavity were compared to the results obtained using a technique involving a *Green* function with two indices. The two methods gave the same results.

The acoustic power radiated from the box was computed using a variational boundary element method. The formulation was implemented with the help of a special integration scheme. The scheme was developed for the calculation of integrals involving integrand having an $1/r$ singularity. For the regular integrands,

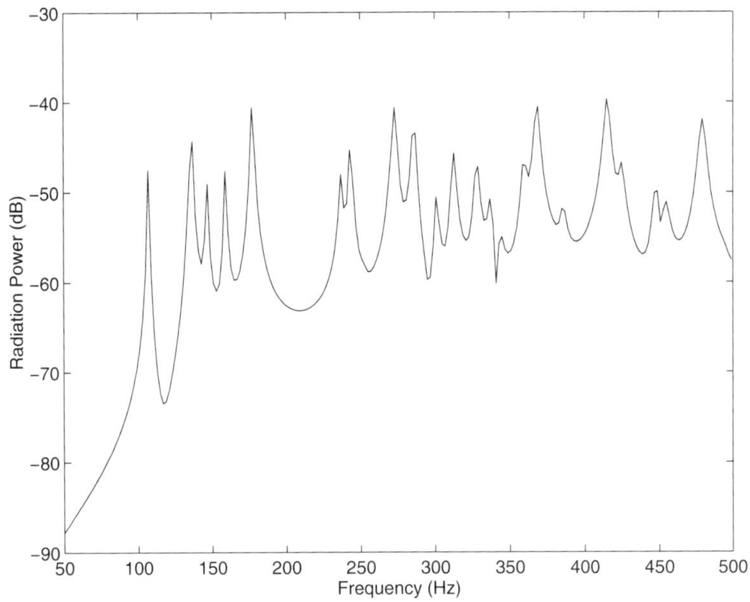

FIGURE 5. Radiation power of a rectangular box.

a selective integration was used in order to increase speed. An important gain in computing time was achieved. The results for a clamped square plate immersed in water were validated by comparison with a technique involving a Ritz method and a semi-analytical integration scheme.

Finally, a typical box problem was considered. A coupled structure-cavity modal analysis was performed and the vibroacoustic response of the system submitted to a point force was calculated.

References

1. H. Bouhioui, *Étude vibroacoustique d'un montage en double paroi de verre*, Ph.D. thesis, Univ. de Technologie de Compiègne, France, 1993.
2. R. R. Jr Craig and C-J Chang, *Substructure coupling for dynamic analysis and testing*, CR-2781, NASA, 1977.
3. M. A. Hamdi, *Formulation variationnelle par équations intégrales pour le calcul de champs acoustiques linéaires proches et lointains*, Ph.D. thesis, Univ. de Technologie de Compiègne, France, 1981.
4. R. H. MacNeal, *A hybrid method of component mode synthesis*, J. Comput. Structures **1** (1971), 581-601.
5. H. Nelisse, O. Beslin, and J. Nicolas, *A generalized approach for the acoustic radiation from a Baffled or unbaffled plate with arbitrary boundary conditions, immersed in a light of heavy fluid*, J. Sound Vibration **211** (1998), no. 2, 207–225.
6. S. Rubin, *An improved component-mode representation*, AIAA Journal **13** (1975), 995-1006.
7. F. Sgard, *Étude numérique du comportement vibroacoustique d'un système plaque-cavité dans un écoulement uniforme pour différents types d'excitation*, Ph.D. thesis, Univ. Sherbrooke, 1995.
8. M. A. Tournour, *Intégration sélective pour le calcul des opérateurs acoustiques dans la formulation variationnelle par équations intégrales*, Internal report of the Acoustics and Vibration Group of Sherbrooke University, Univ. Sherbrooke, 1995.

9. M. A. Tournour and N. Attala, *Vibroacoustic behavior of an elastic box using state-of-the-art* FEM-BEM *approaches*, Noise Control Engrg. J. **46** (1996), no. 3, pp. 83–90.
10. W. Wand and N. Atalla, *A numerical algorithm for double surface integrals over quadrilaterals with* $1/r$ *singularity*, Comm. Numer. Methods Engrg. **13** (1997), 885–890.

DÉPARTEMENT DE GÉNIE MÉCANIQUE, UNIVERSITÉ DE SHERBROOKE, SHERBROOKE, QC, CANADA
E-mail address, M. A. Tournour: `Michel.Tournour@gme.usherb.ca`
E-mail address, N. Atalla: `Noureddine.Atalla@gme.usherb.ca`

Centre de Recherches Mathématiques
CRM Proceedings and Lecture Notes
Volume **21**, 1999

Error Analysis of Mixed Finite Elements for Cylindrical Shells

Geng Yang, Michel C. Delfour, and Michel Fortin

Abstract. In this paper, based on the Naghdi shell model, we analyze the uniform convergence of mixed finite element methods for cylindrical shell problems using macroelement techniques. We show that Taylor-Hood elements $P_2 - P_1$ and P_1 iso P_2 are locking free elements for the model problems. Optimal error estimates are presented with these elements.
Résumé. On analyse la convergence uniforme des méthodes éléments finis mixes pour la coque cylinqrique en utilisant le technique de macro-élément. On montre que les éléments Taylor-Hood $P_2 - P_1$ et P_1 iso P_2 convergent uniforement. Des estimations d'erreurs sont données pour ces éléments.

1. Introduction

As it is well known, standard finite element methods fail to give good approximation for some parameter-dependent problems due to locking phenomena when the parameter is too small. Instead, mixed methods, based on introducing some new variables have been successfully used to avoid locking.

In thin plate and shell theory, the deformation behavior is usually described by mathematical models which contain the thickness t of the plate or shell as a parameter. For moderately thin plate problems, Reissner-Mindlin model is often used. Shear locking caused by the Kirchhoff-Love constraints may occur in the finite element approximation of the model. Some mixed finite element methods are proposed to get uniform error estimates in the plate thickness t (see, for instance, [**2,4,5,8**] and [**14**]). For the Naghdi shell model, both shear locking and membrane locking are involved (cf. [**1,15**]). In [**10**] and [**15**], Leino and Pitkäranta discussed the performance of high-order h-p finite element schemes for cylindrical shell problems. Locking is observed at low value of p and when the finite element grid is not aligned with the axis of the cylinder. They proposed a special strategy of selective reduced integration in order to avoid locking phenomena. Some energy estimates were given in [**12,13**]. For thin spherical shell problems, Kirmse [**9**] considered a finite element method with a penalty term for solving the limit problem of the shell

1991 *Mathematics Subject Classification.* 65N30, 65N12, 73K15, 73V05.
This research was supported in part by National Sciences and Engineering Research Council of Canada Research Grant 8730 and Infrastructure Grant 7939, and by a FCAR grant from the Ministère de l'Éducation du Québec.
This is the final form of the paper.

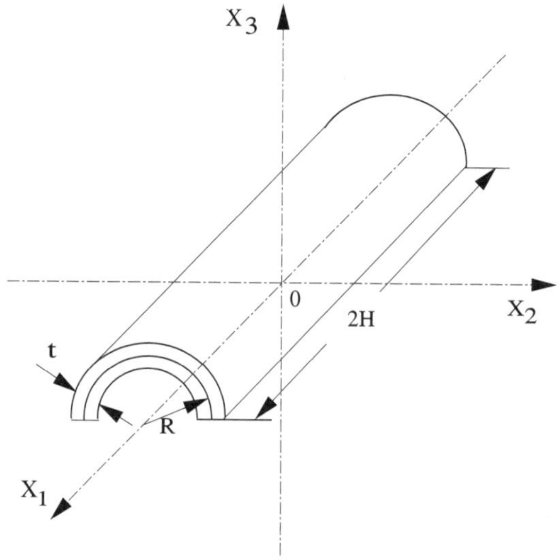

FIGURE 1. The shell geometry.

model. The technique used is based on the so-called discrete Kirchhoff triangle (DKT) elements. Recently, Arnold and Brezzi **[1]** introduced a nonstandard mixed formulation for Naghdi shell model. They proved a uniform convergence in the thickness t for a family of mixed finite elements. The choice of elements, however, is unbalanced from the point of view of approximation order. In **[16]** a family of balanced elements were studied for cylindrical shell problems. Some uniform and optimal error estimates were obtained. Nevertheless, the elements used in **[1]** and **[16]** contain the bubble functions which induce addition cost and cannot improve the approximation order in error analysis.

In this work, we consider some mixed finite elements without the bubble functions for cylindrical shell problems. Using macroelement techniques **[6]** and the mixed formulation introduced by Arnold and Brezzi **[1]**, we analyze the uniform convergence of mixed finite element methods. We examine particularly Taylor-Hood elements $P_2 - P_1$ and P_1 iso P_2. We show that these approximations converge uniformly in the thickness of the shell t. We also give optimal error estimates.

The organization of the paper is as follows. In Section 2 we present the Naghdi model for cylindrical shells and a mixed formulation. In Section 3 we describe mixed finite element methods and introduce so-called macroelement techniques. In Section 4 we examine particularly some elements by using the framework developed in Section 3. We prove the uniform convergence and optimal error estimates.

2. A Mixed Formulation of the Shell Model

We consider a hemicylindrical shell shown in. 1. The shell will occupy in Cartesian coordinates the region

$$\Omega = \left\{ (x_1, x_2, x_3) \mid -H < x_1 < H, x_3 > 0, \left(R - \frac{t}{2}\right)^2 < x_2^2 + x_3^2 < \left(R + \frac{t}{2}\right)^2 \right\}.$$

The middle surface S of the shell is

$$S = \{(x_1, x_2, x_3) \mid -H < x_1 < H, x_3 > 0, x_2^2 + x_3^2 = R^2\}.$$

We map S onto the square

$$\omega = (-H, H) \times (-L, L), \quad L = \pi R/2,$$

by the curvilinear coordinate transformation:

$$\begin{pmatrix} x_1 \\ x_2 \\ x_3 \end{pmatrix} = \begin{pmatrix} x \\ R\sin(y/R) \\ R\cos(y/R) \end{pmatrix},$$

therefore, the shell occupies the region $\omega \times (-t/2, t/2)$ in the curvilinear coordinate system (x, y, z) placed at the center of the middle surface. We denote by v_1, v_2, v_3 and ψ_1, ψ_2 the displacements along the coordinate axes and rotations in the x- and y-directions, respectively. We take Dirichlet boundary conditions on $\partial\omega$ and define

$$V^5 = \{(v, \psi) := (v_1, v_2, v_3, \psi_1, \psi_2) \in \left(H_0^1(\omega)\right)^5\}.$$

Here $H_0^1(\omega)$ denotes Sobolev space. The norm in V^5 is given by

$$\|(v, \psi)\|_k^2 = \sum_{i=1}^{3} \|v_i\|_k^2 + \sum_{i=1}^{2} \|\psi_i\|_k^2 \quad \text{for } k = 0, 1.$$

The energy functional of the shell is given by

$$\Pi = \frac{t^3}{2} a(u, \theta; u, \theta) + \frac{t}{2} b(u, \theta; u, \theta) + \frac{t}{2} c(u, \theta; u, \theta) - t^3 F(u, \theta), \tag{2.1}$$

where $a(u, \theta; u, \theta)$, $b(u, \theta; u, \theta)$, $c(u, \theta; u, \theta)$ and $F(u, \theta)$ correspond to the bending energy, the transverse shear energy, the membrane deformation energy and the potential energy due to external loads. They are defined as: (cf. **[3, 11]**)

$$a(u, \phi; v, \psi) = K \int_\omega \Big\{ \nu \big[\kappa_{11}(u, \phi) + \kappa_{22}(u, \phi)\big]\big[\kappa_{11}(v, \psi) + \kappa_{22}(v, \psi)\big] + (1 - \nu) \sum_{i,j=1}^{2} \kappa_{ij}(u, \phi)\kappa_{ij}(v, \psi) \Big\}\, dx\, dy, \tag{2.2}$$

$$b(u, \phi; v, \psi) = 6K(1 - \nu) \int_\omega \big[\rho_1(u, \phi)\rho_1(v, \psi) + \rho_2(u, \phi)\rho_2(v, \psi)\big]\, dx\, dy, \tag{2.3}$$

$$c(u, \phi; v, \psi) = 12K \int_\omega \Big\{ \nu \big[\beta_{11}(u, \phi) + \beta_{22}(u, \phi)\big]\big[\beta_{11}(v, \psi) + \beta_{22}(v, \psi)\big] + (1 - \nu) \sum_{i,j=1}^{2} \beta_{ij}(u, \phi)\beta_{ij}(v, \psi) \Big\}\, dx\, dy, \tag{2.4}$$

here $K = E/(12(1 - \nu^2))$, E is the Young's modulus, $\nu \in (0, 0.5)$ is the Poisson ratio of the material. κ_{ij}, β_{ij} and ρ_i are respectively the bending shear, membrane strains and transverse shear as follows:

$$\kappa_{11}(u,\theta)=\frac{\partial\theta_1}{\partial x},\quad \kappa_{22}(u,\theta)=\frac{\partial\theta_2}{\partial x}+\frac{1}{R}\frac{\partial u_2}{\partial y}+\frac{u_3}{R^2},$$
$$\kappa_{12}(u,\theta)=\kappa_{12}(u,\phi)=\frac{1}{2}\left(\frac{\partial\theta_1}{\partial y}+\frac{\partial\theta_2}{\partial x}\right)-\frac{1}{2R}\frac{\partial u_2}{\partial x},$$
$$\rho_1(u,\theta)=\theta_1-\frac{\partial u_3}{\partial x},\quad \rho_2(u,\theta)=\theta_2-\frac{\partial u_3}{\partial y}+\frac{u_2}{R},$$
$$\beta_{11}(u,\theta)=\frac{\partial u_1}{\partial x},\quad \beta_{22}(u,\theta)=\frac{\partial u_2}{\partial y}+\frac{u_3}{R},$$
$$\beta_{12}(u,\theta)=\beta_{21}(u,\theta)=\frac{1}{2}\left(\frac{\partial u_1}{\partial y}+\frac{\partial u_2}{\partial x}\right).$$

The shell problems may be formulated as the problem of minimizing the energy functional Π over the space V^5. The minimizer (u,θ) is equivalently determined by the following variational problem:

(2.5) Find $(u,\theta)\in V^5$ such that
$$a(u,\theta;v,\psi)+t^{-2}b(u,\theta;v,\psi)+t^{-2}c(u,\theta;v,\psi)=F(v,\psi),\quad \forall(v,\psi)\in V^5.$$

The problem (2.5) is uniquely solvable whenever F is a bounded linear functional in V^5 (cf. [**3**, p. 34]).

In our analysis we will make use of the mixed formulation suggested by Arnold and Brezzi [**1**], in which the following auxiliary variables are defined:

$$\phi_i=6K(t^{-2}-c_0)(1-\nu)\rho_1(u,\theta),\ \ i=1,2,$$
$$\lambda_1=12K(t^{-2}-c_0)\big[\beta_{11}(u,\theta)+\nu\beta_{22}(u,\theta)\big],$$
$$\lambda_2=12K(t^{-2}-c_0)\big[\nu\beta_{11}(u,\theta)+\beta_{22}(u,\theta)\big],$$
$$\lambda_3=12K(t^{-2}-c_0)(1-\nu)\beta_{12}(u,\theta),$$

where $c_0>0$ is a given constant. Then we have a mixed formulation for the problem (2.5):

Problem M^t: Find $(u,\theta;\phi,\lambda)\in V^5\times W^5$ such that

$$\begin{cases}A(u,\theta;v,\psi)+B(v,\psi;\phi,\lambda)=F(v,\psi), & \forall(v,\psi)\in V^5,\\ B(u,\theta;\eta,\chi)-\frac{t^2}{1-c_0t^2}C(\phi,\lambda;\eta,\chi)=0, & \forall(\eta,\chi)\in W^5.\end{cases}\tag{2.6}$$

Here

$$W=\{(\eta,\chi)\mid \eta_i,\chi_j\in L^2(\omega), i=1,2,\ j=1,2,3\},$$
$$A(u,\theta;v,\psi)=a(u,\theta;v,\psi)+c_0\big[b(u,\theta;v,\psi)+c(u,\theta;v,\psi)\big],\tag{2.7}$$
$$B(v,\psi;\eta,\chi)=\int_\omega\{\rho_1\eta_1+\rho_2\eta_2+\beta_{11}\chi_1+2\beta_{12}\chi_3+\beta_{22}\chi_2\}\,dx\,dy,\tag{2.8}$$
$$C(\phi,\lambda;\eta,\chi)=\frac{1}{E}\int_\omega\{2(1+\nu)\phi_1\eta_1+2(1+\nu)\phi_2\eta_2+(\lambda_1-\nu\lambda_2)\chi_1 +(\lambda_2-\nu\lambda_1)\chi_2+2(1+\nu)\lambda_3\chi_3\}\,dx\,dy.\tag{2.9}$$

In the mixed formulation, the form A will be coercive on the whole space V^5 mentioned in the following lemma. The proof is elementary.

LEMMA 2.1. *The bilinear forms A, B and C are bounded for $t \in (0, \sqrt{c_0})$ on $V^5 \times V^5$, $V^5 \times W^5$ and $W^5 \times W^5$ respectively. Moreover A and C are symmetric and coercive on V^5 and W^5.*

Let k_1, k_2 be two constants such that

$$A(v, \psi; v, \psi) \geq k_1 \big\|(v, \psi)\big\|_1^2, \quad \forall (v, \psi) \in V^5. \tag{2.10}$$

$$C(\eta, \chi; \eta, \chi) \geq k_2 \big\|(\eta, \chi)\big\|_0^2, \quad \forall (\eta, \chi) \in W^5. \tag{2.11}$$

We denote the norms of the forms A, B and C by $\|A\|$, $\|B\|$ and $\|C\|$. We also define a seminorm on W^5 by

$$|||(\eta, \chi)||| = \sup_{0 \neq (v,\psi) \in V^5} \frac{B(v, \psi; \eta, \chi)}{\|(v, \psi)\|_1}, \quad \forall (\eta, \psi) \in W^5,$$

so that $|||(\eta, \chi)||| \leq \|B\| \big\|(\eta, \chi)\big\|_0$. Since A and C are symmetric, Problem M^t corresponds to the saddle point problem (cf. [**6**], p. 44)

$$\inf_{(v,\psi) \in V^5} \sup_{(\eta,\chi) \in W^5} \left\{ \frac{1}{2} A(v, \psi; v, \psi) + B(v, \psi; \eta, \chi) - \frac{t^2}{2(1 - c_0 t^2)} C(\eta, \chi; \eta, \chi) - F(v, \psi) \right\}.$$

Then we have the following theorem (cf. [**1, 6**]):

THEOREM 2.2. *For every bounded linear form $F \in (V^5)' : V^5 \to \mathcal{R}$, and $t \in (0, \sqrt{c_0})$, Problem M^t has a unique solution $(u, \theta; \phi, \lambda)$ in $V \times W$. Moreover there exists a constant α depending on $\|A\|$, $\|B\|$, $\|C\|$, k_1 and k_2 such that*

$$\big\|(u, \theta)\big\|_1 + |||(\phi, \lambda)||| + t\big\|(\phi, \lambda)\big\|_0 \leq \alpha \|F\|_{(V^5)'}. \tag{2.12}$$

3. Finite Element Method and Error Estimation

3.1. General remarks. Our finite element method is based on the mixed formulation in Problem M^t. Let $\mathcal{T}_h$ be a regular sequence of finite element grids of ω indexed by the mesh size h. Assume that to each $\mathcal{T}_h$ there are associated some finite element spaces $V_h \subset V$ and $W_h \subset H^1 \subset W$ (that is, we consider only continuous approximations for the auxiliary variables (ϕ_h, λ_h)). The discrete problem corresponding to Problem M^t has the form:

Problem M_h^t: Find $(u_h, \theta_h; \phi_h, \lambda_h) \in V_h^5 \times W_h^5$ such that

$$\begin{cases} A(u_h, \theta_h; v_h, \psi_h) + B(v_h, \psi_h; \phi_h, \lambda_h) = F(v_h, \psi_h), & \forall (v_h, \psi_h) \in V_h^5, \\ B(u_h, \theta_h; \eta_h, \chi_h) - \dfrac{t^2}{1 - c_0 t^2} C(\phi_h, \lambda_h; \eta_h, \chi_h) = 0, & \forall (\eta_h, \chi_h) \in W_h^5. \end{cases} \tag{3.1}$$

Eq. (3.1) defines a mixed finite scheme for solving shell problem.

For this discrete problem we have two following theorems (see [**1**] for details). One gives the existence and uniqueness of the solution, and nonuniform estimates in t. The other describes uniform estimates under a stability hypothesis on the form B. We suppose that the form F is bounded linear and $t \in (0, \sqrt{t_0})$.

THEOREM 3.1. *Problem M_h^t has a unique solution $(u_h, \theta_h; \phi_h, \lambda_h)$. Moreover there is a constant α depending on $\|A\|$, $\|B\|$, $\|C\|$, k_1 and k_2 such that*

$$\begin{aligned}(3.2)\quad & \big\|(u,\theta)-(u_h,\theta_h)\big\|_1 + t\big\|(\phi,\lambda)-(\phi_h,\lambda_h)\big\|_0 \\ & \le \alpha \inf_{\substack{(v_h,\psi_h)\in V_h^5 \\ (\eta_h,\chi_h)\in W_h^5}} \Big[t^{-1}\big\|(u,\theta)-(v_h,\psi_h)\big\|_1 + |||(\phi,\lambda)-(\eta_h,\chi_h)||| \\ & \qquad\qquad + t\big\|(\phi,\lambda)-(\eta_h,\chi_h)\big\|_0\Big].\end{aligned}$$

THEOREM 3.2. *Assume that there is a constant k_0 such that*

$$\sup_{0\neq(v_h,\psi_h)\in V_h^5} \frac{B(v_h,\psi_h;\eta_h,\chi_h)}{\|(v_h,\psi_h)\|_1} \ge k_0\, |||(\eta_h,\chi_h)||| \quad \forall(\eta_h,\chi_h)\in W_h^5. \tag{3.3}$$

Then there is a constant β depending on $\|A\|$, $\|B\|$, $\|C\|$, k_1, k_2 and k_0 such that

$$\begin{aligned}(3.4)\quad & \big\|(u,\theta)-(u_h,\theta_h)\big\|_1 + |||(\phi,\lambda)-(\phi_h,\lambda_h)||| + t\big\|(\phi,\lambda)-(\phi_h,\lambda_h)\big\|_0 \\ & \le \beta \inf_{\substack{(v_h,\psi_h)\in V_h^5 \\ (\eta_h,\chi_h)\in W_h^5}} \Big[\big\|(u,\theta)-(v_h,\psi_h)\big\|_1 + |||(\phi,\lambda)-(\eta_h,\chi_h)||| \\ & \qquad\qquad + t\big\|(\phi,\lambda)-(\eta_h,\chi_h)\big\|_0\Big].\end{aligned}$$

The hypothesis (3.3) is usually hard to check. We give here another criteria that will be used in our analysis.

THEOREM 3.3. *Assume that there exists an operator π_h from V^5 to V_h^5 satisfying*

$$\begin{cases} B\big(\pi_h(v,\psi)-(v,\psi);\eta_h,\chi_h\big)=0, & \forall(\eta_h,\chi_h)\in W_h^5, \\ \big\|\pi_h(v,\psi)\big\|_1 \le \alpha\big\|(v,\psi)\big\|_1, & \forall(v,\psi)\in V^5, \end{cases} \tag{3.5}$$

then condition (3.3) *holds.*

The proof of this theorem is elementary. In fact, for given $(\eta_h,\chi_h)\in W_h^5$, from the definition of the seminorm, there exists $(v,\psi)\in V^5$ such that

$$\frac{B(v,\psi;\eta_h,\chi_h)}{\|(v,\psi)\|_1} \ge \frac{1}{2}\, |||(\eta_h,\chi_h)||| .$$

Then the conclusion follows

$$\frac{B(\pi_h(v,\psi);\eta_h,\chi_h)}{\|\pi_h(v,\psi)\|_1} = \frac{B(v,\psi;\eta_h,\chi_h)}{\|\pi_h(v,\psi)\|_1} \ge \frac{B(v,\psi;\eta_h,\chi_h)}{\alpha\|(v,\psi)\|_1} \ge \frac{1}{2\alpha}\, |||(\eta_h,\chi_h)||| .$$

3.2. Macroelement techniques. In order to apply Theorem 3.3 in our finite element analysis, we introduce here the macroelement techniques [**6**]. Suppose that the domain ω is partitioned into triangles. A macroelement M is defined by an union of a fixed number m of adjacent elements

$$M = \sum_{i=1}^{m} T_i .$$

Supposing that a mesh of macroelements $\mathcal{M}_h$ is used, we define space

$$V_{h0,M} = \big\{v_h \bigm| v_h\in V_h, v_h = 0 \text{ in } \omega\setminus M\big\}. \tag{3.6}$$

Then V_h can be written as

$$V_h = \widetilde{V_h} \oplus (\oplus_M V_{h0,M}),$$

where $\widetilde{V_h}$ is some space.

With these macroelements we have:

THEOREM 3.4. *Suppose that we have an operator* $\pi_h^1 : V \to V_h$ *satisfying*

$$\sum_{M\in\mathcal{M}_h} h_M^{2r-2}|v-\pi_h^1 v|_{r,M}^2 \le c_1\|v\|_1^2, \quad r=0,1, \quad \forall v\in V. \tag{3.7}$$

Here c_1 *is independent of* h, h_M *denotes the diameter of the macroelement* M. *If the following two conditions are satisfied, then the condition* (3.3) *and the uniform convergence estimations* (3.4) *hold.*

(i) *If on every* M *the matrices associated with*

$$\int_M pq\,dx\,dy, \quad \forall p\in V_{h0,M}, \quad \forall q\in W_h|_M, \tag{3.8}$$

and

$$\int_M pq\,dx\,dy, \quad \forall p\in V_{h0,M}, \quad \forall q\in \frac{\partial W_h|_M}{\partial x}, \tag{3.9}$$

have rank $\dim(W_h|_M)$ *and* $\dim\big((\partial W_h|_M)/\partial x\big)$ *respectively.*

(ii) *For any function* $f\in V$ *and for all* $g\in W_h|_M$, *if* $\int_M f\partial g/\partial x\,dx\,dy=0$, *then* $\int_M f\partial g/\partial y\,dx\,dy = 0$.

PROOF. We will employ Theorem 3.3. We just need to construct an operator π_h satisfying (3.5), which will be done in the following three steps.

First, we take

$$\pi_h^1(v,\psi) := (\pi_h^1 v_1, \pi_h^1 v_2, \pi_h^1 v_3, \pi_h^1\psi_1, \pi_h^1\psi_2), \quad \forall(v,\psi)\in V^5.$$

From (3.7) we have

$$\big\|\pi_h^1(v,\psi)\big\|_1 \le c_2\big\|(v,\psi)\big\|_1, \quad \forall(v,\psi)\in V^5, \tag{3.10}$$

with a constant c_2 independent of h.

Secondly, we define $\pi_h^2 : V^5 \to (\oplus_M V_{h0,M})^5$ such that

$$B\big(\pi_h^2(v,\psi)-(v,\psi);\eta_h,\chi_h\big)=0, \quad \forall(\eta_h,\chi_h)\in W_h^5. \tag{3.11}$$

This operator can be defined by solving five linear systems of equations from expression (2.8) of the form B. Indeed, by integrating by parts, we have

$$\begin{aligned}
B(v,\psi;\eta_h,\chi_h)
&= \sum_M \int_M \bigg\{ \Big(\psi_1 - \frac{\partial v_3}{\partial x}\Big)\eta_{h1} + \Big(\psi_2 - \frac{\partial v_3}{\partial y} + \frac{v_2}{R}\Big)\eta_{h2} \\
&\qquad + \Big(\frac{\partial v_1}{\partial x}\Big)\chi_{h1} + \Big(\frac{\partial v_1}{\partial y} + \frac{\partial v_2}{\partial x}\Big)\chi_{h3} + \Big(\frac{\partial v_2}{\partial y} + \frac{v_3}{R}\Big)\chi_{h2}\bigg\}\,dx\,dy \\
&= \sum_M \int_M \bigg\{ \psi_1\eta_{h1} + \Big(\psi_2 + \frac{v_2}{R}\Big)\eta_{h2} + v_3\Big(\frac{\partial \eta_{h1}}{\partial x} + \frac{\partial \eta_{h2}}{\partial y}\Big) \\
&\qquad - v_1\Big(\frac{\partial \chi_{h1}}{\partial x} + \frac{\partial \chi_{h3}}{\partial y}\Big) - v_2\Big(\frac{\partial \chi_{h3}}{\partial x} + \frac{\partial \chi_{h2}}{\partial y}\Big) + \frac{v_3}{R}\chi_{h2}\bigg\}\,dx\,dy.
\end{aligned}$$

Let $\pi_h^2(v,\psi) = (\tilde v_1, \tilde v_2, \tilde v_3, \widetilde{\psi}_1, \widetilde{\psi}_2)$, from (3.11) we have

$$\begin{aligned}(3.12)\quad \sum_M \int_M \Big\{ &(\widetilde{\psi}_1 - \psi_1)\eta_{h1} + \Big(\widetilde{\psi}_2 - \psi_2 + \frac{\tilde v_2 - v_2}{R}\Big)\eta_{h2} \\ &+(\tilde v_3 - v_3)\Big(\frac{\partial \eta_{h1}}{\partial x} + \frac{\partial \eta_{h2}}{\partial y}\Big) - (\tilde v_1 - v_1)\Big(\frac{\partial \chi_{h1}}{\partial x} + \frac{\partial \chi_{h3}}{\partial y}\Big) \\ &- (\tilde v_2 - v_2)\Big(\frac{\partial \chi_{h3}}{\partial x} + \frac{\partial \chi_{h2}}{\partial y}\Big) + \frac{\tilde v_3 - v_3}{R}\chi_{h2}\Big\}\, dx\, dy = 0.\end{aligned}$$

Therefore, from the arbitrariness of (η_h, χ_h), $\tilde v_1$, $\tilde v_2$ $\tilde v_3$, $\widetilde{\psi}_1$ and $\widetilde{\psi}_2$ can be defined in order by solving the following linear systems

$$(3.13)\quad \int_M (\tilde v_1 - v_1)\frac{\partial \chi_{h1}}{\partial x}\, dx\, dy = 0, \quad \forall M \in \mathcal{M}_h,\ \chi_{h1} \in W_h|_M,$$

$$(3.14)\quad \int_M (\tilde v_2 - v_2)\frac{\partial \chi_{h3}}{\partial x}\, dx\, dy = \int_M (\tilde v_1 - v_1)\frac{\partial \chi_{h3}}{\partial y} dx\, dy, \quad \forall M \in \mathcal{M}_h,\ \chi_{h3} \in W_h|_M,$$

$$(3.15)\quad \int_M (\tilde v_3 - v_3)\chi_{h2}\, dx\, dy = \int_M (\tilde v_2 - v_2)\frac{\partial \chi_{h3}}{\partial y}\, dx\, dy, \quad \forall M \in \mathcal{M}_h,\ \chi_{h2} \in W_h|_M.$$

$$(3.16)\quad \int_M (\widetilde{\psi}_1 - \psi_1)\eta_{h1}\, dx\, dy = -\int_M (\tilde v_3 - v_3)\frac{\partial \eta_{h1}}{\partial x} dx\, dy, \quad \forall M \in \mathcal{M}_h,\ \eta_{h1} \in W_h|_M,$$

$$\begin{aligned}(3.17)\quad \int_M (\widetilde{\psi}_2 - \psi_2)\eta_{h2}\, dx\, dy = -\int_M &\Big(\frac{\tilde v_2 - v_2}{R}\Big)\eta_{h2}\, dx\, dy \\ &- (\tilde v_3 - v_3)\frac{\partial \eta_{h2}}{\partial y} dx\, dy, \quad \forall M \in \mathcal{M}_h,\ \eta_{h2} \in W_h|_M,\end{aligned}$$

Then, for a given $(v_1, v_2, v_3, \psi_1, \psi_2)$ and from the condition (i), $\tilde v_1$ can be well defined by (3.13) (we take a minimal solution if uniqueness fails, note that $(\tilde v_1, \tilde v_2, \tilde v_3, \widetilde{\psi}_1, \widetilde{\psi}_2)\big|_M \in V_{h0,M}^5$). By the condition (ii), the linear system (3.14) becomes

$$(3.18)\qquad \int_M (\tilde v_2 - v_2)\frac{\partial \chi_{h3}}{\partial x}\, dx\, dy = 0, \quad \forall M \in \mathcal{M}_h,\ \chi_{h3} \in W_h|_M,$$

So, by the same reason, $\tilde v_2$ can be found from (3.18). Once we have $(\tilde v_1, \tilde v_2)$, the others $\tilde v_3$, $\widetilde{\psi}_1$ and $\widetilde{\psi}_2$ can be determined in order from (3.15), (3.16) and (3.17) by the condition (i). Clearly, the solutions $(\tilde v_1, \tilde v_2, \tilde v_3, \widetilde{\psi}_1, \widetilde{\psi}_2)$ satisfy the condition (3.11). For this operator a scaling argument gives (cf. **[6]**)

$$(3.19)\qquad \big\|\pi_h^2(v,\psi)\big\|_{1,M} \le c_3\big(h_M^{-1}\big\|(v,\psi)\big\|_{0,M} + \big|(v,\psi)\big|_{1,M}\big),$$

with a constant c_3 independent of h. Combining (3.7) and (3.19), we have

$$(3.20)\qquad \big\|\pi_h^2(I - \pi_h^1)(v,\psi)\big\|_1 \le c_4\big\|(v,\psi)\big\|_1, \quad \forall (v,\psi) \in V^5,$$

where the constant c_4 is independent of h.

Finally, let $\pi_h(v,\psi) = \pi_h^1(v,\psi) + \pi_h^2\big((v,\psi) - \pi_h^1(v,\psi)\big)$, then the condition (3.5) follows from (3.10), (3.11) and (3.20). In fact, π_h is an operator from V^5 to V_h^5.

From the definition of π_h^2, we have

$$B(\pi_h(v,\psi)-(v,\psi);\eta_h,\chi_h)=B(\pi_h^1(v,\psi)+\pi_h^2((v,\psi)-\pi_h^1(v,\psi))-(v,\psi);\eta_h,\chi_h)$$
$$=0,\quad \forall\,(v,\psi)\in V^5,\quad \forall(\eta_h,\chi_h)\in W_h^5.$$

Moreover,

$$\begin{aligned}\|\pi_h(v,\psi)\|_1&=\|\pi_h^1(v,\psi)+\pi_h^2((v,\psi)-\pi_h^1(v,\psi))\|_1\\&\le\|\pi_h^1(v,\psi)|_1+\|\pi_h^2((v,\psi)-\pi_h^1(v,\psi))\|_1\\&\le c_2\|(v,\psi)\|_1+c_4\|(v,\psi)\|_1\\&=(c_2+c_4)\|(v,\psi)\|_1.\end{aligned}$$

Therefore, the desired results follow from Theorem 3.3. □

4. Analysis of Some Elements

In this section, we will use the framework of the previous section to examine some elements. The major task is to verify the hypothesis in Theorem 3.4. The elements considered here are P_2-P_1 and P_2 iso P_1 elements. They also be known as Taylor-Hood conforming elements. We use the notation $\mathcal{P}_k(T)$ for the set of functions on T which are the restrictions of polynomials of degree no greater than $k\ge 0$ and define

$$L_k^s(\mathcal{T}_h)=\{v\in H^s(\mathcal{T}_h)\mid v|_T\in\mathcal{P}_k(T),\ \forall T\in\mathcal{T}_h\}.$$

4.1. (P_2-P_1) elements. For a triangle decomposition $\mathcal{T}_h$ of ω, the finite element spaces are:

$$\begin{cases}V_h=L_2^1(\mathcal{T}_h),\\ W_h=L_1^1(\mathcal{T}_h).\end{cases}\tag{4.1}$$

That is, the variables $(v_1,v_2,v_3,\psi_1,\psi_2)$ are approximated by standard Lagrange finite elements of degree two P_2, while $(\eta_1,\eta_2,\eta_3,\chi_1,\chi_2)$ are interpolated by the piecewise linear P_1 (see Fig. 2). In the sequel, we shall restrict ourselves to special meshes. We start from a mesh of quadrilaterals, each of these being considered as a macro-element M by dividing it into four triangles T_1, T_2, T_3 and T_4 as shown in Fig. 3. Let s_1, s_2, s_3, s_4 denote the areas of T_1, T_2, T_3 T_4 and suppose that $s_i\ne 0$, $1\le i\le 4$. We shall impose on our quadrilaterals a further restriction, that is $y_2\ne y_4$ and $y_1\ne y_3$ where y_i is the coordinate of vertex i on the y axis.

One remarks that a regular rectangular mesh satisfies these assumptions and that it still does after rather severe distorsions. On every M we have five degrees of freedom associated with the internal nodes of P_2 elements. For this choice of elements we have the following two lemmas.

LEMMA 4.1. *The rank of the matrix associated with*

$$\int_M pq\,dx\,dy,\quad \forall p\in V_{h0,M},\ \forall q\in W_h|_M,\tag{4.2}$$

and

$$\int_M pq\,dx\,dy,\quad \forall p\in V_{h0,M},\ \forall q\in\frac{\partial W_h|_M}{\partial x},\tag{4.3}$$

are $\dim(W_h|_M)$ *and* $\dim((\partial W_h/(\partial x)|_M)$ *respectively.*

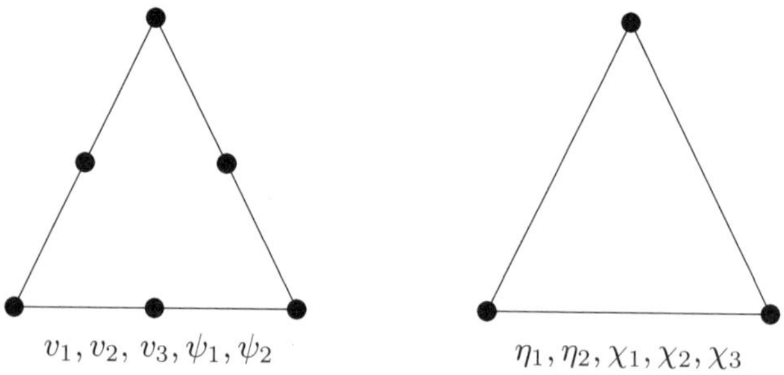

FIGURE 2. Conforming element $P_2 - P_1$.

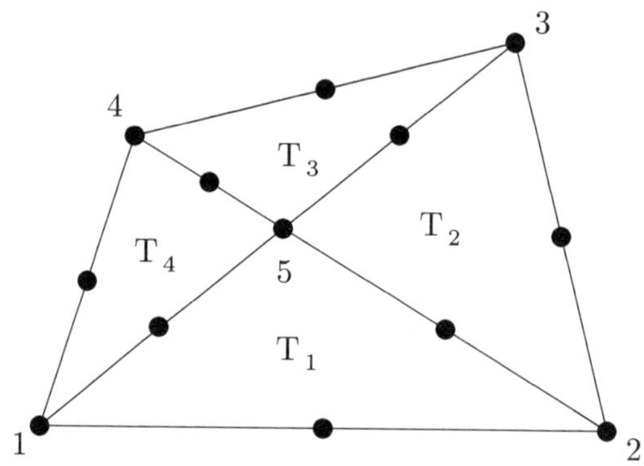

FIGURE 3. Macroelement associated with $P_2 - P_1$ element.

PROOF. Clearly, $\dim(W_h|_M) = 5$ and a straightforward computation gives $\dim\big((\partial W_h|_M)/(\partial x)\big) = 3$.

The matrix associated with (4.2) has the form

$$G_1 = \frac{1}{60}\begin{pmatrix} -(s_1+s_4) & 8(s_1+s_4) & 4s_1 & 0 & 4s_4 \\ -(s_1+s_2) & 4s_1 & 8(s_1+s_2) & 4s_2 & 0 \\ -(s_2+s_3) & 0 & 4s_2 & 8(s_2+s_3) & 4s_3 \\ -(s_3+s_4) & 4s_4 & 0 & 4s_3 & 8(s_3+s_4) \\ 2(s_1+s_2+s_3+s_4) & 8(s_1+s_4) & 8(s_1+s_2) & 8(s_2+s_3) & 8(s_3+s_4) \end{pmatrix}.$$

The matrix G_1 is similar to

$$\begin{pmatrix} 1 & 0 & 0 & 0 & 0 \\ 0 & 1 & 0 & 0 & 0 \\ 0 & 0 & s_1 & -2(s_2+s_3) & s_4 \\ 0 & 0 & s_1+s_2 & 0 & s_3+s_4 \\ 0 & 0 & (s_1+s_2)(s_3+s_4) & s_2s_4 - s_1s_3 & 0 \end{pmatrix}.$$

It implies $\operatorname{rank}(G_1) = 5$.

Another matrix corresponding to (4.3) has the form

$$G_2 = \frac{1}{6}\begin{pmatrix} 0 & y_2 - y_4 & y_2 - y_5 & 0 & y_5 - y_4 \\ 0 & y_2 - y_1 & y_3 - y_1 & y_3 - y_5 & 0 \\ 0 & 0 & y_5 - y_2 & y_4 - y_2 & y_4 - y_5 \end{pmatrix},$$

it is clear that $\operatorname{rank}(G_2) = 3$. □

LEMMA 4.2. *On every macroelement M, for any $f \in V$, if for all $g \in W_h|_M$*

$$\int_M f \frac{\partial g}{\partial x} = 0,$$

then

$$\int_M f \frac{\partial g}{\partial y} = 0.$$

PROOF. Let N_i be the piecewise linear basis funtions at i-th node of the triangulation $\mathcal{T}_h$. On a macroelement M shown in Fig. 3, since $W_h = L_1^1(\mathcal{T}_h)$, $(\partial N_i)/(\partial x)$ and $(\partial N_i)/(\partial y)$ $(i = 1,2,3,4,5)$ are piecewise constant on M. Moreover, $(\partial N_i)/(\partial x)$ and $(\partial N_i)/(\partial y)$ $(i = 1,2,3)$ are basis functions of the spaces $(\partial W_h|_M)/(\partial x)$ and $(\partial W_h|_M)/(\partial y)$ respectively. For any $f \in V$,

$$\int_M f \frac{\partial N_1}{\partial x}\, dx\, dy = \frac{y_2 - y_4}{2(s_1 + s_4)} \int_{T_1 \cup T_4} f dx\, dy,$$

Then $\int_M f(\partial N_1)/(\partial x)\, dx\, dy = 0$ implies $\int_{T_1 \cup T_4} f\, dx\, dy = 0$. So

$$\int_M f \frac{\partial N_1}{\partial y}\, dx\, dy = \frac{x_4 - x_2}{2(s_1 + s_4)} \int_{T_1 \cup T_4} f\, dx\, dy = 0.$$

Similarly,

$$\int_M f \frac{\partial N_i}{\partial y}\, dx\, dy = 0, \text{ if } \int_M f \frac{\partial N_i}{\partial x}\, dx\, dy = 0, \text{ for } i = 2, 3.$$

Therefore, Lemma 4.2 holds. □

We give now uniform convergence estimates which is one of main results of this paper.

THEOREM 4.3. *With the choice of V_h and W_h by (4.1), for every bounded linear form $F \in (V^5)'$ and $t \in (0, \sqrt{c_0})$, there exists a unique solution $(u_h, \theta_h; \phi_h, \lambda_h) \in V_h^5 \times W_h^5$ to Problem M_h^t. Moreover, there is a constant γ_1 independent of h, t and F such that*

$$\text{(4.4)} \quad \|(u,\theta) - (u_h,\theta_h)\|_1 + |||(\phi,\lambda) - (\phi_h,\lambda_h)||| + t\|(\phi,\lambda) - (\phi_h,\lambda_h)\|_0 \\ \le \gamma_1 h^2 \big[\|(u,\theta)\|_3 + \|(\phi,\lambda)\|_2\big].$$

PROOF. The existence and uniqueness of the solution follows Theorem 3.1. We employ Theorem 3.4 to prove the uniform convergence (4.4). We take Clement's interpolation operator as $\pi_h^1 : H_0^1(\omega) \to L_2^1(\mathcal{T}_h) \cap H_0^1(\omega)$ satisfying (cf. **[6, 7]**)

$$\text{(4.5)} \quad \sum_{T \in \mathcal{T}_h} h_T^{2r-2} |q - \pi_h^1 q|_{r,T}^2 \le c_1 \|q\|_1^2, \quad r = 0, 1, \quad \forall q \in V = H_0^1(\omega).$$

Here h_T denotes the diameter of the triangle T. We have then

$$\begin{aligned} \text{(4.6)} \quad \sum_{T \in \mathcal{T}_h} h_T^{2r-2} |q - \pi_h^1 q|_{r,T}^2 &= \sum_{M \in \mathcal{M}_h} \sum_{T \in M} h_T^{2r-2} |q - \pi_h^1 q|_{r,T}^2 \\ &\ge \sum_{M \in \mathcal{M}_h} \sum_{T \in M} h_M^{2r-2} |q - \pi_h^1 q|_{r,T}^2 \quad (\text{since } h_T \le h_M) \\ &= \sum_{M \in \mathcal{M}_h} h_M^{2r-2} |q - \pi_h^1 q|_{r,M}^2, \quad r = 0, 1, \quad \forall q \in V. \end{aligned}$$

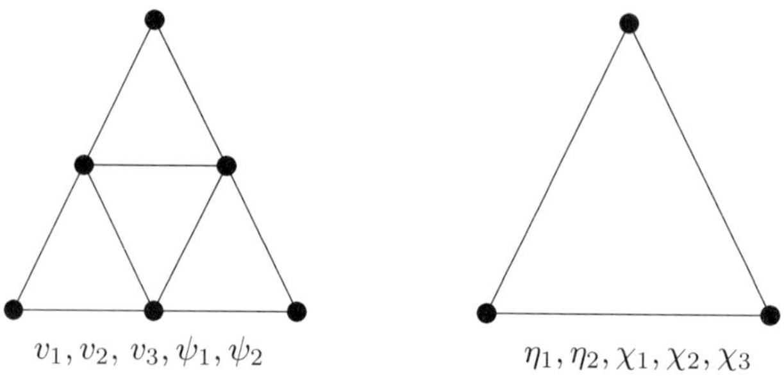

FIGURE 4. Conforming element P_1 iso P_2.

Thus, we have (3.7). On the other hand, the conditions (i) and (ii) follow from Lemma 4.1 and Lemma 4.2. Therefore, Theorem 3.4 holds. So, there is a constant β depending on $\|A\|$, $\|B\|$, $\|C\|$, k_1, k_2 and k_0 such that

$$\begin{aligned} &\|(u,\theta)-(u_h,\theta_h)\|_1 + |||(\phi,\lambda)-(\phi_h,\lambda_h)||| + t\|(\phi,\lambda)-(\phi_h,\lambda_h)\|_0 \\ &\le \beta \inf_{\substack{(v_h,\psi_h)\in V_h^5\\(\eta_h,\chi_h)\in W_h^5}} \Big[\|(u,\theta)-(v_h,\psi_h)\|_1 + |||(\phi,\lambda)-(\eta_h,\chi_h)||| \\ &\qquad + t\|(\phi,\lambda)-(\eta_h,\chi_h)\|_0\Big]. \end{aligned} \tag{4.7}$$

Finally, the error estimates (4.4) can be obtained from (4.7) by using a standard error analysis of finite element method (Note that $|||(\eta,\chi)||| \le \|B\|\|(\eta,\chi)\|_0$). □

4.2. (P_1 iso P_2) elements. The so-called "P_1 iso P_2" elements consist of four piecewise linear elements for the approximations of the variables v_1, v_2, v_3, ψ_1, ψ_2, while the others η_1, η_2, χ_1, χ_2, χ_3 remain linear on the elements (see Fig. 4). The finite element spaces are defined as

$$\begin{cases} V_h = L_1^1(\widehat{\mathcal{T}_h}), \\ W_h = L_1^1(\mathcal{T}_h), \end{cases} \tag{4.8}$$

where $\widehat{\mathcal{T}_h}$ is obtained by dividing $\mathcal{T}_h$ into four subtriangles. For this choice of elements, we can verify that Theorem 3.4 holds. Then it gives us the uniform convergence. Indeed, on a macroelement M shown in Fig. 5, the matrices associated with (3.8) and (3.9) are similar to the following matrices respectively:

$$G_1 = \begin{pmatrix} s_1+s_4 & 10(s_1+s_4) & 4s_1 & 0 & 4s_4 \\ s_1+s_2 & 4s_1 & 10(s_1+s_2) & 4s_2 & 0 \\ s_2+s_3 & 0 & 4s_2 & 10(s_2+s_3) & 4s_3 \\ s_3+s_4 & 4s_4 & 0 & 4s_3 & 10(s_3+s_4) \\ s_1+s_2+s_3+s_4 & 3(s_1+s_4) & 3(s_1+s_2) & 3(s_2+s_3) & 3(s_3+s_4) \end{pmatrix}.$$

$$G_2 = \begin{pmatrix} y_2-y_4 & 3(y_2-y_4) & 3(y_2-y_5) & 0 & 3(y_5-y_4) \\ y_3-y_1 & 3(y_5-y_1) & 3(y_3-y_1) & 3(y_3-y_5) & 0 \\ y_4-y_2 & 0 & 3(y_5-y_2) & 3(y_4-y_2) & 3(y_4-y_5) \end{pmatrix}.$$

Suppose that $y_1 \neq y_5$ and $y_2 \neq y_4$, a straightforward calculus gives $\operatorname{rank}(G_1) = 5$ and $\operatorname{rank}(G_2) = 3$. Moreover, the space $W_h|_M$ chosen here is the same as that in $P_2 - P_1$ element. So, Lemma 4.2 (i.e., the condition (ii)) also holds for P_1 iso P_2

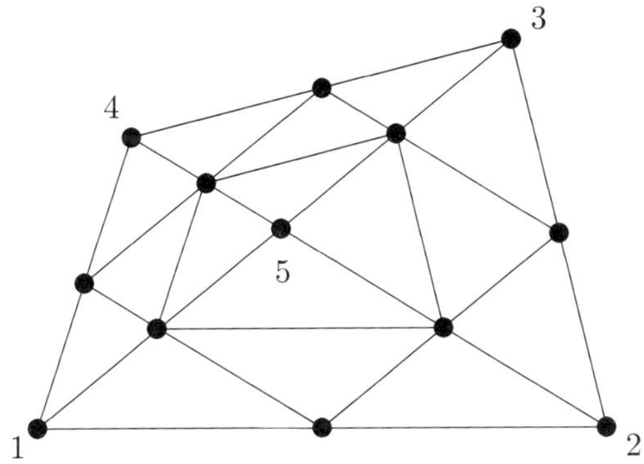

FIGURE 5. Macroelement associated with P_2 iso P_1 element.

element. Therefore, we have the same conclusion like $P_2 - P_1$ elements. The proof is similar to that for Theorem 4.3. We won't repeat it here.

THEOREM 4.4. *With the choice of V_h and W_h by (4.8), for every bounded linear form $F \in (V^5)'$ and $t \in (0, \sqrt{c_0})$, there exists a unique solution $(u_h, \theta_h; \phi_h, \lambda_h) \in V_h^5 \times W_h^5$ to Problem M_h^t. Moreover, there is a constant γ_2 independent of h, t and F such that*

$$(4.9) \quad \left\|(u,\theta) - (u_h,\theta_h)\right\|_1 + |||(\phi,\lambda) - (\phi_h,\lambda_h)||| + t\left\|(\phi,\lambda) - (\phi_h,\lambda_h)\right\|_0 \le \gamma_2 h^2\Big[\left\|(u,\theta)\right\|_3 + \left\|(\phi,\lambda)\right\|_2\Big].$$

References

1. D. N. Arnold and F. Brezzi, *Locking free finite elements for shells*, Research Report AM 127, Pennsylvania State Univ., University Park, PA, 1993.
2. D. N. Arnold and R. S. Falk, *A uniformly accurate finite element method for Reissner-Mindlin plate*, SIAM J. Numer. Anal. **26** (1989), 1276–1290.
3. M. Bernadou, *Méthodes d'éléments finis pour les problèmes de coques minces*, Masson, Paris, 1994.
4. F. Brezzi, K. Bathe, and M. Fortin, *Mixed-interpolated elements for Reissner-Mindlin plates*, Internat. J. Numer. Methods Engrg. **28** (1989), 1787–1801.
5. F. Brezzi and M. Fortin, *Numerical approximation of Mindlin-Reissner plates*, Math. Comp. **47** (1986), 151–158.
6. ______, *Mixed and hybrid finite element methods*, Springer-Verlag, New York, 1991.
7. P. Clement, *Approximation by finite element functions using local regularization*, RAIRO, Anal. Numer. **9** (1975), 33–76.
8. R. Duran and E. Liberman, *On mixed finite element methods for the Reissner-Mindlin plate model*, Math. Comp. **5** (1992), no. 8, 561–573.
9. A. Kirmse, *Bending-dominated deformation of spherical shells: Analysis and finite-element approximation*, SIAM J. Numer. Anal. **30** (1993), 1015–1040.
10. Y. Leino and J. Pitkäranta, *On the membrane locking of $h-p$ finite element in a cylindrical shell problem*, Internat. J. Numer. Methods Engrg. **37** (1994), 1057–1070.
11. P. M. Naghdi, *The theory of shells and plates*, Handbuch der Physik, vol. VI a-2, Springer-Verlag, Berlin, 1972, pp. 425-640.
12. J. Piila and J. Pitkäranta, *Energy estimates relating different linear elastic models of a thin cylindrical shell. I. The membrane-dominated case*, SIAM J. Numer. Anal. **24** (1993), 1–22.
13. ______, *Energy estimates relating different linear elastic models of a thin cylindrical shell. II. The case of free boundary*, SIAM J. Math. Anal. **26** (1995), 820–849.

14. J. Pitkäranta, *Analysis of some low-order finite element schemes for Mindlin-Reissner and Kirchhoff plates*, Numer. Math. **53** (1988), 237–254.
15. ———, *The problem of membrane locking in finite element analysis of cylindrical shells*, Numer. Math. **61** (1992), 523–542.
16. G. Yang, M. C. Delfour, and M. Fortin, *Uniformly convergent mixed finite element methods for cylindrical shells*, CRM-2319, Univ. de Montréal; Math. Models Methods Appl. Sci. (1995) (submitted).

E-mail address: yangg@dns.njupt.edu.cn

Centre de recherches mathématiques et Dép. de mathématiques et de statistique, Université de Montréal, C. P. 6128, Succ. Centre-ville, Montréal (Qc), H3C 3J7, Canada

E-mail address: delfour@crm.umontreal.ca

Dép. de mathématiques et de statistique, Université Laval, Ste-Foy, Québec, Canada, G1K 7P4

E-mail address: mfortin@mat.ulaval.ca